Wheat Improvement

Matthew P. Reynolds • Hans-Joachim Braun
Editors

Wheat Improvement

Food Security in a Changing Climate

 Springer

 CGIAR

 CIMMYT MR
International Maize and Wheat Improvement Center

Editors
Matthew P. Reynolds
Wheat Physiology, Global Wheat program
International Maize and Wheat
Improvement Center (CIMMYT)
Texcoco, Mexico

Hans-Joachim Braun
Global Wheat program
International Maize and Wheat
Improvement Center (CIMMYT)
Texcoco, Mexico

ISBN 978-3-030-90675-7   ISBN 978-3-030-90673-3   (eBook)
https://doi.org/10.1007/978-3-030-90673-3

This Springer imprint is published by the registered company Springer Nature Switzerland AG
The registered company address is: Gewerbestrasse 11, 6330 Cham, Switzerland

This textbook is dedicated to Dr. Sanjaya Rajaram, a titan of the wheat world who succumbed to COVID-19 on 17 February 2021. "Raj," as he was called by those who knew him, carried the mantle of his grand mentors, Norman Borlaug and Glenn Anderson, the driving forces of the wheat revolution of the twentieth century. He took over CIMMYT's bread wheat program in the early 1970s and proceeded to lead a second Green Revolution in wheat production into the early 2000s that continues through today. Raj not only stands as a titan but is also an example for a person, who made it from a poor background to the world leader in his field.

Foreword

For more than 10,000 years, wheat has been the cornerstone of food and nutritional security and is currently the most widely grown crop in the world. Due to its unique processing and quality characteristics, and since it can be easily transported and stored – it is also the world's most traded crop and often the first choice when food aid is needed for famine-struck regions.

Grown on all five continents and in more diverse environments than any other crop, wheat is vulnerable to a wide range of transboundary diseases and abiotic stresses, particularly heat and drought. Resistance to these stresses plays an important role in efforts to breed for yield stability, the most requested trait among wheat farmers across the globe.

Half a century ago, wheat was also one of the most-studied crops. But for reasons related to its biology – wheat is self-pollinated, and thus its seed can be readily saved and shared for the next crop – it has not attracted the same private sector investment in breeding research as crops with a higher financial return on investment, for example hybrid and genetically modified (GMO) crops. Consequently, the public sector remains the largest provider of improved wheat varieties. This is particularly true in the Global South where more than 1.5 billon resource-poor people are dependent on a constant and affordable supply of wheat as a staple food.

Globally, the crop provides about 20% of all human dietary protein and calories. Climate change and consequential periods of extreme heat, cold and drought, combined with disease threats, represent huge challenges. A 2 °C temperature increase will reduce wheat yields in the Global South by 10–15%. At the same time, average yields will need to go up 40% by 2050 to provide enough food for a still growing population. Provision of sufficient calories and protein remains essential. Estimates from 2020 show that around 820 million people still go to bed hungry each night, only a slight decrease from the 2000 estimate of 900 million, indicating that we are unlikely to reach the UN goal to end hunger by 2030.

Furthermore, in addition to calories, other nutritional aspects of diets must be assured, especially for consumers whose dietary options are restricted. Wheat scores well here too, being an important source of dietary fibre, minerals, B vitamins and other micronutrients, as well as an outstanding source of plant protein. Contrary to the 'food-fad' misinformation emanating from the Global North, there is no evidence that intensive breeding has decreased the nutritional quality of wheat, nor that wheat proteins trigger adverse responses in the vast majority of people.

This book covers all aspects of wheat improvement, from utilizing genetic resources to breeding and selection methods, data analysis, biotic and abiotic stress tolerance, yield potential, genomics, quality nutrition and processing, physiological pre-breeding, and seed production. It dedicates a final part to rapidly evolving technologies and their potential to accelerate genetic gains and adaptation.

This is the first book in many years focusing on wheat science in such a wide and comprehensive way. I commend the editors and Springer for bringing out this important publication now. While this textbook focuses on wheat *per se*, its 32 chapters, written by leaders in their disciplinary fields, address cutting-edge issues relevant to many other crops. Considering the remarkable progress made in genetics, molecular breeding, phenomics at breeding scale and bioinformatics, I am sure that this book will be immensely useful to students – the future wheat science leaders – and that it will help scientists, plant breeders, extensionists, agro-industrialists, farmers and policy developers better understand how wheat can remain a pillar for sustainable global food and nutrition security.

Deputy Chair, Board of Trustees   William (Bill) Angus
Wheat Improvement Centre (CIMMYT)
Texcoco, Mexico

Editor of The World Wheat Books (1–3)

Founder of Angus Wheat Consultants Ltd.
Rattlesden, UK

Former Head of Wheat Breeding
Limagrain Rothwell, UK

Former Wheat Breeder, PBI
London, UK

Preface

Springer's suggestion to develop a textbook on wheat improvement, while out of the blue, was also very prescient. The looming challenges to food security are perhaps greater now than at any previous moment in modern history. They are, of course, related to demographics and the environment, but they are also political in terms of who they affect and how they are addressed. We hope this volume will provide a valuable new tool for wheat scientists, policy makers, and farmers – who, for the most part, have remarkably achieved global food sufficiency to date – so that continued benefits of crop research will be both timely and universal in their reach. We especially want to thank all authors who contributed their valuable time and effort to this volume, in spite of busy and often overwhelming schedules. We also recognize the thousands of collaborators around the world who enable wheat improvement – with a special mention of the International Wheat Improvement Network that has supported globally coordinated wheat improvement for more than 50 years – and the millions of farmers who, year in and year out, stoically face the risks and challenges of cultivation.

Texcoco, Mexico
July 2021

Matthew P. Reynolds
Hans-Joachim Braun

Acknowledgments

The editors are indebted to Fatima Escalante for her timely and conscientious technical editing of this volume and for her valuable help and patience in communicating with authors and editors.

This work was supported, in whole or in part, by the Bill & Melinda Gates Foundation [Accelerating Genetic Gains, OPP1215722], the Foundation for Food and Agriculture Research [HeDWIC, DFs-190000000013], the International Wheat Yield Partnership [IWYP Hub Project W0293, MTO 069018], and Mexico's Secretariat of Agriculture and Rural Development (SADER) [Trigo para Mexico Project].

Contents

List of Figures

List of Tables

About the Editors and Contributors

Editors

 **Matthew P. Reynolds** is Distinguished Scientist at the International Maize and Wheat Improvement Centre (CIMMYT) and alumnus of Oxford (BA), Reading (MS), and Cornell (PhD) Universities. He currently leads the Wheat Physiology lab at CIMMYT and serves on the management committee of the CGIAR Platform for Big Data in Agriculture, as well as leading its community of practice for crop modelling. He has been active in developing global collaborations to tap into expertise of plant scientists worldwide –such as the International Wheat Yield Partnership – and the Heat and Drought Wheat Improvement Consortium, with the view to underpinning food security through crop improvement, with a special focus on the Global South. He has published widely in the areas of crop physiology, genomics, and pre-breeding and has been included in the top 1% of the world's researchers since 2018 – most recently across plant and animal sciences – by Web of Science. He has honorary positions at Nottingham, Texas A&M, and Oklahoma State Universities and is board member of the Global Plant Council. He is Fellow of both the Crop Science Society of America and the American Society of Agronomy and was appointed to the Mexican Academy of Science in 2018.

 **Hans-Joachim Braun,** a native of Germany with background in wheat breeding, has led CIMMYT's Global Wheat Program from 2004 until 2020, when he retired. He was responsible for technical direction and implementation of the program and lead and managed 40 internationally recruited scientists, who develop wheat germplasm that is distributed to around 200 cooperators in more than 100 countries and grown of more than 50% of the spring wheat area in developing countries. During his 37 years in international agriculture, he became familiar with all major wheat based systems globally. He lived from 1985 to 2005 in Turkey, leading the Turkey CIMMYT ICARDA International Winter Wheat Improvement Program. He contributed to the development of more than 40 winter wheat varieties released mainly in West and Central Asia, which are grown on more than 2 million ha. Braun was instrumental in recognizing Zn deficiency and soil-borne diseases as a major constraint for winter wheat production in the dryland areas of West Asia. He has published more than 50 peer-reviewed articles, book chapters, and received various awards, including the Friendship Award of China for his contribution to develop disease resistant wheat lines for Gansu province, the Crop Science of America Fellowship Award, and the International Agronomy Award of the American Society of Agronomy. Braun received his PhD from the University of Hohenheim, Germany, in 1983.

Contributors

Eduard Akhunov Kansas State University, Manhattan, KS, USA

Samir Alahmad The University of Queensland, Queensland Alliance for Agriculture and Food Innovation, Brisbane, Australia

Gregorio Alvarado International Maize and Wheat Improvement Center (CIMMYT), Texcoco, Mexico

Juan B. Álvarez Departamento de Genética, Escuela Técnica Superior de Ingeniería Agronómica y de Montes, Universidad de Córdoba, Córdoba, Spain

Jose Luis Araus Integrative Crop Ecophysiology Group, Faculty of Biology, University of Barcelona, Barcelona, Spain

Senthold Asseng Department of Life Science Engineering, Technical University of Munich, Freising, Germany

Michael Ayliffe CSIRO, Canberra, ACT, Australia

Ilse Barrios-Perez Soil and Crop Sciences, Texas A&M University, College Station, TX, USA

Bhoja R. Basnet International Maize and Wheat Improvement Center (CIMMYT), Texcoco, Mexico

Alison R. Bentley International Maize and Wheat Improvement Center (CIMMYT), Texcoco, Mexico

Sridhar Bhavani International Maize and Wheat Improvement Center (CIMMYT), Texcoco, Mexico

Gurcharn Singh Brar Faculty of Land and Food Systems, The University of British Columbia, Vancouver, BC, Canada

Hans-Joachim Braun International Maize and Wheat Improvement Centre (CIMMYT), Texcoco, Mexico

Martina Bruschi Alma Mater Studiorum–Università di Bologna, University of Bologna, Bologna, Italy

Maria Luisa Buchaillot Integrative Crop Ecophysiology Group, Faculty of Biology, University of Barcelona, Barcelona, Spain

Fatima Camarillo-Castillo International Maize and Wheat Improvement Center (CIMMYT), Texcoco, Mexico

J. Jesús Cerón-Rojas International Maize and Wheat Improvement Center (CIMMYT), Texcoco, Mexico

Giovanny Covarrubias-Pazaran International Maize and Wheat Improvement Center (CIMMYT), Texcoco, Mexico

Leonardo A. Crespo-Herrera International Maize and Wheat Improvement Center (CIMMYT), Texcoco, Mexico

José Crossa Colegio de Post-Graduados (COLPOS), Montecillos, Mexico
International Maize and Wheat Improvement Center (CIMMYT), Texcoco, Mexico

Abdelfattah A. Dababat International Maize and Wheat Improvement Center (CIMMYT), Texcoco, Mexico
International Maize and Wheat Improvement Center (CIMMYT), Ankara, Turkey

Jason Donovan International Maize and Wheat Improvement Center (CIMMYT), Texcoco, Mexico

Mustapha El Bouhssini Mohamed VI Polytechnic University, Benguerir, Morocco

Olaf Erenstein International Maize and Wheat Improvement Center (CIMMYT), Texcoco, Mexico

Justin Faris United States Department of Agriculture (USDA), Fargo, ND, USA

John P. Fellers Plant Pathology, USDA-ARS-Hard Winter Wheat Genetics Research Unit, Manhattan, KS, USA

R. A. (Tony) Fischer Agriculture & Food, CSIRO, Canberra, Australia

Richard B. Flavell International Wheat Yield Partnership (IWYP), Texas A&M AgriLife Research, College Station, TX, USA

M. John Foulkes School of Biosciences, University of Nottingham, Leicestershire, UK

Navin C. Gahtyari ICAR-Vivekananda Parvatiya Krishi Anusandhan Sansthan, Almora, Uttarakhand, India

Maraeva Gianella Department of Biology and Biotechnology "L. Spallanzani", University of Pavia, Pavia, Italy

Bikram S. Gill Wheat Genetics Resource Center, Department of Plant Pathology, Kansas State University, Manhattan, KS, USA

Peter Giovannini Global Crop Diversity Trust, Bonn, Germany

Bram Govaerts International Maize and Wheat Improvement Center (CIMMYT), Texcoco, Mexico

Cornell University, Ithaca, NY, USA

Velu Govindan International Maize and Wheat Improvement Center (CIMMYT), Texcoco, Mexico

Surbhi Grewal School of Biosciences, University of Nottingham, Loughborough, UK

Simon Griffiths The John Innes Centre, Norwich, UK

Jose Rafael Guarin Department of Agricultural and Biological Engineering, University of Florida, Gainesville, FA, USA

Center for Climate Systems Research, Columbia University, New York, NY, USA

Carlos Guzmán Departamento de Genética, Escuela Técnica Superior de Ingeniería Agronómica y de Montes, Universidad de Córdoba, Córdoba, Spain

Filippo Guzzon International Maize and Wheat Improvement Center (CIMMYT), Texcoco, Mexico

J. Jefferson Gwyn International Wheat Yield Partnership (IWYP), Texas A&M AgriLife Research, College Station, TX, USA

Marion Harris North Dakota State University, Fargo, ND, USA

Dirk B. Hays Soil and Crop Sciences, Texas A&M University, College Station, TX, USA

Xinyao He International Maize and Wheat Improvement Center (CIMMYT), Texcoco, Mexico

Lee T. Hickey The University of Queensland, Queensland Alliance for Agriculture and Food Innovation, Brisbane, Australia

David P. Hodson International Maize and Wheat Improvement Center (CIMMYT), Texcoco, Mexico

Nora Honsdorf Kiel University, Kiel, Germany

International Maize and Wheat Improvement Center (CIMMYT), Texcoco, Mexico

Julio Huerta-Espino INIFAP, Campo Experimental Valle de Mexico, Texcoco, Mexico

Marla Itria Ibba International Maize and Wheat Improvement Center (CIMMYT), Texcoco, Mexico

Moti Jaleta International Maize and Wheat Improvement Center (CIMMYT), Addis Ababa, Ethiopia

Arun Kumar Joshi International Maize and Wheat Improvement Center (CIMMYT), New Delhi, India

Borlaug Institute for South Asia (BISA), Ludhiana, Punjab, India

Philomin Juliana Borlaug Institute for South Asia (BISA), New Delhi, India

Shawn C. Kefauver Integrative Crop Ecophysiology Group, Faculty of Biology, University of Barcelona, Barcelona, Spain

Zakaria Kehel International Center for Agricultural Research in the Dry Areas (ICARDA), Rabat, Morocco

Ian P. King School of Biosciences, University of Nottingham, Loughborough, UK

Julie King School of Biosciences, University of Nottingham, Loughborough, UK

Margaret R. Krause Department of Plants, Soils and Climate, Utah State University, Logan, UT, USA

Uttam Kumar Borlaug Institute for South Asia (BISA), New Delhi, India

Evans Lagudah CSIRO, Canberra, ACT, Australia

Peter Langridge School of Agriculture Food and Wine, University of Adelaide, Adelaide, SA, Australia

Wheat Initiative, Julius-Kühn-Institute, Berlin, Germany

Ming Luo CSIRO, Canberra, ACT, Australia

Marco Maccaferri Alma Mater Studiorum–Università di Bologna, University of Bologna, Bologna, Italy

Ian Mackay SRUC, Edinburgh, UK

Johannes W. R. Martini International Maize and Wheat Improvement Center (CIMMYT), Texcoco, Mexico

Ky L. Mathews University of Wollongong, Wollongong, NSW, Australia

Kristina D. Michaux HarvestPlus, IFPRI, Washington, DC, USA

Gemma Molero KWS Momont SAS, Mons-en-Pévèle, France

Suchismita Mondal International Maize and Wheat Improvement Center (CIMMYT), Texcoco, Mexico

Boyd A. Mori University of Alberta, Edmonton, AB, Canada

Craig Morris USDA-ARS Western Wheat Quality Laboratory, Pullman, WA, USA

Khondoker Abdul Mottaleb International Maize and Wheat Improvement Center (CIMMYT), Texcoco, Mexico

Thomas S. Payne International Maize and Wheat Improvement Center (CIMMYT), Texcoco, Mexico

Wolfgang H. Pfeiffer HarvestPlus, IFPRI, Washington, DC, USA

Charlotte Rambla The University of Queensland, Queensland Alliance for Agriculture and Food Innovation, Brisbane, Australia

Mandeep Singh Randhawa International Maize and Wheat Improvement Center (CIMMYT), Nairobi, Kenya
International Centre for Research in Agroforestry (ICRAF), Nairobi, Kenya

Greg Rebetzke CSIRO Agriculture and Food, Canberra, ACT, Australia

Matthew P. Reynolds International Maize and Wheat Improvement Centre (CIMMYT), Texcoco, Mexico

Richard A. Richards CSIRO Agriculture and Food, Canberra, ACT, Australia

Jeffrey L. Rosichan Crops of the Future Collaborative, Foundation for Food & Agriculture Research, Washington, DC, USA

Chandan Roy Department of Plant Breeding and Genetics, Bihar Agricultural University, Bhagalpur, Bihar, India

Jessica E. Rutkoski Department of Crop Sciences, University of Illinois at Urbana-Champaign, Urbana, IL, USA

Mark C. Sawkins International Wheat Yield Partnership (IWYP), Texas A&M AgriLife Research, College Station, TX, USA

Pawan Kumar Singh International Maize and Wheat Improvement Center (CIMMYT), Texcoco, Mexico

Ravi P. Singh International Maize and Wheat Improvement Center (CIMMYT), Texcoco, Mexico

Mike Sissons NSW Department of Primary Industries, Tamworth Agricultural Institute, Calala, Australia

Gustavo A. Slafer ICREA (Catalonian Institution for Research and Advanced Studies), AGROTECNIO (Center for Research in Agrotechnology), University of Lleida, Lleida, Spain

Kai Sonder International Maize and Wheat Improvement Center (CIMMYT), Texcoco, Mexico

Mark E. Sorrells College of Agriculture and Life Sciences, Cornell University, Ithaca, NY, USA

Sivakumar Sukumaran International Maize and Wheat Improvement Center (CIMMYT), Texcoco, Mexico

Wuletaw Tadesse International Center for Agricultural Research in the Dry Areas (ICARDA), Rabat, Morocco

Fernando H. Toledo International Maize and Wheat Improvement Center (CIMMYT), Texcoco, Mexico

Richard M. Trethowan Plant Breeding Institute, School of Life and Environmental Sciences, The University of Sydney, Sydney, Australia

Roberto Tuberosa Alma Mater Studiorum–Università di Bologna, University of Bologna, Bologna, Italy

Jelle Van Loon International Maize and Wheat Improvement Center (CIMMYT), Texcoco, Mexico

Mateo Vargas Parasitología Agrícola, Universidad Autónoma Chapingo, Texcoco, Mexico

Nele Verhulst International Maize and Wheat Improvement Center (CIMMYT), Texcoco, Mexico

Stephen H. Visscher Global Institute for Food Security, Saskatoon, SK, Canada

Kai P. Voss-Fels The University of Queensland, Queensland Alliance for Agriculture and Food Innovation, Brisbane, Australia

Wei Wang Kansas State University, Manhattan, KS, USA

Abbreviations

1-MCP	1-methyl cyclopropane
90K SNP chip	90,000 single nucleotide polymorphism array
ABA	Abscisic acid
ABC	ATP-binding cassette
AFLP	Amplified fragment length polymorphisms
AgMIP	Agricultural Modeling Intercomparison and Improvement Project
AgRenSeq	Associated genetics R gene enrichment sequencing
AHEAD	Alliance for Wheat Adaptation to Heat and Drought
ALS	Acetolactate synthase
ANOVA	Analysis of variance analysis
APR	Adult-plant resistance
APSIM	Agricultural production systems modulator
ASR	All-stage resistant
AY	Attainable yield
B2IR	Beds 2 irrigations
B5IR	Beds 5 irrigations
BARI	Bangladesh Agricultural Research Institute
BC_1	Single-backcrosses
BGRI	Borlaug Global Rust Initiative
BLHT	Beds late heat
BLUP	Best linear unbiased prediction
BM	Biomass
BS	Breeder seed
BYDV	Barley yellow dwarf virus
CA	Conservation agriculture
CAPS	Cleaved amplified polymorphism
CBD	Convention on Biological Diversity
CCII	Composite cross II
CCN	Cereal cyst nematode
CCR	Common root rot

CENEB	Centro Experimental Norman E. Borlaug
CESIM	Constrained index ESIM
CIMMYT	International Maize and Wheat Improvement Centre
CLGSI	Constrained linear genomic selection index
CLPSI	Constrained LPSI
COTF	Crops of the Future Consortium
COVID-19	Coronavirus disease
CR	Crown rot
CRISPR	Clustered regularly interspaced short palindromic repeats
CRISPR-SpCas9	CRISPR-Cas9 from *Streptococcus pyogenes*
CRRK	Cysteine-rich receptor kinase
crRNAs	CRISPR RNAs
CS	Certified Seed
CSB	Community seed banks
CT	Canopy temperatures
CTD	Canopy temperature depression
CVD	Cardiovascular disease
CWANA	Central and West Asia and North Africa
CWR	Crop wild relatives D163
DArT	Diversity arrays technology
DEM	Digital elevation model
DGGW	Delivering genetic gains in Wheat
DH	Days to heading
DHs	Doubled-haploids
DMI	Demethylation inhibitors
DON	Deoxynivalenol
DRRW	Durable rust resistance in wheat
DSB	Double strand breaks
DSI	Drought susceptibility index
DUF26	Domain with unknown function
DUS	Distinct, uniformly, stable
Eca	Apparent soil electrical conductivity
EGF_CA	Epidermal growth factor-calcium binding
EIAR	Ethiopian Institute of Agriculture Research
EN	Endangered
Eps	Earliness per se
ESIM	Eigen selection index method
EST-SSR	Expressed sequence tag-derived simple sequence repeat markers
ESWYT	Elite spring wheat yield trial
ETI	Effector triggered immunity
EW	Epicuticular wax
EWGs	Expert Working Groups
F5IR	Flat 5 irrigations
FA	Factor Analytic

FACE	Free-air carbon dioxide enrichment
Faster-RCNN	Faster Regional Convolutional Neural Network
FBS	Food balance sheet
FDRIP	Flat drip
FE	Fruiting efficiency
FFAR	Foundation for Food & Agriculture Research
FHB	Fusarium head blight
FHBSN	Fusarium head blight screening nursery
FIGS	Focused identification of germplasm strategy
Fr	Frost tolerance
FS	Foundation Certified Seed
GA	Gibberellic acid
gBLUP	Genomic best linear unbiased prediction
GBSS	Granule-bound starch synthase
GCA	General combining ability
GCM	Global climate models
GCRMS	Global cereal rust monitoring system
GE	Genetic engineering or genome editing
GEBVs	Genomic estimated breeding values
GEI	Genotype-by-environment interaction
GGT	Take-ALL, *Gaeumannomyces graminis var. tritici*
GM	Genetic modification
GOBii	Genomic open-source breeding informatics initiative
GP-1	Primary gene pool
GP-2	Secondary gene pool
GP-3	Tertiary gene pool
GPR	Ground penetrating radar
GS	Genomic selection
GSII	Glutathione
GUB_WAK	Extracellular galacturonan binding
GWAM	Genome-wide association mapping
GWAS	Genome-wide association study
GWP	Global Wheat Program
GxE	Genotype-by-Environment
GxExM	Genotype x environment x management
GY	Grain yield
HDR	Homology-directed repair
HeDWIC	Heat and Drought Wheat Improvement Consortium
HI	Harvest Index
HI-Edit	Haploid induction editing technology
HMW-Gs	High molecular weight glutenins
HNT	High night temperatures
HRW	Hard red winter
HRWSN	High rainfall wheat screening nursery
HRWYT	High rainfall wheat yield trial

HSI	Hue saturation intensity
HST	Host-selective toxin
HTFP	High-throughput field phenotyping
HTPP	High-throughput phenotyping
HTWYT	Heat Tolerance wheat yield trial
HW-DT-EM	Hard white-drought tolerant-early maturity
HW-DT-NM	Hard white-drought tolerant-normal maturity
HW-HiR-NM	Hard white-high rainfall-normal maturity
HW-HT-EM	Hard white-heat tolerant-early maturity
HW-OE-NM	Hard white-optimum environment-normal maturity
IBP	Integrated Breeding Platform (IBP)
IBWSN	International Bread Wheat Screening Nursery
ICARDA	International Center for Agriculture Research in the Dry Areas
ID	Identifier
IMU	Inertial measurement unit
INDELs	Insertion or deletion of bases
INV	Invertases
ISEPTON	International Septoria Observation Nursery
ISTA	International Seed Testing Association
ITMI	International Triticeae Mapping Initiative
ITPGRFA	International Treaty on Plant Genetic Resources for Food and Agriculture
IUCN	International Union for Conservation of Nature
IWIN	International Wheat Improvement Network
IWYP	Wheat yield partnership
IYPTE	IWYP yield potential trait experiment
KARLO	Kenya Agriculture and Research Organization
KASP	Kompetitive allele-specific PCR
KB	Karnal bunt
KBRSN	Karnal bunt resistance screening nursery
L/LM-ICs	Lower-middle income countries
LBRSN	Leaf blight resistance screening nursery
LD	Linkage disequilibrium
LGSI	Linear genomic selection index
LI	Light interception
Li+B150:C165DAR	Light detection and ranging
LMICs	Low- and middle-income countries
LMMs	Linear mixed models
LMSI	Linear marker selection index
LMW-Gs	Low molecular weight glutenins
LOX	Lipoxygenases
LPSI	Linear phenotypic selection index
LR	Leaf (or brown) rust
LSI	Linear selection index

MABC	Marker-assisted backcrossing
MAPK	Mitogen-activated protein kinase
MARPLE	Mobile and real-time plant disease
MAS	Marker-assisted selection
MAV	*Macrosiphum avenae* virus
ME	Mega-environment
MEF	Managed environment facility
MET	Multi-environmental testing
MGE	Multiplexed gene editing
MoT	*Magnaporthe oryzae* pathotype triticum
MRI	Magnetic resonance imaging
MSP	Major sperm protein
MTAs	Marker-trait associations
MutChromSeq	Mutation chromosome sequencing
MutRenSeq	Mutation resistance gene enrichment sequencing
NAMs	Nested-association mapping populations
NARS	National Agricultural Research Systems
NASA	National Aeronautics and Space Administration
NBS-LRR	Nucleotide Binding Site- Leucine Rich Repeat
NCDs	Non-communicable diseases
NDVI	Normalized difference vegetation index
NE	Necrotrophic effectors
NHEJ	Non-homologous end joining
NILs	Near isogenic lines
NIRS	Near-infrared spectroscopy
NLR	Nucleotide-binding leucine-rich repeat
NLS	Nuclear location signals
NRM	Natural resources management
NS	Nucleus seed
NUE	Nitrogen use efficiency
OCS	Optimum contribution selection
P&D	Pests and diseases
PAM	Protospacer-adjacent motif
PAMPs or DAMPs	Damage-associated molecular patterns
PAS	Publicly available standard
PAV	*Padi avenae* virus
PBC	Pseudo-black chaff
PCD	Plant cell death
PE	Prime editing
pegRNA	Editing gRNA
PGR	Plant genetic resources
PGRFA	Plant genetic resources for food and agriculture
PH	Plant height
PIEs	Plant improvement experiments
PK	Protein kinase

Ppd	Photoperiod
PPO	Polyphenol oxidase
PPP	Precision phenotyping platform
PR-1	Pathogenesis-related protein 1
p-rep	Partially replicated
PRRs	Pattern recognition receptors
PTI	PAMP-triggered immunity
PVS	Participatory varietal selection
PY	Potential yield
PYT	Preliminary yield trial
PYw	Water-limited potential yield
QoI	Quinone outside inhibitor
QTL	Quantitative trait loci
R	Resistance
RAPD	Random amplified polymorphic sequences
RCBD	Randomized complete block design
RCP	Representative concentration pathway
RCRS	Rapid-cycle recurrent selection
RED	Restriction enzyme digestion
REML	Residual maximum likelihood
RenSeq	Resistance gene enrichment sequencing
reps	Replications
RFLP	Restriction fragment length polymorphisms
Rht	Plant height
RLN	Root lesion nematode
RLPSI	Restricted LPSI
RT	Reverse transcriptase
RUE	Radiation-use efficiency
RWA	Russian wheat aphid
S	Susceptible
SA	South Asia
SABWGPYT	South Asian bread wheat genomic prediction yield trials
SAWSN	Semi-arid wheat screening nursery
SAWYT	Semi-arid wheat yield trial
SB	Spot blotch or speed breeding
SBCMV	Soil-borne cereal mosaic virus
SBE	Starch branching enzyme
SBPs	Soil-borne pathogens
SCA	Specific combining ability
SCAR	Sequence characterized amplified regions
SDHI	Succinate dehydrogenase inhibitors
SE	Selection environments
Ser/Thr	Serine/threonine
sgRNA	Single-guide RNA (sgRNA)
SGV	*Schizaphis graminum* virus

SMTA	Standard Material Transfer Agreement
SNB	*Stagonospora nodorum* blotch
SNPs	Single nucleotide polymorphisms
SPS	Sucrose phosphate synthase
SR	Stem rust
SREG	Sites regression
SRI	Spectral reflectance indices
SRRSN	Stem rust resistance screening nursery
SRW	Soft red winter
SS	Starch synthase
SSBs	Single strand breaks
SSD	Single seed descent method
SSR	Simple sequence repeats
STB	*Septoria tritici* Blight
SUC	Sucrose
SUS	Sucrose synthase
SVI	Spectral vegetation indices
T2DM	Type 2 diabetes mellitus
TaGW2	Thousand grain weight (TaGW2)
TALEN	Transcription activator-like effector nuclease
TCA	Tribarbozylic acid
TE	Transpiration efficiency
TFs	Transcription factors
TILLING	Targeting induced local lesions in genomes
TKW	Thousand-kernel weight
TL	Truthful labelled SEED
TOAR	Tropospheric ozone assessment report
TPEs	Target population of environments
tracrRNA	Trans-activating crRNA
TS	Tan spot
UAVs	Unmanned aerial vehicles
UGI	Uracil glycosylase inhibitor
UM/H-Ics	Upper-middle- and high-income countries
UPOV	International Union for the Protection of Plant Varieties
VI	Vegetation indices
Vrn	Vernalization
VU	Vulnerable
WAK	Wall-associated kinase
WB	Wheat blast
WCWR	Wheat crop wild relatives
WGR	Wheat genetic resources
WMS	Wheat microsatellites
WSMV	Wheat streak mosaic virus
WSSF	Wheat stem sawfly
WUE	Water use efficiency

WYCYT	CIMMYT wheat yield collaboration yield trial
X ray CT	X-ray computed tomography
YR	Stripe (yellow) rust
ZFN	Zinc finger nuclease
ZFP	Zinc finger protein

Part I
Background

Chapter 1
Wheat Improvement

Matthew P. Reynolds and Hans-Joachim Braun

> *Almost certainly the first essential component of social justice is adequate food for all*
>
> Norman Ernest Borlaug

Abstract Wheat is a staple for rich and poor alike. Its improvement as a discipline was boosted when statisticians first distinguished heritable variation from environment effects. Many twentieth century crop scientists contributed to the Green Revolution that tripled yield potential of staple crops but yield stagnation is now a concern, especially considering the multiple challenges facing food security. Investments in modern technologies – phenomics, genomics etc. – provide tools to take both translational research and crop breeding to the next level. Herein wheat experts address three main themes: "Delivering Improved Germplasm" outlining theory and practice of wheat breeding and the attendant disciplines; 'Translational Research to Incorporate Novel Traits' covers biotic and abiotic challenges and outlines links between more fundamental research and crop breeding. However, effective translational research takes time and can be off-putting to funders and scientists who feel pressure to deliver near-term impacts. The final section 'Rapidly Evolving Technologies & Likely Potential' outlines methods that can boost translational research and breeding. The volume by being open access aims to disseminate a comprehensive textbook on wheat improvement to public and private wheat breeders globally, while serving as a benchmark of the current status as we address the formidable challenges that agriculture faces for the foreseeable future.

Keywords Breeding precedents · New-technologies · Interdisciplinary research · Proof of concept · Food security · Wheat breeding benchmark

M. P. Reynolds (✉) · H.-J. Braun
International Maize and Wheat Improvement Centre (CIMMYT), Texcoco, Mexico
e-mail: m.reynolds@cgiar.org

1.1   Learning Objectives

• Provide background to the rest of the textbook and crop breeding generally.
• Highlight the need for integration among disciplines.
• Outline factors involved to achieve proofs of concept and impacts.

1.2   Background on Crop Breeding

Wheat is one of approximately 300,000 potentially edible plant species, of which just over 100 are commonly cultivated (Fig. 1.1). Of these just three – maize, rice, and wheat – provide nearly 60% of all human calories [2] and wheat alone provides approximately 20% of all calories and protein [3]. Plant breeding has been evolving since humans first selected among plants and their seed, for whatever purpose. Wallace et al. [4] and Fernie and Yan [5] divided the evolution of breeding into four stages. Stage 1 was phenotypic selection by farmers, stage 2 the era of hybridization. Most current breeding programs are in stage 3, characterized by use of biotechnologies like marker-assisted breeding, genomic selection, transgenics and use of bioinformatics. We are now entering stage 4, breeding by design, i.e. genome editing and precision breeding supported by big data analysis targeted to develop crops

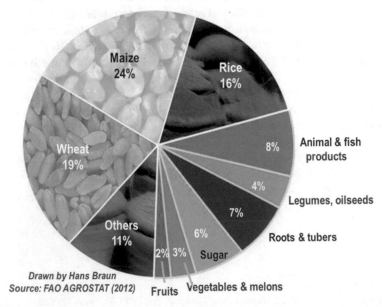

Fig. 1.1 The proportions of crops produced globally as a % of their total dry matter (approximately 3 billion tons annually). (Figure drawn by Hans-Joachim Braun with data from Ref. [1])

that meet farmer and consumer expectations in terms of yield and yield stability, biotic and abiotic stress tolerance, and nutrition and quality.

Interestingly, no new plant domestication has occurred in modern history, clear evidence of the formidable challenges associated with crop 'domestication'. There is one partial exception, namely triticale a relative of wheat [6], but even that was a hybridization of two domesticated species, wheat and rye, and has been quite difficult to commercialize despite its robustness to stress and multiple potential uses.

The principles of breeding are similar across most crops since they are cultivated in similar ways, and new cultivars face similar types of challenges in their respective growing environment. These include resisting or tolerating diseases and pests, and since most crops are field-grown, they must also adapt to variable temperatures, water supply, light and soil conditions, while flowering and maturing within defined time windows. Crop management can optimize the plant's environment to some degree, including for nutrients, control of biotic threats, as well as through choice of sowing dates, crop rotations, and irrigation where feasible. However, significant yield gaps in most annual cropping systems [7, 8], attest to the importance of selecting for heritable traits through plant breeding. Once obtained, a new cultivar can normally be relied on to express desirable traits, including yield and other agronomic and commercial expectations, as well as robustness to seasonal variation that may include a range of abiotic stresses, within a given target population of environments. In other words, guided hybridization and heritable-trait selection is a highly effective way to boost and/or protect crop productivity, since changing cultivars is one of the easiest interventions to achieve at the farm level [9].

1.3 Crop Improvement in Pre-history

Domestication of wild plants to fit agriculture is believed to have started in the Neolithic age at least 12,000 years ago in the fertile crescent, that finally led to the around 100 species that we cultivate today; though in fact a much larger number of plant species (7000) are considered semi-cultivated [10] if we include herbs, spices, medicinal plants etc. Considering the characteristics that have been passed down through history, and in comparison to wild ancestors, it is clear that early plant breeder/farmers selected for three main trait classes: (1) Preferential growth of edible organs to maximize yield; (2) Palatability and nutritional value; (3) Adaptation to a range of biotic and abiotic stresses, a problem challenging breeders to the present day [11, 12]. In short, modern day breeding is qualitatively the same discipline as our ancestors practiced; the principal selection objectives remain much the same though the breeding tools have changed.

1.4   Breeding in the Industrial Age

Mendel's work led to the first scientific proof of hereditary principles and the new discipline of genetics catalyzed crop research with the objective of boosting productivity through breeding. Gartons Agricultural Plant Breeders in the UK was one of the first companies to commercialize higher yielding cultivars. William Farrer in Australia bred the first rust resistant wheat strain. Meanwhile Nazareno Strampelli in Italy bred several high yielding, early maturing, rust resistant and short strawed wheat lines using the *Rht 8* dwarfing gene. Some of his lines made global impact and were exported to the Americas and China [13], and also used decades later as parents by Norman Borlaug. The new discipline of statistics enabled traits to be dissected genetically allowing a quantitative distinction between heritable variation and environment effects on trait [14].

These efforts and the work of Gonjiro Inazuka in Japan created the foundation of the Green Revolution leading to a paradigm shift in plant breeding and crop management. This was kick started by the dissemination of semi-dwarf genes in wheat and other cereals in the 1960s. Before the adoption of shorter lines, cereal yields were limited by lodging if plants became too tall as a result of yield-boosting inputs like N and irrigation water. It took over 10 years to achieve effective introgression of Rht1 from Norin-10, but its pleiotropic effects improved harvest index (HI) and nitrogen use efficiency (NUE), as well as lodging resistance, spearheading the Green Revolution [15]. The new generation of semi-dwarf spring wheat lines were also photoperiodic insensitive which was of paramount importance for their wide adoption; Borlaug himself admitted that this was a case of serendipity – 'an unplanned collateral effect of shuttle breeding'.

The Green Revolution in the 1960s, based in wheat on *Rht1* and *Rht 2* dwarfing genes and breeding genetic backgrounds to suit them, and the biotechnology revolution from the 1980s onwards, have delivered increasingly sophisticated methodologies for crop improvement. In the meantime, breeding programs have been efficiently meeting the demands of a fast-growing global population through steady genetic gains and broad-spectrum resistance to pests and diseases in wheat and other staple crops, with exceptionally high returns on investment documented [16]. Some suggest that this success has led to complacency, and both public and private sectors struggle to achieve the investments needed to match predicted human food demand by mid-century. The situation is especially ironic, given that many breeding programs now struggling for operational funds – have already made initial investments in modern technologies such as phenomics, genomics and informatics that are crucial to further increase genetic gains. In addition to helping increase the efficiency of selection for mainstream traits – yield, and yield stability, abiotic and biotic stress tolerance, phenology, quality and nutrition – these technologies can be powerful tools in translational research aimed at achieving step changes in yield and adaptation to emerging stresses.

1.5 Technologies That Have Impacted Crop Breeding in Recent Decades

This volume attempts to present the most relevant disciplines and research approaches that are likely to impact wheat breeding for the foreseeable future, building on tried and test approaches as well as new and emerging technologies.

Among these the most important effort, at least from the point of view of sustainable crop production and in addition to selecting for incremental yield gains, is breeding for resistance to pathogens and pests (i.e. maintenance breeding), this being a task that only becomes harder as agriculture intensifies. Maintenance breeding is reminiscent of the legend of Sisyphus, whose task started over and over again just as he had nearly finished, and so it is with the constant evolution of new pest and disease races, as well as the periodic emergence of new threats that jump host-species barriers, such as wheat blast [12]. For many diseases the challenge is made even harder since new sources of resistance are mainly found in relatively exotic materials such as landraces and wild relatives. This continuous challenge to find resistance genes against new disease pathotypes follows the same principals as the need to develop new vaccines effective against new CoVID-19 variants [17]. Molecular technologies can now be applied in breeding for resistance to many diseases where the genes are of relatively large effect. With recent advances in gene-cloning and gene-stacking, it is now technically possible for example, to combine stem rust resistance genes so that they do not recombine and are inherited like a single trait [18] and thereby underpin durable resistance. All rust resistance genes used in the stack originate from wheat and closely related genomes (i.e. cisgenics). However, since genetic modification (GM) technology can be used to stack the wheat resistance genes, policy makers and consumers must first accept such products. Then gene stacking technology could be expanded to other diseases, having fundamental impacts in terms of durable and sustainable crop protection and reducing agro-chemical footprints globally.

The approaches and technologies used to deliver new, higher yielding, broadly adapted, disease resistant wheat lines, many with specific quality and nutritional characteristics, are described in Part II of this volume entitled "Delivering Improved Germplasm". This section outlines the theory and practice of wheat breeding and the disciplines it routinely integrates to deliver on farmer and consumer needs. These methods underpin food security, especially in countries where many external inputs such as fungicide or insecticide are out-of-reach for resource poor farmers. Resistance to biotic stresses also helps safeguard farmers, agricultural communities and ultimately consumers from the potential hazards of widescale application of such chemical protectants.

On the other hand, for any complex genetic trait – such as many associated with yield potential and climate resilience – the chances of cloning a causative gene or identifying reliable molecular markers decreases with the numbers of genes involved in its expression. Hence genomic selection for yield involves modelling of largely random markers in order to train QTL-based models of yield prediction; exercises which have underscored the importance of genetic background and environment in determining which alleles impact crop performance. Nonetheless, the process remains largely stochastic and is challenging to apply on all of the complex traits that have been shown – and will be shown – to be involved in yield determination and adaptation to biotic and abiotic stresses. In order for breeding to reach the final 'deterministic' stage and catch up with the technological revolutions that are happening in phenomics, genomics, in silico breeding, etc. an even larger integration of disciplines is required.

1.6 Integration of Disciplines

Crop improvement relies on integration and application of many disciplines and has been exemplary in achieving this, having underpinned global food security since the Green Revolution, during which time human population has more than doubled. During this time frame, namely the last half century, the area sown to cereals globally – has not changed significantly while yields have tripled. It is clear that crop research has achieved outstanding impacts on breeding and crop management, while policy and the adaptability of farmers to embrace new technologies have had life-saving outcomes [19]. Nonetheless, the challenges that agriculture faces now are not just to feed nearly 10 billion people within the next 3 decades, but to achieve it sustainably under a warmer and more unpredictable climate, and often with less water, less N and declining soil quality [20]. Clearly research, breeding and agronomy must become even more effective and responsive to a range of stakeholders.

The explosion in fundamental plant science of recent decades has uncovered the physiological and genetic basis of many traits as well as genetic markers in model species. Nonetheless, many of these outputs have yet to be tested and translated into applied breeding. Clearly, the need for investment in translational research is more critical than ever. Sequencing of the wheat genome, in conjunction with thorough phenotypic characterization of elite material in appropriate field environments, will lead to a comprehensive physiological and biochemical basis of crop yield and adaptation. Such information will enable modelling the effects of and interactions among candidate traits and genes in different target locations, and help inform and refine breeding strategies. Meanwhile, advances in phenomics and genomics have the potential to be mainstreamed in three main areas of crop improvement: (1) Characterizing candidate parents to help design more strategic crosses; (2) Screening progeny at breeding scale to identify genotypes that express the targeted traits; (3) Facilitating the exploration of vast collections of relatively underutilized crop genetic resources. Advanced phenomics approaches – such as use of hand-held androids, drones and plane/satellite mounted sensors – make screening of such collections much more feasible at scale [21]. At the same time, genomics is also mobilizing to the field, with portable genotyping kits that have the potential to revolutionize global disease surveillance, potentially averting pandemics [22]. Such technologies scale readily to mainstream breeding and are equally valid for biotic and abiotic factors.

For these reasons, the volume includes a dedicated section entitled 'Part III Translational Research to Incorporate Novel Traits', covering biotic and abiotic challenges. Translational research in this context, is defined as the application of any scientific knowledge to crop improvement. Translational research of this kind provides an essential link between more fundamental research and crop breeding, adding value to both. The challenge however, is to demonstrate genetic gains using up to date and representative germplasm, in relevant environments. Therefore, translation often takes time and can be off-putting to funders and scientists who feel pressure to deliver near-term impacts. As a result, relatively few scientists occupy the applied research space where proofs of concept for crop improvement hypotheses are rigorously tested in a breeding context. Nonetheless, it can be accelerated with newer tools and technologies and these are discussed in the final part of this volume 'Part IV Rapidly Evolving Technologies & Likely Potential'.

1.7 Networking and Sharing

No matter how advanced the understanding of a component of a problem, holistic understanding is required to solve many cropping-system level challenges. New tools and approaches can help fill knowledge gaps and potentially accelerate genetic gains directly. A recent review involving industry and academia set out to define major knowledge gaps with potential to improve crop productivity across a broad

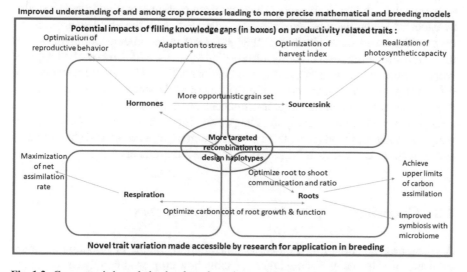

Fig. 1.2 Current trait-knowledge bottlenecks and potential research outcomes on crop productivity. (Reprinted with permission from Ref. [23])

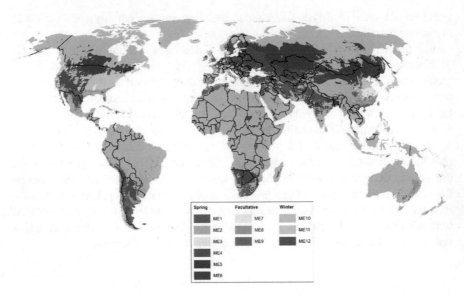

Fig. 1.3 The International Wheat Improvement Network (IWIN) embraces a global collaboration of wheat scientists testing approximately 1,000 new high yielding, stress adapted, disease resistant wheat lines each year. Breeding is directed towards 12 different ME, representing a range of temperature, moisture, and disease profiles. Spring wheat: *ME1* irrigated, high yield, *ME2* high rainfall disease prone environments, *ME3* acid soils, *ME4* water limitation, *ME5* heat stress, *ME6* temperate, high latitude.; Facultative wheat: *ME7* irrigated, moderate cold, *ME8* high rainfall, moderate cold, *ME9* low rainfall, moderate cold.; Winter wheat: *ME10* irrigated severe cold, *ME11* high rainfall/irrigated, severe cold, *ME12* low rainfall, severe cold. (Figure drawn by Kai Sonder and adapted from Ref. [3])

range of crops and environments. These research bottlenecks if addressed can also be expected to complement existing knowledge (Fig. 1.2), thereby also capitalizing on previous investment. However, other gaps exist in our understanding of how to maximize the output and stability of cropping systems. Since many challenges to wheat production are experienced across continents (Fig. 1.3), global collaboration offers many advantages, in terms of efficiency of scale, encompassing representative sites within and among target environments, and by coordinating efforts across a range of stakeholders thereby avoiding costly duplication of effort [24]. In summary, maximizing the impacts from crop research requires cross-stakeholder interaction to share know-how tailored to stakeholder requirements [25].

1.8   Choosing Crop Improvement Approaches

A young crop scientist may be overwhelmed by the volume of scientific literature available, and the many different theories about how crop productivity can or should be boosted. In addition, there are bandwagons in crop science [26] that both funding bodies as well as peer-pressure 'encourage' the science community to board. Joining can be a useful learning experience, positive for the career and possibly lead to impacts. However, the true scientific mind goes where the evidence takes it. Luckily science still upholds its internal standards through the institution of voluntary, anonymous peer-review, helping to maintain the scientific bar high in terms of objectivity and rigor. However, no one is without bias and keeping an open mind is always a worthy challenge. As an example, a recent study challenged a growing movement that believe – with some justification – that an industrial model of agriculture with its intensive farming practices, make society more vulnerable to unpredictable climate and other environmental impacts. The study looked specifically at the impact of winter wheat selection in North Europe under intensive inputs, with respect to its genetic gains across a range of high and low input systems. The results showed that the genetic gains achieved at high input stood up when tested across all levels of input [27], mirroring similar findings in Spring wheat breeding [28]. However, such results, valuable and practical as they are, should be taken at face-value and not be used to make sweeping generalizations about one cropping system over another. For example, while crop yields tripled over the last 60 years, Nitrogen (N) application increased tenfold [29]. Only research conducted objectively can provide the answers we need as contributors to food security; and proofs of concept can only come from outputs of research that are tested directly in the appropriate plant breeding and crop management contexts, before they can be scaled to meet the challenges that agriculture must face in the future.

The future of food security will depend on a combination of the ecological prudence of the past and the technological advances of today (M.S. Swaminathan)

1.9 Main Objectives of the Textbook 'Wheat Improvement – Food Security in a Changing Climate'

While the scientific context for each main section of this volume has been presented already, outlines of individual chapters are not listed here, as the information is readily accessible in the Table of Contents and in the Abstracts of each chapter. However, it is worth mentioning the why. The textbook was developed with three main objectives. One was to put together in a single volume, a compendium of knowledge about the theory and practice of wheat improvement to serve as a guide to full-time students of the field as well as scientists from a given discipline wishing to brief themselves on areas outside of their own expertise. Among the authorship are world authorities in their respective fields which certainly lends weight to the content. There is a CIMMYT bias in authorship, partly a tactic to ensure timely delivery of the book as a whole, but also reflecting the paramount role that CIMMYT has played on wheat breeding globally for more than half a century, currently impacting around 70% of all wheat grown globally and generating an estimated extra revenue of $2–3 billion dollars annually for farmers in the Global South alone [16].

Nonetheless, readers should not assume this volume to be a definitive, last word on wheat improvement. Authors of all chapters were asked to cite the literature in a selective way, so as to give readers access to other sources that complement understanding and in many cases provide alternative perspectives. Furthermore, while the attempt was made to cover key disciplines, it is recognized that what may be priorities from a global perspective (reflecting the professional background and bias of the editors), can allow important challenges and disciplines to be overlooked. For example, there is no chapter in this volume on chilling and freezing stress tolerance which are especially important for winter wheat. The fall sown winter wheat crop must survive harsh frosts and snow cover without incurring irreversible tissue damage caused by internal ice crystals. They must also be able to fix carbon on cold but sunny winter mornings when chilling can be an important factor that causes photooxidative damage; readers are referred to Muhammad et al. [30] for an up to date review on cold stress acclimation in wheat. While micronutrients are addressed in the chapter covering microelement deficiency and toxicity, macronutrients are not covered in this volume. Despite wheat being a good nitrogen scavenger, there is much interest in breeding for nitrogen and other macronutrient use efficiency, for example [31], while a body of literature on the impact of the microbiome on crop nutrition is starting to accumulate, including possible genotype effects [32]. Neither was a chapter on roots commissioned but readers are referred to "Wheat root systems as a breeding target for climate resilience" just published [33]. Lodging resistance is missing despite its persistent negative impact on wheat (and other crops), but readers are referred to a comprehensive review on the subject for cereals [34] and more recent efforts to identify genetic bases in wheat [35].

A second objective is to disseminate the information in this book as much as it can be useful since (i) wheat is the most widely grown crop globally, (ii) many wheat colleagues – particularly in the Global South – work with very restricted budgets, so access to costly literature is therefore limited, and (iii) potentially to serve as a technical reference point for the many stakeholders involved in wheat improvement. Through a grant from the Bill and Melinda Gate's Foundation, the cost for publishing this volume as open access is covered, so the whole volume can be shared electronically, printed locally, and even translated to other languages if desired without restrictions.

Finally, as with any textbook, this volume benchmarks the state-of-the-art in wheat breeding, but at a key moment in the history of agriculture. Decisions and actions that are taken now will be pivotal to future food security for a number of reasons, for which crop breeding – if adequately resourced – can provide at least partial solutions. The factors are well known and have already shown global impacts: a less predictable and generally harsher climate; declining water resources; widescale attrition and disappearance of arable soils; a burgeoning population with increased demands for wheat products; grave concerns about the evolution of new pests and disease races and the threat of crop pandemics looming closer as some diseases are already jumping species barriers; an imperative to reduce the environmental footprint of agriculture to help avert devastating sea level rises for example, associated with global warming; a need to produce more on the same land to decelerate encroachment of agriculture into precious natural ecosystems, and the list goes on. These are significant challenges not only for breeding per se, but also to the way agriculture – the widely recognized cornerstone of civilization – will be conducted in the future. However, if you are reading this you have already embraced the challenge.

1.10 Key Concepts

Wheat breeding has a long history and excellent precedents Many new technologies can be applied to emerging problems; interdisciplinary approaches applied through collaborative research are likely to be more efficient than working in silos, assuming objectivity; proof of concept need to be achieved in the appropriate context before breeding pipelines are changed.

1.11 Conclusions

Wheat breeding has been extremely successful especially since the Green Revolution and much of the progress made was due to the open sharing of germplasm and knowledge among wheat scientists, which holds up until today. As long as hybrid wheat does not become a widely accepted reality, wheat research is likely to remain a critical activity in the public domain, in particular in the Global South where most wheat is produced. In order to match predicted demand and adapt the crop to a more challenging environment, crop scientists must demonstrate objectivity and rigor, in order to combine technologies – both old and new – that will deliver reliable productivity gains. We trust, this book will help to generate interest among young scientists to enter the exciting field of crop and in particular wheat improvement.

Nobody is qualified to become a statesman who is entirely ignorant of the problems of wheat (Socrates/Plato)

References

1. Food and Agriculture Organization of the United Nations (2012) FAO statistics
2. FAO (2004) What is Agrobiodiversity? Fact sheet. Building on Gender, Agrodiversity and Local Knowledge. Rome. https://www.fao.org/3/y5609e/y5609e02.htm
3. Braun H, Atlin G, Payne T (2010) Multi-location testing as a tool to identify plant response to global climate change. In: Reynolds M (ed) Climate change and crop production. CIMMYT, Mexico City, pp 115–138
4. Wallace JG, Rodgers-Melnick E, Buckler ES (2018) On the road to breeding 4.0: Unraveling the good, the bad, and the boring of crop quantitative genomics. Annu Rev Genet 52:421–444. https://doi.org/10.1146/annurev-genet-120116-024846
5. Fernie AR, Yan J (2019) De novo domestication: an alternative route toward new crops for the future. Mol Plant 12:615–631. https://doi.org/10.1016/j.molp.2019.03.016

6. Ammar K, Mergoum M, Rajaram S (2004) The history and evolution of triticale. In: Mergoum M, Gomez-Macpherson H (eds) Triticale improvement and production. FAO, Rome, pp 1–9
7. Lobell DB, Cassman KG, Field CB (2009) Crop yield gaps: their importance, magnitudes, and causes. Annu Rev Environ Resour 34:179–204. https://doi.org/10.1146/annurev.environ.041008.093740
8. Fischer RA, Byerlee D, Edmeades GO (2014) Crop yields and global food security: will yield increase continue to feed the world? ACIAR monograph No. 158. Australian Centre for International Agricultural Research, Canberra
9. Atlin GN, Cairns JE, Das B (2017) Rapid breeding and varietal replacement are critical to adaptation of cropping systems in the developing world to climate change. Glob Food Sec 12:31–37. https://doi.org/10.1016/j.gfs.2017.01.008
10. Smýkal P, Nelson MN, Berger JD, Von Wettberg EJB (2018) The impact of genetic changes during crop domestication. Agronomy 8. https://doi.org/10.3390/agronomy8070119
11. Singh RP, Hodson DP, Jin Y, Lagudah ES, Ayliffe MA, Bhavani S, Rouse MN, Pretorius ZA, Szabo LJ, Huerta-Espino J, Basnet BR, Lan C, Hovmøller MS (2015) Emergence and spread of new races of wheat stem rust fungus: continued threat to food security and prospects of genetic control. Phytopathology 105:872–884. https://doi.org/10.1094/PHYTO-01-15-0030-FI
12. Ristaino JB, Anderson PK, Bebber DP, Brauman KA, Cunniffe NJ, Fedoroff NV, Finegold C, Garrett KA, Gilligan CA, Jones CM, Martin MD, MacDonald GK, Neenan P, Records A, Schmale DG, Tateosian L, Wei Q (2021) The persistent threat of emerging plant disease pandemics to global food security. Proc Natl Acad Sci 118:e2022239118. https://doi.org/10.1073/pnas.2022239118
13. Salvi S, Porfiri O, Ceccarelli S (2013) Nazareno Strampelli, the 'Prophet' of the green revolution. J Agric Sci 151:1–5. https://doi.org/10.1017/S0021859612000214
14. Fisher RA, Mackenzie WA (1923) Studies in crop variation. II. The manurial response of different potato varieties. J Agric Sci 13:311–320. https://doi.org/10.1017/S0021859600003592
15. Borlaug NE (2007) Sixty-two years of fighting hunger: personal recollections. Euphytica 157:287–297. https://doi.org/10.1007/s10681-007-9480-9
16. Lantican MA, Braun H-J, Payne TS, Singh RP, Sonder K, Michael B, van Ginkel M, Erenstein O (2016) Impacts of international wheat improvement research, pp 1994–2014
17. Smale M, Reynolds MP, Warburton M, Skovmand B, Trethowan R, Singh RP, Ortiz-Monasterio I, Crossa J (2002) Dimensions of diversity in modern spring bread wheat in developing countries from 1965. Crop Sci 42:1766–1779
18. Luo M, Xie L, Chakraborty S, Wang A, Matny O, Jugovich M, Kolmer JA, Richardson T, Bhatt D, Hoque M, Patpour M, Sørensen C, Ortiz D, Dodds P, Steuernagel B, Wulff BBH, Upadhyaya NM, Mago R, Periyannan S, Lagudah E, Freedman R, Lynne Reuber T, Steffenson BJ, Ayliffe M (2021) A five-transgene cassette confers broad-spectrum resistance to a fungal rust pathogen in wheat. Nat Biotechnol 39:561–566. https://doi.org/10.1038/s41587-020-00770-x
19. Pingali PL (2012) Green revolution: Impacts, limits, and the path ahead. Proc Natl Acad Sci 109:12302–12308. https://doi.org/10.1073/pnas.0912953109
20. Stewart BA, Lal R (2018) Increasing world average yields of cereal crops: it's all about water. In: Sparks DL (ed) Advances in agronomy, vol 151, pp 1–44
21. Reynolds M, Chapman S, Crespo-Herrera L, Molero G, Mondal S, Pequeno DNL, Pinto F, Pinera-Chavez FJ, Poland J, Rivera-Amado C, Saint-Pierre C, Sukumaran S (2020) Breeder friendly phenotyping. Plant Sci 295:110396. https://doi.org/10.1016/j.plantsci.2019.110396
22. Radhakrishnan GV, Cook NM, Bueno-Sancho V, Lewis CM, Persoons A, Debebe Mitiku A, Heaton M, Davey PE, Abeyo B, Alemayehu Y, Badebo A, Barnett M, Bryant R, Chatelain J, Chen X, Dong S, Henriksson T, Holdgate S, Justesen AF, Kalous J, Kang Z, Laczny S, Legoff J-P, Lesch D, Richards T, Randhawa HS, Thach T, Wang M, Hovmøller MS, Hodson DP, Saunders DGO (2019) MARPLE, a point-of-care, strain-level disease diagnostics and surveillance tool for complex fungal pathogens. BMC Biol. https://doi.org/10.1186/s12915-019-0684-y
23. Reynolds M, Atkin OK, Bennett M, Cooper M, Dodd IC, Foulkes MJ, Frohberg C, Hammer G, Henderson IR, Huang B, Korzun V, McCouch SR, Messina CD, Pogson BJ, Slafer GA, Taylor NL, Wittich PE (2021) Addressing research bottlenecks to crop productivity. Trends Plant Sci 26:607–630. https://doi.org/10.1016/j.tplants.2021.03.011

24. Reynolds MP, Braun HJ, Cavalieri AJ, Chapotin S, Davies WJ, Ellul P, Feuillet C, Govaerts B, Kropff MJ, Lucas H, Nelson J, Powell W, Quilligan E, Rosegrant MW, Singh RP, Sonder K, Tang H, Visscher S, Wang R (2017) Improving global integration of crop research. Science 357:359–360. https://doi.org/10.1126/science.aam8559
25. Cornelissen M, Malyska A, Nanda A, Lankhorst R, Parry M, Rodrigues V, Pribil M, Nacry P, Inzé D, Baekelandt A (2020) Biotechnology for tomorrow's world: scenarios to guide directions for future innovation. Trends Biotechnol. https://doi.org/10.1016/j.tibtech.2020.09.006. (in press)
26. Bernardo R (2016) Bandwagons I, too, have known. Theor Appl Genet 129:2323–2332. https://doi.org/10.1007/s00122-016-2772-5
27. Voss-Fels KP, Stahl A, Wittkop B, Lichthardt C, Nagler S, Rose T, Chen T-W, Zetzsche H, Seddig S, Baig MM, Ballvora A, Frisch M, Ross E, Hayes BJ, Hayden MJ, Ordon F, Leon J, Kage H, Friedt W, Stützel H, Snowdon RJ (2019) Breeding improves wheat productivity under contrasting agrochemical input levels. Nat Plants 5:706–714. https://doi.org/10.1038/s41477-019-0445-5
28. Reynolds M, Braun H (2019) Benefits to low-input agriculture. Nat Plants 5:652–653. https://doi.org/10.1038/s41477-019-0462-4
29. FAO (2017) The future of food and agriculture. Trends and challenges. Rome. 163 pp. https://www.fao.org/3/i6583e/i6583e.pdf
30. Hassan MA, Xiang C, Farooq M, Muhammad N, Yan Z, Hui X, Yuanyuan K, Bruno AK, Lele Z, Jincai L (2021) Cold stress in wheat: plant acclimation responses and management strategies. Front Plant Sci 12:1234. https://doi.org/10.3389/fpls.2021.676884
31. Salim N, Raza A (2020) Nutrient use efficiency (NUE) for sustainable wheat production: a review. J Plant Nutr 43:297–315. https://doi.org/10.1080/01904167.2019.1676907
32. Azarbad H, Constant P, Giard-Laliberté C, Bainard LD, Yergeau E (2018) Water stress history and wheat genotype modulate rhizosphere microbial response to drought. Soil Biol Biochem 126:228–236. https://doi.org/10.1016/j.soilbio.2018.08.017
33. Ober ES, Alahmad S, Cockram J, Forestan C, Hickey LT, Kant J, Maccaferri M, Marr E, Milner M, Pinto F, Rambla C, Reynolds M, Salvi S, Sciara G, Snowdon RJ, Thomelin P, Tuberosa R, Uauy C, Voss-Fels KP, Wallington E, Watt M (2021) Wheat root systems as a breeding target for climate resilience. Theor Appl Genet 134:1645–1662. https://doi.org/10.1007/s00122-021-03819-w
34. Berry PM, Sterling M, Spink JH, Baker CJ, Sylvester-Bradley R, Mooney SJ, Tams AR, Ennos AR (2004) Understanding and reducing lodging in cereals. Adv Agron 84:217–271. https://doi.org/10.1016/S0065-2113(04)84005-7
35. Piñera-Chavez FJ, Berry PM, Foulkes MJ, Sukumaran S, Reynolds MP (2021) Identifying quantitative trait loci for lodging-associated traits in the wheat doubled-haploid population Avalon × Cadenza. Crop Sci. https://doi.org/10.1002/csc2.20485

Chapter 2
History of Wheat Breeding: A Personal View

R. A. (Tony) Fischer

Abstract For more than a century, breeding has delivered huge benefits as a major driver of increased wheat productivity and of stability in the face of inevitable disease threats. Thus, the real cost of this staple grain has been reduced for billions of consumers. Steady breeding progress has been seen across many important traits of wheat, currently for potential yield averaging about 0.6% p.a. This yield progress continues to rely of extensive multilocational yield testing but has, however, become more difficult, even as new breeding techniques have improved efficiency. Breeding will continue to evolve as new approaches, being proposed with increasing frequency, are tested and found useful or not. High throughput phenotyping (HTPP), applying modern crop physiology, and molecular markers and genomic selection (GS) are in this phase right now. Such new techniques, along with pre-breeding for new traits, will likely play a larger role in this future improvement of wheat. New tools will also include genetic engineering (GE), as society's need for its benefits become more urgent. The steady privatization of breeding seems unlikely to cease in the developed world but will continue to struggle elsewhere. It would seem wise, however, that a significant portion of the world's pre-breeding research remains in the public sector, while maintaining close and equitable contact with those delivering new varieties.

Keywords Yield progress · Plant pathology · Grain quality · Biometrics · Pre-breeding · Privatization

2.1 Learning Objectives

- To know about and be proud of the past achievements of wheat breeders
- To understand the successful techniques making for this progress and the importance of breeding × agronomy interactions
- To be aware of the new breeding technologies but mindful of the need for validation in the real world

R. A. (Tony) Fischer (✉)
Agriculture & Food, CSIRO, Canberra, Australia
e-mail: tony.fischer@csiro.au

© The Author(s) 2022
M. P. Reynolds, H.-J. Braun (eds.), *Wheat Improvement*,
https://doi.org/10.1007/978-3-030-90673-3_2

- To appreciate the evolution towards larger multidisciplinary breeding teams and the continuing key role of teamwork and strong leadership.
- To recognize the ongoing place for public research in wheat breeding which is steadily privatizing.

2.2   Introduction

I am not a wheat breeder, rather I have been a crop physiologist/agronomist specializing in wheat for most of my long career in Australia and in Mexico at CIMMYT. Therefore, it is both an honour and a special challenge to contribute to this book targeting young scientists, many early in wheat-breeding careers. The challenge is to tell you something of past and present wheat breeding that is of value for your future career in agriculture. I say agriculture because many of us finish in other often-related fields than where we start. This is not bad, for I am firmly believe in scientific breadth, as well as depth in some speciality, likely to be breeding in your cases. What I have in mind is commonly described as the T-trained person, the "jack-of-all trades and master of one".

The inspiration that one derives from being amongst leading wheat breeders is important. In my case, in the early 1960s it was Albert Pugsley and Jim Syme at Wagga Wagga (where William Farrer Australia's famous first wheat breeder had worked), then from 1970 to 1975 at CIMMYT, Norman Borlaug, Frank Zillinsky, Glenn Anderson and Sanjaya Rajaram, and all the US and Canadian breeders who were regularly in NW Mexico to attend to their winter nurseries of spring cereals. My second period at CIMMYT (1988–1995) as Wheat Program Director again put me in touch with wheat breeding around the world. For you, there will be others, your contemporaries, but I recommend that you read about your predecessors, especially Borlaug (e.g., Vietmeyer's 2011 book [1], see also the vintage Borlaug 1968 IWGS presentation below). Successful wheat breeders of my vintage were very dedicated to breeding, hardworking, spending long hours in the nurseries and field plots, very focussed on their breeding goals and prepared to persist decades to achieve them. They led small teams of scientists and technicians with a firm hand and changed successful breeding strategies reluctantly. As a young scientist at the International Wheat Genetics Symposium in 1968 in Canberra, I witnessed crop physiologist Professor Colin Donald deliver for the first time his radical concept of a wheat ideotype [2]; it did not go down very well with the assembled breeders from around the world. Borlaug's description of his already remarkably successful breeding program in Mexico, with its unique emphasis on efficiency and broad adaptation, along with his fiery ridicule of bureaucrats and "band wagons" for hampering scientific progress in agriculture, was much more popular [3]!

Since that congress when my wheat career was just beginning, many things in wheat breeding around the world have gradually changed, while breeding progress in key traits has been maintained almost uninterrupted. Lessons have been learnt, supporting technologies have advanced in almost unimaginable ways, and the organization of breeding has altered notably. Some things have however not changed., nor should they as we look to the near future. The rest of this Chapter will deal with these issues, briefly given the space available and since many will reappear in detail in later Chapters.

2.3   Past Wheat Improvement at the Farm Level and in the Breeders' Plots

World *wheat yield* has increased remarkably linearly at about 40 kg/ha/y over the last 60 years (Fig. 2.1); for projection to meet future demand, the key number is this slope relative to today's yield of 3.5 t/ha, namely 1.16%. Fischer and Connor [4] argue that while this rate of increase is probably adequate to balance world wheat demand growth, a greater rate would help poor consumers by reducing pressure on prices, would protect against negative contingencies, and would reduce the pressure for greater wheat area (including clearing new land to achieve this). Yields in most wheat-growing countries and regions reveal similar close-to-linear increases at various rates clustered around the world figure [5] (also see Chap. 4). For example the irrigated Yaqui Valley of NW Mexico, where CIMMYT's major yield testing and selection is undertaken, shows one of the higher rates of absolute increase (Fig. 2.1,

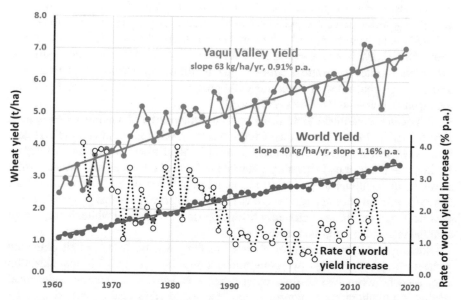

Fig. 2.1 Annual wheat yields from 1961 to 2019 for the world and the Yaqui Valley of northwest Mexico, also, relative rate of increase of world wheat yield with time based on 7 year moving average and plotted against the middle year. Note % p.a. slopes are expressed relative to the yield at the end of any period. (Sources: World yields (fao.org/faostat/en/#data/QC, accessed 17 September 2020); Yaqui Valley yields (various official sources in the State of Sonora, Mexico))

63 kg/ha/year, but currently only 0.91% p.a.), reflecting this breeding effort and the concurrent modernization of crop management by the Mexican farmers.

The percentage rate of increase in yield is a reasonable measure of productivity progress. Although it has fallen steadily with time – in % not in kg/ha – as shown for the world in Fig. 2.1, it is still strongly positive. I say wheat improvement because the yield progress has involved new varieties, new agronomy (or crop management), and the positive interaction between the two (G × M). The key agronomic changes include mechanized planting giving better plant populations and more timely sowing, increased fertilizer use, more irrigation, and improved weed and disease control. There has been endless discussion as to whether breeding or agronomy has played the greater role, but neither discipline alone could have achieved even half this progress; they have been complementary throughout, with agronomy continuing to create new challenges and opportunities for breeding.

The primary target of modern wheat breeding has been increased yield, through eliminating yield-related deficiencies such as lodging and shattering, fixing optimum height and flowering date, and seeking to raise inherent yield. *Breeding progress for yield* is commonly measured in vintage trials [6]. If management is excellent, water adequate, and diseases are absent or controlled, this measures potential yield (PY) under the best agronomy and weather of the time, thus including progress due to G × M. If water supply is inadequate as in many rainfed regions, we have water-limited potential yield (PYw). There are now new ways to measure such progress for multilocational multiyear national trials [6]. Throughout it is argued that progress is most usefully expressed, as above, relative to yield of the most recent cultivars in any series. Recent reports of breeding yield progress in wheat from around the world have been complied [7]: from 34 case studies the average rate of progress was 0.58 ± 0.034 % p.a., ranging from 0.2% to 1.1% p.a. There was no significant difference in rates of progress between spring and winter wheats, nor between PY and PYw. Recent rates of breeding yield progress with rice, maize and soybean averaged also around 0.6% to 0.7% p.a. [7].

Wheat breeders made yield progress under a *variety of breeding schemes* suitable for self-pollinated crops (see Chaps. 5 and 7). What is common to all systems is massive investment in yield measurement in field plots, beginning as early as F5 in home fields, then in steadily reduced numbers of advanced lines at increasing numbers of locations representing the target population of environments (TPEs). Since the middle of last century there has been no big change in this general scheme; new developments have continually been proposed, and if worthwhile, incorporated into the scheme to improve breeding efficiency.

Borlaug's unique strategy of shuttle breeding was controversial: it delivered greater efficiency through two generations a year but was novel in that selection alternated between two distinct environments in Mexico, preceding widespread testing in collaborating programs around the world. This testing was adopted by CIMMYT when it began in 1966 which, along with ICARDA starting in the 1970s, and building on early efforts by USDA and FAO, gave rise to the extensive and unique international network for testing and germplasm distribution [8], which continues to this day. The strategy of selecting and testing in many environments has been vindicated with the production of a number of superior cultivars having *broad adaptation,* meaning good performance across locations and years (e.g., Siete Cerros 66, Pavon 76, Anza (via California), Seri 82, Attila or PBW343 (via India), Borlaug100). Other breeding efforts have also delivered a small number of varieties which have dominated over large and seemingly diverse regions (e.g., Florence Aurore from Tunisia, Gabo from Australia, Bezostaya from Krasnodar in Russia, Capelle Deprez from France). As mentioned, the relative yield progress seen in PYw, or at lower soil fertility levels, is unabated compared to that under PY conditions and what's more, the cultivar ranking generally changes little across a large range of such resource inputs (e.g., recent references [9, 10]). In fact, after allowing for differences in flowering date, which can be important in particular years, amongst low latitude spring wheats there are few significant crossover interactions in the absence of disease; the characteristic fan pattern of variety yield response to site mean yield, popularized long ago by Finlay and Wilkinson [11] in barley, remains valid even today. Finally, some advocate intrinsic yield stability, which is of limited value since it tends to mean low average yield; yield responsiveness (to good years and management) is what modern farmers need!

In the case of wheat, the second target for breeding, taking up to one half of the breeding effort, is aimed at *biotic stress resistances,* strengthening, and then maintaining genetic resistance to diseases (see Chaps. 8, 9, 19 and 20). This is adding a useful type of stability but is rarely related to PY. Also included is a smaller investment directed against insect pests and nematodes (see Chap. 20) for which biocides are less effective and more dangerous. Plant pathology has been the discipline most closely linked with wheat breeding since its outset, indeed many pathologists have become successful breeders. The first single genes identified were major rust resistance genes, and many years later, in 2003, the first wheat rust gene was sequenced. Being a serious disease that knows no borders, rust was the reason for the first international screening nurseries, as already mentioned. Since then, this collaboration has grown and a host of major and minor rust resistance genes have been identified, catalogued, sometimes sequenced, and freely shared and utilized by breeders around the world. Early warning systems and ongoing deployment of new major genes and more durable minor ones have meant that wheat yield losses due to rust are lower now than ever, notwithstanding the apparent uniformity of modern wheats. This is a powerful tribute to unfettered international collaboration amongst wheat breeders and rust pathologists; the current Borlaug Global Rust Initiative (BGRI) is the latest iteration in this process.

Since there are so few yield losses due to rust these days, rust breeding is now directed more at maintaining resistance with the deployment of more durable resistance genes including GE solutions. Attention has also passed to the multitude of other diseases of wheat, for many of which host plant resistance can be effective. There is, however, never any room for complacency, with new diseases and new virulences likely to threaten wheat at any time and readily spread in our interconnected world. The latest wheat example is the occurrence of a wheat blast (*Magnaporte oryzae* pathotype *Triticum*), first seen in Brazil in 1985, and now in Asia (Bangladesh in 2016) and Africa (Zambia in 2018). Genetic biotic stress resistance, including GE solutions, will probably become even more important if societies only partly justifiable fear of biocides continues to grow.

The *industrial quality of wheat* (its suitability for products for human consumption, in particular the many forms of bread and noodles) has been the third major target for breeders (see Chap. 11) and an important element of productivity gain not captured in yield statistics. Rapid low-cost tests for various quality traits were widely used for screening early generations from the 1960s onwards, and overall industrial quality has generally been improved even in the face of the inevitable decline in grain protein concentration with yield improvement. (e.g., [12, 13]). Market price premiums for desirable quality are essential, so farmers as well as consumers see the benefit.

In the last 30 years, concern has grown for the *nutritional and health qualities* of wheat, especially its inherently low concentrations of iron and zinc, values which had tended to fall as wheat PY had been lifted by breeding. A variety have recently been released in India with improved grain zinc levels, and there are genotypes in the pipeline with other health-giving properties (e.g., high iron, soluble fibre, fructans). These issues are likely to receive more attention in the future (see Chap. 12).

2.4   Past Activities Associated with Greater Breeding Success and Efficiency

Genetic variation is essential to breeding success: especially since the middle of the twentieth century there has been a big increase in collection and conservation of *wheat genetic resources*, ranging from wild wheat ancestors through land races to named varieties and genetic stocks (see Chaps. 16, 17 and 18). Fortunately, genetic resource scientists moved quickly to collect such materials as modern varieties began to replace land races in farmers' fields. However, because of linkage drag, the utilization of these materials by breeders has been slow. Disease resistance and some quality genes are the best examples of useful introgression into modern cultivars. Also, some accessions have now been identified as sources for increase in yield, this includes 1B/1R, 1A/1R, 2NS, and the LR19 translocation from *Agropyron*. This process has been helped partly through the large effort that CIMMYT has put into creating synthetic wheats with new accessions of *Triticum tauschii*, and incorporating them into its breeding program where they have demonstrated both increased PY and PYw; success with such material in Sichuan, China, is a recent example [14]. Wheat genetic resources are safely conserved but their exploitation in breeding, which is a form of pre-breeding, remains too slow due to underfunding.

Breeding efficiency over the last 50 years has been greatly facilitated by *allied fields of technology and science* combining with breeder ingenuity. This includes the mechanization of seeding and harvesting (see Chap. 15), the acceleration of generation advance (see Chap. 30), and automation of all repetitive tasks, including NIR-based measurement of quality traits and molecular markers for difficult to measure qualitative traits (see Chap. 28). Biometrics has brought large *advances in trial design and computing* for processing of data and applying complex algorithms to field measurements correcting for spatial variation in ever more efficient plot designs (see Chap. 13). This progress is probably now reaching the limits imposed by measurement error and soil spatial variation. This is a special problem as the relative yield gains being sought become smaller (note 0.6% p.a. is only a 3% jump every 5 years). Finally, the ever-present G × E (genotype × environment) driven by both locational and annual variation in E remains a special challenge. Many statistical models have been applied over the years, with factor analytics the most recent (see Chap. 3). Also, crop simulation modelling is valuable for characterizing environments, especially rainfed ones (e.g., [15]. Such modelling is now based on sufficiently-sound physiological knowledge to also allow the exploration *in silico* across TPEs of the effect of changes in some key traits (e.g., phenological ones), but such modelling is very unlikely to be a substitute for accurate multilocational yield testing (see Chap. 31). The past failure of many breeders to adequately measure their environments (soil, weather), and thus facilitate a better understanding of the basis of G × E, has always been a weakness, but national and global weather services are now filling this gap.

In the late 1960s it was expected that *physiology* would help breeders accelerate yield progress, explaining why CIMMYT first hired me, a disciple of the physiological thinking of Lloyd Evans and UK physiologists, especially Roger Austin, and breeder John Bingham. Much is now known about the crop physiological changes behind the yield progress since 1960: generally flowering date is unchanged or slightly earlier, height is substantially reduced (from >120 cm to <90 cm), harvest index has increased as has grain number (/m^2), but not necessarily spike number (/m^2). Stomatal conductance and leaf photosynthetic rate have increased along with leaf erectness, and lately biomass is also increasing, as is grain weight in some places. Apart from earliness and height reduction, and with a few exceptions such as erect leaves, almost nowhere in the world were the other changes either pre-emptively identified by crop physiologists, and/or deliberately selected by wheat breeders. There are lessons in this observation: maybe physiology should not have been so focussed on retrospective studies, missing opportunities for testing traits in breeders' populations and in early generation indirect yield selection, some of which such as harvest index, fruiting efficiency and stomatal conductance/canopy temperature are discussed in depth recently in Fischer and Rebetzke [16]. One constraint was that physiological studies often paid little attention to the crowded crop situation in which yield is to be delivered. Donald [2] in 1968 pointed out how much smaller than the isolated wheat plant was the plant under heavy competition in the crop and argued that for higher yield the crop plant needed traits that made it less competitive and more "communal"; lately this neglected notion has received solid support in retrospective studies of yield progress. Another constraint with early physiology was that trait measurement was too slow/expensive for use in selection by breeders, and a final constraint, physiology often did not work sufficiently closely to and cooperatively with real breeding programs. HTPP has been proposed lately as one way of dealing with the trait measurement constraint (see Chap. 27), but there still needs to be an intimate link with open-minded and well-resourced breeders.

These days widespread *pre-breeding* aims to transfer to elite materials (and validate) potentially useful physiological (and morphological) traits, for their subsequent easier incorporation into better varieties by other breeders who generally don't have the resources for risky pre-breeding (see Chap. 25). Dwarfing genes, alternatives to those which catalysed the Green Revolution, are a potentially useful target for such exploitation. Another current use of physiological knowledge, undertaken in CIMMYT Wheat Fisiologia, is in the selection of parents with measured physiological traits which are likely to be complementary for yield [17].

Over the last century, other new techniques to aid crop genetic improvement have, like physiology, been highlighted but have often failed to realize their early claims of success. Simmonds [18] disparagingly called them "*band wagons*" and his list includes induced polyploidy, mutation breeding, physiology (again), and somaclonal variation; F1 hybrids for wheat could also be added, but that effort continues in several breeding programs, encouraged by successes with hybrid rice since the 1980s. The lesson for the breeder regarding band wagons, and they appear with regularity, seems to be to hasten slowly, change currently successful programs gradually and only after solid evidence of efficiency gains has been gathered. We shall return to this, for Simmonds also included biotechnology in his bandwagon list!

2.5   Some Future Considerations for Breeding

History is of little use if it doesn't guide the future. Field grown wheat will be around for your lifetime and field testing of yield in crop-like plots will remain paramount. But what may change are the breeding tools, the natural environment, and the agronomy. Indeed. innovation never ceases and wheat breeding is now engaging with a suite of new tools (band wagons if you like) proposed to improve the efficiency and effectiveness of the breeding, as described in the Chaps. 16 to 32 on translational research. Unfortunately, space limits attention to these issues here.

The first consideration which must be emphasized, however, is an ongoing problem with field testing, namely *bias in plot trials*. In small plots (< say 3 m^2) which are harvested without trimming, yield can easily be biased by as much as the breeding progress expected to be achieved in 5 years (only 3% at best). There is little doubt that cultivars can perform differently in plot ends and edge rows than in inside bordered-rows, and that where paths are narrow (<50 cm) plants in edge rows can compete for light and nutrients (and moisture if rainfed) with adjoining plots; all this distorts or even negates their performance relative to inner rows [19]. Larger sown plots and/or edge trimming is essential, while certain simple measurements (e.g., path NDVI) can help detect and perhaps correct for such bias.

New tools offer help with the biggest specific challenge facing wheat breeding, the need *to continue to lift potential yield*. After 100 years of success in this area, relative rates of breeding progress for yield have, as we have seen, slowed to currently around 0.6% p.a., yet breeding investment in real terms has probably increased. Does this herald an approach to the biological limit for yield? Probably it does. But can new tools and pre-breeding lift rate of progress and/or ultimately push back this limit? Is greater progress to be achieved by focussing now more on specific adaptation, better exploiting the locational component of G × E which is so often noted in multilocational trails (see Chap. 3)? Will seed production and heterosis be improved enough for F1 hybrid wheat to become a reality? These are exciting questions which will be resolved one way or another in the next 20 years of your breeding careers.

HTPP and *GS* have already been mentioned for predicting yield advance; together they could be even better (e.g., [20]). GS allows the shortening of the generation cycle: while HTPP must be applied to segregating populations if it is to be truly useful (e.g., [16]). The new environments predicted by climate change modellers could be another target, but this needs to proceed cautiously because of the uncertainties. Besides the best way to adapt to climate change is to be field testing widely, due to the simple fact that a significant proportion of years across locations in any decade predict better than any model those of the next decade!

GE (often less usefully abbreviated to GM) and gene editing must be part of the near future for wheat breeding, but they will have great difficulty raising potential yield simply because of the genetic complexity of this quantitative trait, the product of millennia of evolution and over a century of breeding. The numerous promising reports on GE crop plants in controlled environments, where mainly photosynthetic, partitioning and drought resistance traits were targeted, have so far failed to deliver extra grain yield in the field [21, 22]. However the first GE event to enhance wheat yield (HB4, see [23] has now been approved in Argentina: substantial yield increases (>20%) have been measured in multi-year large plots and fields when dryness has restricted yields to less than 2 t/ha, while there are no yield penalties at higher levels. Another promising wheat GE event has been the modification of pericarp expansins to give larger grains apparently without the compensatory negative genetic trade-off commonly seen in crops between grain weight and grain number ($/m^2$) [24]. In the meantime, we desperately need *GE to enhance other traits* besides potential yield: the scientific prospects are much better because many such traits are less complex than yield, and there is now often precedent from other crops. Such traits in wheat could include improved nutritional value, such as high iron wheat [25], better resistance to rust (see Chap. 19), or environmentally desirable traits like biological nitrification inhibition. Regulatory barriers to GE traits will fade as society accepts their proven safety and realizes it cannot do without their manifest benefits.

Passing to the *changing natural environment of cropping*: CO_2 is rising inexorably (currently about 2 ppm p.a.), related to this climate is changing (largely warming but maybe drying in middle latitudes, and greater frequency of extreme heat events). Atmospheric pollution (aerosols, ozone in particular) is rising (and declining in some regions where pollution controls are enforced). Finally, water scarcity in irrigated systems is increasing, especially in Asia, due to overextraction of aquifers. The optimal genetic makeup of cultivars will interact with all these changes.

Related to natural environments changes are those in *wheat agronomy, and the cropping, farming and social systems* within which the wheat crop is grown, the input and product prices, and, ultimately, our social licence to farm, which relates to the increasing regulation of cropping practices. Breeders need always to be alert to these developments and hence remain in contact with agronomists and farmers, policy makers and ultimately the public. One example suffices: in southern Australia, Flohr and colleagues [26] recently describe a striking G × M change. Conservation agriculture had improved fallow storage of moisture; along with a gradual shift in rainfall patterns (probably linked to climate change), this opened opportunities for earlier than normal planting of wheat (April instead of May-June). Planting date could be advanced 4–8 weeks, but the optimum flowering date in the spring remained unchanged. Only new combinations of the wheat phenology alleles could deliver cultivars giving optimal flowering dates when being planted much earlier; essentially this meant a switch from spring wheats to fast winter types. The longer crop cycle (sowing to anthesis) had the bonus of bringing deeper roots; in many situations yield improved notably. This new system often requires deeper seeding hence it needed wheats with longer coleoptiles (= alternative dwarfing genes to the Norin10 ones) which was enabled by pre-breeding. Since the early planted winter wheat can deliver substantial winter forage to grazing sheep or cattle without grain yield loss, the whole transformation is aided by the notable rise in the ratio of meat/wheat prices on world markets. Of course, the wheat farming system must have access to grazing animals, which is the case in Australia (and West Asia-North Africa). This serves to remind us that wheat is part, not only of a cropping, but also a farming system.

2.6 Organization and Funding of Wheat Breeding

Ultimately the success of plant breeding (and your jobs as breeders) depends on how this complex task is organized and financed. The roots of modern breeding lie in the late nineteenth century, just before the rediscovery of Mendel's notions of genes and inheritance in 1900. Even then there were *private and public breeding organizations,* although wheat breeding has rarely had the protection of secrecy provided by commercial F1 hybrids (as with maize for example). Notwithstanding this, as time passed, the private wheat breeders became gradually stronger, especially in Europe. Plant variety protection under UPOV rules and seed sale royalties gave greater income security to the private sector, which had become formalized into farmer-owned cooperatives and companies. Following the 1964 Plant Variety and Seeds Act in 1964, a milestone was the full privatization of wheat breeding in the UK in 1987, in accord with the free market concepts of the time; there are valuable lessons in this disruptive experience [27]. Outside of Europe, apart from Argentina, the privatization of breeding was slower. However, this has now accelerated in the New World, especially USA and Australia but less so in Canada, and lastly has begun in Asia. Uniquely, in USA wheat breeding is supported by utility patents and licensing, accompanying check-off fees and royalties on seed sales, and in Australia support is entirely from end-point royalties on grain sales [28]. Payment for private varieties has always been a challenge with wheat since seed can and commonly is saved on-farm without fear of genetic change. Provided there is reasonable adherence to the relevant laws and regulations, the various schemes mentioned here have generally been successful in returning just rewards to the breeder and better varieties for growers.

Around 2020, the biggest multinational wheat breeding efforts are found in traditional breeding companies like Limagrain (French) and KWS (German) in Europe, where also there are several smaller ones such as RAGT Semances (France) and Staaten-Union, the latter uniquely strong in F1 hybrid wheat. Multinational life companies have, through mergers and takeovers in the last 25 years or so, also become significant players in wheat breeding: firstly Syngenta, then relative newcomers Bayer, BASF, and Corteva Agriscience: combining breeding and agricultural chemicals has both synergistic and, unfortunately, perverse elements. All these companies are moving cautiously into the developing world and the ex-Soviet Union, where there were only a few smaller home-grown private breeding companies (e.g., Mahyco (India), SeedCo (East Africa), Buck and Klein (both on Argentina)). Here the public system continues to take major breeding responsibility in the form of state and national wheat breeding institutions and some Universities; this will probably remain the case until and if F1 hybrid wheat becomes feasible. CIMMYT and ICARDA's wheat breeding which targets the developing world has, of course, remained public since its inception around half a century ago, with support from many governments, non-profit organizations and institutions. These two centers continue to play a vital role in supplying international trials of advanced breeding lines and facilitating collaboration amongst all of the worlds' wheat breeders, with germplasm and performance results distributed free of change to all *bona fide* breeders, whether public or private. Their impact has been huge [29]. With competing breeding entities in most countries, another very desirable component is publicly controlled independent testing of candidate varieties for yield and other important attributes, and the associated registration of new varieties. The final critical step in the breeding process is the national seed systems for getting new varieties to farmers (see Chap. 14).

Along with greater privatization and consolidation, wheat breeding has become obviously a big team effort, with the inevitable involvement of associated disciplines such as pathology, cereal quality and biometrics, aided often these days by service providers for routine trials and testing work. *Pre-breeding research* has emerged as a vital supporting activity, but generally is separated from breeding and still publicly funded, essentially because it is a long-term high-risk activity with potential benefits for society which are maximized if its fruits are widely shared. As mentioned, these activities range from genetic resource conservation, the discovery of novel useful genes (traits) in this material (or creation through GE or gene editing), and the incorporation of new traits into lines and populations having relevant modern genetic backgrounds for utilization by all breeders, commercial and public. Also included is strategic plant science aimed at understanding the physiological and functional molecular basis of important wheat traits, with a view to more efficient manipulation of the traits in breeding and selection strategies (see Chap. 28). It suffices here to emphasize that because of "market failure" pre-breeding research merits public investment, and this includes funding from major non-profit organizations as we have seen lately with the International Wheat Yield Partnership (IWYP) and the Heat and Drought Wheat Improvement Consortium (HeDWIC). In the future, more traits may be protected as intellectual property, as is usually the case with GE ones, but meeting equity goals will remain important to maximize benefit and societal acceptance.

The smooth transfer of products of pre-breeding in an equitable way so that all commercial breeders benefit is a challenge yet to be solved. Europe seems to have made most progress in imbedding independently funded pre-breeding research into the private breeding process in a mutually beneficial manner. CIMMYT and ICARDA's wheat improvement teams are rare in that they have had for many years carried out pre-breeding alongside their breeding of advanced lines for variety release by NARS, but efficient in-house collaboration can still be challenging. How much further along would CIMMYT be if the early promise for yield advance seen with cumbersome stomatal conductance measurements on F2 plants [30] had been pursued a little longer, thereby encountering the huge efficiency gains in conductance measurement coming from infrared thermometry. This demands open and enlightened leadership, multidisciplinary teamwork, and adequate long-term stable financial support. Balancing this with the need to consider the endless stream of breeding innovations being proposed is a critical challenge: effective breeding programs should only be adopting new technologies when these have been tested in pilot mode and found to deliver!

2.7 Key Concepts

- The goals of wheat breeding have changed little, increased potential yield and host plant resistance remain paramount
- The technology and science of breeding has changed gradually but reliance on multilocational yield testing remains essential
- Genetic engineering and gene editing are starting to deliver valuable trait opportunities for breeding, as is innovative agronomy (examples in Sect. 2.4)
- Multidisciplinary breeding teams have become more important and their effective leadership remains a challenge
- Privatization of wheat breeding grows steadily, but there remains an essential role for the public sector breeding research and pre-breeding and a challenge linking it closely to variety production.

Acknowledgements The author thanks the Editors, and John Passioura and Richard Richards for their input, and especially Stephen Baenziger and William Angus for detailed comments on the manuscript.

References

1. Vietmeyer N (2011) Our daily bread; the essential Norman Borlaug. Bracing Books, Lorton
2. Donald CM (1968) The breeding of crop ideotypes. Euphytica 17:385–403. https://doi.org/10.1007/BF00056241
3. Borlaug NE (1968) Wheat breeding and its impact on world food supply. In: Proceeding of the 3rd international wheat genetics symposium Canberra 1968. Canberra, Australia, pp 1–36
4. Fischer RA, Connor DJ (2008) Issues for cropping and agricultural science in the next 20 years. Field Crop Res 222:121–142. https://doi.org/10.1016/j.fcr.2018.03.008
5. Fischer RA, Byerlee D, Edmeades GO (2014) Crop yield and global food security: will yield increase continue to feed the world, Monograph 158. Australian Centre for International Agricultural Research, Canberra. https://aciar.gov.au/publication/mn158
6. Fischer RA (2015) Definition and determination of crop yield, yield gaps, and the rates of change. Field Crop Res 182:9–18. https://doi.org/10.1016/j.fcr.2014.12.006
7. Fischer RA (2020) Advances in the potential yield of grain crops. In: Gustafson JP, Raven PH, Ehrlich PR (eds) Population, agriculture and biodiversity: problems and prospects. University of Missouri Press, Columbia, pp 150–180
8. Byerlee D, Lynam JK (2020) The development of the international center model for agricultural research: a prehistory of the CGIAR. World Dev 135:105080. https://doi.org/10.1016/j.worlddev.2020.105080
9. Voss-Fels KP, Stahl A, Wittkop B, Lichthardt C, Nagler S, Rose T, Chen T-W, Zetzsche H, Seddig S, Baig MM, Ballvora A, Frisch M, Ross E, Hayes BJ, Hayden MJ, Ordon F, Leon J, Kage H, Friedt W, Stützel H, Snowdon RJ (2019) Breeding improves wheat productivity under contrasting agrochemical input levels. Nat Plants 5:706–714. https://doi.org/10.1038/s41477-019-0445-5
10. Mondal S, Dutta S, Crespo-Herrera L, Huerta-Espino J, Braun HJ, Singh RP (2020) Fifty years of semi-dwarf spring wheat breeding at CIMMYT: grain yield progress in optimum, drought and heat stress environments. Field Crop Res 250:107757. https://doi.org/10.1016/j.fcr.2020.107757
11. Finlay KW, Wilkinson GN (1963) The analysis of adaptation in a plant breeding programme. Aust J Agric Res 14:742–754. https://doi.org/10.1071/AR9630742
12. Laidig F, Piepho H-P, Rentel D, Drobek T, Meyer U, Huesken A (2017) Breeding progress, environmental variation and correlation of winter wheat yield and quality traits in German official variety trials and on-farm during 1983–2014. Theor Appl Genet 130:223–245. https://doi.org/10.1007/s00122-016-2810-3
13. Guzmán C, Autrique E, Mondal S, Huerta-Espino J, Singh RP, Vargas M, Crossa J, Amaya A, Peña RJ (2017) Genetic improvement of grain quality traits for CIMMYT semi-dwarf spring bread wheat varieties developed during 1965–2015: 50 years of breeding. Field Crop Res 210:192–196. https://doi.org/10.1016/j.fcr.2017.06.002
14. Tang Y, Wu X, Li C, Yang W, Huang M, Ma X, Li S (2017) Yield, growth, canopy traits and photosynthesis in high-yielding, synthetic hexaploid-derived wheats cultivars compared to non-synthetic wheats. Crop & Pasture Sci 68:115–125. https://doi.org/10.1071/CP16072
15. Chenu K, Porter J, Martre P, Basso B, Chapman S, Ewert F, Bindi M, Asseng S (2017) Contribution of crop models to adaptation in wheat. Trends Plant Sci 22:472–490. https://doi.org/10.1016/j.tplants.2017.02.003
16. Fischer RA, Rebetzke GJ (2018) Indirect selection for potential yield in early-generation, spaced plantings of wheat and other small-grain cereals: a review. Crop & Pasture Sci 69:439–459. https://doi.org/10.1071/CP17409
17. Reynolds M, Langridge P (2016) Physiological breeding. Curr Opin Plant Biol 31:162–171. https://doi.org/10.1016/j.pbi.2016.04.005
18. Simmonds NW (1991) Bandwagons I have known. Trop. Agric. Assocn. Newsl. 11:7–10

19. Rebetzke G, Fischer RA, van Herwaarden AF, Bonnett DG, Chenu K, Rattey AR, Fettell NA (2014) Plot size matters: interference from intergenotypic competition in plant phenotyping studies. Funct Plant Biol 41:107–118. https://doi.org/10.1071/FP13177

20. Cooper M, Technow F, Messina C, Gho C, Totir LR (2016) Use of crop growth models with whole-genome prediction: application to a maize multienvironment trial. Crop Sci 56:2141–2156. https://doi.org/10.2135/cropsci2015.08.0512

21. Passioura JB (2020) Translational research in agriculture. Can we do it better? Crop & Pasture Sci 71:517–528. https://doi.org/10.1071/CP20066

22. Nuccio ML, Paul M, Bate NJ, Cohn J, Cutler SR (2018) Where are the drought tolerant crops? An assessment of more than two decades of plant biotechnology effort in crop improvement. Plant Sci 273:110–119. https://doi.org/10.1016/j.plantsci.2018.01.020

23. Gonzalez FG, Rigalli N, Miranda PV, Romagnoli M, Ribichich KF, Trucco F, Portapila M, Otegui E, Chan RL (2020) An interdisciplinary approach to study the performance of second-generation genetically modified crops in field trials: a case study with soybean and wheat carrying the sunflower HaHB4 transcription factor. Front Plant Sci 11:178. https://doi.org/10.3389/fpls.2020.00178

24. Calderini DF, Castillo FM, Arenas-M A, Molero G, Reynolds MP, Craze M, Bowden S, Milner MJ, Wallington EJ, Dowle A, Gomez LD, McQueen-Mason SJ (2020) Overcoming the trade-off between grain weight and number in wheat by the ectopic expression of expansin in developing seeds leads to increased yield potential. New Phytol. https://doi.org/10.1111/nph.17048

25. Beasley JT, Bonneau JP, Sánchez-Palacios JT, Moreno-Moyano LT, Callahan DL, Tako E, Glahn R, Lombi E, Johnson AAT (2019) Metabolic engineering of bread wheat improves grain iron concentration and bioavailability. Plant Biotechnol 17:1514–1526. https://doi.org/10.1111/pbi.13074

26. Flohr BM, Hunt JR, Kirkegaard JA, Evans JR, Trevaskis B, Zwart A, Swan A, Fletcher AL, Rheinheimer B (2018) Fast winter wheat phenology can stabilise flowering date and maximize grain yield in semi-arid Mediterranean environments. Field Crop Res 223:12–25. https://doi.org/10.1016/j.fcr.2018.03.021

27. Galushko V, Gray R (2018) Twenty five years of private wheat breeding in the UK: lessons for other countries. Field Crop Res 223:12–25

28. Alston JM, Gray RS (2013) Wheat research in Australia: the rise of public-private-producer partnerships. EuroChoices 12:30–35. https://doi.org/10.1111/1746-692X.12017

29. Alston JM, Pardey PG, Rao X (2020) The payoff to investing in CGIAR research. SoAR (Supporters of Agricultural Research). Virginia, pp. 156

30. CIMMYT (1978) CIMMYT report on wheat improvement 1978. 117–128. CIMMYT, Mexico, DF.

Chapter 3
Defining Target Wheat Breeding Environments

Leonardo A. Crespo-Herrera, José Crossa, Mateo Vargas, and Hans-Joachim Braun

Abstract The main objective of a plant breeding program is to deliver superior germplasm for farmers in a defined set of environments, or a target population of environments (TPE). Historically, CIMMYT has characterized the environments in which the developed germplasm will be grown. The main factors that determine when and where a wheat variety can be grown are flowering time, water availability and the incidence of pests and diseases. A TPE consists of many (population) environments and future years or seasons, that share common variation in the farmers' fields, it can also be seen as a variable group of future production environments. TPEs can be characterized by climatic, soil and hydrological features, as well as socioeconomic aspects. Whereas the selection environments (SE) are the environments where the breeder does the selection of the lines. The SE are identified for predicting the performance in the TPE, but the SE may not belong to the TPE. The utilization of advanced statistical methods allows the identification of GEI to obtain higher precision when estimating the genetic effects. Multi-environmental testing (MET) is a fundamental strategy for CIMMYT to develop stable high grain yielding germplasm in countries with developing economies. An adequate MET strategy allows the evaluation of germplasm in stress hotspots and the identification of representative and correlated sites; thus, breeders can make better and targeted decisions in terms of crossing, selection and logistic operations.

Keywords Mega-environments (ME) · Target population of environments (TPEs) · Selection environments (SEs) · Genotype-by-environment interaction (GEI)

L. A. Crespo-Herrera (✉) · J. Crossa · H.-J. Braun
International Maize and Wheat Improvement Center (CIMMYT), Texcoco, Mexico
e-mail: l.crespo@cgiar.org; j.crossa@cgiar.org; h.j.braun@cgiar.org

M. Vargas
Universidad Autonoma Chapingo, Texcoco, Mexico

3.1 Learning Objectives

- Identify the factors that drive wheat adaptation for the classification of target environments.
- Identify the statistical methods that can be used for defining TPEs.
- Identify the importance of a multi-environmental testing strategy for wheat breeding with a global scope.

3.2 Introduction: Wheat Mega-environments in History and the Context of Global Wheat Breeding

The success of a breeding program and, particularly, of a program with international dimensions such as CIMMYT's wheat breeding program, depends heavily on the characterization of the environments where the germplasm will be grown.

Historically, since its earliest efforts to breed wheat in the 1940s, CIMMYT has characterized the environments in which the developed germplasm will be grown. At that time, this characterization was restricted to the geographical areas in Mexico. However, as soon as CIMMYT's mandate became global in the 1970s, 15 agroecological zones were defined, for instance, the region that encompasses India, Pakistan, Bangladesh and Nepal (South Asia), or the Nile Valley zone (Egypt and Sudan).

Those agroecological zones were redefined in the late 1980s. As it became evident that the sole geographical description was inadequate, given the diversification of the production systems, the need for high-yielding and stable germplasm and the simple fact that for specific conditions, certain traits were needed, particularly those related to stress tolerance and quality requirements. Hence, this redefinition led to the concept of Mega-Environment (ME), described by Rajaram et al. [1] as a "broad" but not necessarily "contiguous area" with similar biotic and abiotic constraints, cropping systems, consumer preferences and production volume. By 1992, twelve ME had been conceived, six for each spring and winter growth habit. Here we present only those corresponding to spring wheat (Table 3.1).

After CIMMYT's target environment classification, and thanks to the historical data collected by the International Wheat Improvement Network (IWIN), this was followed by several reports on germplasm adaptation and performance in the context of ME (see Chap. 7). A study published by DeLacy et al. [4] demonstrated that the major discrimination factors for these ME were latitude and the presence/absence of stresses, plus the agreement between the ME-based locations and the location groups obtained from pattern analysis. Another study reported by Hodson and White [3] defined ME classification with the aid of GIS tools, and hence, a more quantitative and specific classification was proposed. In such a study, it was demonstrated that long term environmental variables, mainly temperature and precipitation in the coolest and wettest quarter of the year, were effective to separate environments based on abiotic stresses and growth habits, i.e., spring vs. winter growth type.

Table 3.1 Spring bread wheat mega-environments, land area, characteristics and required traits

Mega environment	Area (ha)[a]	Climate conditions[b]	Biotic stresses	Abiotic stresses	Key agronomic traits
1	47.2	Favorable, irrigated, low rainfall. Coolest quarter mean min temp ≥3 °C <11 °C	Rusts (leaf, yellow and stem rust)	Lodging	Water and nutrient use efficiency
2	5.9	a. High rainfall. Highland summer rain. Wettest quarter (three consecutive wettest months) mean min temp ≥3 °C <16 °C, wettest quarter precipitation ≥250 mm, elevation ≥1400 m b. High rainfall. Lowland winter rain. Coolest quarter mean min temp ≥3 °C <16 °C, coolest quarter precipitation ≥150 mm, elevation <1400 m	Yellow rust, stem rust, STB & FHB	Lodging	Sprouting resistance
3	1.3	Same as ME2, topsoil pH < 5.2	Same as ME2	Acid soils	
4	13.5	a. Low rainfall. Coolest quarter mean min temp ≥ 3 °C < 11 °C, wettest quarter precipitation ≥100 mm <400 mm c. Low rainfall, stored moisture. Coolest quarter mean min temp ≥3 °C <16 °C, wettest quarter precipitation ≥100 mm <400 mm	Rusts, STB, tan spot, root diseases	Drought	Water use efficiency
5	2.1	Coolest quarter mean min temp >11 °C <16 °C, High rainfall or irrigated	Rusts and spot blotch in low rainfall areas, Rusts and fusarium in high rainfall areas	High temperature	Early maturity
6	21	High latitude (>45 °N or S). Coolest quarter mean min temp less than −13 °C, warmest quarter mean min temp ≥9 °C	Rusts	High temperature and drought	Photosensitivity

[a]Data from Lantican et al. [2]; [b]Climatic conditions according to Hodson and White [3]

Additional studies derived from the historical data provided by the IWIN have shown how CIMMYT wheat germplasm performs and adapts throughout the locations within each ME [5–7]. These studies also demonstrated that CIMMYT's main yield testing site located in northwest Mexico correlates positively with the locations that belong to each ME for spring wheat.

3.3 Major Factors That Broadly Impact the Definition of Target Environments

3.3.1 Flowering Time: Photoperiod and Vernalization

Flowering time is a fundamental adaptive trait, as it determines where and when a variety can be grown, and, in general, largely determines the reproductive success of a plant. Flowering must occur during an optimal environmental period that permits the full development of the reproductive organs. This period should also be long enough to allow optimal grain filling.

One factor that highly determines flowering time in wheat is photoperiod. In wheat there is a series of dominant genes (*Ppd*) located on chromosomes 2A, 2B and 2D that induce an insensitive reaction to photoperiod [8–10]. Photoperiod insensitivity means that plants reach flowering even under short days, provided that any vernalization requirements have been met. One characteristic of the wheat cultivars derived from the Green Revolution is their insensitivity to photoperiod, which along with their short stature and disease resistance, significantly contributed to their adaptation to a broad range of environments.

Various studies have been conducted to determine the advantages of photoperiod-sensitive (PS) and photoperiod-insensitive germplasm (PI). For high latitude locations, evidence indicates that PS germplasm may have an advantage over PI germplasm [11, 12]. High GEI in Northern Europe [11], North America [13], and other high latitude locations in Asia [14], indicates that regional adaptation plays a major role in breeding spring wheat.

The geographical division suggested by Worland et al. [11] in Europe where PI and PS spring wheat germplasm displays better adaptation is 45–46° N. For practical purposes, wheat grown north of Paris is frequently PS, while south of that latitude the germplasm that better adapts is PI due to the summer conditions in Southern Europe [11].

Another factor that largely determines flowering time is *vernalization*. In this context, vernalization is the exposure to cold temperatures after germination to acquire or accelerate the ability to flower [15]. In northern latitudes where winters are cold, vernalization sensitivity is required, as it delays floral initiation which consequently protects ear development when low temperatures can severely damage it [16], hence conferring adaptability to northern latitudes.

The distinction between spring and winter growth habits is determined by a series of genes that can express both sensitivity and insensitivity to vernalization. Of these series of genes—*Vrn1, Vrn2* and *Vrn3*—*Vrn1* and *Vrn3* on the homoeologous groups 5 and 7, respectively, are dominant for spring growth habit, while *Vrn2* on chromosome 5A is dominant for winter growth habit [17]. In winter wheat sown and germinated in the autumn, *Vrn2* suppresses *Vrn3*, which in turn impedes the expression of *Vrn1*; then, as winter approaches, lower temperatures downregulate *Vrn2*, facilitating the upregulation of *Vrn3*, which in turn promotes Vrn1 transcription for the induction of flowering [18].

In geographical regions where wheat is grown during the winter and harvested late in the spring, the presence of *Vrn1* dominant genes confers adaptability to those lower latitude regions. The *Vrn-A1* and *Vrn-D1* genes of the *Vrn1* series are the most common, although all three (*Vrn-A1, Vrn-B1* and *Vrn-D1*) are present in CIMMYT's germplasm either alone or in combination [19].

The two previously mentioned factors—photoperiod and vernalization—alone, broadly determine the target breeding environments (high and low latitude regions) and, consequently, the type of germplasm that is required for each environment/s, since they guideline the planting and harvesting times.

The paradigm until the early years (1940s) of wheat breeding in Mexico dictated that breeding must be conducted in the environment where the future varieties will be cultivated [20]. However, given the need to accelerate the development of high-yielding and stem rust resistant germplasm, two generations per year started to be grown—using shuttle breeding—with the sole objective of speeding up the breeding process (Chap. 30 describes new technologies to speed up breeding). This paradigm shift took place years before any deep knowledge on the photoperiod in/sensitivity in wheat was available [21]. As germplasm exchange happened through the assembling of the first international yield trials during 1960s, the daylength effect on the materials became evident, since those shuttled-bred wheats developed in Mexico would adapt in most places in latitudes lower than 45° N [22].

3.3.2   Water Availability and Temperature

Water availability for the wheat crop is paramount to determine key traits in breeding. Water availability can favor optimal growing conditions, in the absence of high temperatures. However, drought stress is a constraint for wheat production in locations where water access is limited, either because of the lack of irrigation equipment or because the climate is dry (low rainfall).

Drought is one of the most severe factors that reduce wheat productivity (see Chap. 23 for details). In meteorological terms, it is defined as the absence of rain for a certain period, during which plants suffer from the lack of water in the soil. Yield losses of 20% can occur if plants are grown with 40% less water than required to avoid the stress [23]. This loss varies depending on the phenological stage at which the stress occurs, for instance, it can be larger if water is limited at the reproductive stage than if it occurs only at the vegetative stage [23].

Plants are drought stressed when water for the roots is limited and when the transpiration rate becomes higher. Drought can affect germination and plant establishment, growth, biomass accumulation, leaf senescence and, consequently, grain yield, but at the cellular level, it affects membrane integrity, pigment content, osmotic adjustment, photosynthetic activity, gas exchange and cell elongation [24].

Regions that are typically considered prone to drought stress are North Africa, some regions in West and Central Asia, and some locations in South America. Regions that are considered optimal in terms of water availability are the Nile Valley in Africa, the Northwestern Gangetic Plains and Northwestern Mexico.

Temperature is considered a stress factor that drastically influences wheat productivity once vernalization requirements—if any—have been met (see Chap. 22). Temperatures above optimum thresholds take high relevance, particularly in the context of climate change, since it determines the traits that the plants must carry to cope with the stress, such as earliness to avoid terminal heat stress [25]. It is estimated that for every °C increase above the a base temperature (13 °C) grain yield decreases by 6% [26]. Higher temperatures modify wheat phenology by reducing the number of days to reach flowering and maturity, consequently reducing the number of days in which plants can intercept light for photosynthesis, which leads to a reduction in biomass and grain yield. Larger yield reductions are expected in tropical and subtropical regions where wheat is grown, such as regions in India, which is a major wheat producer in the world [27].

3.3.3 Diseases

Following the fundamental paradigm in plant pathology (disease triangle), a disease outbreak occurs if there is (1) an adequate (susceptible) host, (2) a virulent pathogen, and if (3) favorable environmental conditions are present (see Chap. 19 for details). Hence, diseases tend to follow specific distribution patterns depending on the whether their environmental requirements are met.

While rusts, as a group of diseases, are found in all wheat growing areas, other leaf diseases occur in certain environments and crop management conditions (see Chaps. 8, 9 and 19 for details). Disease distribution and occurrence are dependent on both temporal and spatial variation, and these factors determine the resistance traits that cultivars must carry for certain environments. For instance, tan spot (caused by *Pyrenophora tritici-repentis*) incidence is linked to an expansion of zero-tillage practices (in Brazil, Argentina, Paraguay) or in places where climate does not allow fast stubble decomposition (Central Asia) and monocropping is common [28]. Septoria tritici blotch (caused by *Mycosphaerella graminicola*) is most common in temperate (15–20 °C) and humid wheat growing regions. Powdery mildew (caused by *Blumeria graminis* f. sp. *tritici*) is common in highly productive areas with maritime or semi-continental climate, particularly in China and South America [28].

3.4 Target Population of Environments

The main objective of a plant breeding program is to deliver superior germplasm for farmers in a defined set of environments, or a target population of environments (TPE). A TPE consists of many (population) environments and future years or seasons. A TPE is also a variable group of future production environments. Climatic (seasonal) variation in the same farmer's field might change drastically year after year causing the exacerbation of GEI. GEI can have two components: (1) static and predictable (repeated) variability due to the location (site) where the trial has been established, and (2) dynamic and unpredictable variability due to the year effect.

Target environments should be characterized by climatic, soil and hydrological characteristics as well as by socioeconomic characteristics. There are different ways to group trials and environments into TPE. One is to group together sites where line means are highly correlated. A standard methodology is to use *stratified hierarchical cluster analyses* of the sites based on climatic variables and production traits [29].

The selection environments (SE) are the environments where the breeder does the selection of the lines. The SE are identified for predicting the performance in the TPE, but the SE may not belong to the TPE. If the lines in the SE predict those in the TPE, then (1) it is important to compute the genetic correlations between the lines in the SE versus the same (or related lines) in the TPE and show some relatively high correlations between lines in SE and in TPE; (2) for screening lines in the SE, the repeatability (broad-sense heritability) in the SE should be high; (3) SE should allow a large number of lines to be screened at a low cost, so SE should allow high selection intensity (i).

3.5 Multi-environmental Testing and Genotype-by-Environment Interactions

As CIMMYT's mandate became international, the observations made between 1944 and the 1960s established the bases for the definition of target environments on a global scale. Along with this, the implementation of a breeding strategy based on ME targeted breeding, a diverse gene pool for crossing, shuttle breeding, selection under optimal conditions and multilocation testing have led to the enhancement of the adaptability and stability that characterize CIMMYT spring wheat germplasm to date.

Multi-environmental testing (MET) is a paramount strategy for CIMMYT to develop stable high grain yielding germplasm in countries with developing economies (see Chap. 7). An adequate MET strategy allows the evaluation of germ-plasm in stress hotspots and the identification of representative and correlated sites; thus, breeders can make better and targeted decisions in terms of crossing, selection and logistic operations. Another highly important aspect for CIMMYT's MET strategy is that collaborators can directly evaluate CIMMYT's elite germplasm, and hence they can make line selections for further evaluation and variety release, as well as utilize the germplasm as parental lines in their breeding programs to improve local adaptation.

Every year CIMMYT undertakes significant efforts to distribute international nurseries that comply with global and local seed health regulations, to collaborators within the IWIN, with the only request of returning the data to CIMMYT, for breeders to analyze them in a global context and support breeding decisions. The nurseries are of three different types: yield trials, observation nurseries (prior to yield trials), and trait specific nurseries (Table 3.2), Chap. 7 describes in detail international yield trials for bread wheat. Between 2013 and 2017, CIMMYT's wheat germplasm was distributed to 350 collaborators in 80 countries per year.

Despite the large variability between MEs, it is possible to simulate the most significant ones at CIMMYT's main testing site in northwest Mexico, Ciudad Obregon, a semi-arid location with suitable infrastructure for irrigation and available machinery for establishing the planting systems common around the globe. At this location in the Yaqui Valley, it is possible to mimic optimal, drought and heat stressed environments by applying the water management system corresponding to each ME, in combination with different sowing dates. This MET at one single location that is highly correlated with representative international locations [6, 7, 30] allows the breeders to select the elite germplasm that will most likely have an outstanding performance in international yield trials, and will consequently provide National Agricultural Research Centers a selection of CIMMYT's best materials every year.

Analysis of these international trials requires the utilization of advanced statistical methods that are able to parsimoniously model the GEI, obtain higher precision when estimating the genetic effects, and allow the identification of GEI patterns, for instance, the Factor Analytic (FA) and Sites Regression (SREG) models [31–33]. The FA model utilizes the leading principal components of the GEI covariance matrix in a mixed model framework, and accounts for the maximum amount of variation with a reduced number of parameters [32]. In the SREG model, the genotype and the GEI are estimated together, which is useful for evaluating METs, as its first and second principal components account for the non-crossover and the crossover interaction, respectively [34]. This property allows a visual examination to discriminate genotypes and sites with and without crossover interactions [34]. The FA and SREG models have been used to identify the trend of genetic gains and site correlations in CIMMYT's international nurseries [30, 35].

Table 3.2 International nurseries annually distributed by CIMMYT within the International Wheat Improvement Network

Nursery type	Trial/nursery	Abbreviation	Target environment	Grain Color	BW/DW[a]
Yield trials	Elite Spring Wheat Yield Trial	ESWYT	ME1	White	BW
	Harvest Plus Yield Trial	HPYT	ME1	White	BW
	Heat Tolerant Wheat Yield Trial	HTWYT	ME5	White	BW
	High Rainfall Wheat Yield Trial	HRWYT	ME2	Red	BW
	High Rainfall Wheat Screening Nursery	HRWSN	ME2	Red	BW
	Int. Durum Yield Nursery	IDYN	ME1, ME4, ME5		DW
	South Asia Bread Wheat Genomic Prediction Yield Trial	SABWGPYT	ME1, ME4, ME5	White	BW
Observation	Int. Bread Wheat Screening Nursery	IBWSN	ME1, ME4, ME5	White	BW
	Int. Durum Screening Nursery	IDSN	ME1, ME4, ME5		DW
	Semi Arid Wheat Screening Nursery	SAWSN	ME4	White	BW
	Semi Arid Wheat Yield Trial	SAWYT	ME4	White	BW
	Wheat Yield Consortium Yield Trial	WYCYT			BW
Trait specific	Fusarium Head Blight Screening Nursery	FHBSN			BW
	Harvest Plus South Asia Screening Nursery	HPAN	ME1, ME4, ME5	White	BW
	Heat Tolerance Screening Nursery	HTSN	ME5	White/Red	BW
	Helmithosporium Leaf Blight Screening Nursery	HLBSN			
	Int. Septoria Observation Nursery	ISEPTON			BW
	Karnal Bunt Screening Nursery	KBSN		White	BW
	Stem Rust Resistance Screening Nursery	SRRSN		White/Red	BW
	Stress Adtaptive Trait Yield nursery	SATYN			BW

[a]*BW* Bread Wheat, *DW* Durum wheat

3.6 Example of TPE Definition

We applied the mentioned methodology to a set of locations in India with data from the Elite Spring Wheat Yield Trials that are distributed internationally by CIMMYT, upon request. Daily meteorological data for these locations in India were obtained from the NASA Langley Research Center POWER Project funded through the NASA Earth Science/Applied Science Program. Then we implemented a principal component analysis to infer the number of groups (TPEs) that would explain most of the variation and then perform hierarchical clustering with the Euclidean distance matrix of data. From our analysis we obtained three main TPEs for India, in agreement with the wheat producing zones determined by the Indian government (Table 3.3 and Fig. 3.1), and that together account for more than 97% of India's wheat producing area. Finally, we obtained the correlated response to selection,

Table 3.3 Agroecological zones for wheat production in India and CIMMYT's breeding target population of environments (TPE)

Zone	Area covered	Estimated area (m ha)	Estimated productivity (kg/ha)	Estimated production (mt)	TPE
Northern Hills Zone (NHZ)	Western Himalayan regions of Jammu & Kashmir (except Jammu & Kathua dist.); Himachal Pradesh (except Una & Paonta Valley); Uttarakhand (except Tarai area); Sikkim & hills of West Bengal & North Eastern States	0.82	2203	1.81	
North Western Plains Zone (NWPZ)	Punjab, Haryana, Delhi, Rajasthan (except Kota & Udaipur divisions), western Uttar Pradesh (except Jhansi divison), parts of Jammu & Kashmir (Jammu & Kathua District), parts of Himachal Pradesh (Una district & Paona valley) and Uttarakhand (Tarai region)	12.33	4527	55.82	TPE1
North Eastern Plains Zone (NEPZ)	Eastern Uttar Pradesh, Bihar, Jharkhand, Odisha, West Bengal, Assam and plains of North Eastern States	8.85	2509	22.20	TPE2
Central Zone (CZ)	Madhya Pradesh, Chhattisgarh, Gujrat, Rajasthan (Kota & Udaipur divisions), Uttar Pradesh (Jhansi division)	6.84	2978	20.37	TPE3
Peninsular Zone (PZ) & Southern Hills Zone (SHZ)	PZ: Maharashtra, Karnatka, Andhra Predesh, Telengana, Goa, plains of Tamil Nadu. SHZ: Hilly areas of Tamil Nadu & Kerla comprising the Nilgiri & Palni hills of southern plateau	0.71	1404	1.00	TPE3
All Zones		**29.55**	**3424**	**101.20**	

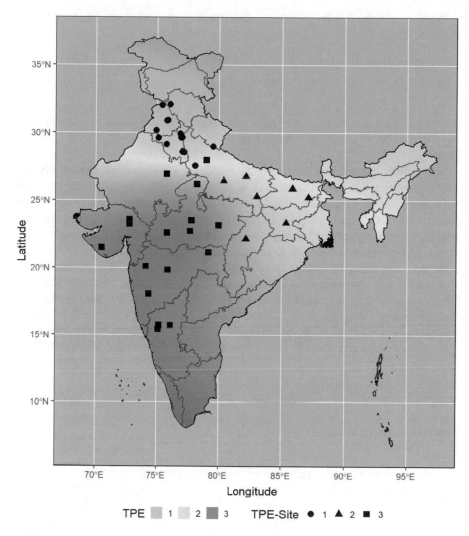

Fig. 3.1 TPE classification in India, obtained from environmental data

between each TPE and SE in Mexico, by first obtaining the genetic correlations with Eq. 3.1.

$$r_A = \frac{p_{i,j}}{\sqrt{h_i^2 * h_j^2}}$$

(3.1)

where r_A is the genetic correlation, $p_{1,2}$ represents the phenotypic correlation between site i and j, and h_i^2 and h_j^2 are the heritability of sites i and j, respectively.

Then, assuming that the same selection intensity applied in the SE is applied in the TPE (Eq. 3.2):

$$CR = \overline{r_A} * \sqrt{\frac{h^2_{SE}}{h^2_{TPE}}} \tag{3.2}$$

where CR is the correlated response to selection, $\overline{r_A}$ is the genetic correlation averaged over years of testing, h^2_{SE} is the heritability of the SE and h^2_{TPE} is the heritability of the TPE.

Our results indicate that the centralized breeding efforts in combination with the MET can give a selection efficiency (CR) as high as in the TPE, assuming the same selection intensity (Table 3.4). However, several factors are in place to obtain this result: CIMMYT's main yield testing site allows the simulation of various environments, the high heritability usually observed in the testing phase, a relatively stable (semi-arid) climate with favorable temperatures, water availability, irrigation infrastructure and mechanized operations. Furthermore, this result does not consider the fact that the selection intensity can be several times higher in the SE than in the TPE, given that several thousands of lines (~9000) are tested annually under optimal conditions (Stage 1 testing), from which ~1000 are selected to be tested in the SE (Stage 2), and ~300 are evaluated in the SE in the Stage 3 of testing, to finally distribute 46–48 new elite lines in each international yield trial nursery (Table 3.2).

Table 3.4 Average heritability (H²), genetic correlations and correlated selection response between SE and TPE in India

			Genetic correlation			Correlated selection response		
		Average H²	1	2	3	1	2	3
SE[a]	B2IR	0.62	0.40	0.62	0.45	0.46	1.05	0.63
	B5IR	0.61	0.63	0.55	0.31	0.73	0.93	0.43
	BLHT	0.85	0.32	0.54	0.25	0.43	1.08	0.42
	F5IR	0.65	0.51	0.50	0.40	0.61	0.87	0.56
	FDRP	0.59	0.23	0.10	0.34	0.26	0.16	0.47
TPE	1	0.46						
	2	0.21						
	3	0.32						

[a]Selection environments (SE) are: Beds 5 irrigations (B5IR): Trials conducted on raised beds with full irrigation management (optimal), 500 mm of available water. Flat 5 irrigations (F5IR): Trials planted on flat land with full irrigation (optimal), 500 mm of available water. Beds 2 irrigations (B2IR). Trials conducted on raised beds with partial irrigation, 260 mm of available water. Flat Drip (FDRIP): Trials planted on flat land with severe drought, 180 mm of available water. Beds late heat (BLHT): Trials planted in February, subject to heat stress and fully irrigated, 500 mm of available water

3.7 Key Concepts and Conclusions

Characterizing TPEs is critical for any plant breeding endeavor to succeed. Determining the main factors that may limit wheat productivity in a determined set environments (TPEs) is fundamental to incorporate key traits in breeding. Such limitations include, but are not limited to: water availability, temperature and incidence of pests and diseases. Additionally, for a breeding program to succeed, it is important that the SE display relatively high correlations with the TPEs, allow a higher selection intensity and accuracy of selection (higher repeatability). At CIMMYT's main testing location in northwest Mexico, it is possible to mimic optimal, drought and heat stressed environments to artificially create SE that, at one single location, are highly correlated with representative international locations to allow breeders the selection of elite germplasm with potential outstanding performance in international yield trials, and in so doing, to provide National Agricultural Research Centers a selection of CIMMYT's best materials every year.

For a refined definition of TPEs, statistical methods such as the FA model and SREG coupled with the climatic, soil, hydrological and socioeconomic characteristics of the environments can be applied to allow the identification of GEI patterns. These models (FA and SREG) have the advantage of being parsimonious and allow to measure the extend of non-crossover and crossover GEI.

Multi-environmental testing is paramount to identify high yielding and climate reliance germplasm, as well as to determine the GEI patterns that conform potential TPEs. The CIMMYT MET strategy has the benefit of evaluating the germplasm in stress hotspots, identification of representative and correlated sites, rapid response to new constraints (see Chap. 9 for examples) and direct access to germplasm for CIMMYT collaborators, so the materials can be used as parents or directly released as varieties.

References

1. Rajaram S, Van Ginkel M, Fischer RA (1994) CIMMYT's wheat breeding mega-environments (ME). In: Li ZS, Xin ZY (eds) Proceedings of the 8th International wheat genetic symposium. Beijing, pp 1101–1106
2. Lantican MA, Braun H-J, Payne TS, Singh RP, Sonder K, Michael B, van Ginkel M, Erenstein O (2016) Impacts of international wheat improvement research, pp 1994–2014
3. Hodson DP, White JW (2007) Use of spatial analyses for global characterization of wheat-based production systems. J Agric Sci 145:115
4. DeLacy IH, Fox PN, Corbett JD, Crossa J, Rajaram S, Fischer RA, van Ginkel M (1993) Long-term association of locations for testing spring bread wheat. Euphytica 72:95–106. https://doi.org/10.1007/BF00023777
5. Braun H-J, Pfeiffer WH, Pollmer WG (1992) Environments for selecting widely adapted spring wheat. Crop Sci 32:1420–1427. https://doi.org/10.2135/cropsci1992.0011183X003200060022x

6. Lillemo M, Van Ginkel M, Trethowan RM, Hernandez E, Rajaram S (2004) Associations among international CIMMYT bread wheat yield testing locations in high rainfall areas and their implications for wheat breeding. Crop Sci 44:1163–1169
7. Lillemo M, Van Ginkel M, Trethowan RM, Hernandez E, Crossa J (2005) Differential adaptation of CIMMYT bread wheat to global high temperature environments. Crop Sci 45:2443–2453
8. Mohler V, Lukman R, Ortiz-Islas S, William M, Worland AJ, Van Beem J, Wenzel G (2004) Genetic and physical mapping of photoperiod insensitive gene Ppd-B1 in common wheat. Euphytica 138:33–40. https://doi.org/10.1023/B:EUPH.0000047056.58938.76
9. Beales J, Turner A, Griffiths S, Snape JW, Laurie DA (2007) A pseudo-response regulator is misexpressed in the photoperiod insensitive Ppd-D1a mutant of wheat (Triticum aestivum L.). Theor Appl Genet 115:721–733. https://doi.org/10.1007/s00122-007-0603-4
10. Bentley AR, Turner AS, Gosman N, Leigh FJ, Maccaferri M, Dreisigacker S, Greenland A, Laurie DA (2011) Frequency of photoperiod-insensitive Ppd-A1a alleles in tetraploid, hexaploid and synthetic hexaploid wheat germplasm. Plant Breed 130:10–15. https://doi.org/10.1111/j.1439-0523.2010.01802.x
11. Worland AJ, Borner A, Korzun V, Li WM, Petrovic S, Sayers EJ (1998) The influence of photoperiod genes on the adaptability of European winter wheats. Euphytica 100:385–394. https://doi.org/10.1023/a:1018327700985
12. Dyck JA, Matus-Cádiz MA, Hucl P, Talbert L, Hunt T, Dubuc JP, Nass H, Clayton G, Dobb J, Quick J (2004) Agronomic performance of hard red spring wheat isolines sensitive and insensitive to photoperiod. Crop Sci 44:1976–1981. https://doi.org/10.2135/cropsci2004.1976
13. Lanning SP, Hucl P, Pumphrey M, Carter AH, Lamb PF, Carlson GR, Wichman DM, Kephart KD, Spaner D, Martin JM, Talbert LE (2012) Agronomic performance of spring wheat as related to planting date and photoperiod response. Crop Sci 52:1633–1639. https://doi.org/10.2135/cropsci2012.01.0052
14. Trethowan RM, Morgunov A, He Z, De Pauw R, Crossa J, Warburton M, Baytasov A, Zhang C, Mergoum M, Alvarado G (2006) The global adaptation of bread wheat at high latitudes. Euphytica 152:303–316. https://doi.org/10.1007/s10681-006-9217-1
15. Chouard P (1960) Vernalization and its relations to dormancy. Annu Rev Plant Physiol. https://doi.org/10.1146/annurev.pp.11.060160.001203
16. Law CN, Worland AJ (1997) Genetic analysis of some flowering time and adaptive traits in wheat. New Phytol 137:19–28. https://doi.org/10.1046/j.1469-8137.1997.00814.x
17. Kamran A, Iqbal M, Spaner D (2014) Flowering time in wheat (Triticum aestivum L.): a key factor for global adaptability. Euphytica 197:1–26. https://doi.org/10.1007/s10681-014-1075-7
18. Distelfeld A, Li C, Dubcovsky J (2009) Regulation of flowering in temperate cereals. Curr Opin Plant Biol 12:178–184
19. Beem J, Mohler V, Lukman R, Ginkel M, William M, Crossa J, Worland AJ (2005) Analysis of genetic factors influencing the developmental rate of globally important CIMMYT wheat cultivars. Crop Sci 45:2113–2119. https://doi.org/10.2135/cropsci2004.0665
20. Borlaug NE (1983) Contributions of conventional plant breeding to food production. Science 219(80):689–693
21. Mckinney HH, Sando WJ (1935) Earliness of sexual reproduction in wheat as influenced by temperature and light in relation to growth phases. J Agric Res 5:621–541
22. Rajaram S (1994) Wheat germplasm improvement: historical perspectives, philosophy, objectives, and missions_. In: Rajaram S, Hettel GP (eds) Wheat breeding at CIMMYT: commemorating 50years of research in mexico for global wheat improvement. CIMMYT, Texcoco, pp 1–10
23. Daryanto S, Wang L, Jacinthe P-A (2016) Global synthesis of drought effects on maize and wheat production. PLoS One 11:e0156362. https://doi.org/10.1371/journal.pone.0156362
24. Anjum SA, Xie X, Wang L, Saleem MF, Man C, Lei W (2011) Morphological, physiological and biochemical responses of plants to drought stress. Afr J Agric Res 6:2026–2032. https://doi.org/10.5897/ajar10.027

25. Mondal S, Singh RP, Crossa J, Huerta-Espino J, Sharma I, Chatrath R, Singh GP, Sohu VS, Mavi GS, Sukuru VSP, Kalappanavar IK, Mishra VK, Hussain M, Gautam NR, Uddin J, Barma NCD, Hakim A, Joshi AK (2013) Earliness in wheat: a key to adaptation under terminal and continual high temperature stress in South Asia. Field Crop Res 151:19–26. https://doi.org/10.1016/j.fcr.2013.06.015

26. Asseng S, Ewert F, Martre P, Rotter RP, Lobell DB, Cammarano D, Kimball BA, Ottman MJ, Wall GW, White JW, Reynolds MP, Alderman PD, Prasad PVV, Aggarwal PK, Anothai J, Basso B, Biernath C, Challinor AJ, De Sanctis G, Doltra J, Fereres E, Garcia-Vila M, Gayler S, Hoogenboom G, Hunt LA, Izaurralde RC, Jabloun M, Jones CD, Kersebaum KC, Koehler A-K, Muller C, Naresh Kumar S, Nendel C, O'Leary G, Olesen JE, Palosuo T, Priesack E, Eyshi Rezaei E, Ruane AC, Semenov MA, Shcherbak I, Stockle C, Stratonovitch P, Streck T, Supit I, Tao F, Thorburn PJ, Waha K, Wang E, Wallach D, Wolf J, Zhao Z, Zhu Y (2015) Rising temperatures reduce global wheat production. Nat Clim Chang 5:143–147

27. Asseng S, Cammarano D, Basso B, Chung U, Alderman PD, Sonder K, Reynolds M, Lobell DB (2017) Hot spots of wheat yield decline with rising temperatures. Glob Chang Biol 23:2464–2472. https://doi.org/10.1111/gcb.13530

28. Duveiller E, Singh RP, Nicol JM (2007) The challenges of maintaining wheat productivity: pests, diseases, and potential epidemics. Euphytica 157:417–430. https://doi.org/10.1007/s10681-007-9380-z

29. Cooper M, Hammer GL (1996) Plant adaptation and crop improvement. CAB International, Wallingford

30. Crespo-Herrera LA, Crossa J, Huerta-Espino J, Autrique E, Mondal S, Velu G, Vargas M, Braun HJHJ, Singh RPRP (2017) Genetic yield gains in CIMMYT's international elite spring wheat yield trials by modeling the genotype × environment interaction. Crop Sci 57:789–801. https://doi.org/10.2135/cropsci2016.06.0553

31. Burgueño J, Crossa J, Cornelius PL, Yang R-C (2008) Using factor analytic models for joining environments and genotypes without crossover genotype × environment interaction. Crop Sci 48:1291. https://doi.org/10.2135/cropsci2007.11.0632

32. Meyer K (2009) Factor-analytic models for genotype × environment type problems and structured covariance matrices. Genet Sel Evol 41:21. https://doi.org/10.1186/1297-9686-41-21

33. Crossa J, Vargas M, Joshi AK (2010) Linear, bilinear, and linear-bilinear fixed and mixed models for analyzing genotype × environment interaction in plant breeding and agronomy. Can J Plant Sci 90:561–574. https://doi.org/10.4141/CJPS10003

34. Crossa J, Vargas M, Cossani CM, Alvarado G, Burgueño J, Mathews KL, Reynolds MP (2015) Evaluation and interpretation of interactions. Agron J 107:736. https://doi.org/10.2134/agronj2012.0491

35. Crespo-Herrera L, Crossa J, Huerta-Espino J, Vargas M, Mondal S, Velu G, Payne TSS, Braun H, Singh RPP (2018) Genetic gains for grain yield in CIMMYT's semi-arid wheat yield trials grown in suboptimal environments. Crop Sci 58:1890–1898. https://doi.org/10.2135/cropsci2018.01.0017

Chapter 4
Global Trends in Wheat Production, Consumption and Trade

Olaf Erenstein, Moti Jaleta, Khondoker Abdul Mottaleb, Kai Sonder, Jason Donovan, and Hans-Joachim Braun

Abstract Since its domestication around 10,000 years ago, wheat has played a crucial role in global food security. Wheat now supplies a fifth of food calories and protein to the world's population. It is the most widely cultivated crop in the world, cultivated on 217 million ha annually. This chapter assesses available data on wheat production, consumption, and international trade to examine the global supply and demand conditions for wheat over the past quarter century and future implications. There is continued urgency to enhance wheat productivity to ensure global food security given continued global population growth and growing popularity of wheat based processed foods in the Global South. To enhance productivity while staying within planetary boundaries, there is a need for substantive investments in research and development, particularly in support of wheat's role in agri-food systems in the Global South.

Keywords Wheat · Food security · Demand · Supply · Trade · Staple cereals

4.1 Learning Objectives

This chapter highlights the continued importance of wheat for global food security over the past quarter century. It aims to illustrate:

- The need to not only consider global wheat supply, but also demand and trade conditions.
- The continued need to invest in wheat productivity enhancement while staying within planetary boundaries.

O. Erenstein (✉) · K. A. Mottaleb · K. Sonder · J. Donovan · H.-J. Braun
International Maize and Wheat Improvement Center (CIMMYT), Texcoco, Mexico
e-mail: o.erenstein@gmail.com; k.mottaleb@cgiar.org; k.sonder@cgiar.org;
j.donovan@cgiar.org; H.J.Braun@outlook.com

M. Jaleta
International Maize and Wheat Improvement Center (CIMMYT), Addis Ababa, Ethiopia
e-mail: m.jaleta@cgiar.org

© The Author(s) 2022
M. P. Reynolds, H.-J. Braun (eds.), *Wheat Improvement*,
https://doi.org/10.1007/978-3-030-90673-3_4

4.2 Introduction

Wheat is one of the world's oldest and most widely used food crops, domesticated more than 10,000 years ago in the Near East's Fertile Crescent. Its domestication took place roughly around the time of rice and somewhat prior to that of maize [1]. Together, the three big global staple cereals – wheat, rice, maize – comprise a major component of the human diet, accounting for nearly half of the world's food calorie and two-fifths of protein intake. Wheat alone plays a particularly crucial role in ensuring global food/nutrition security [2, 3], supplying a fifth of global food calories and protein.

This chapter examines the global wheat supply and demand conditions over the past quarter century and explores future implications. In the subsequent sections we briefly present data and methods and then assess the state of wheat production, consumption, and international trade at the global and regional levels, before concluding.

4.3 Data and Methods

We assess available secondary data on wheat production, consumption and international trade from FAOStat [4] and complementary indicators from other sources and review associated literature.

A modified approach to calculate and map wheat calorie production and demand based on Kinnunen et al. [5] was utilized to produce Figs. 4.2 and 4.5. On the production side the SPAM 2010 [6] wheat production grid was utilized in combination with a calorie value per ton [7] to calculate wheat based energy per 10×10 km^2 pixel. Using raster calculator in ArcMap 10.8.1 this value grid was adjusted based on available data for regional production and postharvest losses [8]. Calorie allocation fractions for food use were applied on country basis, subtracting wheat used for feed and other purposes [7]. For countries without data a value of 1 was assumed for wheat utilization as food.

On the demand side population was represented by the 2017 Landscan data set [9]. This was multiplied with the country specific annual wheat calorie use by person and year [4]. For few countries without current values older FAO data and secondary sources were utilized, and for remaining gaps neighbouring countries were utilized. These values were adjusted upwards for losses related to processing, packaging, and transport as well as consumer food waste [8].

4.4 Trends in Global Wheat Production

By 2018, wheat was cultivated on an estimated 217 million (M) ha of land globally (Triennium Ending – TE2018), making it the most widely grown crop in the world. In comparison, maize has nearly 200 M ha and rice 165 M ha (Table 4.1). In terms

Table 4.1 Global cereal production indicators

		1992–1994 (TE1994)	2016–2018 (TE2018)	Relative change (%)
Wheat	Area (M ha)	220	217	−1.1%
	Production (M mt)	552	752	36.3%
	Yield (mt/ha)	2.5	3.5	37.8%
Rice (Paddy)	Area (M ha)	147	165	13%
	Production (M mt)	532	768	44%
	Yield (mt/ha)	3.6	4.6	28%
Maize	Area (M ha)	135	196	44%
	Production (M mt)	527	1146	118%
	Yield (mt/ha)	3.9	5.9	51%

With data from Ref. [4]

of production, wheat's 752 M tons globally (TE2018) is slightly less than rice (768 M tons paddy – Table 4.1), although both crops are overtaken by maize (1146 M tons, with some 57% used as feed). The divergence reflects the substantially higher maize yields, linked to widespread hybrid and input use and the higher rice yields linked to widespread irrigation. It is of interest to note that of the three main cereal staples, wheat was the only staple recording a slight area decline over the last quarter century (−1% since TE1994), whereby the substantive yield increase (+38%) was the main driver for the similarly substantive increase in production (+36%). In the case of maize, the more than doubling of production over the period was supported by both substantive yield increases and area expansion. Increases in rice production also relied on a combination of yield and area increases (Table 4.1).

Since 1961, the global area under wheat production has oscillated between 200 and 240 M ha (Fig. 4.1). Wheat area peaked around 1980 and has slowly oscillated downwards towards the current 217 M ha (TE2018). Given the relative stability of wheat area (including a modest decline over the last half century) the increase in global wheat production is explained by consistent increases in wheat yield (Fig. 4.1). Yields have steadily increased from a global average of only just over 1 ton/ha in the early 1960s to the current 3.5 tons/ha, nearly quadrupling global wheat production over the period (Fig. 4.1).

Over 120 countries distributed across Europe, Africa, the Americas, Asia and Oceania cultivate wheat [4], spanning both emerging economies and the developed world. From an agronomic perspective, wheat performs better in temperate environments. It can withstand frost and some 150 M ha of wheat is grown in areas where freezing temperatures occur during the wheat growing season. Such frost prevents many other crops from being cultivated since they are frost susceptible, except for some frost tolerant minor crops such as rye, triticale, barley, canola and some legumes. Consequently, in areas with below zero temperatures during the crop cycle

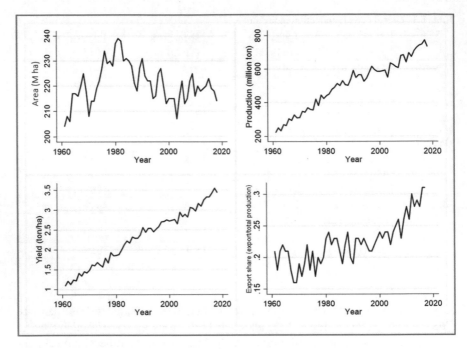

Fig. 4.1 Dynamics of key wheat indicators 1961–2018: wheat area (M ha), production (M ton), yield (ton/ha) and export share (export/total production). (Figure prepared with data from Ref. [4])

wheat thereby is the only biologically and economically feasible crop. Wheat cultivation is spread across the northern and southern latitudes, as well as in highlands and irrigated winter seasons in the lower latitudes (Fig. 4.2).

Asia contributes the most to global wheat production (44%, TE2018), followed by Europe (34%) and the Americas (15% – Fig. 4.3), with small but similar shares for Oceania and Africa (3.4–3.5%). Over the last quarter century, the relative production shares by region have remained largely similar, albeit with a 5% point decline for the Americas (down from 20% in TE1994, associated with the expansion of maize and soybean) and slight increases of 1–2% points in each of the other regions. There is a substantive heterogeneity within each of the continent's regions. For instance, about half of Asia's near 100 M ha are in South Asia, the remainder about equally split between west/central and east/south east Asia. Within South Asia wheat cultivation is concentrated in the Indo-Gangetic plains and within east/southeast Asia in north-eastern China (Fig. 4.2). There are also marked divergences in productivity (e.g., being low in west/central and high in east/southeast Asia), translating in varying regional shares in production.

Roughly 29% of the global wheat area is in low and lower-middle income countries (L/LM-ICs), contributing some 25% to the global wheat production (TE2018 – Table 4.2). This reflects somewhat lower yields (3.1 ton/ha TE2018) than the average yields of upper-middle- and high-income countries (UM/H-ICs, 3.6 ton/ha). Interestingly, the yield growth rate in the two income groups has been similar (1.4% pa – Table 4.2). A wheat area decrease (−0.3% pa) was only observed for

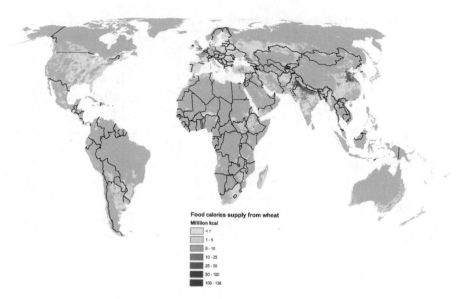

Fig. 4.2 Geography of wheat production (estimated M kcal energy produced by wheat per pixel, ca 10 × 10 km²). Prepared using SPAM 2010 and other sources (see Sect. 4.3 for details)

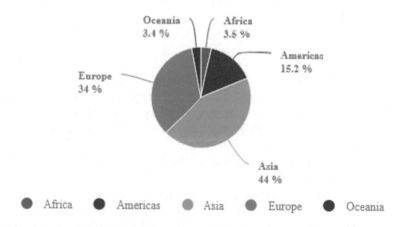

Fig. 4.3 Production shares of wheat by region, TE2018. (Figure prepared with data from Ref. [4])

UM/H-ICs, whereby production increased by only 1.1% pa. L/LM-ICs had a wheat area increase (+0.8% pa), whereby production increased by 2.2% pa (Table 4.2).

Wheat production has been dominated by a handful of countries. In TE2018, 53% of global production came from China, India, Russia, United States and France. Interestingly, these same countries have dominated wheat production since the 1960s, but their order has changed over time – with production in both China and India increasing ten-fold from some 10+ M tons each in the early 1960s to each surpassing 100 M tons currently, and becoming the top 2 global producers. India's

Table 4.2 Regional wheat production indicators

Region	Average TE2018			Average annual growth rate (TE1994–2018)		
	Area (M ha)	Production (M mt)	Yield (mt/ ha)	Area (% pa)	Production (% pa)	Yield (% pa)
Asia	99.3	330.6	3.3	−0.1	1.5	1.6
South	48.9	144.3	3.0	0.6	2.2	1.5
West & central	25.3	51.9	2.2	−0.5	0.8	1.3
East & SE	25.2	134.4	5.3	−0.8	1.2	2.0
Africa	10.1	26.4	2.6	1.0	3.0	1.8
Northern	7.4	19.4	2.6	0.9	2.8	1.8
Sub-Saharan	2.7	7.1	2.6	1.7	3.5	2.0
Americas	34.7	113.9	3.3	−1.1	0.2	1.4
Northern	25.6	85.1	3.3	−1.5	−0.2	1.4
Central & South	9.1	28.8	3.2	0.6	2.0	1.5
Europe	61.7	255.6	4.1	0.4	1.6	1.1
Oceania	11.5	25.4	2.2	1.6	4.4	2.3
L/LM-IC[a]	62.0	191.4	3.1	0.8	2.2	1.4
UM/H-IC[b]	155.3	560.6	3.6	−0.3	1.1	1.4
World	217.3	752.0	3.5	0.0	1.3	1.4

With data from Ref. [4]
[a]Low & lower-middle income countries
[b]Upper-middle & High income countries

rise has been linked to the Green Revolution which combined high-yielding wheat varieties with fertilizer and irrigation and policy support. The expansion of wheat cultivation and intensification in upcoming wheat producing countries has so far not dented the dominance of the traditional top producers, albeit increased the importance of wheat production in the Global South.

The diverse environments where wheat is cultivated have led to the distinction of various wheat mega-environments (ME, see Fig. 1.3), which spread from winter production in northern latitudes to production in the warm and humid environments of Bangladesh and eastern India. The mega-environments have implications for the types of wheat grown (e.g. spring, winter; bread, durum; hardness, colour) and the relevance of associated traits (e.g. heat and drought tolerance; maturity; biotic stress tolerance – [10]; see Chap. 3). The mega-environments are associated with the prevailing wheat production systems, including intensive irrigated systems with high and stable yield potential to extensive rainfed systems with low and variable yield potential and associated input use and mechanization. The wheat production systems and productivity do not always reflect the income categorization of a particular country, with Australia's variable rainfed wheat production being a case in point.

Climate change is set to gradually shift wheat mega-environments, including increased cultivation prospects in the northern and southern latitudes, while increased stress may lead to reduced production in sub-tropical environments (including heat, drought and biotic stresses – e.g., [3, 11]). Over time new pests and diseases or new races of existing diseases, have emerged with far reaching consequences, including wheat rusts [12] and wheat blast [13]. There are also likely trade-offs between climate change adaptation and implications for wheat, with higher CO_2 levels potentially increasing yields (starch) but lowering protein content.

Wheat is far from 'a rich man's crop' for large swathes of the Global South, but pivotal to poor rural producers/consumers (and urban resource poor consumers, see Sect. 4.5). Millions of smallholders in Asia, Africa and South America are engaged in wheat cultivation for their own consumption and income generation. A range of smallholder wheat production systems exist, from rainfed with low and variable productivity in the Central-West Asia and Northern Africa region to smallholder commercial intensified areas such as the NW Indo-Gangetic plains. There is also a contrast between traditional and non-traditional wheat growing areas with implications for the role of wheat for food security and rural livelihoods and implications for innovation and system dynamics (e.g., crop-livestock interactions and use of wheat straw as animal feed). Table 4.3 summarizes the major differences for wheat between L/LM-ICs and UM/H-ICs. There is a remarkable divergence on the reliance on irrigated wheat: some three-fifths of the wheat area in L/LM-ICs being irrigated, whereas irrigated wheat accounts for only 15% in UM/H-ICs – wheat being a relatively low value crop there. The very small farm sizes in the Global South also limit options for variety choice and risk management. Such smallholders cannot grow several varieties to cover risk – they need varieties with yield stability no matter what the weather brings.

Earlier in this section we noted that since the 1960s wheat production increases have been attributed more to intensification than extensification. The challenge

Table 4.3 Global wheat dichotomy

	Low & lower-middle income countries (L/LM-IC)	Upper-middle & High income countries (UM/H-IC)
Wheat area (M ha)	62	155
Wheat production (M mt)	191	561
Wheat yield (mt/ha)	3.1	3.6
High average yields (mt/ha)[a]	6.7 (Egypt, spring wheat)	9 (Ireland, winter wheat)
Average farm size[a]	1–3 ha	Up to 5000 ha
Irrigated wheat area, %[a]	59	15
Wheat consumption (%)		
Food	79	60
Feed/seed/other	21	40

Most data from [4]
[a]Compiled by authors

remains to continue to do so in the coming decades while staying within planetary boundaries [14]. Population growth alone implies the need to produce an additional 132 M mt annually by 2050 to meet wheat food needs at current average consumption levels (see Sect. 4.5). At the same time the demands of increasing land and water scarcity with the added context of climate change are increasingly recognized.

Increasing land pressure implies the continued need to close yield gaps [15]. Wheat also adds to global water demand - with some 1000 litre per kg of grain. The unsustainable portion of the blue water footprint is particularly large in the Indus and Ganges river basins in India and Pakistan and in the north-eastern part of China [16]. Increasing water scarcity implies the continued need to improve water use and water productivity through policy, crop improvement and management (e.g., laser land leveling, drip irrigation). The intensification of wheat production also has raised concerns on environmental externalities (beyond water scarcity), including the heavy doses of chemical fertilizers used in intensive systems (particularly nitrogen). The persistently low nutrient use efficiency (stagnant at around a third over the last quarter century, particularly in China and India, [17]) has led to a quest for improvement, including sustainable intensification and climate change mitigation.

4.5   Trends in Global Wheat Consumption

Wheat has an average annual per capita food consumption of 65.6 kg globally (TE2017), which amounts to 37% of the average annual cereal consumption of 175 kg globally (TE2017, excluding beverages – [4]). Wheat is the second most consumed cereal (as food) after rice (81 kg, 46%). Wheat is consumed in 173 countries, with consumption levels exceeding 50 kg/capita/year in 102 countries [4]. In countries with strong wheat dietary traditions, to include those of in Northern Africa, West/Central Asia and Europe, per capita wheat consumption is particularly high (Table 4.4). As a group, UM/H-ICs consume 68% of global wheat, aided by above average per capita wheat consumption (Table 4.4). Asia stands out as the main aggregate consumer, with 53% of global wheat consumption followed by Europe (26%) and some 10% each in the Americas and Africa (Table 4.4).

Over the past quarter century global per capita wheat consumption has shown a slight decline in most regions and globally (Table 4.4). However, this masks a significant earlier surge in global per capita wheat consumption (55 kg – TE1963 to 70 kg – TE1993), driven by increases in Africa and Asia since the 1960s (Fig. 4.4). In Africa, per capita annual wheat consumption increased from 30 kg – TE1963 to 47 kg – TE1993 and to 49 kg – TE2017 (Fig. 4.4). In Asia, where rice is the major staple crop, corresponding figures were 29 kg – TE1963, increasing to 67 kg – TE1993 and declining slightly to 63 kg – TE2017 (Fig. 4.4), driven by changes in South and South-East Asia. In Asia, the role of China and India stands out. The two countries, which together contain more than 36% (2.8 billion) of the global population (7.7 billion – 2019, [18]), have experienced a drastic increase in wheat consumption: per capita annual wheat consumption in China and India increased from

Table 4.4 Regional wheat consumption indicators

Region	Average 2014–2017		Aggregate consumption (av % pa)		Per capita food consumption (av % pa)	
	Aggregate consumption[a] (M mt/year)	Per capita food consumption[b] (kg/year)	TE1994-13	2014–2017[c]	TE1994-13	2014–2017[c]
Asia	375.1	62.8	1.2	2.8	−0.5	0.5
South	146.4	67.6	1.8	2.9	0.0	0.7
West & Central	61.9	132.6	1.7	−1.3	−0.2	−0.5
East & SE	166.8	49.6	0.7	11.8	−0.6	0.3
Africa	70.7	49.5	3.3	1.0	0.1	−0.7
Northern	45.2	143.8	2.6	−0.1	−0.1	−0.2
SSA	25.5	25.2	5.1	1.0	1.9	−0.7
Americas	78.7	61.4	0.7	2.3	−0.1	−0.2
Northern	41.4	80.4	0.05	4.4	−0.2	0.4
Central & South	37.2	50.5	1.5	−0.1	0.1	−0.6
Europe	186.8	110.5	0.2	2.5	0.1	0.2
Occania	8.4	75.4	3.5	3.0	0.1	0.4
L/LM-IC[d]	225.5	56.4	2.0	1.9	−0.1	−0.2
UM/H-IC[e]	486.0	72.7	0.6	2.7	−0.4	−0.1
World	711.5	65.5	1.0	2.4	−0.4	−0.1

With data from Ref. [4]
[a]Domestic supply quantities in Food Balance Sheet (FBS), across uses
[b]Food supply quantity (kg/capita/year) in FBS (i.e. net of non-food uses)
[c]New FBS method since 2014 [4]
[d]Low and Lower-Middle income countries
[e]Upper-Middle and High income countries

23/29 kg – TE1963 to 78/60 kg – TE1993 and now averaging 63/61 kg – TE2017. Still, given the sheer size of China and India there is a marked within country heterogeneity of wheat consumption. In NW India per capita wheat consumption surpasses 100 kg pa; whereas in SE India rice consumption prevails. Similarly, wheat consumption is more pronounced in NE China and E Pakistan. Figure 4.5 visualizes the heterogeneity in wheat consumption globally and highlights some of the within country variations.

The role of wheat in diets around the globe has been particularly dynamic up to the 1990s affected by income growth, urbanization, and associated life-style changes. Globally GDP per capita has increased by 3.5% per annum (1961–2019, from US$3.9k pc TE1963, to US$11.1k pc in 2019 at 2010 constant prices – [19]). Urbanization has increased from a little more than 34% of global population in 1961 to nearly 56% off late [19]. On the one hand, the nutrition transition posits non-cereal food consumption to increase with increasing GDP per capita and

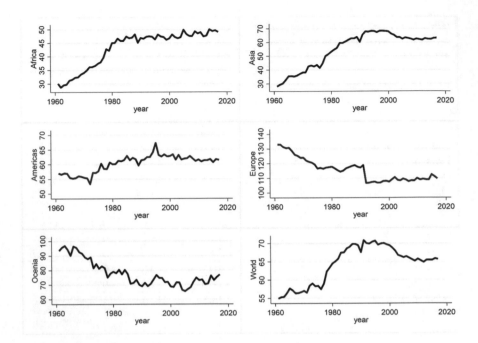

Fig. 4.4 Per capita wheat consumption trends by major regions, 1961–2017. (Figured prepared with data from Ref. [4])

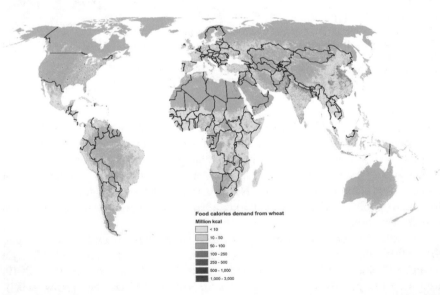

Fig. 4.5 Geography of wheat consumption (estimated M kcal food energy consumed from wheat per pixel, ca 10 × 10 km²). Prepared using data from various sources (see Sect. 4.3 for details)

urbanization. On the other hand, wheat is a special case, with its numerous derived processed food products. Diversification of traditional diets as a result of growing economies, increased global trade, 'modernization' of tastes and consumer fads has boosted per capita consumption of wheat flour in several Asian and SSA countries [1]. An increasing number of empirical studies have highlighted wheat consumption to be increasing particularly in the global South associated inter alia with increasing incomes, rapid urbanization and the allied changes in the lifestyle including India [20], Bangladesh [21] and SSA [22]. This is further aided by changes in the food processing sector and the ability to generate cheap processed wheat-based foods and making them available across the globe.

In addition to dietary change, population growth will continue to add to wheat demand. The global population is set to increase by 2 billion from 7.7 billion currently to a projected 9.7 billion by 2050 (8.9–10.7 billion depending on low and high-fertility assumption rates – [18]). Assuming a constant annual per capita consumption, this implies a potential annual increase of 132 M mt of wheat as food by 2050 (106–224 M mt depending on fertility assumption).

Wheat is primarily produced for food (66% of global production) but a fifth of the grain is used as feed. Over time the feed use share has been steadily increasing globally, from 9% in TE1963 to 18% in TE1993 and the latest 21% (TE2017). Conversely, the food use share declined from 74% in TE1963 to 69% in TE1993 and the latest 66% (TE2017). High seed rates imply 5% of production is used as seed, with remainder divided between losses, processing and non-food uses (Table 4.5). Wheat use presents a marked divergence between country income status. In L/LM-ICs food use predominates (79%), with feed use only 10%; and relatively lower seed and processing use, and higher losses (5%). In contrast, in UM/H-ICs food use drops to 60%, with feed use increasing to 26%; and relatively higher seed, processing and non-food uses, and lower losses (Table 4.5). Furthermore, feed use is concentrated in selected geographies, being particularly high in Australia, followed by Europe and Eastern Asia. Food use is particularly high in South Asia, SSA and Latin America (Table 4.5). In SSA much of the wheat consumption is concentrated among urban populations with supplies derived largely from imports [22].

Various classes of wheat exist with varying use, including various bread wheats (such as Hard Red Winter (HRW), Hard Red Spring, Soft Red Winter (SRW), Hard White and Soft White) and Durum wheat (for pasta and couscous). These bread wheat classes vary in their milling characteristics for bread flour; baking characteristics for pan breads; and processing characteristics for Asian noodles, hard rolls, flat breads, cakes, cookies, snack foods, crackers and pastries; and improving blending (also see Chap. 11). There is often a preference for white wheat in relation to flour extraction and the colour for whole grain products like chapati. Other uses include industrial production of starch, malt, dextrose, gluten, and alcohol.

Wheat alone contributes 18% of the total dietary calories and 19% of proteins globally (Table 4.6). On average, the daily dietary energy intake per capita was 530 kcal from wheat, similar to rice (550 kcal, 19%) compared to a total intake of 2907 kcal (of which 1216 kcal from cereals). The average energy from wheat was

Table 4.5 Wheat utilization, by region, average 2014–2017

Region	Average use (% of domestic supply)					
	Food	Feed	Seed	Losses	Processing	Other uses (non-food)
Asia	73.5	16.1	4.1	3.6	0.4	2.2
South	84.7	4.3	4.1	5.3	0.0	1.7
West & Central	64.9	17.4	7.2	4.5	1.7	3.5
East & SE	67.0	25.9	3.0	1.9	0.3	2.1
Africa	74.9	12.2	1.6	5.6	0.2	4.8
Northern	69.8	17.1	2.0	7.8	0.0	3.3
SSA	83.9	3.7	0.8	1.9	0.6	7.4
Americas	75.6	12.3	4.5	5.4	0.2	4.1
Northern	67.7	19.1	6.3	7.3	0.4	2.4
Central & South	84.7	4.6	2.5	3.3	0.0	6.1
Europe	43.8	36.0	9.4	2.1	4.3	4.7
Oceania	28.3	48.6	7.8	3.4	2.2	11.2
L/LM-IC[a]	79.0	9.5	3.4	5.1	0.2	2.7
UM/H-IC[b]	59.5	26.1	6.3	2.9	1.9	3.7
World	65.7	20.9	5.3	3.6	1.4	3.4

With data from Ref. [4]
[a]Low and Lower-Middle income countries
[b]Upper-Middle and High income countries

however more than double its global average in West/Central Asia and Northern Africa (>1000 kcal), representing over a third of their total energy intake. As a group, the wheat calorie share is similar for UM/H-ICs and L/LM-IC's (Table 4.6).

On average, globally the daily protein intake per capita from wheat was 16 g or 19% of the daily protein intake (82 g, and half the proteins provided by cereals, 32 g). The average protein intake from wheat was somewhat higher in UM/H-ICs (18 g). However, wheat intake represented 20% of protein intake in L/LM-IC's, reflecting more intensive use of wheat as food and lower food intakes in L/LM-IC's. The contribution of wheat as a source of daily protein intake is again substantively higher for regions like Northern Africa and West/Central Asia (38%). Wheat also provides a modest source of daily fat (2.5 g representing 3% of daily intake – Table 4.6).

Wheat's nutritional contribution plays of an important role in addressing the triple burden of undernutrition (hunger), micronutrient malnutrition and overnutrition (overweight, obesity). Indeed, in addition to being a major source of dietary energy and proteins, wheat also provides essential micronutrients and diverse non-nutrient bioactive food components (24). Still, various avenues remain to strengthen wheat's nutritional contribution, including active work on bio- and industrial fortification (e.g. see Chap. 12). There is also considerable potential in improving processing and intake forms (also see Chap. 11). For instance, current intakes of whole grain foods should at least double compared with national dietary guidelines (except in N America, [23]). Whole grain wheat is an important source of dietary fibre with associated health benefits for controlling non-communicable diseases [24].

Table 4.6 Regional wheat food supply indicators

Region	Wheat and products in food supply (TE2017)			Wheat share in total food supply (%/capita/day, TE2017)		
	Food supply (kcal/ capita/day)	Protein supply quantity (g/ capita/day)	Fat supply quantity (g/ capita/day)	Kcal share (%)	Protein supply share (%)	Fat supply share (%)
Asia	527	15.8	2.6	18.7	19.8	3.5
South	567	16.3	2.7	22.6	25.3	4.9
West & Central	1084	32.7	5.2	36.3	37.7	5.6
East & SE	342	10.4	1.6	11.5	12.3	2.1
Africa	392	11.7	1.7	15.0	17.2	3.2
Northern	1136	34.4	5.0	35.5	37.6	7.5
SSA	248	7.4	1.1	9.8	11.2	1.9
Americas	464	13.7	2.1	14.1	14.6	1.7
Northern	608	19.5	3.3	16.4	17.4	2.0
Central & South	343	9.5	1.3	11.2	11.1	1.4
Europe	846	26.3	3.5	25.1	25.7	2.7
Oceania	602	19.1	2.6	18.6	18.6	1.8
L/LM-IC[a]	464	13.4	2.1	18.3	20.2	3.8
UM/H-IC[b]	583	18.0	2.8	18.2	18.9	2.6
World	530	16.0	2.5	18.2	19.4	3.0

With data from Ref. [4]
[a]Low and Lower-Middle income countries
[b]Upper-Middle and High income countries

4.6   Wheat Prices and Trade

Some of the world's biggest wheat producers – China, India – are largely self-sufficient. Nonetheless, wheat is the most widely globally traded cereal, with 25% of global wheat production being exported (TE2018),[1] up from 19% a decade earlier [1]. The global trade reflects a marked spatial disparity between where wheat is produced and where it is consumed (Figs 4.2 and 4.5; [5]). This underpins an active global wheat trade linking surplus production areas in the northern and southern latitudes (net exporters in Europe, Northern America and Australia) to the deficit areas in the lower latitudes (net importers in Africa, Asia and Latin America – Table 4.7). This leads to a marked disparity between countries by income groups: L/LM-ICs being net importers and UM/H-ICs net exporters (Table 4.7). Top exporting

---

[1] 190 M mt of wheat being traded internationally against a global production of 768 M mt in the TE2018 (average import-export). This compares to 161 M mt traded for maize the 2nd most widely traded – but representing 14% given larger production of 1146 M mt (also see Table 4.1; [4]) and primarily exported for feed).

Table 4.7 Regional wheat import/export indicators

Region	Wheat net imports		Top net importing countries (net import M mt/year, TE2018)
	Annual average (M mt/year, TE2018)	Annual growth rate (av % pa, TE1994-18)	
Asia	68.0	2.7	
South	9.8	6.9	Pakistan (−0.6); Sri Lanka (+1.1); India (+2.2); Bangladesh (+5.7);
West & Central	17.3	5.3	Kazakhstan (−4.9); Uzbekistan (+1.6); Saudi Arabia (+2.1); Yemen(+3.0); Turkey (+5.0);
East & SE	41.0	2.6	Rep Korea (+4.3); Vietnam (+4.7); Japan (+5.7); Philippines (+5.7); *Indonesia (+10.3)*
Africa	46.4	4.8	
Northern	28.4	4.0	Tunisia (+2); Sudan (+2.1); Morocco (+4.7); *Algeria (+8.3); Egypt (+10.3)*
SSA	17.9	6.9	Cameroon (+0.7); Ethiopia (+1.5); South Africa (+1.7); Nigeria (+5.0)
Americas	−34.9	−0.7	
Northern	−43.3	−0.7	**Canada (−21.6); US (−21.9)**
Central & South	8.4	3.3	Argentina (−11.7); Colombia (+1.9); Peru (+2.0); Mexico (+4.1); *Brazil (+6.1)*
Europe	−69.0	−25.3	**Russia (−32.9); Ukraine (−17.3); France (−17.0);** *Spain (+6.0)*; *Italy (+7.3)*
Oceania	−15.9	−4.2	**Australia (−16.7)**; Papua New Guinea (+0.2); New Zealand (+0.5)
L/LM-IC[a]	52.7	3.0	
UM/H-IC[b]	−58.1	−3.5	

With data from Ref. [4]
[a]Low and Lower-Middle income countries
[b]Upper-Middle and High income countries

countries include Russia, United States, Canada, Ukraine, France and Australia, each exporting 16–33 M mt/year (TE2018, Table 4.7). Off late (TE2018), top importers include Indonesia, Egypt, Algeria, Italy, Brazil and Spain; each importing 6–10 M mt/year; with a number of other countries also importing substantive amounts (e.g., Bangladesh, Japan, Philippines, Turkey, Nigeria 5–6 M mt – Table 4.7). Imports thereby are spread far and wide across a range of countries, but still creating substantive foreign exchange outlays for annual imports and import dependence for food security, particularly given its prevalence across L/LM-ICs.

There is still substantial heterogeneity in each region. For instance, the Mediterranean region includes major wheat deficit areas, both for Europe (Italy, Spain) and northern Africa; Latin America includes major import reliance with exports from the southern cone; and Oceania's net exports hinges on Australia's

harvests. More worrying is that import dependence continues to increase for the L/
LM-ICs, whereas exports increase for the UM/H-ICs, particularly Europe and
Australia (Table 4.7). The highest growth in wheat imports were observed for sub-
Saharan Africa and South Asia (Bangladesh), followed by West and Central Asia
(Table 4.7). The increased burden of wheat imports has reignited interests in self-
sufficiency, from traditional wheat producers like Ethiopia to across Africa. Indeed,
the current high levels of spatial decoupling between production and consumption
are set to increase further over the coming decades [25].

Underlying the global wheat trade are also some wheat processing and consumption considerations. Turkey is a case in point, being notionally self-sufficient in
terms of domestic wheat production and consumption. Still, Turkey is a major
importer of wheat grain and the leading exporter of wheat flour, with imports not
being taxed if equivalent of flour exported. Australia is a main wheat provider for
Asian noodles, with e.g., a preference for Australian Standard white being relatively
low in protein. Adding to global trade is a decoupling between the production of red
wheat in the Americas and Europe and a preference to consume white wheat products, with much of red wheat being exported, including for industrial uses.

Over the last decades wheat prices nominally increased by 37%, from US$143/
ton (average US HRW/SRW) in TE1994 to US$197/ton in TE2019. HRW tended to
have a somewhat higher price compared to SRW, but the prices have converged in
recent years (Fig. 4.6). The highest prices over the period were observed in
2008–2009 (US$240–250/ton) linked to the global food crisis and in 2012–2014
with even higher nominal prices of up to USD 300/ton (2013). The price oscillations
over the last 10+ years largely track the pattern of urea fertilizer (Fig. 4.6), albeit
that urea prices increased by 139% over the quarter century (from US$99/ton in
TE1994 to US$236 in TE2019). The ratio of wheat-to-urea prices thereby decreased
from 1.45 to only 0.84 over the quarter century. Other staple cereals saw somewhat

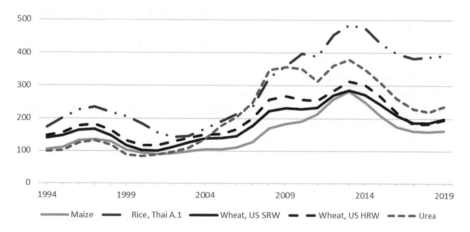

Fig. 4.6 Selected cereal and urea prices (nominal US$/ton, TE1994-2019). (Figure prepared with
data from Ref. [26])

similar price trends over the period, albeit that the increases in nominal prices were somewhat more pronounced for maize (+56%) and particularly for rice (+127%).

The global food crisis and subsequent Arab Spring made their mark on the supply and demand dynamics. Prior to these there were concerns of a longer term decline of wheat prices due to continued wheat productivity growth [2]. The surge in global prices and import dependence sparked social unrest and the Arab Spring when these were passed on to domestic price increases and increased scarcity of wheat and food supply instability (e.g. [27]). The Arab Spring has been associated with the failure of the wheat crop in China [28], illustrating how production shocks can contribute to wheat price spikes. Such concerns have been growing in the context of climate change with increased weather shocks (e.g., heat, droughts, excessive water) and biotic shocks (diseases and pests). Such concerns are also not limited to wheat, as extreme weather conditions can affect global agricultural production across 'breadbaskets' and crops at the same time, leading to synchronized global breadbasket failures and fallout thereof [29]. Global wheat stocks could help buffer shocks, but outside China, global wheat stocks have been oscillating around the current 150 M tons. China has however progressively grown its wheat stock since 2006 after setting a guaranteed floor price to ensure food security and stability [30]. China now has more than half (52%) of global stocks [31]. China thereby is well placed to buffer domestic shocks, albeit less likely to release any onto the global market, also as Chinese domestic prices are relatively high and not internationally competitive [30].

International trade brings into play potential distortions and (dis-)incentives. Domestic (grain) price support (e.g., floor prices, taxes/subsidies, import barriers) can increase domestic relative to world market prices and can boost domestic production as observed in China and India [32, 33]. China's grain subsidy program has been labelled as the largest food self-sufficiency project in the Global South [34]. Agricultural input subsidies provide additional distortions – typically incentivizing intensification but also creating environmental externalities with their excessive use (e.g., nitrogen; irrigation – [32]). Export support (e.g., subsidies) aids domestic surplus producers but undermines producers in importing countries. Removal of agricultural supports globally would raise international wheat prices and potentially increase the cost for many net-importing countries, although also increase incentives for domestic production and import substitution. In addition to the competitive distortions induced by agricultural supports, concerns have also been raised by the underlying resource demands, e.g., the virtual water implicit in the global wheat trade and environmental externalities.

4.7 Key Concepts

This chapter highlighted the continued importance of wheat for global food security over the past quarter century and future implications. It thereby highlighted the need to not only consider global wheat supply, but also demand and trade conditions.

Finally, it highlighted the continued need to invest in wheat productivity enhancement while staying within planetary boundaries.

4.8   Conclusion

Wheat plays a crucial role for global food security and is a critical component in agri-food systems across the globe. It is the most widely grown crop in the world in terms of area. It provides a fifth of food calories and protein to the world's population. It is the most widely internationally traded cereal reflecting the marked spatial disparity between supply and demand. Global wheat production has shown steady growth, mainly propelled by wheat yield increases and wheat cultivation intensification rather than extensification in the form of land expansion.

The paper summarized the state of wheat production, consumption, and international trade at the global and regional levels. It provides a broad-brush appraisal, focusing on the last quarter century. Still, the analysis could be strengthened by improved data to allow for more detailed spatial, dynamic, and political analysis. The sheer size, heterogeneity and evolution of the global wheat economy calls for more detailed analysis about wheat and its role in food systems, including enhanced insights into the associated drivers and modifiers. The political economy of wheat also merits more attention, given the vested interests of subsidized production and the export and processing industries and increasing wheat consumption/production by poor consumers/farmers in the Global South.

With continued global population growth and growing popularity of wheat based processed foods in the Global South there is continued urgency to ensure further transformation of wheat agri-food systems, including sustainable intensification of wheat production to stay within planetary boundaries. The further transformation of wheat production calls for a tripartite contribution of improved germplasm, improved crop management and improved policy. Improved germplasm is particularly needed to continue to raise the wheat yield frontier (yield potential), make it more resilient and address emerging challenges and opportunities (also see Chaps. 7 and 21). This clearly includes climate change which is set to influence wheat production systems and aggravate biotic and abiotic stresses. But it also includes increased attention to quality and demands from consumers and the processing industry (also see Chap. 11). Improved crop management is particularly needed to close yield gaps and stay within planetary boundaries, including reduced environmental externalities linked to water, land and nutrients (also see Chap. 31). Improved policy is particularly needed to create the enabling environment, including value chains, markets and prices and support services and should be dynamic considering the general economic transformation context.

These improvements call for substantive investments in public research and development, particularly in support of wheat agri-food systems in the Global South. There is a general misalignment between private and public interests in wheat germplasm improvement and therefore a need to strengthen the enabling environment and to maintain public support. In the context of the Global South, there is no royalty collection system linked to varietal use (like e.g., Australia's) nor immediate prospects for proprietary varieties like hybrid wheat. Wheat germplasm improvement and research in the Global South will remain in the public domain for the coming decades, including the need for continued funder and CGIAR support; with the role of the private sector mainly limited to contracted seed multiplication (also see Chap. 2). Less obvious too is that much of the needed investments imply maintenance research: the need to keep running as standing still is not an option. Much of crop improvement indeed relies on maintenance breeding – maintaining the yield potential against the evolving biotic and abiotic stresses (also see Chaps. 8, 9, 10, 19, 20, 22 and 23). In much the same way crop management includes doing more with less and reducing environmental externalities. And much of improved policy should ensure the incentives and societal needs are aligned. Much of this may thus not result in visible productivity increases – but should reduce productivity erosion and externalities over time. But taken together the tripartite approach should go a long way to raise global wheat security and stay within planetary boundaries over the coming decades.

References

1. Awika J (2011) Major cereal grains production and use around the world. In: Awika J, Piironen V, Bean S (eds) Advances in cereal science: implications to food processing and health promotion. American Chemical Society, Atlantic City, pp 1–13
2. Dixon J (2007) The economics of wheat: research challenges from field to fork. In: Buck H, Nisi J, Salomon N (eds) Wheat production in stressed environments. Springer, Dordrecht, pp 9–22
3. Shiferaw B, Smale M, Braun H, Duveiller E, Reynolds MP, Muricho G (2013) Crops that feed the world 10. Past successes and future challenges to the role played by wheat in global food security. Food Sci 5:291–317. https://doi.org/10.1007/s12571-013-0263-y
4. FAOStat (2020) FAO Stat. http://www.fao.org/faostat
5. Kinnunen P, Guillaume JHA, Taka M, D'Odorico P, Siebert S, Puma MJ, Jalava M, Kummu M (2020) Local food crop production can fulfil demand for less than one-third of the population. Nat Food 1:229–237. https://doi.org/10.1038/s43016-020-0060-7
6. International Food Policy Research Institute (2019) Global spatially-disaggregated crop production statistics data for 2010 version 2.0. In: International food policy research I. Harvard Dataverse
7. Cassidy ES, West PC, Gerber JS, Foley JA (2013) Redefining agricultural yields: from tonnes to people nourished per hectare. Environ Res Lett 8:34015. http://doi.org/10.1088/1748-9326/8/3/034015
8. FAO (2011) Global food losses and food waste: extent, causes and prevention. FAO, Rome
9. Bright EA, Rose AN, Urban ML, McKee J (2018) LandScan 2017 high-resolution global population data set. Computer software. Version 00. https://www.osti.gov/biblio/1524426
10. Braun HJ, Atlin G, Payne T (2010) Multi-location testing as a tool to identify plant response to global climate change. In: Reynolds M (ed) Climate change and crop production, CABI climate change series. CABI Publishing, Wallingford, pp 115–138
11. Xiong W, Asseng S, Hoogenboom G, Hernandez-Ochoa I, Robertson R, Sonder K, Pequeno D, Reynolds M, Gerard B (2020) Different uncertainty distribution between high and low

latitudes in modelling warming impacts on wheat. Nat Food 1:63–69. https://doi.org/10.1038/s43016-019-0004-2

12. Singh RP, Hodson DP, Huerta-Espino J, Jin Y, Njau P, Wanyera R, Herrera-Foessel SA, Ward RW (2008) Will stem rust destroy the world's wheat crop? Adv Agron 98:271–309. https://doi.org/10.1016/S0065-2113(08)00205-8

13. Mottaleb KA, Singh PK, Sonder K, Gruseman G, Tiwari TP, Barma NCD, Malaker PK, Braun HJ, Erenstein O (2018) Threat of wheat blast to South Asia's food security: an ex-ante analysis. PLoS One 13. https://doi.org/10.1371/journal.pone.0197555

14. Willett W, Rockström J, Loken B, Springmann M, Lang T, Vermeulen S, Garnett T, Tilman D, DeClerck F, Wood A, Jonell M, Clark M, Gordon LJ, Fanzo J, Hawkes C, Zurayk R, Rivera JA, De Vries W, Majele Sibanda L, Afshin A, Chaudhary A, Herrero M, Agustina R, Branca F, Lartey A, Fan S, Crona B, Fox E, Bignet V, Troell M, Lindahl T, Singh S, Cornell SE, Srinath Reddy K, Narain S, Nishtar S, Murray CJL (2019) Food in the Anthropocene: the EAT Lancet Commission on healthy diets from sustainable food systems. Lancet 393:447–492. https://doi.org/10.1016/S0140-6736(18)31788-4

15. Fader M, Rulli MC, Carr J, Dell'Angelo J, D'Odorico P, Gephart JA, Kummu M, Magliocca N, Porkka M, Prell C, Puma MJ, Ratajczak Z, Seekell DA, Suweis S, Tavoni A (2016) Past and present biophysical redundancy of countries as a buffer to changes in food supply. Environ Res Lett 11:55008. https://doi.org/10.1088/1748-9326/11/5/055008

16. Mekonnen MM, Gerbens-Leenes W (2020) The water footprint of global food production. Water 12:2696. https://doi.org/10.3390/w12102696

17. Omara P, Aula L, Oyebiyi F, Raun WR (2019) World cereal nitrogen use efficiency trends: review and current knowledge. Agrosyst Geosci Environ 2:5. https://doi.org/10.2134/age2018.10.0045

18. UN-DESA World population prospects (2019). https://population.un.org/wpp/

19. WorldBank (2020) Data Bank. https://databank.worldbank.org

20. Mittal S (2007) What affects changes in cereal consumption? Econ Polit Wkly 42:444–447. http://www.jstor.org/stable/4419216

21. Mottaleb KA, Rahut DB, Kruseman G, Erenstein O (2018) Changing food consumption of households in developing countries: a bangladesh case. J Int Food Agribus Mark 30:156–174. https://doi.org/10.1080/08974438.2017.1402727

22. Mason NM, Jayne TS, Shiferaw B (2015) Africa's rising demand for wheat: trends, drivers, and policy implications. Dev Policy Rev 33:581–613. https://doi.org/10.1111/dpr.12129

23. Springmann M, Spajic L, Clark MA, Poore J, Herforth A, Webb P, Rayner M, Scarborough P (2020) The healthiness and sustainability of national and global food based dietary guidelines: modelling study. BMJ 370:m2322. https://doi.org/10.1136/bmj.m2322

24. Poole N, Donovan J, Erenstein O (2021) Agri-nutrition research: revisiting the contribution of maize and wheat to human nutrition and health. Food Policy 100:101976. https://doi.org/10.1016/j.foodpol.2020.101976

25. Fader M, Gerten D, Krause M, Lucht W, Cramer W (2013) Spatial decoupling of agricultural production and consumption: quantifying dependences of countries on food imports due to domestic land and water constraints. Environ Res Lett 8:14046. https://doi.org/10.1088/1748-9326/8/1/014046

26. WorldBank (2020) World Bank commodities price data (The pink sheet). http://www.worldbank.org/commodities

27. D'Amour C, Anderson W (2020) International trade and the stability of food supplies in the Global South. Environ Res Lett 15:074005. https://doi.org/10.1088/1748-9326/ab832f

28. Sternberg T (2012) Chinese drought, bread and the Arab Spring. Appl Geogr 34:519–524. https://doi.org/10.1016/j.apgeog.2012.02.004

29. Gaupp F, Hall J, Hochrainer-Stigler S, Dadson S (2020) Changing risks of simultaneous global breadbasket failure. Nat Clim Chang 10:54–57. https://doi.org/10.1038/s41558-019-0600-z

30. Hunt N (2018) Global wheat supply to crisis levels; big China stocks won't provide relief. Reuters

31. Jamieson C (2020) Canada markets: a look at USDA's growing global wheat stocks estimates. https://www.dtnpf.com/agriculture/web/ag/blogs/canada-markets/blog-post/2020/05/13/look-usdas-growing-global-wheat
32. Gulati A, Narayanan S (2003) The subsidy syndrome in Indian agriculture. Oxford University Press, New Delhi
33. Qian J, Ito S, Zhao Z, Mu Y, Hou L (2015) Impact of agricultural subsidy policies on grain prices in China. J Fac Agric Kyushu Univ 60:273–279
34. Yi F, Sun D, Zhou Y (2015) Grain subsidy, liquidity constraints and food security – impact of the grain subsidy program on the grain-sown areas in China. Food Policy 50:114–124. https://doi.org/10.1016/j.foodpol.2014.10.009

Part II
Delivering Improved Germplasm

Chapter 5
Breeding Methods: Line Development

Jessica E. Rutkoski, Margaret R. Krause, and Mark E. Sorrells

Abstract In order to produce successful varieties, wheat breeding programs must develop several strategies that fall under one of the following topics: line development, population improvement, and selection methods. Part I of this chapter focuses on breeding activities related to line development, while Part II discusses population improvement and selection methods. Line development refers to the process of obtaining homozygous inbreds derived from crosses between parental lines. A wide variety of line development methods have been proposed in pursuit of greater efficiency and effectiveness. This chapter aims to provide basic knowledge on line development methods in relation to wheat breeding, describe how and why they came about, and synthesize the results of empirical studies that have evaluated them in order to foster critical thinking and innovation in breeding strategy design.

Keywords Breeding strategies · Line development · Pedigree breeding · Bulk breeding · Single seed descent · Doubled-haploids

5.1 Learning Objectives

- To provide background information on line development approaches in relation to wheat breeding.

J. E. Rutkoski (✉)
Department of Crop Sciences, University of Illinois at Urbana-Champaign, Champaign, IL, USA
e-mail: jrut@illinois.edu

M. R. Krause
Department of Plants, Soils and Climate, Utah State University, Logan, UT, USA
e-mail: margaret.krause@usu.edu

M. E. Sorrells
College of Agriculture and Life Sciences, Cornell University, Ithaca, NY, USA
e-mail: mes12@cornell.edu

© The Author(s) 2022
M. P. Reynolds, H.-J. Braun (eds.), *Wheat Improvement*,
https://doi.org/10.1007/978-3-030-90673-3_5

- To facilitate critical thinking around the role of line development in the design of wheat breeding programs.

5.2   Introduction

Wheat breeding programs that aim to develop varieties must first develop inbred breeding lines so that they can be reproduced for further testing and variety release. Pedigree, bulk, single seed descent, and doubled-haploids are the four main line development methods, while backcross breeding is generally considered to be a useful adjunct to these approaches. Regardless of the line development method being used, the first step is typically to make crosses between different parental plants in order to generate new genetic combinations. If the two parents used in crossing are themselves inbred, then the F_1 progeny will be identical. If one or both of the crossing parents are not inbred, then there will be genetic and phenotypic variability, referred to as 'segregation' in the F_1 progeny. As an alternative to crossing, a breeder can generate novel genetic variation by mutagenizing one or a few plants to induce genetic mutations. The next steps after F_1 seed or mutagenized plants are generated depends on the line development method employed. Following successive generations of line development, a breeder may choose to release one or more lines as varieties or to release a multiline variety composed of more than one selected inbred line.

5.3   Pedigree Breeding

The pedigree method of line development, developed in the 1840s by Vilmorin [1] and rediscovered by Hallett [2] and Nilsson-Ehle [3], allows selection among individual plants and whole families at every inbreeding generation. The process tends to emphasize visual selection among individual plants in the field over successive years as the plants approach homozygosity. To initiate the pedigree breeding process, F_1s from a single cross are space-planted to maximize seed production and to clearly identify individual plants. If there is segregation among the F_1 progeny, selection among F_1s may be imposed. The F_1 plants are harvested individually or bulk harvested, and the resulting F_2s are sown in rows according to the pedigree such that individual plants within families can be identified and harvested individually. Selection is imposed among the F_2 plants, and only the selected plants are carried forward. Each selected F_2 is given a unique identifier (ID) that is recorded along with its pedigree. F_2 plants are harvested individually for their F_3 seed. F_3s that originate from a single F_2 plant are referred to as F_2-derived F_3 families (F_2:F_3). The F_2:F_3 families are typically space planted in rows to enable selection of one or more single plants or single spikes from different plants within each family. The F_2:F_3 families may also be evaluated for yield or quality in order to more accurately select among families. As in the previous generation, selected F_3 plants are given IDs which are recorded along with the ID of their F_2 plant of origin. F_4 seed is harvested from the selected F_3 plants and is planted as F_3:F_4 families in rows. Because most of the variability at this stage is among as opposed to within families, many whole families may be discarded either based on visual assessments of traits such as disease resistance or flowering time and/or quantitative data on traits such as grain yield. The best individual F_4 plants from within the best families are selected and given an ID. F_5 seed is harvested from the selected F_4 plants. The F_4:F_5s are now referred to as 'inbred lines' or 'fixed lines'. F_4:F_5s are expected to be 87.5% homozygous; therefore, they should be phenotypically uniform and stable across generations. The F_4:F_5s are planted as rows, and bulk harvests of each row produce F_4:F_6 seed, which is then used to establish multi-environment yield trials and disease nurseries. Because multiple generations of selection have already been imposed, the F_4-derived lines are expected to be better than the average of their F_1 parents for the traits selected during pedigree breeding, assuming selection during line development was effective. An advantage of the pedigree breeding method is that phenotypic information from related families can be considered during among-family selection to help improve selection accuracy.

While once a popular approach, the pedigree breeding method in its original form is now seldom used in wheat breeding due to its inefficiency. With the pedigree breeding method, a large number of resources must be invested in selection among single plants in early generations. This requires evaluating the selection criteria, performing selection, maintaining seed purity of individual pedigrees, and keeping detailed records of each lineage. In return, a marginal amount of gain from selection is achieved. Although genotypic effects of early generation families are theoretically predictive of their late-generation derivatives [4], early generation selection, as reviewed by Fischer and Rebetzke [5], is particularly ineffective for yield and other low-heritability traits with large genotype-by-environment (GxE) effects. In the case of yield, single plants or families in early generations experience low intragenotypic competition and high intergenotypic competition. Therefore, space planting of single plants or families is not representative of an actual production environments, and meaningful selections for yield performance cannot be made. This point has been demonstrated by empirical studies, which have found low or zero correlation between grain yield measured on single plants [6] or early-generation bulks [7] and grain yield measured in yield plots in later generations. In practice, yield is not typically measured on individual plants for selection during the pedigree breeding process. Instead, breeders often conduct visual selection of plants that appear to be higher yielding. While this visual selection approach is less costly than measuring yield, it is also largely ineffective. A selection experiment conducted by McKenzie and Lambert [8] found that, in barley (*Hordeum vulgare*), visual selection for overall appearance in the F_3 did not improve yield in the F_6, and it led to F_6 lines that were significantly taller and later maturing.

Even if selection in early generations could be conducted in a meaningful way, very little genetic gain would be realized for traits with large GxE effects unless families are evaluated across different locations and selection among families is performed. A study which evaluated both early- and late-generation selection for grain yield or harvest index evaluated in a single environment showed that realized gains in grain yield were little better than random selection [9]. On the other hand, early-generation selection may be effective for traits of high heritability. A study examining the effectiveness of early-generation selection for yield and baking quality in wheat found that selection for protein content and thousand kernel weight in the F_3 generation was effective, but selection for other quality traits and yield was ineffective [10]. Because low-heritability traits like yield tend to be the primary targets of selection, the possible benefits of imposing selection in early generations often do not outweigh the costs. Today, many breeding programs are not conducting early-generation selection and instead employ line development schemes that aim to rapidly generate fixed lines that can be phenotyped accurately for yield and other traits of interest.

5.4   Bulk and Composite Breeding

In the early 1900s, Nilsson-Ehle developed the bulk breeding method [11] which greatly simplified the line development process and enabled breeders to generate lines from many different hybrid combinations with limited resources. In bulk methods of line development, early-generation families are planted and harvested as bulk populations. To begin the process, F_1 plants are harvested in bulk according to their pedigree. In the following season, the F_2 seed from each F_1 bulk is planted as a single row or small plot. Each selected F_2 family is harvested in bulk, producing F_3 seed. In the following season, each F_3 family is again planted as a plot, and selection among plots may be imposed. The process is repeated again until a desired level of uniformity and homozygosity is reached, at which point single spikes within the bulk plots are harvested in order to derive fixed lines. The fixed lines are given IDs and then planted as rows in the following season during which selection is often imposed among rows. Seed harvested from the selected rows is then used to evaluate yield and other traits. In the bulk breeding method employed by Nilsson-Ehle, mass selection (see Sect. 6.3 of Chap. 6) within bulk populations was considered to be an important feature. The idea was to 'assist nature in eliminating the delicate and in conserving the hardy' [11] by relying on abiotic and biotic stresses to aid the culling of poorly-adapted individuals within bulk populations over generations of inbreeding.

Several variations of the bulk breeding method have been suggested to further simplify or improve the process. Harlan and Martini [12] proposed to bulk progeny from multiple cross combinations, creating what is referred to as a 'composite cross population'. This approach enables the sampling of progeny from many diverse cross combinations and then allows natural selection to be imposed among the progeny. A bulk method which derives bulk families from selected F_2 plants and imposes selection among plants in bulk populations was described by Lupton and Whitehouse [13]. This approach was used extensively at CIMMYT, where it was referred to as a 'modified pedigree bulk' method [14]. The CIMMYT wheat program is currently using a 'selected bulk' [15] method in which selection within bulk populations is imposed, but selected plants within F_2 families are bulked rather than harvested individually.

The main advantage of bulk breeding methods is that they are simple and cost efficient because individual plants do not need to be harvested and documented individually. This cost savings can then be invested in the evaluation of fixed lines in multiple environments, which is much more effective for the improvement of low heritability traits like grain yield. At one time, natural selection within bulk populations was believed to be a useful feature of bulk breeding, but several experiments have demonstrated that natural selection within bulk populations often favors genotypes that do not perform well in realistic production environments [16]. The potential for natural selection to favor traits that may be advantageous in natural populations but undesirable in agronomical production systems is, in fact, the main disadvantage of bulk breeding methods.

5.5 Single Seed Descent

In light of the negative impacts of natural selection on bulk breeding populations, the single seed descent method (SSD) was proposed as a way to efficiently generate lines without allowing natural selection to take place [17]. This revolutionary idea enabled the use of off-season nurseries and controlled environments for generation advancement because selection for adaptation to these irrelevant environments could be avoided. In the SSD method, F_1 plants are assigned IDs and harvested for their F_2 seed. Many individual F_2 seeds from each F_1 are sown to generate F_2 plants. From each F_2, lines are derived by planting a single seed each generation. Specifically, one spike is harvested from each F_2 plant, and a single seed is planted to produce F_2:F_3 seed, which is then sown by family. One F_3 spike is harvested from each family, and a single seed is planted to produce F_3:F_4 seed. As in the previous generation, one spike is harvested from each F_4 plant and a single seed is planted. The process is repeated until the lines reach the desired level of homozygosity. For traits conferred by additive effects, the phenotypic distribution of the F_2 population will be the same as the phenotypic distribution of the F_2-derived inbred lines [18]. Thus, transgressive segregates will be preserved, although some anomalies, selection, or attrition is expected [19]. Concerns about missed opportunities for selecting during generation advancement are often raised. However, for yield improvement, the SSD and pedigree methods have been found to perform similarly [20], which is expected because selection for yield in early generations results in very little or no genetic gain [9].

The main advantage of the SSD method is that lines can be rapidly generated in a greenhouse or off-season nursery. Rapid generation advancement in greenhouses [21], also referred to as 'speed breeding', is a technique that is becoming increasingly popular for accelerating line development via SSD or bulk methods. These accelerated breeding methods impose stresses that accelerate plant growth and development. Interestingly, with rapid generation advancement, breeders are reverting back to the random bulk method to simplify the process [21]. Tee and Qualset [22] suggested that under accelerated growth conditions, each plant produces only a few seeds and genetic differences in productivity are not apparent. If this is the case, then SSD is not necessary and bulking whole populations will not alter the genetic composition of the population. To test this hypothesis, Tee and Qualset [22] compared SSD and bulk methods in accelerated growth conditions in two populations. They found that in one population, taller genotypes were favored under bulk selection compared to SSD, while in the other population, there was no difference between bulk- and SSD-derived lines in terms of height, days to heading, and yield. The authors concluded that inadvertent selection in bulk populations was not enough of a concern to warrant using SSD. However, a simulation study by Muehlbauer et al. [19] found that when the standard deviation in the number of seeds produced per plant was greater than 25, progeny from 75% of the original F_2 plants were no longer represented in the population after four generations of bulk breeding. For any given breeding program, the relative merits of SSD and bulk breeding under rapid generation advancement will undoubtedly depend on the germplasm and the nature of the crosses being made. Intuitively, populations derived from parents that are phenotypically very different will experience greater intergenotypic competition effects and reduced between-line versus within-line variation in bulk breeding.

5.6 Doubled-Haploids

Doubled-haploids (DHs) allow breeders to develop homozygous genotypes from heterozygous genotypes in a single generation from the F_1 or in two generations from the F_2. For winter wheat, which can require eight or more weeks of vernalization, DH methods are often used for rapid line development. DHs in wheat can be produced using anther culture or via chromosome elimination, the latter of which is more reliable for wheat breeding. The chromosome elimination method of DH production in wheat begins by hybridizing F_1 wheat plants with maize (*Zea mays*) plants followed by embryo rescue and chromosome doubling using colchicine. For an extensive review of DH production methods in cereals, refer to Humphreys and Knox [23].

Successful DH production results in completely homozygous plants that then undergo seed increase and phenotypic evaluation. In theory, even in the absence of selection, means and variance of DH populations derived from F_1s may differ from those of equivalent SSD populations depending on the linkage phases and interaction effects of favorable loci [24]. The phenotypic distribution of DH populations can have greater kurtosis compared to SSD populations, which would make identifying individuals better than the population mean more difficult unless population sizes are increased [24]. However, empirical studies comparing DH and SSD populations have found little to no differences between them in terms of their phenotypic distributions [25]. To allow greater opportunity for recombination and minimize differences between DH and SSD populations, producing DHs from F_2s or F_3s rather than from F_1s has been suggested [24].

While DHs are being used in some applied wheat breeding programs, their use has been limited by the cost of DH production and the difficulty of either establishing a specialized DH production laboratory in-house or finding a suitable DH service provider. The cost in 2020 of DH production charged by a popular DH service provider in the United States is between $35 and $50 USD per line depending on the details of the order (Heartland Plant Innovations, http://www.heartlandinnovations.com/). In the case of spring wheat breeding at CIMMYT, where two generations of line development can be conducted each year, using DH to develop fixed lines was not advantageous [26]. In winter wheat, off-season nurseries or rapid generation advancement in the greenhouse are alternatives to DH methods that could potentially deliver lines within the same timeframe and at lower cost. It is critical to remember that an established breeding program is producing new populations and lines every year, and accelerating line advancement is only advantageous when new lines are recycled as parents. As DH and generation advancement methods continue improving, breeders should continually reevaluate their options for rapid line development and select the most efficient method available.

5.7   Backcross Methods

Backcross breeding approaches can be employed to transfer a specific trait of interest from a parental donor into another breeding line referred to as the 'recurrent parent'. In this method, the parental donor line is repeatedly crossed to a recurrent parent with the goal of obtaining a line that is nearly genetically identical to the recurrent parent except for the addition of one or a few genes from the donor parent conferring the trait of interest. In practice, linkage drag can result in undesirable linked genes being transferred as well, especially in crosses in which exotic germplasm is the donor parent.

Backcrossing is sometimes referred to as a defensive or conservative breeding strategy because it involves the transfer up to three genes (limited by population size) conferring a simply-inherited trait of interest to correct a defect or otherwise improve a successful variety. It is therefore considered to be a useful adjunct to pedigree, bulk, SSD, and DH, which are typically employed to recover superior combinations of numerous alleles from both parents to improve quantitative traits such as yield. Self-fertilization and backcrossing produce parallel rates of inbreeding but very different genotypes (Table 5.1). An early example used backcrossing to develop 'Baart', a wheat cultivar resistant to common bunt (*Tilletia tritici*) [27]. The author noted that backcross-derived varieties should require less extensive testing prior to release and that the improved variety could then be used in future backcrossing programs rather than using the original exotic line.

Table 5.1 Approach to homozygosity and percent homozygosity at two loci of interest from self-fertilization or backcrossing

| Generations | Self-fertilized | | Backcrossed | |
	% homozygous	% homozygous at 2 desired loci	% homozygous	% homozygous at 2 desired loci
1	25.00	6.25	25.00	25.00
2	56.25	14.06	56.25	56.25
3	76.56	19.14	76.56	76.56
4	87.79	21.97	87.89	87.89
5	93.84	23.46	93.84	93.84

Reprinted with permission from Ref. [28]

The general protocol for backcrossing depends on whether the trait being transferred follows dominant or recessive inheritance. If the trait is dominant, then the plants expressing the trait in each generation are heterozygous and are chosen for crossing to the recurrent parent. However, if the trait is recessive, it will not be obvious which plants carry the recessive allele. This can be remedied by making a test cross to the donor parent (or a self-pollination) at the same time the plant is crossed to the recurrent parent. The progeny of the test cross or self-pollination will segregate if the plant was heterozygous. The crosses made with the heterozygous plant are then advanced to the next backcross. With each backcross generation, the percentage of the recurrent parent genome recovered increases by half (Table 5.1). Population sizes required to have a certain probability of recovering individuals with the desired trait have been published in Sedcole [29]. However, in practice, it may be more efficient to process progeny in batches so that once the desired number of individuals carrying the trait is attained, the next round of backcrosses can be initiated and the entire population need not be evaluated.

A potential drawback to backcross methods is that newer varieties developed using breeding methods such as pedigree, bulk, SSD, or DH may surpass the performance of a backcross-derived variety by the time it is released and available for commercial use. Also, unforeseen problems such as a new race of a pathogen can cause a long-time recurrent parent to become obsolete. In practice, it is recommended to introduce advanced lines into the backcrossing program as early as possible and to carry along several backcross families concurrently so that there can be selection among the families at the end of the program for traits other than the one(s) transferred.

5.8 Mutation Breeding

All genetic variation observed in living organisms has been generated by mutation, structural rearrangements, and recombination. Whereas the aforementioned breeding methods rely on recombination through crossing to develop new genetic combinations and derive breeding lines, mutation breeding represents an alternative approach that does not require crossing and may be useful for improving traits that may lack natural genetic variation. Most natural mutations are deleterious, rare, and recessive. However, plant breeders have sought to generate potentially useful genetic variation by inducing mutations through various means. Because the mutagens are generally not selective, plant breeders are faced with the task of sorting out useful mutations from undesirable ones. Any individual mutagenized plant can have many hundreds or thousands of mutations, creating complications when a deleterious mutation obscures a useful one. Generating large segregating populations is therefore important for identifying useful genetic variants.

The first step in designing a mutation breeding program is to calibrate the dose of the chosen mutagen so that the frequency of mutations is maximized but lethality is limited. A dose/response calibration is required for each mutagen, species, and seed lot. The dose is adjusted by varying the intensity or time for radiation or by varying the concentration of a chemical mutagen. Radiation treatments can be applied to pollen, seeds, seedlings, buds, or whole plants, whereas chemical treatments are used for ungerminated seeds. Polyploids generally tolerate higher doses because of genetic redundancy, though that benefit may be offset by homeologous or duplicate genes masking the effects of recessive mutations. Following mutagenesis, the screening method for desirable mutations depends on the species and whether it is clonally propagated, outcrossing, or inbreeding. For an inbreeding, seed-propogated species such as wheat, mutations can be dominant or recessive with the latter being revealed through selfing. Mutations must be transmitted in the pollen or eggs in order to be transferred across generations.

Mutagenesis impacts the entire genome, producing a large number of undesirable mutants that require an efficient screening technique. Even if a desired variant is found, it is likely to be associated with undesirable mutations that will require elimination through outcrossing or backcrossing. Consequently, mutation breeding should only be considered for certain traits or applications. Examples of plant varieties developed using mutagenesis can be found in the Joint FAO/IAEA Mutant Varieties Database, which compiles information on more than 3200 officially released mutant varieties of over 200 plant species worldwide (https://mvd.iaea.org).

Oladosu [30] reviews multiple examples of mutation breeding for targeting a variety of traits. Assessment of the value of mutation breeding has to consider if the product has been proven to not be a result of outcrossing, recombination, or natural mutations. TILLING (targeting induced local lesions in genomes), which combines chemical mutagenesis and high-throughput screening for point mutations, has been used to create mutant populations for wheat [31]. In summary, mutation breeding can be a useful tool for improving certain traits that may lack natural genetic variation, but an efficient evaluation protocol is required such that large populations can be screened for desirable variants.

5.9   Multilines

Once inbred lines are produced through one or more of the aforementioned line development methods, the breeder may choose to create a multiline variety. It is important to distinguish multilines from blends or mixtures at the outset. Blends can be mixtures of existing varieties or species in various proportions and are sometimes referred to as 'multiblends' [32]. Varieties may be blended for many reasons such as to capture the performance of different varieties or to reduce seed inventory. In constrast to multiline varieties, the development of multiblends does not necessarily require research on the performance of different combinations of mixtures.

The concept of a multiline variety was proposed by Jensen [33] and defined as a combination of pure lines chosen from a breeding program for uniformity of appearance, especially for height and maturity, but also for other characteristics important for a desirable agronomic type. The purpose is to combine different genotypes that have desirable attributes but do not reduce the phenotypic uniformity. Performance data on the components are necessary so that only compatible lines are blended. The individual component lines are maintained separately so that the original blend can be recreated by mixing the seed stocks in the correct proportions, and the breeder has the option of adding or removing individual lines over time. Theoretically, a multiline variety could have a longer life because of enhanced yield stability, broader adaptation, and resistance to diseases. The component lines could have resistance to different races of the pathogen, thus avoiding a potentially devasting disease outbreak that could occur if they were released individually. In summary, the advantages of multiline cultivars include (1) they provide a method to quickly develop a well-buffered, disease-resistant cultivar that can employ several resistance genes; (2) the useful life of a disease resistance gene is extended while a conventional breeding program is ongoing; (3) reduced losses due to disease should stabilize the cultivars deployed; and (4) an individual breeding program can distribute cultivars over a wide area without risk of homogenizing the pathogen population. Disadvantages include (1) the utility of multilines is limited to high-risk regions for disease outbreaks; (2) usually there is no genetic improvement for yield or agronomic traits; (3) substantial labor is required to produce and maintain the component lines; and (4) release of an improved recurrent variety is delayed until the components are produced.

To quantify the performance of multilines, Jensen and Federer [34] applied the concepts and computation of combining ability to competitive ability in wheat. In this application, general combining ability (GCA) refers to the average performance of a line in combinations, and specific combining ability (SCA) refers to the deviations in the expected average performance of combinations. Jensen [32] outlined four different examples for forming a multiline: (1) using a single backcross to generate lines for use in the multiline; (2) crossing unrelated lines; (3) crossing to different selected recurrent parents; and (4) making double crosses where each single cross has a common parent.

   Marshall and Brown [35] used statistical models to determine the effect of intra-populational genetic diversity on the stability of performance of mixtures as estimated by their variance in yield across environments. Their models suggested that, in the absence of intergenotypic interactions, the yield of a multiline will vary less than the least variant component when the component lines perform differently in different environments. Conversely, when there are intergenotypic interactions, the stability of a mixture will be more stable than the best line only when each component responds differently to different environments. They also predicted that, when mixtures are compared to their pure line components, it is expected that improved stability is more easily attained than improved yield because improvements in yield require net positive intergenotypic interactions whereas stability does not.

   In addition to stability, the use of multilines has the potential to improve resistance to disease. Borlaug and Gibler [36] developed wheat lines for multiline cultivars at CIMMYT using 'donor parents' selected from the International Wheat Rust Nursery and backcrossed to recurrent parents. A number of studies have examined possible mechanisms for the observed enhancements in disease resistance within multilines, and generally agree that the reduced inoculum load results from both a lower frequency of initial infection when a spore lands on a resistant component of the multiline and a lower rate of increase in inoculum. A review of multilines for disease control was published by Mundt [37].

   Given the changing climate and the need for greater protection of natural ecosystems and sustainable agricultural practices, multilines and multiblends will likely play an important role in the future agricultural production systems. Further research is warranted on durability of resistance in multilines, experimental design, and design of mixtures.

5.10   Key Concepts

Pedigree, bulk, single seed descent, and doubled-haploids are the four main line development methods. Backcross, mutation and multiline breeding methods are useful supplements to line development. Pedigree breeding is rarely used in wheat breeding because a large number of resources must be invested in selection. Bulk breeding methods are simple and cost efficient. The main advantage of the SSD method is that lines can be rapidly generated in a greenhouse or off-season nursery. Doubled-haploids allow breeders to develop homozygous genotypes from heterozygous genotypes in a single generation but are limited by the high cost. The choice of line development method(s) depends on resources available and selection methods chosen.

5.11 Conclusion

Given the array of options for developing inbred lines, the main challenge is to determine how to build a coherent and efficient breeding strategy given a fixed budget and other resources. It is common for wheat breeding programs to implement multiple line development methods at different stages of the breeding pipeline. Breeders must also consider how to strategically combine line development approaches with population improvement and selection methods, which are described in Chap. 6, in order to produce superior varieties.

References

1. Gayon J, Zallen DT (1998) The role of the Vilmorin company in the promotion and diffusion of the experimental science of heredity in France. J Hist Biol 31:1840–1920
2. Hallett FF (1861) On 'pedigree' in wheat as a means of increasing the crop. J R Agric Soc Engl 22:371–381
3. Nilsson-Ehle H (1908) Einige Ergebnisse von Kreuzungen bei Hafer und Weizen. Berlingska Boktryckeriet
4. Bernardo R (2003) On the effectiveness of early generation selection in self-pollinated crops. Crop Sci 43:1558–1560. https://doi.org/10.2135/cropsci2003.1558
5. Fischer RA, Rebetzke GJ (2018) Indirect selection for potential yield in early-generation, spaced plantings of wheat and other small-grain cereals: a review. Crop Pasture Sci 69:439–459. https://doi.org/10.1071/CP17409
6. Syme JR (1972) Single-plant characters as a measure of field plot performance of wheat cultivars. Aust J Agric Res 23:753–760. https://doi.org/10.1071/AR9720753
7. Fowler WL, Heyne EG (1955) Evaluation of bulk hybrid yests for predicting performance of pure line selections in hard red winter wheat. Agron J 47:430–434. https://doi.org/10.2134/agronj1955.00021962004700090010x
8. McKenzie RIH, Lambert JW (1961) A comparison of F_3 lines and their related F_6 lines in two barley crosses. Crop Sci 1:246–249. https://doi.org/10.2135/cropsci1961.0011183x000100040005x
9. Whan BR, Knight R, Rathjen AJ (1982) Response to selection for grain yield and harvest index in F_2, F_3 and F_4 derived lines of two wheat crosses. Euphytica 31:139–150. https://doi.org/10.1007/BF00028316
10. Briggs KG, Shebeski LH (1971) Early generation selection for yield and breadmaking quality of hard red spring wheat (*Triticum aestivum L.* EM THELL.). Euphytica 20:453–463. https://doi.org/10.1007/BF00035673
11. Newman LH (1912) Plant breeding in Scandinavia. The Canadian Seed Grower's Association, Ottawa
12. Harlan HV, Martini ML (1929) A composite hybrid mixture. Agron J 21:487–490. https://doi.org/10.2134/agronj1929.00021962002100040014x
13. Lupton FGH, Whitehouse RNH (1957) Studies on the breeding of self-pollinating cereals – I. Selection methods in breeding for yield. Euphytica 6:169–184. https://doi.org/10.1007/BF00729886
14. van Ginkel M, Trethowan R, Cukadar B (1998) A guide to the CIMMYT bread wheat program. Wheat special report no.5. CIMMYT, Mexico
15. Singh RP, Rajaram S, Miranda A et al (1997) Comparison of two crossing and four selection schemes for yield, yield traits, and slow rusting resistance to leaf rust in wheat. In: Braun HJ,

Altay F, Kronstad WE, Beniwal SPS, McNab A (eds) Wheat: prospects for global improvement. Springer, Dordrecht, pp 93–101

16. Suneson CA, Wiebe GA (1942) Survival of barley and wheat varieties in mixtures. Agron J 34:1052–1056. https://doi.org/10.2134/agronj1942.00021962003400110010x

17. Goulden CH (1939) Problems in plant selection. In: Proceedings of the seventh international genetics congress. Cambridge University Press, pp 132–133

18. Snape JW, Riggs TJ (1975) Genetical consequences of single seed descent in the breeding of self-pollinating crops. Heredity 35:211–219. https://doi.org/10.1038/hdy.1975.85

19. Muehlbauer FJ, Burnell DG, Bogyo TP, Bogyo MT (1981) Simulated comparisons of single seed descent and bulk population breeding methods. Crop Sci 21:572–577

20. Knott DR, Kumar J (1975) Comparison of early generation yield testing and a single seed descent procedure in wheat breeding. Crop Sci 15:295–299. https://doi.org/10.2135/cropsci1975.0011183x001500030004x

21. Collard BCY, Beredo JC, Lenaerts B et al (2017) Revisiting rice breeding methods – evaluating the use of rapid generation advance (RGA) for routine rice breeding. Plant Prod Sci 1008:1–16. https://doi.org/10.1080/1343943X.2017.1391705

22. Tee TS, Qualset CO (1975) Bulk populations in wheat breeding: comparison of single-seed descent and random bulk methods. Euphytica 24:393–405. https://doi.org/10.1007/BF00028206

23. Humphreys DG, Knox RE (2016) Doubled haploid breeding in cereals. In: Al-Khayri JM, Jain SM, Johnson DV (eds) Advances in plant breeding strategies: breeding, biotechnology and molecular tools. Springer, pp 241–290

24. Jinks JL, Pooni HS (1981) Properties of pure-breeding lines produced by dihaploidy, single seed descent and pedigree breeding. Heredity 46:391–395. https://doi.org/10.1038/hdy.1981.47

25. Park SJ, Walsh EJ, Reinbergs E et al (1976) Field performance of doubled haploid barley lines in comparison with lines developed by the pedigree and single seed descent methods. Can J Plant Sci 56:467–474. https://doi.org/10.4141/cjps76-077

26. Li H, Singh RP, Braun H et al (2013) Doubled haploids versus conventional breeding in CIMMYT wheat breeding programs. Crop Sci 53.74–83. https://doi.org/10.2135/cropsci2012.02.0116

27. Briggs FN (1930) Breeding wheats resistant to bunt by the back-cross method. Agron J 22:239–244

28. Briggs FN (1935) The backcross method in plant breeding. Agron J 27:971–973

29. Sedcole JR (1977) Number of plants necessary to recover a trait. Crop Sci 17:667–668

30. Oladosu Y, Rafii MY, Abdullah N et al (2016) Principle and application of plant mutagenesis in crop improvement: a review. Biotechnol Biotechnol Equip 30:1–16. https://doi.org/10.1080/13102818.2015.1087333

31. Krasileva KV, Vasquez-Gross HA, Howell T et al (2017) Uncovering hidden variation in polyploid wheat. Proc Natl Acad Sci U S A 114:E913–E921. https://doi.org/10.1073/pnas.1619268114

32. Jensen NF (1988) Plant breeding methodology. Wiley, New York

33. Jensen NF (1952) Intra-varietal diversification in oat breeding 1. Agron J 44:30–34. https://doi.org/10.2134/agronj1952.00021962004400010009x

34. Jensen NF, Federer WT (1965) Competing ability in wheat 1. Crop Sci 5:449–452

35. Marshall DR, Brown AHD (1973) Stability and performance of mixtures and multilines. Euphytica 22:405–412

36. Borlaug NE, JW Gibler (1953) The use of flexible composite wheat varieties to control the constantly changing stem rust pathogen. Agronomy Abstracts 81

37. Mundt CC (2002) Use of multiline cultivars and cultivar mixtures for disease management. Annu Rev Phytopathol 40:381–410. https://doi.org/10.1146/annurev.phyto.40.011402.113723

Chapter 6
Breeding Methods: Population Improvement and Selection Methods

Jessica E. Rutkoski, Margaret R. Krause, and Mark E. Sorrells

Abstract In order to produce successful varieties, wheat breeding programs must develop several strategies that fall under one of the following topics: line development, population improvement, and selection methods. This chapter focuses on breeding activities related to population improvement and selection methods, while Chap. 5 discusses line development. The objective of population improvement is to enhance the entire genetic base of the breeding program, while selection methods aim to identify breeding lines with superior potential or performance. As with line development approaches, numerous population improvement and selection methods have been developed in order to enhance breeding program efficiency and achieve genetic improvement. This chapter will provide an overview of population improvement and selection methods in the context of wheat breeding, discuss their advantages and disadvantages, and summarize empirical studies that have evaluated them in order to inform breeding program design.

Keywords Breeding strategies · Population improvement · Recurrent selection · Best Linear Unbiased Prediction (BLUP) · Marker-assisted selection · Genomic selection

J. E. Rutkoski (✉)
Department of Crop Sciences, University of Illinois at Urbana-Champaign, Urbana, IL, USA
e-mail: jrut@illinois.edu

M. R. Krause
Department of Plants, Soils and Climate, Utah State University, Logan, UT, USA
e-mail: margaret.krause@usu.edu

M. E. Sorrells
College of Agriculture and Life Sciences, Cornell University, Ithaca, NY, USA
e-mail: mes12@cornell.edu

M. P. Reynolds, H.-J. Braun (eds.), *Wheat Improvement*,
https://doi.org/10.1007/978-3-030-90673-3_6

83

6.1   Learning Objectives

- To provide background information on population improvement and selection methods in relation to wheat breeding.
- To facilitate critical thinking around roles of population improvement and selection methods in the design of wheat breeding programs.

6.2   Population Improvement

Chapter 5 introduced various line development methods in the context of wheat breeding programs. While representing a key component of the breeding pipeline, line development alone will not lead to the production of successful varieties year after year. To achieve ongoing variety improvement, a population improvement strategy should be implemented in order to enhance the entire genetic base of the breeding program. Recurrent selection is the predominant approach to population improvement in wheat breeding, but evolutionary breeding has also received some attention as an alternative approach.

6.2.1   Evolutionary Breeding

The idea for evolutionary plant breeding [1] grew out of the desire to improve the efficiency of the bulk breeding method. The evolutionary breeding method involves the bulking of F_1 progeny followed by many generations of prolonged natural selection (and incidental artificial selection) in successive natural environments. The concept is simple: let nature select the best adapted genotypes over time, thus minimizing the effort required for traditional selection and testing of individual genotypes.

The exploitation of natural selection represents a similarity between the bulk and evolutionary breeding methods, but there is an important distinction. Bulk breeding is considered a line development method because it is used to inbreed a population during a limited number of generations of self-fertilization following the initial cross until a desired level of homozygosity is reached. Conversely, evolutionary breeding represents a population improvement strategy because it relies on natural outcrossing and selection to generate new genetic combinations and lead to incremental improvement in the population over time. Natural selection acts on these heterogeneous mixtures over many generations to produce populations with superior environmental adaptation.

As with bulk breeding for line development, this reliance on natural selection provides a logistical advantage to the population improvement process, but a potential disadvantage is that nature may select for traits that are undesirable in agronomic settings. For example, when evolutionary breeding populations are replanted, a subset, typically 1/30th to 1/50th or less, of the seed is sampled for use in the next generation. The plants producing the most seeds have the greatest likelihood of representing the next generation, and finite resources available for reproduction may produce a negative correlation between seed size and seed number. Another example is undesired selection for increased lodging. Taller plants may shade out shorter plants, thereby producing more seeds, but tall plants may be prone to lodging.

Most empirical studies testing the utility of the evolutionary breeding method were performed in barley. Hockett et al. [2] evaluated agronomic traits in Composite Cross II (CCII), which was developed by crossing 28 barley lines and sowing the subsequent generations under natural selection. The resulting populations were found to have higher yield than a mixture of the original parents, suggesting that improvement due to natural selection had been made. However, various CCII populations which had been developed in different environments were not significantly different from one another, even when tested in the environment in which they were developed. This suggests that natural selection had not produced local environmental adaptation. Furthermore, contemporary varieties yielded significantly more than the best CCII populations, suggesting that greater genetic progress had been made from conventional breeding methods.

Heterozygosity is expected to be reduced by half in each successive generation after an initial cross is made to a produce segregating population or after a natural outcrossing event. However, empirical studies have demonstrated that natural selection can sometimes preserve the level of heterozygosity in the population when there is an advantage for the heterozygote with respect to relative fitness. For example, Hockett et al. [2] observed a high level of variability remaining in the F_{19} generation of CCII, which authors attributed to the adaptive advantage of the heterozygote. The promotion of outcrossing can further maintain or increase heterozygosity in the population.

An important disadvantage of evolutionary breeding as a population improvement approach is the amount of time required to observe a benefit. Following an initial cross or crosses, long-term progress in evolutionary breeding populations is dependent on natural outcrossing, which occurs at relatively low levels in wheat. This disadvantage can be partially offset by the ability to plant and maintain several populations simultaneously, though it is difficult to predict which populations will produce superior progeny, and this will not be known for several years. Promoting greater levels of outcrossing may further expedite population improvement.

6.2.2   Recurrent Selection

Recurrent selection, reviewed by Rutkoski [3], is a population improvement method which aims to enhance the breeding population as a whole through crossing and recombination. Compared with evolutionary breeding, which largely relies on natural outcrossing, recurrent selection more readily facilitates recombination through successive intermating, and incremental genetic improvement occurs in the population over time as the frequency of favorable alleles increases. The recurrent selection process is a cycle that consists of four sequential activities: (1) crossing to recombine breeding materials; (2) generation of new breeding individuals which are non-inbred plants, families, or inbred lines; (3) evaluation of the breeding individuals by phenotyping and/or genotyping; and (4) identification of the best breeding individuals to use for the next round of crossing. Selection may be imposed on a single quantitative trait of interest or on an index of multiple traits (see Chap. 32). Given this process, the breeding population mean is expected to improve by $R=$ $kr_{xg}\sigma_g/L$ units each year where k is the selection intensity in standard deviation units, r_{xg} is the accuracy of selection, σ_g is the genetic standard deviation, and L is the duration of the breeding cycle defined as the time which elapses between crossing and evaluation of the derived progeny.

In its simplest form, recurrent selection in wheat consists of selecting and intermating (i.e. 'recycling') the best new breeding lines developed by the program each year to generate F_1s that then enter the line development process. Because the breeding germplasm is improving, a different set of 'best' breeding lines will be selected for intermating each year. The rate of improvement each year is heavily affected by the length of the breeding cycle, with shorter breeding cycles leading to faster rates of gain. As a way to shorten the breeding cycle, recurrent selection in wheat can be imposed among S_1 families (equivalent to F_2 families) based on phenotypic or genomic selection [4, 5]. Each cycle, individual plants can be selected out of rapid recurrent selection populations and put through a line development process like bulk or SSD to generate lines for further testing as potential varieties. For traits like yield, which should be evaluated as a uniform stand to eliminate intra- and inter-genotypic competition effects, genomic selection [6] (see Sect. 6.3.4) is a promising solution. A recent simulation study demonstrated that, assuming a fixed budget, rapid genomic recurrent selection in winter wheat has the potential to increase the rate of genetic gain for yield by nearly 2.5-fold compared to a conventional breeding strategy [7].

Generating crosses in wheat involves a labor-intensive process of removing the anthers within each floret with forceps or scissors prior to anthesis and later introducing pollen from another plant. The reproductive biology of wheat therefore acts as a limiting factor to rapid cycling for recurrent selection. Male sterility greatly facilitates outcrossing between individuals, and both recessive and dominant male-sterility genes are available in wheat. Figure 6.1 shows several recurrent selection schemes using a dominant male-sterile gene. Those schemes include phenotypic recurrent selection (method A), variations of half-sib selection (methods B, C, and D), S_1 selection (method E) and combined S_1 and half-sib selection (method F). Line development methods can be readily applied to dominant male-sterile populations to obtain pure lines. Selection of male-fertile plants provides F_2 progeny that will breed true for fertility and can be used directly in bulk, SSD (single seed descent method), DH (Doubled-haploids), or backcross breeding line development schemes.

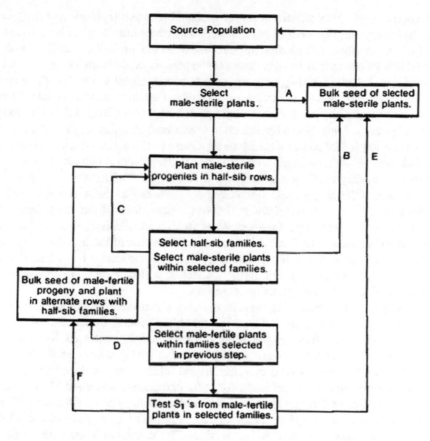

Fig. 6.1 Recurrent selection schemes using a dominant male-sterile allele. (Reprinted with permission from Ref. [8])

6.3   Selection Methods

Selection in wheat breeding programs may be imposed based on rudimentary visual assessments, sophisticated genomic prediction models, or anything in between. Although the appropriate selection method for a given situation depends on the traits of interest, resources available, and the breeding germplasm, results of many years of research in this area have shed light on which are the most promising methods for wheat breeding.

6.3.1   Mass Selection Systems

Mass selection involves selection among single plants based on their single-plant phenotypes, and it is a common practice during line development via pedigree or bulk methods and male sterile-facilitated recurrent selection. Although mass selection is a relatively popular selection method because it is simple and inexpensive to implement, the merit of mass selection in wheat breeding has been controversial. An important limitation on the effectiveness of mass selection is the heritability of the trait of interest. The genetic gain expected from a single generation of mass selection is $kh\sigma_g$, where k is the selection intensity, h is the square root of the heritability, and σ_g is the square root of the additive genetic variance. For quantitative traits like yield, h on a single-plant basis will be low, although as long as h is not zero, some gain from selection is expected. Another major criticism of mass selection is that most traits of interest cannot be measured on single plants in realistic production conditions. Thus, the trait under selection will likely be a correlated 'secondary trait', and selection will be indirect. The relative efficiency of indirect versus direct mass selection is $k_j r_g h_j / k_i h_i$, where k_i is the selection intensity of the trait of interest, h_i is the square root of the heritability of the trait of interest, k_j is the selection intensity of the secondary trait under selection, r_g is the genetic correlation between traits i and j, and h_j is the square root of the heritability of the trait under selection. Thus, if the secondary trait under selection has a heritability level similar to the primary trait, indirect selection will be even less effective than direct selection with equal selection intensities.

Empirical studies evaluating the effectiveness of mass selection in wheat tend to confirm what would be expected based on theory. In general, direct mass selection tends to be effective, although the effect is often small. For example, Redden and Jensen [9] found that direct mass selection for early tillering, a low heritability trait, was effective, and more cycles of selection and intermating were associated with greater genetic gain. Mass selection for traits that cannot be measured on single plants must be done through indirect means. For example, grain yield measured on individual space-planted plants is an indirect measure of grain yield measured in uniform plots under realistic planting densities. The success of indirect mass selection will depend on the nature of the primary and secondary traits and the genetic correlation between them; therefore, results are expected to vary widely. An experiment evaluating mass selection for grain size using mechanical sorting [10] found that selection was effective at improving grain yield per spike and kernel weight, however improvement in grain yield was not evaluated. Thakare and Qualset [11] evaluated indirect selection for yield and found that visual selection of single plants was effective. However, random selection was also similarly effective at improving grain yield in this experiment, indicating that inadvertent or natural selection may have been the cause of the observed yield gain rather than visual selection. Grid selection, or subdividing the field into a grid and performing selection within the grid units, can improve mass selection by reducing the environmental variation among plants in close proximity [12]. Indirect mass selection for grain yield based on yield measurements taken on single, widely-spaced plants in a honeycomb pattern has been found to produce small but significant genetic gains [13].

Although the effectiveness of mass selection has been clearly demonstrated for some traits and populations, the more important question is how well mass selection performs compared to alternative methods on a gain per unit time and cost basis. The major drawback to mass selection is that when performed in pedigree or bulk breeding schemes, as it is often done, mass selection can be at odds with rapid generation advancement, which can have an impact on increasing rates of genetic gain through reducing the breeding cycle duration. Thus, the benefit of mass selection may not be large enough to outweigh its opportunity cost if it precludes reducing the breeding cycle duration. Mass selection in rapid cycle recurrent selection is much more promising; however, it is worth remembering that it is often not possible to measure and select for all traits of interest on a single plant.

6.3.2   Selection Based on Best Linear Unbiased Prediction (BLUP)

While mass selection can be inexpensive and feasible for large numbers of candidates, selection based on data replicated within and/or across environments is more accurate, especially for low heritability traits like grain yield. Typically, wheat breeding programs phenotypically evaluate inbred lines replicated within and across environments, and phenotypic observations on lines are often combined using simple arithmetic means or least-squares. However, the statistical procedure Best Linear Unbiased Prediction (BLUP) is the most effective approach to combine multiple phenotypic observations on a breeding individual into a single value which represents its genetic value. These estimates of individuals' genetic values based on BLUP are referred to as 'BLUPs' or 'random effects'. The BLUP procedure was developed by Henderson [14] for animal breeding as a way of maximizing selection accuracy, and therefore genetic improvement, given all data available while also accounting for non-genetic effects such as effects of environment. BLUPs are referred to as 'shrinkage estimators' because they are compressed towards zero depending upon the degree of uncertainty in the estimate. For example, BLUPs for individuals which appear outstanding based on very little data will be shrunk heavily towards the population mean to reflect what is more likely to be their true genetic value. In contrast, BLUPs for individuals which appear outstanding based on many observations within and across environments will be shrunk towards the population mean only very slightly since their performance is known with a high degree of certainty. In this way, it is possible to accurately rank and compare individuals that have been evaluated over different numbers of environments and/or replications.

Another major advantage of BLUP is that it supports the utilization of data from multiple traits and/or multiple environments and on related individuals in an optimal way through pedigree BLUP and multi-trait pedigree BLUP [15, 16] or genomic BLUP [17] and multi-trait genomic BLUP [18]. The latter are commonly used for genomic selection models. A useful feature of multi-trait BLUPs is that they can be multiplied by a vector of economic weights to easily estimate an optimal selection index [19]. While the use of BLUP for selection is not controversial, many research groups have empirically demonstrated the effectiveness of BLUP methods for wheat breeding, especially for selection based on yield in multiple environments [20].

With the rise in popularity of genomic selection in recent years, BLUP methods are becoming increasingly more important in wheat breeding.

6.3.3 Marker-Assisted Selection

Marker-assisted selection (MAS) is based on the premise that selection based on DNA markers can be more effective or efficient than selection based on phenotypes. Here we introduce conventional MAS, which includes MAS methods other than genomic selection, while more in-depth coverage of MAS can be found in Chap. 28. Conventional MAS in plant breeding, reviewed by Collard and Mackill [21], involves (1) detecting diagnostic markers closely linked to genes affecting the traits of interest; (2) validating those markers in the germplasm where MAS is to be applied; and (3) routine selection based on the validated markers during the breeding process. MAS was a revolutionary idea because it implied that breeders could select on alleles directly without phenotyping [22]. Although MAS never replaced phenotyping and conventional selection methods, it now plays an important role in backcross introgression, gene pyramiding and line development in wheat breeding.

The earliest use of molecular markers in plant breeding was in the backcross method. The ability to identify recombinants close to one or more genes or QTL (Quantitative trait loci) from a donor parent and to simultaneously select for the elite background genotype transformed molecular markers into a modern breeding approach for crop improvement. This strategy, referred to as 'marker-assisted backcrossing' (MABC), greatly improved the efficiency of backcrossing alleles that are recessive, epistatic, or affecting traits that cannot be easily measured on a single plant basis.

Tanksley et al. [22] first proposed the use of markers in backcrossing to introgress target genes of interest and select for the genome of the recurrent parent. Hospital et al. [23] was among the first to investigate, through simulation, the optimization of molecular markers to simultaneously perform 'background selection', where selections are made against donor alleles at non-target loci, and 'foreground selection', in which the target loci are selected. The authors evaluated variables such as time and intensity of selection, population size, and number and position of markers and found that MABC led to a gain of about two generations to recover the recurrent parent genome compared to conventional backcrossing without the use of markers. This reduction in the amount of time required for backcrossing is substantial when considering that MABC can be conducted year-round in the greenhouse because phenotyping is not necessary. Their simulation also showed that three markers per non-carrier chromosome (100 cM) were adequate to select for the elite background genotype in early generations because few recombination events have occurred. In later generations, most of the recurrent parent genome has been recovered, so few donor parent segments remain to be eliminated.

Empirical studies have demonstrated that MABC is effective for wheat breeding for traits that are conferred by few large-effect loci. For example, Randhawa et al.

[24] used MABC with foreground and background selection to introgress a yellow rust (*Puccinia striiformis* f. sp. *tritici*) resistance gene into an elite background in only two backcross generations while recovering 97% of the recurrent parent genome. An important consideration for the application of MABC is that it can be less effective than phenotypic selection if the trait of interest provided by the recurrent parent is conferred by multiple QTL that are not tightly linked to the markers used for foreground selection. Furthermore, as discussed in Chap. 5, backcrossing is a conservative breeding strategy because it cannot not produce lines that are superior to the best parent for quantitative traits such as grain yield. Thus, the cultivars that result from backcrossing may be difficult to commercialize unless the trait introgressed is of high economic value.

In addition to MABC, gene or QTL pyramiding using markers has been proposed in order to achieve an ideal genotype containing two or more genes or QTL originating from different parents. The simplest pyramiding strategy, demonstrated by Liu et al. [25] with two powdery mildew (*Erysiphe graminis* f. sp. *tritici*) resistance genes, relies on crossing two near isogenic lines (NILs), where each NIL contains different alleles in the same genetic background. The resulting genotypes are then self-pollinated, and lines homozygous for both genes are selected. Unfortunately, for most desired gene pyramids, appropriate NILs are not readily available. In this case, a crossing and selection strategy which combines genes or QTL from two different parents is needed. Obtaining the desired gene or QTL combination in early generations requires large population sizes. Selection among inbred lines can reduce the number of lines needing to be screened; however, resources must be spent to generate inbred lines that will ultimately get discarded. Bonnett et al. [26] discusses strategies to efficiently pyramid multiple genes in wheat using molecular markers.

Although pyramiding is useful in some cases, most traits of interest in wheat are complex and cannot be improved sufficiently through pyramiding. To breed for quantitative traits as well as traits conferred by major-effect loci, an approach referred to as 'forward breeding' is preferred. With forward breeding, major-effect alleles are first introgressed into elite breeding lines which are then used in crosses. Populations which segregate for the major-effect alleles are subject to MAS in early or late generations, and then lines are derived and evaluated for all traits of interest. Anderson et al. [27] described the application of a MAS forward breeding strategy for improving Fusarium Head Blight resistance where MAS is applied for multiple loci in the F_2 and F_3 generations on a few breeding populations and then lines derived from these populations are phenotypically evaluated.

6.3.4 Genomic Selection

Genomic selection (GS) [6] is a form of MAS which is vastly different from conventional MAS in its approach. Unlike many other MAS strategies, the goal of GS is to improve the breeding germplasm as a whole for all traits of interest over multiple cycles of population improvement. GS is based on genomic estimated breeding values (GEBVs), which are estimates of individuals' values as parents based on genomic markers. Accurate estimation of GEBVs requires phenotypic and genotypic data on a 'training population'. This information is fed into a genomic prediction model, and GEBVs are predicted for selection candidates which have been genotyped but not necessarily phenotyped. For an ongoing wheat breeding program, breeding lines developed over the past few years can serve as the training population, as long as phenotypic and genome-wide marker data are available for these lines.

Selection based on a GEBV can be more effective than selection based on a phenotype or BLUP estimated without genomic relationships. However, the main advantage of selection based on GEBVs is that they can be estimated for individuals that have not yet been phenotyped. This allows breeders to identify parents to be used in crossing much earlier in the breeding process. For example, in a conventional wheat breeding program, selection is typically imposed among breeding lines that have undergone 2–3 years of line development and 2–3 years of testing. In a typical wheat breeding program implementing a conservative GS strategy, selection is imposed among breeding lines that have undergone 2–3 years of line development and 0–1 years of testing. This reduction in the breeding cycle duration is what leads to faster rates of genetic gain [28]. Counterintuitively, the time and effort devoted to phenotyping may remain unchanged because breeding lines will continue to undergo testing even after they are selected as parents in order to gather phenotypic data that will be needed to make variety release decisions and to update the training population.

Prior to implementing GS in a wheat breeding program, several conditions should be met. First, the breeding program should be able to routinely obtain inexpensive genome-wide marker data within 1–6 months of tissue sampling and DNA extraction. Increasing the speed of genome-wide marker data acquisition and reducing the cost of genotyping relative to phenotyping can improve the potential for using GS to shorten the breeding cycle duration and accelerate rates of genetic gain [29]. Second, a breeding program should be selecting parents largely from within the breeding program itself. This recycling of elite lines within the program is needed so that genetic gain can be achieved over cycles and so that phenotypic and marker data generated on the breeding materials can contribute to training an accurate prediction model for GS in future generations. Finally, the breeding program should be collecting and carefully managing high-quality phenotypic data on all traits of interest and on traits highly correlated with the traits of interest, which are referred to as 'secondary traits'. Breeders may cull plants or lines based on traits that are visually observed in the field but not systematically phenotyped. In order to continue selecting for these traits using GS, it will be necessary to record and manage phenotypic data for all target traits. Data on secondary traits, while not essential, can help increase GS accuracies when used together with data on the traits of interest in multi-trait GS models [30]. For example, multi-trait GS models for predicting grain yield in wheat were shown to be more accurate than single-trait GS models when using secondary trait data in the form of aerial imagery [31].

To begin using GS, it is recommended that breeding programs start by genotyping all lines that are being phenotyped for yield and other important traits and use all available phenotypic and marker data for selection. Endelman et al. [29] showed that this strategy is ideal if the cost of genotyping is similar to or higher than the cost of phenotyping. Another advantage of genotyping all lines being phenotyped is that it also allows genotypic and phenotypic data to be accumulated. This then becomes the GS model training population for future generations. After multiple years of phenotyping and genotyping, it may be possible to accurately estimate GEBVs on lines that have not yet been phenotyped to enable selection in early stages of line development in order to reduce the breeding cycle duration and increase the rate of genetic gain.

Ideally, GS will be implemented in a two-part strategy in which selection will be performed among breeding lines as well as among individual plants in a rapid-cycle recurrent selection program. The effectiveness of this strategy can be further enhanced by integrating optimum contribution selection (OCS) [32]. OCS, reviewed by Woolliams et al. [33], is a method which optimizes how much selected parents participate in crosses in order to control how fast the population loses genetic variability. GS can lead to more rapid loss of genetic variability compared conventional selection methods largely because it enables many cycles of breeding to be performed in a short time period. It has also been observed that selected individuals are more likely to be close relatives of one another as GS accuracy decreases, leading to faster rates of inbreeding and faster losses in genetic variance per cycle [34]. Because the strategy of rapid cycle GS is based on achieving many cycles of low accuracy GS in a short period of time, genetic variance loss is expected to be especially severe in rapid cycle GS programs. Empirically, faster losses in genetic variance under GS compared to PS have been observed in a short-term recurrent experiment for quantitative stem rust resistance in wheat [4]. Veenstra et al. [5] found that GS for fructan content in wheat seeds was effective using both truncation selection and OCS but inbreeding was significantly reduced using OCS. Because populations with more genetic variability can achieve higher rates of genetic gain compared to those with less, it is important that the loss of genetic variability in GS-based breeding programs be managed, especially if a rapid cycle strategy is adopted. Given that marker genotypes and estimates of genomic relationships are available to breeders implementing GS, managing loss of genetic variability over time is feasible using estimates of genomic relationship in OCS and/or by placing a higher weight on the effects of favorable, low-frequency marker alleles [34] in the GEBV estimation procedure.

Although the complexities of implementing GS in a wheat breeding program may seem daunting, even a simple GS strategy can be useful. Most wheat breeding programs in the United States and at CIMMYT are implementing some form of GS to help improve line advancement decisions during testing and to improve parent selection. Most of these programs are not yet routinely using rapid cycle recurrent GS, multi-trait GS models, or OCS. Over time, as more analytical tools are developed and as breeders become more skilled in GS methods and analytical techniques, breeding programs will be able to evolve further to take advantage of the potential of GS to maximize rates of genetic gain in wheat using selection procedures that are increasingly complex and data-driven.

6.4   Key Concepts

The goal of population improvement is to enhance the genetic base of the breeding program, while selection methods aim to identify breeding lines with superior potential or performance. Recurrent selection is the predominant approach to population improvement in wheat breeding and aims to enhance the breeding population as a whole through crossing and recombination. Generating crosses in wheat is a labor-intensive process, however, both recessive and dominant male-sterility genes are available in wheat that can greatly facilitate intermating. Mass, best linear unbiased prediction, marker-assisted selection and genomic selection are commonly used selection methods in wheat breeding.

6.5   Conclusions

Considering the multitude of approaches to population improvement and selection methods as well as the many line development methods described in Chap. 5, breeders are tasked with determining how to best combine these into an effective breeding strategy given fixed financial resources and infrastructure. Rather than take this challenge head-on, most successful breeding programs repeat what has traditionally been done with minor modifications. Complete redesign of a breeding program is rarely undertaken, as it can be disruptive to the ongoing variety development process. However, it should be recognized that methods like GS are inherently disruptive when used to their full potential. Thus, flexibility and an open mind will be needed in order for a breeding program to develop and deploy an optimal strategy using the latest methodological advancements.

References

1. Suneson CA (1956) An evolutionary plant breeding method. Agron J 48:188–191. https://doi.org/10.2134/agronj1956.00021962004800040012x
2. Hockett EA, Eslick RF, Qualset CO et al (1983) Effects of natural selection in advanced generations of barley composite cross II. Crop Sci 23:752–756
3. Rutkoski JE (2019) A practical guide to genetic gain. In: Advances in agronomy. Academic, pp 217–249. https://doi.org/10.2135/cropsci2018.09.0537
4. Rutkoski J, Singh RP, Huerta-Espino J et al (2015) Genetic gain from phenotypic and genomic selection for quantitative resistance to stem rust of wheat. Plant Geno 8:1–10. https://doi.org/10.3835/plantgenome2014.10.0074

5. Veenstra LD, Poland J, Jannink J, Sorrells ME (2020) Recurrent genomic selection for wheat grain fructans. Crop Sci 60:1499–1512. https://doi.org/10.1002/csc2.20130
6. Meuwissen TH, Hayes BJ, Goddard ME (2001) Prediction of total genetic value using genome-wide dense marker maps. Genetics 157:1819–1829
7. Gaynor RC, Gorjanc G, Bentley AR et al (2017) A two-part strategy for using genomic selection to develop inbred lines. Crop Sci 57:2372–2386. https://doi.org/10.2135/cropsci2016.09.0742
8. Sorrells ME, Fritz SE (1982) Application of a dominant male-sterile allele to the improvement of self-pollinated crops. Crop Sci 22:1033–1035
9. Redden RJ, Jensen NF (1974) Mass selection and mating systems in cereals. Crop Sci 14:345–350. https://doi.org/10.2135/cropsci1974.0011183x001400030001x
10. Derera NF, Bhatt GM (1972) Effectiveness of mechanical mass selection in wheat (Triticum Aestivum L.). Aust J Agric Res 23:761–768. https://doi.org/10.1071/AR9720761
11. Thakare RB, Qualset CO (1978) Empirical evaluation of single-plant and family selection strategies in wheat. Crop Sci 18:115–118. https://doi.org/10.2135/cropsci1978.0011183x001800010030x
12. Gardner CO (1961) An evaluation of effects of mass selection and seed irradiation with thermal neutrons on yield of corn. Crop Sci 1:241–245
13. Lungu DM, Kaltsikes PJ, Larter EN (1987) Honeycomb selection for yield in early generations of spring wheat. Euphytica 36:831–839. https://doi.org/10.1007/BF00051867
14. Henderson CR (1975) Best linear unbiased estimation and prediction under a selection model. Biometrics 31:423–447. https://doi.org/10.2307/2529430
15. Henderson CR (1975) Best linear unbiased estimation and prediction under a selection model. Biometrics 31:423–447. https://doi.org/10.2307/2529430
16. Henderson CR, Quaas RL (1976) Multiple trait evaluation using relatives' records. J Anim Sci 43:1188–1197. https://doi.org/10.2527/jas1976.4361188x
17. Hayes BJ, Visscher PM, Goddard ME (2009) Increased accuracy of selection by using the realized relationship matrix. Genet Res 91:47–60. https://doi.org/10.1017/S0016672308009981
18. Calus MPL, Veerkamp RF (2011) Accuracy of multi-trait genomic selection using different methods. Genet Sel Evol 43:26. https://doi.org/10.1186/1297-9686-43-26
19. Henderson CR (1973) Sire evaluation and genetic trends. J Anim Sci 1973:10–41
20. Crossa J, Burgueño J, Cornelius PL et al (2006) Modeling genotype × environment interaction using additive genetic covariances of relatives for predicting breeding values of wheat genotypes. Crop Sci 46:1722–1733. https://doi.org/10.2135/cropsci2005.11-0427
21. Collard BCY, Mackill DJ (2008) Marker-assisted selection: an approach for precision plant breeding in the twenty-first century. Philos Trans R Soc B Biol Sci 363:557–572
22. Tanksley SD (1983) Molecular markers in plant breeding. Plant Mol Biol Report 1:3–8. https://doi.org/10.1007/BF02680255
23. Hospital F, Chevalet C, Mulsant P (1992) Using markers in gene introgression breeding programs. Genetics 132:1199–1210
24. Randhawa HS, Mutti JS, Kidwell K et al (2009) Rapid and targeted introgression of genes into popular wheat cultivars using marker-assisted background selection. PLoS One 4:e5752. https://doi.org/10.1371/journal.pone.0005752
25. Liu J, Liu D, Tao W et al (2000) Molecular marker-facilitated pyramiding of different genes for powdery mildew resistance in wheat. Plant Breed 119:21–24. https://doi.org/10.1046/j.1439-0523.2000.00431.x
26. Bonnett DG, Rebetzke GJ, Spielmeyer W (2005) Strategies for efficient implementation of molecular markers in wheat breeding. Mol Breed 15:75–85
27. Anderson JA, Chao S, Liu S (2007) Molecular breeding using a major QTL for Fusarium Head Blight resistance in wheat. Crop Sci 47:S-112. https://doi.org/10.2135/cropsci2007.04.0006IPBS
28. Heffner EL, Lorenz AJ, Jannink J-L, Sorrells ME (2010) Plant breeding with genomic selection: gain per unit time and cost. Crop Sci 50:1681–1690. https://doi.org/10.2135/cropsci2009.11.0662

29. Endelman JB, Atlin GN, Beyene Y et al (2014) Optimal design of preliminary yield trials with genome-wide markers. Crop Sci 54:48–59. https://doi.org/10.2135/cropsci2013.03.0154
30. Jia Y, Jannink JL (2012) Multiple-trait genomic selection methods increase genetic value prediction accuracy. Genetics 192:1513–1522. https://doi.org/10.1534/genetics.112.144246
31. Rutkoski J, Poland J, Mondal S et al (2016) Canopy temperature and vegetation indices from high-throughput phenotyping improve accuracy of pedigree and genomic selection for grain yield in wheat. G3 Genes Geno Genet 6:2799–2808. https://doi.org/10.1534/g3.116.032888
32. Gorjanc G, Gaynor RC, Hickey JM (2018) Optimal cross selection for long-term genetic gain in two-part programs with rapid recurrent genomic selection. Theor Appl Genet 131:1953–1966. https://doi.org/10.1007/s00122-018-3125-3
33. Woolliams JA, Berg P, Dagnachew BS, Meuwissen THE (2015) Genetic contributions and their optimization. J Anim Breed Genet 132:89–99. https://doi.org/10.1111/jbg.12148
34. Jannink J (2010) Dynamics of long-term genomic selection. Genet Sel Evol 42:35. https://doi.org/10.1186/1297-9686-42-35

Chapter 7
Achieving Genetic Gains in Practice

Ravi P. Singh, Philomin Juliana, Julio Huerta-Espino, Velu Govindan,
Leonardo A. Crespo-Herrera, Suchismita Mondal, Sridhar Bhavani,
Pawan Kumar Singh, Xinyao He, Maria Itria Ibba,
Mandeep Singh Randhawa, Uttam Kumar, Arun Kumar Joshi,
Bhoja R. Basnet, and Hans-Joachim Braun

Abstract Accelerating the rate of genetic gain for grain yield together with key
traits is pivotal for delivering improved wheat varieties. The key strategies of
CIMMYT's spring bread wheat improvement program to continuously increase
genetic gains and deliver elite wheat lines to national partners in the target countries
include: breeding for product profiles that prioritize selection traits; robust choice of
diverse parents by leveraging all phenotypic and genotypic data; effective crossing
schemes with an optimal proportion of different types of crosses; early-generation
advancement using the selected-bulk breeding scheme that reduces operational
costs; the two generations/year field based "shuttle-breeding" that reduces the
breeding cycle time while selecting breeding populations in contrasting environ-

R. P. Singh (✉) · V. Govindan · L. A. Crespo-Herrera · S. Mondal · S. Bhavani · P. K. Singh ·
X. He · M. I. Ibba · B. R. Basnet · H.-J. Braun
International Maize and Wheat Improvement Center (CIMMYT), Texcoco, Mexico
e-mail: r.singh@cgiar.org; velu@cgiar.org; l.crespo@cgiar.org; s.mondal@cgiar.org; s.
bhavani@cgiar.org; pk.singh@cgiar.org; x.he@cgiar.org; m.ibba@cgiar.org; b.r.basnet@
cgiar.org; h.j.braun@cgiar.org

P. Juliana · U. Kumar
Borlaug Institute for South Asia (BISA), New Delhi, India
e-mail: p.juliana@cgiar.org

J. Huerta-Espino
INIFAP, Campo Experimental Valle de Mexico, Texcoco, Mexico
e-mail: j.huerta@cgiar.org

M. S. Randhawa
International Maize and Wheat Improvement Center (CIMMYT), Nairobi, Kenya

International Centre for Research in Agroforestry (ICRAF), Nairobi, Kenya
e-mail: m.randhawa@cgiar.org

A. K. Joshi
Borlaug Institute for South Asia (BISA), New Delhi, India

International Maize and Wheat Improvement Center (CIMMYT), New Delhi, India
e-mail: a.k.joshi@cgiar.orgr

ments with diverse biotic and abiotic stresses; making advancement decisions for elite lines using data from intensive multi-trait, multi-year and multi-environment phenotyping; integrating new methods like genomic selection; utilizing yield and phenotypic data from international yield trials and screening nurseries generated by worldwide partners for identifying and utilizing superior lines; and maintaining effective partnerships with the National Agricultural Research Systems who serve as key leaders in developing, releasing, and disseminating varieties to farmers. In addition to these strategies, new breeding schemes to reduce the cycle time and recycle parents in 2–3 years are being piloted and optimized to further accelerate genetic gain.

Keywords Product-profile · Crossing · Selection · Advancement · Phenotyping

7.1   Learning Objectives

- Product profile-based breeding.
- Parental selection and crossing strategies.
- Early-generation advancement and selection strategies.
- Advancement decisions for elite lines and phenotyping strategies.
- International screening nurseries and yield trials.
- Integration of genomic selection.
- Partnerships with national programs.

7.2   Introduction

Wheat, the world's second largest food crop and it's largest primary commodity is grown on over 215 million hectares annually and is consumed by over 2.5 billion people in 89 countries. With a global production of about 760 million metric tons, wheat provides 20% of the world's calories and protein [1]. However, with an increasing global population and changing diets, the current global average rate of wheat yield increase (0.9%) is insufficient to meet the projected rising demands by 2050 [2]. Moreover, other escalating challenges like the evolution and spread of new biotypes of diseases and pests, climate change including weather variabilities, temperature fluctuations, increased frequencies of drought and heat stresses etc. [3–6], necessitate continuous efforts to accelerate the rate at which biotic and abiotic stress resilience is built into new wheat varieties, along with higher grain yield (GY), market-preferred traits and nutritional quality. While this can be achieved through a combination of genetic gains, improved agronomy, and policy changes, increasing genetic gains is a highly effective and feasible intervention for delivering improved wheat varieties to farmers.

The International Maize and Wheat Improvement Centre's (CIMMYT's) spring bread wheat breeding program that is widely recognized as the main source of new varieties and elite lines, especially for the developing world, delivers over 1% annual genetic gain for GY, while ensuring diverse resistance to rusts and other important diseases and building climate resilience [7, 8]. The elite spring bread wheat lines developed by CIMMYT's Global Wheat Program (GWP) are released by national programs and private companies and grown on about 40 million ha in developing countries. In addition, another 20 million ha are sown to varieties that are derived from national breeding programs by using a CIMMYT bred line as parent. This makes CIMMYT's GWP by far the biggest provider of spring bread wheat germplasm globally and driver of genetic gain by deploying many successful breeding strategies to continuously deliver improved germplasm to the target countries in Asia, Africa and Latin America with impacts reaching beyond these targeted geographies. In addition, CIMMYT also breeds winter/facultative wheat (Box 7.1), adapted to West and Central Asia, from Ankara, Turkey and for China from Beijing. In this chapter, we discuss some of the main strategies deployed by the wheat breeding program to achieve genetic gains for GY together with other key biotic, abiotic, nutrition and quality traits while enhancing genetic diversity for relevant traits in elite breeding germplasm.

Box 7.1: Winter and Facultative Wheat

Winter and facultative wheat varieties cover around one third (80 million ha) of the global wheat area [9]. The biggest winter wheat producers are China, Russia, USA, France and Ukraine. Except in China, basically all winter wheat produced is rainfed. The terms winter, facultative and spring refer to sowing time and therefore define the adaptation of wheat to low temperatures. While this makes sense for countries that experience winter, it is misleading for the global south, where more than 90% of all wheat is of spring type and sown in fall/winter. For facultative wheat, there exists no clear definition and is compared to true winter wheats, in general less cold tolerant, has a shorter vernalization period, starts growth in spring earlier, flowers earlier and is grown in areas with milder winters, late fall rains or when late sowing is required due to tight crop rotations. Central and West Asian countries and South America grow facultative wheats on large areas. The Turkey-CIMMYT-ICARDA International Winter Wheat Improvement Program based in Turkey is developing winter and facultative wheats for these regions.

The main genetic difference between winter and spring wheat is the allelic combination for vernalization (*Vrn*) and Photoperiod (*Ppd*) and presence of alleles for frost tolerance (*Fr*). The combination of these alleles is a major determinant for the adaptation of a wheat cultivar [10]. Winter wheats are also considered to have a better tillering capacity. The knowledge of the genetics of adaptation in wheat has revealed that the distinction between what is a winter and a spring wheat is now very blurred and it really depends on the environment. For example, in the UK, some recently released spring wheats can be sown in autumn and some winter wheats are adapted to early spring sowing. By definition, winter wheats require vernalization, frost tolerance,

(continued)

Box 7.1: (continued) and are generally photoperiod sensitive while 'true' spring wheats are vernalization and photoperiod insensitive and susceptible to freezing temperatures. But, there are combinations of alleles that defy this broad characterization.

Winter wheat can survive freezing temperatures after they had gone through a hardening process. (see [11]) The biological limit to survive low temperature in wheat is −20 °C for 12 h without snowcover. Only rye and Triticale (up to −23 °C) are more tolerant. Vernalization is a widespread temperature control mechanism in the plant kingdom that assures that plants do not enter generative stages prior to winter. Vernalization is fastest when wheat is exposed to temperatures between 4 and 5 °C for 4–8 weeks. Wheat vernalizes between −2 and 16 °C, but lower or higher temperatures extend the time to vernalize significantly. The 4–8 weeks required to vernalize winter wheat defines the limit for rapid cycling and doubled haploids. Winter wheat breeding programs using single seed descent on a commercial scale are therefore limited to 3 cycles per year.

For other traits, the genes in winter and spring wheat are similar, though spring and winter wheat gene pools are distinctly different. To exploit these genetic differences, winter and spring wheats are often crossed. Spring type is dominant over winter type. Derivatives of Spring × Winter crosses developed in the 70s at CIMMYT represented a breakthrough in yield stability and yield potential and were released worldwide in the global south. Today, WxS crosses are made to exploit heterosis in hybrid programs and in particular to raise the yield potential. While WxS derivatives have excellent yield potential, many lines still need to be improved for grain plumpness i.e. less grain shrivelling. It is foreseen, that winter × spring crosses and their derivatives become increasingly important for global wheat improvement.

7.3   Product Profile-Based Breeding

To deliver client-oriented improved germplasm with high potential for adoption by farmers, the spring bread wheat program has recently adopted the approach of breeding for product profiles, which is similar to the mega-environment (ME) targeted breeding approach used for decades [12]. A product profile is defined as a set of targeted attributes that a new plant variety, or animal breed, is expected to meet in order to be successfully released onto a market segment [13]. It is essentially a combination of basic or must-have traits and value-added traits targeted in a new variety that can replace a current market-leading product in a target production zone. Hence, through consultations with national partners, considering the requirement of target countries for releasing varieties, knowledge accumulated over years of international collaboration, knowledge of the consumption of wheat-based products (end-use quality) by rural and urban populations, consideration of agroecological conditions, and market segmentation, the GWP has prioritized six market segments/product profiles (Table 7.1) along with the key selection traits (Table 7.2)

Table 7.1 Spring wheat market segments for CIMMYT wheat breeding pipelines

Market segment (broad) – bread wheat	Mega-environment	Representative wheat countries/regions	Area (million ha)	Average grain yield (t/ha)	Total wheat production (million tons)	Value of crop (billion USD)[a]	Farming family (million)	Food for people in market segment (million)
1. Hard White-Optimum Environment-Normal Maturity (HW-OE-NM)	1	Northwestern Plain Zone of India, Central and Northwestern Pakistan, irrigated mid hills of Nepal, Afghanistan, Iran, Iraq, Libya, Turkey, Egypt, Uzbekistan, Mexico, Chile, Zambia, Zimbabwe.	26.25	4.3	112.9	25.96	19.69	940
2. Hard White – Heat Tolerant-Early Maturity (HW-HT-EM)	5a	Northeastern Plain Zone of India, Bangladesh, Terai of Nepal, in addition to some wheat growing areas in Myanmar, Egypt, Sudan, recently opened irrigated areas of Ethiopia, Bolivia and Mexico.	10.46	2.46	25.7	5.92	9.92	322
3. Hard White – Drought Tolerant Normal Maturity (HW-DT-NM)	4a	Afghanistan, Iran, Iraq, Lebanon, Turkey, Syria, Morocco, Algeria, Tunisia, Libya, Mexico and Central American countries.	8.9	2.16	19.2	4.42	2.94	107
4. Hard White-Drought Tolerant Early Maturity (HW-DT-EM)	4c	Central and Peninsular zones of India, and Southern Punjab of Pakistan.	7	2.84	19.9	4.57	4.6	248
5. Hard White – High Rainfall – Normal Maturity (HW-HiR-NM)	2	Highlands of Ethiopia, Nepal, India, Mexico and other countries.	2.08	3.3	6.9	1.58	2.3	113

(continued)

Table 7.1 (continued)

Market segment (broad) – bread wheat	Mega-environment	Representative wheat countries/regions	Area (million ha)	Average grain yield (t/ha)	Total wheat production (million tons)	Value of crop (billion USD)[a]	Farming family (million)	Food for people in market segment (million)
6. Hard Red-High Rainfall – Normal Maturity (HR-HiR-NM)	2/4b	Highlands of Kenya, Uganda, Tanzania, Rwanda, Bolivia, Turkey, Iran and Mexico (spillover benefitting countries in southern Africa and South America with 9.8 million ha).	0.31	1.9	0.6	0.14	0.15	3
Total			55.0	3.37	185.2	42.59	39.59	1732

[a]Additional USD 10 billion estimated value of wheat straw, essential for livestock in the targeted market segments

Table 7.2 Selection traits and their priorities in the product profiles

Key traits	Product profile/market segment				
	1. Hard White-Optimum Environment-Normal Maturity (HW-OE-NM)	2. Hard White – Heat Tolerant – Early Maturity (HW-HT-EM)	3. Hard White-Drought Tolerant – Normal Maturity (HW-DT-NM)	4. Hard White-Drought Tolerant – Early Maturity (HW-DT-EM)	5. Hard White – High Rainfall-Normal Maturity (HW-HiR-NM) & 6. Hard Red-High Rainfall-Normal Maturity (HR-HiR-NM)
High and stable yield potential	XXX	XXX	XXX	XXX	XXX
Water use efficiency/ Drought tolerance	X	X	XXX	XXX	XX
Heat tolerance	XX	XXX	XX	XXX	X
End-use quality	XXX	XXX	XXX	XXX	XXX
Enhanced grain Zn (and Fe) content	XXX	XXX	XXX	XXX	XXX
Stem rust (Ug99 & other)	XX	XX	XX	XXX	XXX
Stripe rust	XXX	XX	XXX	XX	XXX
Leaf rust	XXX	XXX	XXX	XXX	XX
Septoria tritici blotch	–	–	XXX	–	XXX
Spot blotch	X	XXX	–	X	–
Fusarium – head scab and myco-toxins	–	–	–	–	XX
Wheat blast	X	XXX	X	X	X
Maturity	Normal-late	Early	Normal	Early	Normal
	Importance: X= low, XX= moderate, XXX= high				

Common agronomic traits: Plant height, stem strength, leaf health, spike fertility, grain size, grain plumpness, etc for targeting wheat area of approximately 55 million ha (spillover benefit reaching to another 9.8 million ha) in Asia, Africa and Latin America. Chapters 8, 9, 10, 11 and 12 highlight the importance of biotic and abiotic stresses, end-use and nutrition quality for wheat breeding. Moreover Chap. 3 describes the targeted breeding environments.

7.4 Parental Selection and Crossing Strategies

Chapters 5 and 6 describe the breeding methods. The selection of parents for crossing is one of the most important steps for improvement of GY and other traits [14]. Hence, all the available phenotypic data for GY, agronomic traits, disease resistance, end-use quality, and molecular marker data are leveraged to select the best parents. Since diversity is also a key criterion, parental selection each year is done simultaneously from four different cohorts of elite breeding lines that are at different stages of GY testing. This includes about 1% (~100) of lines in the stage 1 or first-year GY trials, 15% (~150) of lines in the stage 2 or second-year yield trials, 30% (~80) of lines in the stage 3 or third-year yield trials, and 10% (~20 most outstanding) of lines in the international trials are recycled as parents. In addition, elite lines from pre-breeding programs, national partners (including private sector), newly released varieties, targeted synthetics, donors for new genes/traits are also used and about 10% of the crosses made are with these parents.

An optimal proportion of simple crosses, top crosses (three-way) and single-backcrosses (BC_1) is made each year [15] to obtain superior progenies and increase the genetic gain for multiple traits simultaneously. A simple cross is made among elite parents or between an elite parent and another parent that is a donor for a trait that the elite parent lacks. Currently, about 1500 simple crosses are made annually in the summer season (May-October) in the field at CIMMYT's headquarters at El Batan to include parents immediately from the main crop season in Ciudad Obregon (Cd. Obregon or Obregon) and other field sites worldwide (November-May). In cases where both the parents are not elite, a top cross is made using a third elite parent. Similarly, when one parent lags in GY, or when non-CIMMYT parents are crossed, a BC_1 is made using the elite parent as the recurrent parent. The BC_1 approach was initiated in the early 1990s, to transfer 3–4 quantitative trait loci (QTL) based adult plant resistance to leaf rust (LR), and to introduce some major rust resistance genes from winter wheat [16]. However, it was observed that the frequency of BC_1-derived advanced lines with the same or higher GY than the checks was 6–7 times higher than the lines derived from simple and top crosses. Hence the BC_1 approach was very advantageous, because in addiction to shifting the mean GY of the progeny towards the higher yielding parent, it was also possible to simultaneously utilize the useful GY QTL from the donor parent. But we observed that making a second backcross to obtain BC_2 was not very useful, as the GY shifted towards the recurrent parent's GY. Nonetheless, BC_2 is required for transferring traits from distant sources, e.g. high zinc content, where land races and synthetics are used, for the incorporation of resistance genes through marker-assisted back-crossing etc. Each year, about 1200 top crosses and BC_1s (about half each) are made in the field in Obregon. Double crosses (4 way, between two F_1s) were used in the 1970s [12], but were discontinued in the early 1980s, because of too much variation in the F_2s and until today, no CIMMYT line derived from a double cross has been released.

7.5 Early-Generation Advancement and Selection Strategies

A combination of two effective breeding strategies is used to achieve genetic gains for key traits in early generations. One of them is the selected-bulk breeding scheme, where all segregating early generations until the F_4 or the F_5 stages are selected visually for agronomic features, phenology, LR, stem rust (SR), stripe rust (YR), spike fertility and tillering capacity; spikes from the selected plants are harvested and threshed in bulk; grains sieved to retain only larger, plump and healthy grains [16]. This scheme proved to be highly effective considering the operational costs, as it permitted retaining large numbers of selected plants in each population at a low cost.

The other successful strategy involving the growing of two generations per year, thereby reducing the breeding cycle time by half (five years for obtaining the stage 1 GY trial results) is the Obregon-Toluca field-based shuttle breeding program, where germplasm is shuttled between these two contrasting environments in Mexico. Obregon, which is located at 39 m.a.s.l in the Sonora desert of Northwestern Mexico has CIMMYT's main wheat breeding and GY phenotyping research station, the Centro Experimental Norman E. Borlaug (CENEB), where wheat is sown in November and harvested in late April/May. The desert conditions in Obregon along with insignificant to no rainfall during the crop season facilitate screening for extreme drought stress under drip irrigation, as well as for GY potential under full irrigation. In addition, screening for tolerance to early and terminal heat stresses is feasible by altering the planting time. Besides, Obregon also favors screening and selecting segregating populations, head-rows and advanced lines for biotic stresses like SR and LR. On the other hand, Toluca located at 2640 m.a.s.l in the highlands of the State of Mexico has CIMMYT's research station, where wheat is sown in May and harvested in September/October. It serves as an ideal site for screening diseases like YR, STB (Septoria tritici Blight) and FHB (Fusarium Head Blight), because of cooler temperatures and high rainfall (>1000 mm) favoring epidemics of these diseases. The adaptation and GY stability of CIMMYT's elite lines in a range of targeted environments has been attributed to the response from selection under these highly contrasting shuttle-breeding environments, with diverse day-length, temperature regimes, rainfall patterns and biotic stresses during the breeding cycles [17].

A description of the 'Obregon-Toluca shuttle' (Fig. 7.1) for early-generation advancement and selection strategies for simple crosses, top crosses and BC_1s is described in Table 7.3. In addition to the Obregon-Toluca shuttle, the Mexico-Kenya shuttle breeding program (Fig. 7.1) was initiated in 2008 to increase the frequency

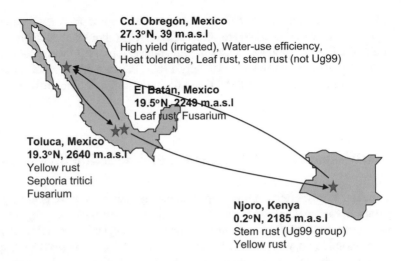

Fig. 7.1 The two-generations/year Ciudad Obregon-Toluca shuttle breeding implemented by Norman E. Borlaug in Mexico in 1945 to reduce breeding cycle time, and the Mexico-Kenya shuttle-breeding initiated in 2008 to rebuild resistance to stem rust in CIMMYT wheat germplasm

of resistance to the Ug99 race group of SR fungus [18]. This shuttle has permitted to drastically increase the frequencies of alleles involved in quantitative resistance to SR in combination with high GY potential, as well as the enrichment of breeding lines with some race-specific resistance genes. In this shuttle, the F_3 and F_4 selected-bulk populations (about 1000 plants/population) that are harvested in Toluca in September are shipped to Njoro, Kenya, selected for two consecutive seasons and the resulting F_5 and F_6 selected-bulk populations are brought back to Obregon for individual plant selection. To reduce a year in breeding cycle time, the procedure was modified in 2020 by shipping F_2s (from F_1-top and BC_1) to Kenya for shuttle breeding.

About 70,000 individual plant-derived plots (2-rows, 0.2 m²), which comprise the F_5 to F_7s (from the Obregon-Toluca shuttle and the Mexico-Kenya shuttle) are grown in Toluca during the months of May to October (Table 7.3). The best lines are selected for agronomic traits, YR and STB resistance. This is followed by planting of about 25,000 selected plots (2-rows, 0.56 m²) in Obregon during the months of November to April, which are further selected for agronomic traits and LR resistance. The selected plots are then cut to obtain enough seed, selection for grain characteristics is done, and seeds from the selected lines are retained for the first year of GY trials. About 13,000 selected plots (2 rows, 0.3 m²) of advanced lines are then grown in Toluca (and El Batan for seed multiplication for international shipment) during May to October, selected for agronomic traits, YR and STB resistance. Finally, about 9000 elite lines are selected for evaluation in the first year of GY trials, using the seed retained previously in Obregon. Simultaneously, a parallel set of these plots are also grown for SR and YR phenotyping in Kenya.

Table 7.3 Description of the early-generation advancement and selection strategies for simple crosses, top crosses and single-backcrosses (BC_1) in the shuttle breeding scheme

Year	Activities	Field	Duration	Details
1	Crossing	El Batan	May–October	Parents sown as crossing block on 3 sowing dates. About 1500 simple crosses are made.
	F_1s (simple crosses) grown	Obregon	November–April	The 1500 F_1 (simple crosses) are sown (2 rows/1 m long) and about 1200 top/BC_1 are made. About 750 selected F_1 plots are harvested to obtain the F_2 populations.
2	F_2s (simple crosses) & F_1s (top/BC_1) grown	Toluca	August–October	About 1200 F_2 plants per simple cross (~700 crosses) and 400 per top/BC_1 cross (~600 crosses each) are grown with 8–10 cm spacing; and selected for agronomic and disease resistance traits. The spikes from the selected plants are harvested in bulk.
	F_3s (simple crosses) & F_2s (top/BC_1) grown	Obregon	November–April	About 400 F_3 plants per simple cross (~700 crosses) and 1200 F_2 plants per cross from F_1 top/BC_1 (~600 crosses each) grown with 8–10 cm spacing; and selected for agronomic and disease resistance traits. The spikes from the selected plants are harvested in bulk.
3	F_4s (simple crosses) & F_3s (top/BC_1) grown	Toluca	May–October	About 400 F_4 plants per simple cross (~700 crosses) and 400 F_3 plants per cross from F_1 top/BC_1 (~600 crosses each) grown with 8–10 cm spacing; and selected for agronomic and disease resistance traits. The spikes from the selected plants are harvested in bulk.
	F_5s (simple crosses) & F_4s (top/BC_1) grown	Obregon	November–April	About 300 F_5 plants per simple cross (~700 crosses) and 300 F_4 plants per top/BC_1 (~600 crosses each) grown with 8–10 cm spacing; and selected for agronomic traits and leaf rust resistance. Selected plants are harvested individually and selection for grain characteristics is done after threshing.
4	F_6s (simple crosses) and F_5s top/BC_1 grown	Toluca	May–October	About 70,000 individual-plant derived plots (2-rows, 0.2 m^2) are selected for agronomic traits, uniformity, yellow rust and Septoria tritici blight resistance. Bulk harvesting several thousands of lines is not possible in Toluca due to rainy conditions hence the left-over seed kept in Obregon is used after culling the discarded lines.
	F_6s (simple crosses) and F_5s top/BC_1 grown	Obregon	November–April	About 25,000 plots (2-rows, 0.56 m^2) are selected for agronomic traits, uniformity and leaf rust resistance. The selected plots are cut to obtain sufficient seed, grain selection is done, and seeds from the selected lines are retained for the first year of yield trials and for phenotyping/multiplication.

(continued)

Table 7.3 (continued)

Year	Activities	Field	Duration	Details
5	Advanced lines grown	Toluca	May–October	About 13,000 plots (2 rows, 0.3 m²) are selected for agronomic traits, yellow rust and Septoria tritici blight resistance. About 9000 lines are selected for evaluation in the stage 1 (first year) yield trials using the seed retained in Obregon.
		El Batan	May–October	Seed multiplication under chemical control for stem rust and yellow rust phenotyping in Kenya is done.

7.6 Advancement Decisions for Elite Lines and Phenotyping Strategies

Advancement decisions for elite lines that are to be included in the international nurseries are made using phenotyping data for several traits evaluated in multiple locations/environments, as described in Table 7.4. About 9000 elite lines that enter the stage 1 of GY testing in Obregon in year 5 from crossing are phenotyped for GY at the research station in Obregon during the months of November-April. Traits like days to heading, days to maturity, plant height, lodging and agronomic scores are also recorded. Simultaneous screening for SR and YR is done in Njoro and about 1400 lines are selected, using the stage 1 GY, SR and YR data, in addition to an acceptable range for agronomic traits like plant height, heading and maturity. These selected lines are then phenotyped for resistance to several diseases and end-use quality in Toluca, El Batan and Njoro. Simultaneously, seed multiplication is done in El Batan, as required by the Mexican quarantine system, before large-scale seed multiplication in Mexicali (Karnal bunt free site in Mexico) can be done for international nurseries.

All the agronomic, GY, disease resistance and quality data generated during the summer season is used to select about 1000 lines from the 1400 lines, for inclusion in the stage 2 GY trials. These lines are evaluated for: GY under six simulated selection environments (SEs) in Obregon, LR and Karnal bunt resistance in Obregon, SR and YR resistance in Njoro, YR resistance in Ludhiana, SB (Spot Blotch) resistance in Agua Fria. Simultaneously, larger-scale seed multiplication of these lines also takes place in Mexicali. Finally, considering the multi-trait data generated for the lines comprising stages 1 and 2 GY trials, a strong selection criteria is applied to select about 280 white grained lines, which comprise the stage 3 GY trials, that are then evaluated for GY in three simulated environments in Obregon while a much larger scale seed multiplication is being done for international trials.

The phenotyping strategies for different traits evaluated in the elite lines are described for grain yield and other traits in Boxes 7.2 and 7.3, respectively.

Table 7.4 Traits phenotyped in the elite lines from different grain yield testing stages, the location or the environment where they are phenotyped, the time of phenotyping, the number of lines phenotyped and the number of replications (reps.) for phenotyping

Trait evaluated	Location/environment	Season/months	Number of lines	Reps.
Stage 1 of yield testing				
Stem and stripe rusts	Njoro-field	January–May (Off-season)	~9000	1
Grain yield	Obregon-field – Raised bed-5 irrigations	November–April		2
Days to heading	Obregon-field – Raised bed-5 irrigations			1
Days to maturity	Obregon-field – Raised bed-5 irrigations			1
Plant height	Obregon-field – Raised bed-5 irrigations			1
Quality traits	El Batan – quality laboratory	June–September	Selected (~1400)	1
Stem & stripe rusts	Njoro-field	June–October (Main-season)		1
Stripe rust and Septoria tritici blight	Toluca-field	May–October		1
Fusarium head blight	El Batan-field	May–October		1
Leaf rust				1
Stage 2 of yield testing				
Grain yield	Obregon-field – Raised bed-5 irrigations	November–April	~1000	2
	Obregon–Flat-5 irrigations		~1000	2
	Obregon-field – Raised bed-2 irrigations – moderate drought stress		~1000	2
	Obregon-field–Flat-drip managed – high drought stress		~1000	2
	Obregon-field – Early-sown heat stress-5 irrigations		~1000	2
	Obregon-field – Late-sown heat stress-5 irrigations		~1000	2
Days to heading	All the above six environments		~1000	1
Days to maturity	All the above six environments		~1000	1
Plant height	All the above six environments		~1000	1
Leaf rust	Obregon-field		~1000	1
Karnal bunt	Obregon-field		~1000	2
Stem & stripe rusts	Njoro-field	Off – and main-seasons	~1000	1
Stripe rust	Ludhiana and Karnal-field		~1000	1
Spot blotch	Agua Fria-field	November–March	~1000	1
Fusarium head blight	El Batan-field	May–October	~300	1

(continued)

Table 7.4 (continued)

Trait evaluated	Location/environment	Season/months	Number of lines	Reps.
Leaf rust and yellow rust	Seedling tests: Greenhouse, El Batan		~1000	1
Stagonospora nodorum blotch and tan spot	Seedling tests: Greenhouse, El Batan		~700	2
All quality traits			~500	1
Stage 3 of yield testing				
Grain yield	Obregon – Raised bed-5 irrigations	November–April	280	2
	Obregon – Flat-drip managed – high drought stress		280	2
	Obregon – Late-sown heat stress-5 irrigations		280	2

Box 7.2: Phenotyping for Grain Yield in Simulated Managed Environments at Ciudad Obregon, Mexico to Identify High and Stable Yielding, Drought Stress and Heat Stress Tolerant Elite Wheat Lines
Stage 1 GY trials. Lines are grown on raised beds at the optimal planting time (late November to first week of December) and irrigated optimally with five irrigations in total and with about 500 mm water. The GY evaluation plot size is 4.8 m² and the lines are sown as three rows over each of the two beds that are of 80 cm width.

Stage 2 GY Trials. About 1000 lines are evaluated for GY in six simulated environments, as follows: (a) **Raised bed-5 irrigations** – The lines are sown on raised beds at the optimal time and receive 500 mm water in five irrigations similar to stage 1 trials. (b) **Flat-5 irrigations** – The lines are sown in the flat planting system, as most of the irrigated wheat in developing countries is grown. Sowing is during the optimal planting time and the lines receive 500 mm of water in five irrigations. The plot size is 5.46 m², and the lines are sown in six rows that are 18 cm apart and 4.2 m in length. (c) **Raised bed-2 irrigations-moderate drought stress** – The lines are sown during the optimal planting time on raised beds in a moderately drought stressed environment that receives 250 mm of water in two irrigations. The plot size is 4.8 m², and the lines are sown in three rows over each of the two beds that are of 80 cm width. (d) **Flat-drip managed-high drought stress** – The lines are sown during the optimal planting time in the flat planting system, with about 180 mm of water supplemented through drip irrigation. The plot size is 5.85 m², and the lines are sown in six rows that are 18 cm apart and 4.5 m in length. (e) **Early-sown heat stress-5 irrigations** – The lines are sown on raised beds, about 3 weeks before the optimum planting time (early November) and receive optimal irrigation. The plot size is 4.8 m² and the lines are sown in three rows over each of the two beds that are of 80 cm width. (f) **Late-sown heat stress-5 irrigations** – The lines are sown on raised beds about 90 days after the optimal time (last week of February) and exposed to high-temperature stress during entire crop cycle, with optimal irrigation. The plot size is 4.8 m² and the lines are sown in three rows over each of the two beds that are of 80 cm width.

Stage 3 GY trials: Lines are evaluated in the raised bed-5 irrigations, flat-drip managed – high drought stress and late sown heat stress-5 irrigations environments, with similar conditions as for the stage 2 environments. The lines in stages 1, 2 and 3 of GY testing are sown in 300+ trials, 39 trials and 10 trials, respectively with each trial comprising 28 lines and two high-yielding check varieties in six blocks.

Box 7.3: Phenotyping of Elite Lines for Resistance to Wheat Diseases and End-Use Quality Traits
Field and Greenhouse Responses to Leaf Rust, Stem Rust, and Stripe Rust. Field response to LR (caused by *Puccinia triticina* Eriks.) is evaluated at CIMMYT's research stations in El Batan and Obregon, SR (caused by *Puccinia graminis* Pers. f. sp. *tritici*) is evaluated at the Kenya Agricultural and Livestock Research Organization, Njoro and YR (caused by *Puccinia striiformis* West. f. sp. *tritici*) is evaluated in Toluca, Njoro and Ludhiana (India). For all the rust evaluations in Mexico, the lines are sown in 0.7 m long paired rows over raised beds that are 30-cm-wide, whereas in Kenya and India, the lines are sown in the flat planting system. Appropriate rust spreaders that are artificially inoculated with a mixture of urediniospores of the most relevant races of the pathogen in the phenotyping fields are sown around the experimental fields, as well as on the hills that are on one side of the plot, in the midst of the pathway. Uredinospores are sprayed on the spreaders four-six weeks after sowing, depending on the field sites in Mexico. In Kenya, uredin-iospores of the SR pathogen races belonging to the Ug99 lineage are sprayed to create an artificial rust epidemic. The plants within the border rows are also inoculated by injecting a suspension of freshly collected urediniospores in water using a hypodermic syringe, twice prior to booting. However, YR infection in Kenya is from natural infection as the main phenotyping is targeted to SR. Susceptible and resistant checks are sown every 20–30 lines in nurseries and serve as indicators of disease pressure. Rust response is scored twice or thrice between the early and late-dough stages at weekly to 10-days intervals after the severity of the susceptible checks reaches 80–100%. The percentage of infected tissue (0–100%) is assessed using the modified Cobb Scale, in addition to the disease reaction. The lines in stage 2 GY trials are also phenotyped for resistance to LR and YR in the seedling stage at CIMMYT's greenhouses in El Batan, using the standard inoculation method with the most appropriate races.

Field Response to Septoria tritici Blight and Spot Blotch. Field response to STB (caused by *Zymoseptoria tritici* (Desm.) Quaedvlieg & Crous) is evaluated at Toluca. The inoculum for STB consists of a mixture of six aggressive strains, that are used to inoculate the plants 45 days after sowing using an ultra-low volume applicator. In addition, two more applications are made at weekly intervals. A border row of a susceptible spreader and a resistant variety is planted around the field. Disease evaluation is done using the double-digit scale (00–99) which is slightly modified from the Saari-Prescott 0–9 scale for rating foliar diseases. After three to four evaluations, the double-digit scores are used to calculate the disease severity percentages, from which the area under the disease progression curve is obtained. Field response to SB (caused by *Bipolaris sorokiniana* Sacc.) is evaluated at CIMMYT's research station in Agua Fria, Mexico. The lines are sown during November and harvested in March. A mixture of virulent races that occur naturally in Agua Fria are collected from leaves and used for inoculation. Disease evaluation is done similar to STB.

Field Response to Fusarium Head Blight. Field response to FHB (caused by *Fusarium graminearum*) is evaluated at the El Batan experimental station, during the summer season (May to October). The lines are planted in 1-m double rows and five checks that represent a range of resistance to susceptibility responses are included for every 50 entries. A mixture of five aggressive *Fusarium graminearum* isolates are used for field inoculation, which comprise isolates collected from naturally infected wheat spikes in different places at the State of Mexico. Spray inoculation targeted to each line's anthesis stage is done using an inoculum of 50,000 spores/ml and is repeated two days later. From anthesis to the early dough stage, the lines are misted for 10 min each hour, from 9 am to 8 pm, thereby creating a humid environment that is favorable for FHB development. Response to FHB is scored three times at 20, 25, and 30 days post-inoculation, on 10 spikes that had been tagged at anthesis. The FHB index is calculated using the total numbers of infected spikes and spikelets of each spike using the formula: FHB index (%) = severity × incidence, where severity is the averaged percentage of diseased spikelets, and incidence is the percentage of symptomatic spikes.

Field Response to Karnal Bunt. Field response to Karnal bunt (caused by *Tilletia indica*) is evaluated at Obregon. The lines are sown in two planting dates and artificial inoculation is done from January to March during the booting stage, by injecting a sporidial suspension of the fungus with a hypodermic syringe into the boot, when the awns emerge. Overhead sprinklers are used during the inoculation period for five times a day, with 20 min of misting each time, to maintain humidity via intermittent misting. When the plants mature, five inoculated heads are harvested and threshed, and the number of infected and uninfected grains per head is counted. Disease severity is then calculated as the percentage of infected grains in each head and the average infection from five spikes is obtained.

Greenhouse Response to Stagonospora nodorum Blotch and Tan Spot. Seedling resistance to Stagonospora nodorum blotch (SNB, caused by *Parastagonospora nodorum* (Berk.) Quaedvlieg, Verkley & Crous) and tan spot (caused by *Pyrenophora tritici-repentis* (Died.) Drechsler) is evaluated in CIMMYT's greenhouses in El Batan. Inoculum production and inoculation for SNB are done as described in [19] and check varieties Erik, Glenlea, 6B-662, and 6B-365 are planted every 20 rows. Reaction to SNB is scored on the second leaf of each seedling 7 days post-inoculation, with the 1–5 lesion rating scale. For tan spot seedling response evaluation, race 1 or isolate *Ptr*1 of the pathogen is used. Inoculum production and checks are similar to that for SNB evaluation. The seedlings are then rated for tan spot response, seven days post-inoculation, using a 1–5 lesion rating scale.

End-use Quality Traits. In all the quality analyses, grains from the high yielding environment with reduced protein content are used, which allows for better discrimination of lines for quality. Some of the end-use quality traits evaluated in the elite lines include: (a) Mixing time (minutes) that is obtained from a mixograph (National Mfg. Co.) according to the American Association of Cereal Chemists (AACC) method 54-40A. (b) Alveograph W or the work value under the curve and Alveograph P/L (mm mm^{-1}), which is the tenacity vs. extensibility, or the ratio of the height to the length of the curve, both of which are obtained from the Chopin Alveograph (Tripette and Renaud, AACC method 54-30A) and used to analyze dough rheological properties. (c) Flour sodium-dodecyl sulfate sedimentation volume (mL) that is measured using 1 g of flour. (d) Bread loaf volume (cm^3) that is assessed by the rapeseed displacement method according to the AACC method 10-05.01, using pup loaves that are baked as pan bread with the slightly modified AACC method 10-09. (e) Grain protein content which is measured on a 12.5% moisture basis. (f) Grain hardness or the particle size index and moisture content, that are measured using Near-infrared spectroscopy (NIR System 6500, Foss) according to the methods AACC 39-10, 39-70A, and 39-00, respectively. The Brabender Quadrumat Jr. (C. W. Brabender OHG) is used to mill grain samples, that are optimally tempered to 13–16.5%, based on the hardness. (g) Flour protein and moisture content are determined with the Antaris II FT-NIR analyzer (Thermo). Calibration for moisture (AACC Method 44-15A), and protein content (AACC Method 46-11A) are done in the NIRS instruments. (h) Flour yield is obtained as the percentage recovered from milling. (i) Test weight (kg hL^{-1}) is obtained by weighing a 37.81 mL sample. (j) Thousand kernel weight (g) is obtained by weighing the kernels, that were counted using the digital image system SeedCount SC5000 (Next Instruments). (k) Grain color is scored visually as red or white.

Table 7.5 International screening nurseries and yield trials derived from the spring bread wheat improvement program, their abbreviations, the number of entries, the target mega-environment (ME) and the grain color requirement for that environment

Trial/Nursery	Abbreviation	Entries (No.)	Target mega-environment (ME)	Grain color
Screening nurseries				
International Bread Wheat Screening Nursery	IBWSN	250–300	ME1, ME2, ME5	White
Semi-arid Wheat Screening Nursery	SAWSN	250–300	ME4	White
High Rainfall Wheat Screening Nursery	HRWSN	150–200	ME2, ME4	Red
Disease based nurseries				
International Septoria Observation Nursery	ISEPTON	100–150	ME2, ME4	White/Red
Leaf Blight Resistance Screening Nursery	LBRSN	100–150	ME4, ME5	White/Red
Stem Rust Resistance Screening Nursery	SRRSN	100–150	All MEs	White/Red
Fusarium Head Blight Screening Nursery	FHBSN	50–100	ME2, ME4	White/Red
Karnal Bunt Resistance Screening Nursery	KBRSN	50–100	ME1	White/Red
Yield trials				
Elite Spring Wheat Yield Trial	ESWYT	50	ME1, ME2, ME5	White
Semi-arid Wheat Yield Trial	SAWYT	50	ME4	White
Heat Tolerance Wheat Yield Trial	HTWYT	50	ME1, ME4, ME5	White
High Rainfall Wheat Yield Trial	HRWYT	50	ME2, ME4	Red

7.7 International Screening Nurseries and Yield Trials for Identifying Superior Lines from Multi-environment Phenotyping

CIMMYT's spring bread wheat breeding program has continued to develop and deliver germplasm to best serve the targeted major wheat growing countries through the partnership called International Wheat Improvement Network (IWIN). The spring bread wheat IWIN partners include over 200 public and private sector institutions distributed worldwide in about 80 countries, who currently test approximately 700 new elite CIMMYT spring bread wheat lines annually at over 200 field sites, resulting in a massive exchange of germplasm and valuable phenotypic datasets. These lines targeted to different mega-environments (MEs) with specific biotic and abiotic stresses [16] are distributed through several international screening nurseries and yield trials, that are described in Table 7.5 and details of the yield trials are provided in Box 7.4.

Box 7.4: The Annually Distributed International Yield Trials Distributed by the Spring Bread Wheat Breeding Program
Elite Spring Wheat Yield Trial (ESWYT): The ESWYTs comprise lines with high and stable GY relative to checks in optimally irrigated trials using three years of GY testing data. Parental diversity is also used to ensure that most lines are derived from different crosses. These lines also show good to moderate drought and heat tolerance (most lines with around 90% or higher yields compared to the checks under drought and late-sown heat stressed environments) and have normal heading and maturity. They are targeted to the irrigated environments with mostly favorable temperatures during the crop season (ME1 mega-environment) which include the Northwestern Gangetic Plains of South Asia, most of Egypt, northwestern Mexico (Obregon), various spring wheat growing areas of Turkey, Afghanistan, Iran, etc. Furthermore, the lines with heat tolerance adapt in Sudan, Nigeria, etc. and lines with STB resistance adapt in Ethiopia.

Semi-Arid Wheat Yield Trial (SAWYT): The SAWYTs comprise lines with high and stable GY relative to checks under drought stressed environments and are mostly from different crosses. The GY in optimally irrigated and late-sown heat stressed environments is generally 90% over the checks and the lines have normal heading and maturity. They are targeted to the semi-arid or rainfed or partially irrigated areas in South Asia, West Asia, North Africa and low to moderate rainfall areas of East Africa. In years of good rainfall, STB resistance is also required for these lines.

Heat Tolerant Wheat Yield Trial (HTWYT): The HTWYTs comprise early maturing lines showing high and stable GY among the early maturing group in optimally irrigated trials, with heat tolerance (similar or higher GY than early maturing check Baj#1 in the late-sown heat stressed environment) and drought tolerance (90% or higher yields than the checks under drought). These are for irrigated or partially irrigated environments where temperatures rise fast post-flowering and often the night time temperatures are slightly warmer throughout the crop season (ME5). The main target areas include the Eastern Gangetic Plains of South Asia, Southern Pakistan, Central and peninsular India, etc. These areas require wheat lines that are early heading and maturing (7–9 days in Obregon compared to lines in ESWYTs) to avoid grain shriveling due to hot temperatures.

High Rainfall Wheat Yield Trial (HRWYT): The HRWYTs are targeted for rainfed areas requiring red grain color wheat, e.g. Kenya and some other East African highlands (excluding Ethiopia where white grained wheat is required), Central and South American countries.

Considering the screening nurseries, the phenotyping data generated for the lines in stage 2 GY trials including GY, agronomic traits, resistance to diseases and quality are used in selecting about 250–300 white grained lines for including in the international bread wheat screening nurseries, 250–300 white grained lines for the semi-arid wheat screening nurseries, and about 150–200 red grained lines for the high rainfall wheat screening nurseries. In addition, other disease-based screening nurseries like the International Septoria Observation Nursery, Leaf Blight Resistance Screening Nursery, Stem Rust Resistance Screening Nursery, Fusarium Head Blight Screening Nursery and the Karnal Bunt Resistance Screening Nursery are prepared for screening these specific diseases in MEs where they are important.

The GY data generated during the three stages of testing and data from all other phenotyped traits are used in finalizing the international yield trial nurseries, which are prepared in summer and distributed worldwide for sowing in November or later depending on the hemisphere. The international yield trials are replicated (two replications) and comprise 50 entries including one local check added by the cooperator (different for each cooperator), three common CIMMYT checks (consistent over trials and change over years by maintaining overlapping checks) and 46 different entries each year. The 50 entry international yield trials were considered by the most partners to be of adequate size, based on their phenotyping capacity and the frequency of lines retained for subsequent testing, leading to varietal release or use as parents in the local breeding programs. Each year partners request trials and nurseries using https://www.cimmyt.org/resources/seed-request/ and the data returned (recovery rate of 60%) is well-maintained in the database and made publicly available at http://orderseed.cimmyt.org/iwin/iwin-results-1.php.

7.8   Integration of Genomic Selection

The spring bread wheat program constantly evaluates and integrates new breeding methods to increase the rate of genetic gain for key traits. One such promising method that has been integrated in the breeding pipeline is genomic selection (GS) that leverages genome-wide molecular marker information to select individuals based on their predicted genetic merit [20]. While GS has transformed animal breeding by increasing the accuracy of selections, reducing cycle time and phenotyping cost, its application in wheat breeding needs a better understanding of its fit in different stages of the breeding cycle and its comparative advantage over conventional breeding strategies.

Hence, GS research at CIMMYT primarily focusses on: (i) evaluating genomic prediction models for traits with different heritabilities and genetic architectures [21–25]. While the within-nursery cross-validation accuracies were moderate to high for most traits, forward predictions (using a previous nursery/year to predict the next nursery/year) were challenging for low-heritable traits like GY [25] (ii) comparing different marker densities, marker platforms and training population designs for optimizing GS schemes [25–27] (iii) comparing genomic and pedigree-based predictions in populations with different family-structures to understand the relative advantage of genomic predictions over the pedigree [28] (iv) comparing selections made from GS and the baseline phenotypic selections [29] (v) understanding the potential of GS for predicting the performance of lines in the target environments including South Asia [30, 31]. The genomic-estimated breeding values of most traits for all the lines evaluated in stages 1 and 2 of GY testing are routinely obtained each year and integrated in selection decisions.

7.9 Partnerships with National Programs for Variety Identification, Release, and Dissemination

CIMMYT maintains effective partnerships with the National Agricultural Research Systems (NARS) who are leaders in developing, releasing, and disseminating varieties to the farmers. Their responsibilities include managing the required multisite yield trials before variety release, providing seed of new varieties at the time of release, promoting new varieties and maintaining basic seed. The NARS partners' local and regional breeding programs also develop new varieties, derived from elite CIMMYT lines. Targeted NARS partners in India, Pakistan, Bangladesh and Nepal also have an early access to a larger set (540 lines and checks) of spring bread wheat lines called the South Asian Bread Wheat Genomic Prediction Yield Trials (SABWGPYTs) that was initiated as part of the U.S. Agency for International Development's Feed the Future project, for local phenotyping, selection and use in breeding. Similarly, NARS partners in Kenya and Ethiopia have early access to elite lines combining good yields, quality and resistance to three rusts and STB.

As an outcome of the very successful partnerships between CIMMYT and NARS, at least 183 direct CIMMYT-derived spring bread wheat varieties have been released by 24 partner countries during 2015–2021 (Table 7.6), replacing older lesser productive and disease susceptible varieties and ensuring adequate wheat production and affordable food for low income wheat consumers.

7.10 Outlook to Further Accelerate Genetic Gain

The annual genetic gains reported in several studies from the evaluation of CIMMYT's international nurseries in the target environments, serve as good indicators of the progress made from the current breeding strategies at CIMMYT. For example: in the optimally irrigated ME1, annual GY genetic gains of 1.63% and 0.72% (compared to the long-term check PBW343, and local checks that are continuously updated with new varieties by NARS partners, respectively) have been reported, using the ESWYTs evaluated from 2006 to 2014 [8]. Similarly, in the low yielding rainfed or partially irrigated ME4, an annual GY genetic gain of 1.8%

Table 7.6 The 183 direct CIMMYT-derived spring bread wheat varieties released by 25 partner countries during 2015 to March 2021

Country	Name of variety
Afghanistan	Daima-17, Lalmi-17, Shamal-17, Garmser-18, Pakita 20, Jowzjan 20, Nasrat 20
Algeria	Ain El Hadjar, Bordj Mehis, El Hachimia, Nif Encer
Argentina	BIOCERES 1008, MS INTA 815
Australia	Borlaug100, SEA Condamine
Bangladesh	BARI Gom 31, BARI Gom 33, WMRI Gom 3
Bhutan	Bumthang kaa Drukchu
Bolivia	Cupesi CIAT, INIAF Tropical, Yotau, INIAF Okinawa
Egypt	Misr 3
Ethiopia	Amibara 2, Deka, Kingbird, Lemu, Wane, Bondena, Hadis, Hibist, Ga'ambo 2, Balcha, Boru, Dursa, Adet 1
India	Ankur Shiva, DBW107, DBW110, DBW168, DBW93, HI1612, HI1605, HS562, PBW658, PBW677, PBW1Zn, Pusa Kiran, Pusa Vatsala, Super 252, Super 272, Super 404, WB2, WH1142, DBW187, HI1620, DBW222, NIAW3170, HI1628, HD3249, DBW252, HI1621, HUW711, Mucut, Tarak, VL Gehun 967, DBW303, WH1270
Iran	Baharan, Barat, Ehsan, Mehrgan, Rakhsahn, Sarang, Talaei, Tirgan, Torabi, Mearaj, Kelateh, Paya, Kabir, Sahar, Farin, Araz, Arman
Kenya	Kenya Deer, Kenya Falcon, Kenya Hornbill, Kenya Peacock, Kenya Pelican, Kenya Songbird, Kenya Weaverbird, Kenya Kasuku, Kenya Jakana
Jordan	Ghweir 1
Mexico	Bacorehuis F2015, Conatrigo F2015, Ñipal F2016, Ciro NL F2016, RSI Glenn, Noroeste F2018, Noeheli F2018, Hans F2019
Nepal	Chyakhura, Danphe, Munal, Tilottama, Zinc Gahun 1, Zinc Gahun 2, Bheri-Ganga, Himganga, Khumal-Shakti, Borlaug 2020
Nigeria	Lacriwhit 9, Lacriwhit 10
Pakistan	Anaaj-17, Barani-17, Borlaug 2016, Ihsan-16, Israr-shaheed-2017, Khaista-17, Kohat-17, NIFA-Aman, Pakhtunkhwa-15, Pasina-2017, Pirsabak-15, Shahid-2017, Sindhu-16, Ujala-16, Wadaan-2017, Zincol 2016, Ghazi 19, Markaz 19, Bhakkar 19, Gulzar 19, Fahim 19, NIFA Awaz, Aghaz 2019, Umeed-e-Khass 2019, Akbar 19, MH-2020, Subhani 20, MA 2020, Bhakkar20, AZRC Dera 2020, IV-2, Swabi 1, Zarghoon 2021, Pirsabak 2021, NIA Zarkhiaz 2020
Paraguay	Caninde 31, Itapua 90
Peru	INIA 440 K'ANCHAREQ
Rwanda	Cyumba, Gihundo, Keza, Kibatsi, Majyambere, Mizero, Nyangufi, Nyaruka, Reberaho, Rengerabana
Spain	Tujena, Santaella, Montemayor, Setenil
Sudan	Ageeb, Akasha
Tajikistan	Haydari, Roghun
Turkey	Altinoz, Ekinoks, Kayra, Koc 2015, Nisrat, Polathan, Karmen, Kirve, Sahika, Simge
Zambia	Falcon

(compared to the mean of four long-term checks) has been reported, using the SAWYTs evaluated between 2002 and 2013 [32]. Furthermore, in the high-rainfall and low rainfall environments of ME2, annual GY genetic gains of 1.17% and 0.73% (compared to the local checks), respectively were reported using the HRWYTs evaluated between 2007 and 2016 [33]. All these studies clearly indicate that continuous genetic gain for GY is achieved in the target environments, where the international spring bread wheat nurseries distributed by CIMMYT are evaluated.

Chapter 30 describes the methods for accelerating breeding cycles. In the current CIMMYT breeding program, it takes a minimum of five years from making simple crosses to obtaining stage 1 GY trial results and six years to obtaining stage 2 GY trial results, which contributes most of the parents for recycling. There are opportunities to accelerate generation advancement by growing 4 generations/year in a greenhouse/screenhouse/speed breeding facility, as well as expand stage 1 trials to multiple selection environments and shortening the breeding cycle time has the potential to accelerate genetic gain. Hence, the spring bread wheat breeding program has initiated piloting and optimization of two breeding schemes that will permit 3- and 2-years breeding cycle time for simple crosses and an additional year for top/BC$_1$. These schemes will attempt to ensure that the loss from not selecting in early segregating generations can be compensated by selection in later generations. Intensification of data-driven decisions for choosing parents by incorporating the genomic-estimated breeding values of parents and using them to eliminate populations and advanced lines at an earlier stage are also considered useful to accelerate genetic gain in the new breeding schemes. The two breeding schemes are briefly described in Sects. 7.10.1 and 7.10.2.

7.10.1 'Rapid Bulk Generation Advancement (RBGA) Scheme (Three-Year Breeding Cycle Time)

In RBGA scheme simple crosses will be made in a field screenhouse in Toluca with sowing of parents initiated in late May, soon after the completion of Obregon season, and F1–F3 generations advanced as bulk in the same screenhouse within one year. In year 2, the F4 populations will be grown in Toluca field for the selection of space-sown plants having the required agronomic traits and disease resistance (YR and STB). Individual spikes will then be harvested and selected for grain characteristics, and head rows will be sown in Obregon in November for selection as small plots for agronomic traits and resistance to diseases (LR, SR). In year 3, the harvested advanced lines with good grain traits will be sown in El Batan and Toluca for seed multiplication. Phenotyping for resistance to LR, YR and STB, while simultaneous genotyping will permit genomic selection, thus, advancing fewer lines to stage 1 trials in Obregon. Seed produced in El Batan will be used to conduct stage 1 trials in 4–5 selection environments in Obregon and phenotyping for resistance to rusts, spot blotch and other diseases in Mexico, Kenya and South Asia. All data will be used for selecting elite parents for recycling using breeding values.

7.10.2 'Rapid-Cycle Recurrent Selection (RCRS)' Scheme (Two-Year Breeding Cycle Time)

Although RBGA is potentially a powerful scheme, opportunities exist to further reduce breeding cycle time by growing F3 derived F4 head rows in Toluca field and then using the seed from selected harvested plots to grow stage 1 yield trials in Obregon in 2–3 selection environments. LR phenotyping and genotyping for estimating breeding values using all data will be used for selecting the best parents for recycling. The 2-year RCRS scheme is especially useful in decoupling population improvement from elite lines (product) extraction and has the potential to simultaneously accelerate genetic gain for a few traits such as grain yield and grain zinc.

7.11 Key Concepts

Delivering genetic gain in farmers fields requires a well-targeted breeding program that needs to select high value parents for hybridization, maintain and add new genetic diversity for relevant traits in breeding populations, conduct accurate phenotyping and select for a range of relevant traits to build the trait package for the development of farmers and market preferred varieties. New methods, such as speed breeding, genomic selection and gene-editing are expected to further enhance the current rates of genetic gains by improving the selection accuracy and reducing the breeding cycle length.

7.12 Conclusion

In this chapter, we have provided an overview of the CIMMYT spring wheat breeding program and discussed several successful breeding strategies like effective parental choice, the selected-bulk breeding scheme, the shuttle-breeding program, rigorous multi-environment phenotyping, international nurseries, and partnerships with national programs. These strategies and their optimization over time, have been instrumental in building a strong spring wheat breeding program at CIMMYT that continuously delivers genetic gains for GY along with other key traits. Moreover, we have also provided descriptions of new breeding schemes that offer promise to accelerate genetic gain by shortening the breeding cycle time, while delivering superior varieties.

References

1. Braun HJ, Atlin G, Payne T (2010) Multi-location testing as a tool to identify plant response to global climate change. In: Climate change and crop production. CABI, Wallingford, pp 115–138
2. Ray DK, Mueller ND, West PC, Foley JA (2013) Yield trends are insufficient to double global crop production by 2050. PLoS One 8. https://doi.org/10.1371/journal.pone.0066428
3. Singh RP, Singh PK, Rutkoski J, Hodson DP, He X, Jørgensen LN, Hovmøller MS, Huerta-Espino J (2016) Disease impact on wheat yield potential and prospects of genetic control. Annu Rev Phytopathol 54:303–322. https://doi.org/10.1146/annurev-phyto-080615-095835
4. Wheeler T, von Braun J (2013) Climate change impacts on global food security. Science 341:508–513. https://doi.org/10.1126/science.1239402
5. Zampieri M, Ceglar A, Dentener F, Toreti A (2017) Wheat yield loss attributable to heat waves, drought and water excess at the global, national and subnational scales. Environ Res Lett 12:064008
6. Shiferaw B, Smale M, Braun HJ, Duveiller E, Reynolds M, Muricho G (2013) Crops that feed the world 10. Past successes and future challenges to the role played by wheat in global food security. Food Secur 5:291–317. https://doi.org/10.1007/s12571-013-0263-y
7. Lantican MA, Braun H-J, Payne TS, Singh RP, Sonder K, Michael B, van Ginkel M, Erenstein O (2016) Impacts of international wheat improvement research: 1994–2014. CIMMYT, Mexico
8. Crespo-Herrera LA, Crossa J, Huerta-Espino J, Autrique E, Mondal S, Velu G, Vargas M, Braun HJ, Singh RP (2017) Genetic yield gains in CIMMYT's international Elite Spring Wheat Yield Trials by modeling the genotype × environment interaction. Crop Sci 57:789–801. https://doi.org/10.2135/cropsci2016.06.0553
9. Braun HJ, Saulescu NN (2002) Breeding winter and facultative wheat. In: Curtis BC, Rajaram S, Gómez Macpherson H (eds) Bread wheat improvement and production, FAO Plant Production and Protection Series No. 30. Food and Agriculture Organization of the United Nations, Rome
10. Kiss T, Balla K, Veisz O, Láng L, Bedő Z, Griffiths S, Isaac P, Karsai I (2014) Allele frequencies in the VRN-A1, VRN-B1 and VRN-D1 vernalization response and PPD-B1 and PPD-D1 photoperiod sensitivity genes, and their effects on heading in a diverse set of wheat cultivars (Triticum aestivum L.). Mol Breed 34:297–310. https://doi.org/10.1007/s11032-014-0034-2
11. Saulescu NN, Braun H-J (2001) Breeding for adaptation to environmental factors. Cold tolerance. In: Reynolds MP, Ortiz-Monasterio I, McNab A (eds) Application of physiology in wheat breeding. Mexico, CIMMYT, pp 111–123
12. Rajaram S, Borlaug NE, Van Ginkel M (2002) CIMMYT international wheat breeding. In: Bread wheat improvement production. FAO, Rome, pp 103–117
13. Ragot M, Bonierbale M, Weltzein E, Genetics NF (2018) From market demand to breeding decisions: a framework. Work Pap 2:1–53
14. Singh RP, Rajaram S, Miranda A, Huerta-Espino J, Autrique E (1998) Comparison of two crossing and four selection schemes for yield, yield traits, and slow rusting resistance to leaf rust in wheat. Euphytica 100:35–43. https://doi.org/10.1023/a:1018391519757
15. Singh RP, Huerta-Espino J, Bhavani S, Herrera-Foessel SA, Singh D, Singh PK, Velu G, Mason RE, Jin Y, Njau P, Crossa J (2011) Race non-specific resistance to rust diseases in CIMMYT spring wheats. Euphytica 179:175–186. https://doi.org/10.1007/s10681-010-0322-9
16. Singh RP, Trethowan R (2008) Breeding spring bread wheat for irrigated and rainfed production systems of the developing world. In: Breeding major food staples. Wiley, New York
17. Braun HJ, Rajaram S, Van Ginkel M (1996) CIMMYT's approach to breeding for wide adaptation. Euphytica 92:175–183. https://doi.org/10.1007/BF00022843
18. Singh RP, Herrera-Foessel S, Huerta-Espino J, Singh S, Bhavani S, Lan C, Basnet BR (2014) Progress towards genetics and breeding for minor genes based resistance to Ug99 and other rusts in CIMMYT high-yielding spring wheat. J Integr Agric 13:255–261. https://doi.org/10.1016/S2095-3119(13)60649-8

19. Singh PK, Mergoum M, Ali S, Adhikari TB, Elias EM, Hughes GR (2006) Identification of new sources of resistance to Tan Spot, Stagonospora Nodorum Blotch, and Septoria Tritici Blotch of wheat. Crop Sci 46:2047–2053. https://doi.org/10.2135/cropsci2005.12.0469

20. Meuwissen THE, Hayes BJ, Goddard ME (2001) Prediction of total genetic value using genome-wide dense marker maps. Genetics 157:1819–1829

21. Battenfield SD, Guzmán C, Gaynor RC, Singh RP, Peña RJ, Dreisigacker S, Fritz AK, Poland JA (2016) Genomic selection for processing and end-use quality traits in the CIMMYT spring bread wheat breeding program. Plant Genome 0:0. https://doi.org/10.3835/plantgenome2016.01.0005

22. Juliana P, Singh RP, Singh PK, Crossa J, Huerta-Espino J, Lan C, Bhavani S, Rutkoski JE, Poland JA, Bergstrom GC, Sorrells ME (2017) Genomic and pedigree-based prediction for leaf, stem, and stripe rust resistance in wheat. Theor Appl Genet 130. https://doi.org/10.1007/s00122-017-2897-1

23. Crossa J, Pérez P, Hickey J, Burgueño J, Ornella L, Ceró N-Rojas J, Zhang X, Dreisigacker S, Babu R, Li Y, Bonnett D, Mathews K (2013) Genomic prediction in CIMMYT maize and wheat breeding programs. Heredity (Edinb) 112:48–60. https://doi.org/10.1038/hdy.2013.16

24. Velu G, Crossa J, Singh RP, Hao Y, Dreisigacker S, Perez-Rodriguez P, Joshi AK, Chatrath R, Gupta V, Balasubramaniam A, Tiwari C, Mishra VK, Sohu VS, Mavi GS (2016) Genomic prediction for grain zinc and iron concentrations in spring wheat. Theor Appl Genet 129:1595–1605. https://doi.org/10.1007/s00122-016-2726-y

25. Juliana P, Poland J, Huerta-Espino J, Shrestha S, Crossa J, Crespo-Herrera L, Toledo FH, Govindan V, Mondal S, Kumar U, Bhavani S, Singh PK, Randhawa MS, He X, Guzman C, Dreisigacker S, Rouse MN, Jin Y, Pérez-Rodríguez P, Montesinos-López OA, Singh D, Mokhlesur Rahman M, Marza F, Singh RP (2019) Improving grain yield, stress resilience and quality of bread wheat using large-scale genomics. Nat Genet 51:1530–1539. https://doi.org/10.1038/s41588-019-0496-6

26. Juliana P, Singh RP, Singh PK, Crossa J, Rutkoski JE, Poland JA, Bergstrom GC, Sorrells ME (2017) Comparison of models and whole-genome profiling approaches for genomic-enabled prediction of Septoria Tritici Blotch, Stagonospora Nodorum Blotch, and Tan Spot resistance in wheat. Plant Genome 10. https://doi.org/10.3835/plantgenome2016.08.0082

27. Juliana P, Montesinos-López OA, Crossa J, Mondal S, Pérez LG, Poland J, Huerta-Espino J, Crespo-Herrera L, Govindan V, Dreisigacker S, Shrestha S, Pérez-Rodríguez P, Pinto Espinosa F, Singh RP (2019) Integrating genomic – enabled prediction and high – throughput phenotyping in breeding for climate – resilient bread wheat. Theor Appl Genet 132:336546. https://doi.org/10.1007/s00122-018-3206-3

28. Juliana P, Singh RP, Braun HJ, Huerta-Espino J, Crespo-Herrera L, Govindan V, Mondal S, Poland J, Shrestha S (2020) Genomic selection for grain yield in the CIMMYT wheat breeding program—status and perspectives. Front Plant Sci 11. https://doi.org/10.3389/fpls.2020.564183

29. Juliana P, Singh RP, Poland J, Mondal S, Crossa J, Montesinos-López OA, Dreisigacker S, Pérez-Rodríguez P, Huerta-Espino J, Crespo-Herrera L, Govindan V (2018) Prospects and challenges of applied genomic selection—a new paradigm in breeding for grain yield in bread wheat. Plant Genome 11:1–17. https://doi.org/10.1136/bmj.2.1403.1129-a

30. Pérez-Rodríguez P, Crossa J, Rutkoski J, Poland J, Singh R, Legarra A, Autrique E, de los Campos G, Burgueño J, Dreisigacker S (2017) Single-step genomic and pedigree genotype × environment interaction models for predicting wheat lines in international environments. Plant Genome 10:1–15. https://doi.org/10.3835/plantgenome2016.09.0089

31. Juliana P, Singh RP, Braun H-J, Huerta-Espino J, Crespo-Herrera L, Payne T, Poland J, Shrestha S, Kumar U, Joshi AK, Imtiaz M, Rahman MM, Toledo FH (2020) Retrospective quantitative genetic analysis and genomic prediction of global wheat yields. Front Plant Sci 11:1328. https://doi.org/10.3389/fpls.2020.580136

32. Crespo-Herrera LA, Crossa J, Huerta-Espino J, Vargas M, Mondal S, Velu G, Payne TS, Braun H, Singh RP (2018) Genetic gains for grain yield in CIMMYT's semi-arid wheat yield

trials grown in suboptimal environments. Crop Sci 58:1890–1898. https://doi.org/10.2135/cropsci2018.01.0017

33. Gerard GS, Crespo-Herrera LA, Crossa J, Mondal S, Velu G, Juliana P, Huerta-Espino J, Vargas M, Rhandawa MS, Bhavani S, Braun H, Singh RP (2020) Grain yield genetic gains and changes in physiological related traits for CIMMYT's high rainfall wheat screening nursery tested across international environments. F Crop Res 249:107742. https://doi.org/10.1016/j.fcr.2020.107742

Chapter 8
Wheat Rusts: Current Status, Prospects of Genetic Control and Integrated Approaches to Enhance Resistance Durability

Sridhar Bhavani, Ravi P. Singh, David P. Hodson, Julio Huerta-Espino, and Mandeep Singh Randhawa

Abstract The three rusts are the most damaging diseases of wheat worldwide and continue to pose a threat to global food security. In the recent decades, stem rust races belonging to the Ug99 (TTKSK) and Digalu (TKTTF) race group resurfaced as a major threat in Africa, the Middle East and Europe threatening global wheat production. In addition, the evolution and migration of new aggressive races of yellow rust adapted to warmer temperatures into Europe and Asia from Himalayan region are becoming a significant risk in several wheat production environments. Unique and complex virulence patterns, continuous evolution to overcome effective resistance genes in varieties, shifts in population dynamics, transboundary migration have resulted in localized/regional epidemics leading to food insecurity threats. This underscores the need to identify, characterize, and deploy effective rust resistant genes from diverse sources into pre-breeding lines and future wheat varieties. The use of genetic resistance and deployment of multiple race specific and pleiotropic adult plant resistance genes in wheat lines can enhance resistance durability. Recent advances in sequencing annotated wheat reference genome with a detailed analysis of gene content among sub-genomes will not only accelerate our understanding of the genetic basis of rust resistance bread wheat, at the same time wheat breeders can now use this information to identify genes conferring rust resistance.

S. Bhavani (✉) · R. P. Singh · D. P. Hodson
International Maize and Wheat Improvement Center (CIMMYT), Texcoco, Mexico
e-mail: s.bhavani@cgiar.org; r.singh@cgiar.org; d.hodson@cgiar.org

J. Huerta-Espino
INIFAP, Campo Experimental Valle de Mexico, Texcoco, Mexico
e-mail: j.huerta@cgiar.org

M. S. Randhawa
International Maize and Wheat Improvement Center (CIMMYT), Nairobi, Kenya

International Centre for Research in Agroforestry (ICRAF), Nairobi, Kenya
e-mail: m.randhawa@cgiar.org

© The Author(s) 2022
M. P. Reynolds, H.-J. Braun (eds.), *Wheat Improvement*,
https://doi.org/10.1007/978-3-030-90673-3_8

Progress in genetic mapping techniques, new cloning techniques and wheat transformation methods over the last two decades have not only resulted in characterizing new genes and loci but also facilitated rapid cloning and stacking multiple genes as gene cassettes which can be future solution for enhancing durable resistance.

Keywords Rust resistance · Race specific genes · Adult plant resistance genes · Breeding technologies

8.1 Learning Objectives

- Geographical distribution of three rust diseases, impact, management strategies and briefly address the new molecular tools in the current era to enhance resistance breeding and opportunities for wheat improvement.

8.2 Economic Importance, Historical Impacts, Status of Rust Diseases

Pests and diseases (P&D) have historically affected food production either directly through losses in crop production or quality. Currently, these losses are exacerbated by the changing climate threatening food security and rural livelihoods across the globe. Nearly 200 wheat pests and diseases in wheat have been documented, of which fifty are considered economically important because of their potential to cause substantial yield losses. Two studies estimated potential grain yield losses due to disease at 18% and 21.5% at global level and (10.1–28.1%) per hotspot for wheat [1], however, losses can be significantly higher in areas where susceptible wheat varieties are still grown. Rust pathogens are present in all wheat growing environments and have constantly hindered global wheat production since domestication and still continue to threaten the global wheat supplies. It is estimated that global annual losses to wheat rust pathogens can be around 15 million tons valued at US\$ 2.9 billion [2]. Documented evidence suggesting wheat rusts could be one of the earliest pathogens wherein spores of SR dating back to 1300 BC were detected in Israel, rust was reported as serious disease of cereals in Italy and Greece more than 2000 years ago and festival called "Robigalia" was celebrated to protect the crops from rusts and smuts [3]. The continuous effort to increase genetic gains is not possible without overcoming several of the current barriers such as climate change coupled with a variety of unpredictable abiotic and biotic stresses that pose significant threat to wheat production both locally and globally (see Chap. 7). Genetic uniformity of wheat in the quest of developing high-performing cultivars, has also contributed pathogen resurgence to the point wherein diseases threaten global wheat production. This review considers the three rust diseases affecting wheat productivity, and the emerging threats considering the geographical distribution, impacts, and management strategies and briefly address the new molecular tools in the current era to enhance resistance breeding and deployment opportunities for wheat improvement.

There are three wheat rust diseases, namely stem (black) rust, stripe (yellow) rust and leaf (brown) rust, all belonging to the members of the Basidiomycete family, genus *Puccinia*, and named *P. graminis* f. sp. *tritici* (*Pgt*), *P. striiformis* f. sp. *tritici* (*Pst*) and *P. triticina* (*Pt*), respectively.

8.2.1   Stem Rust

Stem rust (SR), or black rust is common where wheat plants are exposed to warmer environments at later stages of crop growth. SR has the potential to completely annihilate a healthy looking crop when an epidemic occurs and linear yield losses have been observed, early infections can result in no grain fill and panicles can be reduced to chaff [4]. SR epidemics have been significantly curtailed by eliminating its alternate host (barberry species) between 1918 and 1980 in the USA and in the UK, adoption of semi-dwarf, early maturing rust resistant varieties developed by CIMMYT (International Maize and Wheat Improvement Center), and the use of fungicides.

Wheat growing environments of East Africa are unique epidemiological regions that favor wheat production all-round the year in different regions providing continuous green bridge for pathogen evolution and survival resulting in frequent localized epidemics. Even though it was under control for over three decades the recent re-emergence of SR race "Ug99" in East Africa posed a serious threat to global wheat production [5].

The stem rust race Ug99 (TTKSK) caused widespread damage in Kenya [6] carrying unique virulence as it was able to overcome over 50% of the known SR resistance genes including widely deployed genes *Sr31* and *Sr38*. Following the spread of race Ug99, resistant cultivar "Kenya Mwamba" was released in 2001 (known to carry gene *Sr24*) which became a popular variety with farmers however; in 2006, race TTKST (Ug99+*Sr24* virulence) was detected in Kenya, resulting in severe localized epidemics in Kenya [7]. Through sustained breeding efforts of CIMMYT, several new varieties with resistance to TTKSK and TTKST were released post 2009, of which "Kenya Robin" became a leading variety combining high yield potential and SR resistance covering 40% of the wheat area in Kenya by 2014. However, in the same year the breakdown of resistance in Robin and two variants of Ug99 race group with virulence to resistance gene *SrTmp* were identified, viz. race TTKTK (Ug99+*SrTmp*), and TTKTT (Ug99+*Sr24*+*SrTmp*) [8]. These two genes (*Sr24* and *SrTmp*) were quite important in conferring effective resistance to SR races in the USA, CIMMYT, South America and Australian wheat germplasm increasing the vulnerability of varieties to Ug99 race group [9] not only for East Africa but predicted migration paths threatening production in other wheat growing environments [9]. In 2018, another new race with virulence to *Sr8155B* gene was identified in Kenya in 2018 (unpublished data) and currently, seven of the fourteen variants within the Ug99 race group have evolved in Kenya, making it the hot spot for evolution of Ug99 race group.

Stripe rust epidemics in Ethiopia in 2010, prompted release of varieties carrying good levels of stripe rust and SR resistance of which cultivar "Digalu" (carrying high yield potential and rust resistance to both YR and Ug99 race group) became a popular variety with farmers by 2013–2014 occupying approximately 31% of wheat area under production. However, in 2013, devastating localized epidemics of SR were reported on Digalu caused by race TKTTF, a SR race unrelated to the Ug99 race group [10]. This race was able to overcome resistance gene *SrTmp* present in Digalu and was later detected in Kenya, this race was previously reported in Turkey, Lebanon and Iran. Airborne dispersal models also indicated a migration route into East Africa from sources in the Middle East. Race TKTTF has also now been detected in Germany, UK, Sweden, Denmark. In addition to Digalu race group, diverse SR races with rare combination of virulence to *Sr9e* and *Sr13* have been found in the central highlands of Ethiopia [11], which have been quite important for durum wheat as these genes are deployed in both North America and Australia.

Widespread eradication of the alternate host, common barberry, had resulted in effective control of SR in Western Europe until 2013. However, unusual SR infections on winter and spring wheat in 2013 and race analyses identified six SR races, similar to Digalu race with additional virulence to *Sr7a, Sr45,* and *SrTt*-3 [12]. Common barberry is now being implicated as a source of new stem rust race diversity in Georgia and Western Siberia and SR epidemics on oats in Sweden. Since 2014, several large-scale stem rust outbreaks have been reported and virulent races are spreading rapidly. Race TKKTP with virulence combination for *Sr24, Sr36, Sr1A.1R* and *SrTmp* [13], races TRTTF and TKKTF (virulence to *Sr1A.1R*) have also been identified. Race TKKTF is spreading rapidly and now detected in 17 countries across Europe, North Africa, the Middle East and East Africa. Similarly, race TTRTF caused epidemics on durum wheat in Sicily since 2015 and is now detected in 10 countries in Europe, North Africa, the Middle East and East Africa [14].

8.2.2   Stripe Rust

Stripe (yellow) rust (YR) is a common disease in wheat and well adapted to temperate areas with humid and cool weather, aggressive races adapted to warmer temperatures have migrated and spread across geographies since 2000 [15]. Race shifts towards higher rates of mutation for virulence within the *Pst* pathogen has resulted in vulnerability of widely deployed cultivars. Global yield losses to YR is estimated at 5.5 million tons per year [16]. Production losses in North America alone since 2000 exceeded over one million tons, and in China, losses over 1.8–6.0 million tons were observed under epidemic conditions. Similar reports of yield losses to YR in

Europe in the recent decade has been attributed largely to the race shifts derived from the Himlayan region [17]. Historically, impact of newly evolved YR races on wheat productivity have been occasional, however, new incursions have often resulted in widespread damage, e.g. incursion of YR races from Europe into eastern Australia in 1978, western Australia in early 2002. Exotic incursions of YR races replaced the existing populations in the USA since 2000 and race shifts in the European *Pst* populations in 2011 and 2012 by races from the Himalayan region [17] are very good examples of exotic races with different genetic *Pst* lineages causing significant impact on host susceptibility. A recent study linking both virulence and race structure with recent YR epidemics in different geographies [18] suggested different *Pst* races in distinct genetic lineages, where aggressive strains adapted across diverse environments were spreading across continents, including the more recent outbreak of YR in Argentina.

8.2.3 Leaf Rust

Leaf (or brown) rust (LR), is the most common rust disease in both winter wheat and spring wheat growing areas, as well as on durum wheat. Yield losses due to LR can be substantial if susceptible varieties are infected at early stages coupled with favourable temperatures and moisture conditions resulting in rapid progress in short time span. Yield losses are largely due to the reduction of kernels per spike and lower kernel weights.

Populations of *Pt*, are specifically adapted to either tetraploid durum wheat or hexaploid common wheat and races conferring virulence to several of the *LR* genes are prevalent throughout the world [19]. Since the early 2000s, races of *Pt* that are highly virulent on durum wheat cultivars have spread across South America, Mexico, Europe, the Mediterranean basin, and the Middle East. These races confer virulence to *Lr71* gene, widely present in durum wheat, however, are avirulent to many of the LR genes that are found in common wheat. In Ethiopia another group of *Pt* races have been found that are highly virulent on durum wheat yet avirulent to the highly susceptible common wheat cultivars such as "Thatcher" and "Little Club" [20] and these isolates are unique to Ethiopia.

On a global scale, most populations of *Pt* are unique in their virulence and molecular genotypes. Even though the most common mode of evolution is through mutation and selection in a given environment, there is evidence for recent migration of *Pt* races between different continental regions. Since the mid 1990s isolates of *Pt* with virulence to *Lr1*, *Lr3a*, and *Lr17a*, and avirulence to *Lr28*, have increased and spread across the U.S. and Canada. These isolates also had a unique molecular genotype, which indicated that these were likely recently introduced to North America. Since the early 2000s these isolates with identical or highly similar virulence and molecular genotypes have been found in Europe, South America, Ethiopia, Turkey and Pakistan [21]. Similarly isolates of *P. triticina* with virulence to durum wheat that also have identical or highly related molecular genotypes have been found in the Middle East, South America, Europe, Ethiopia, Tunisia, Mexico and the U.S [22].

8.3 Global Rust Phenotyping Network – Critical Tool to Understand Host Resistance and Pathogenic Diversity on a Global Scale

A global network of precision field-based wheat disease phenotyping platforms of the CGIAR Program WHEAT (see http://wheat.org), were developed with the support of national agricultural research institutes. The objective is to generate multi-location disease phenotypic data, under defined management practices, and fostering germplasm exchange. The selected locations also represent hotspots for specific diseases and future-climate analogue sites. This model opens opportunities to increase coordination in wheat phenotyping, avoiding duplications, and building on efficiency and capacity for research. The global wheat phenotyping network (Fig. 8.1) has eight regional hubs/hotspot sites that facilitate screening and selection for diseases, viz. SR and YR (Kenya, Ethiopia, Turkey, India), Fusarium (China, Uruguay), wheat blast (Bolivia and Bangladesh), leaf blight (Nepal and India), soil borne diseases (Turkey), Septoria (Kenya, Ethiopia, Uruguay and Tunisia) alongside CIMMYT, HQ stations of Toluca (YR, Septoria), El Batan (leaf rust), Agua Fria (leaf blight) and Obregon (leaf rust) (Mexico).

In the last decade, effective partnership between CIMMYT, KALRO (Kenya Agriculture and Research Organization), EIAR (Ethiopian Institute of Agriculture Research) and BGRI (Borlaug Global Rust Initiative) through DRRW (Durable rust

International wheat phenotyping network

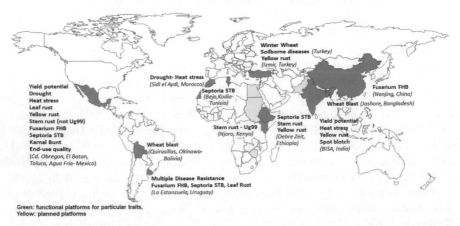

Fig. 8.1 International wheat phenotyping hubs spread across several countries led by NARS in collaboration with CIMMYT/ICARDA

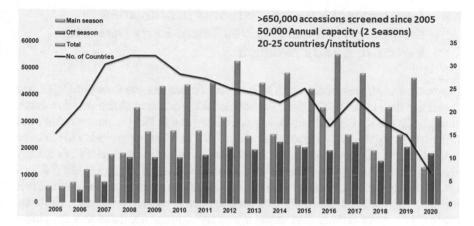

Fig. 8.2 Wheat accessions phenotyped during 2005–2020 for Ug99 resistance at Njoro (Kenya) and participating countries, in partnership with Kenya Agriculture Livestock Research Organization, Kenya

resistance in wheat) and DGGW (Delivering Genetic Gains in Wheat) projects have established functional SR phenotyping platforms which have made a significant impact to the global wheat research in addressing the threat of SR. International SR phenotyping platforms established at Njoro (KALRO, Kenya) and Debrezeit (EIAR, Ethiopia) play key roles in evaluating global wheat germplasm from several countries and institutions. Over 650,000 wheat accessions have been screened against Ug99 and derived races since 2005, and the screening capacity at KALRO, Njoro has increased to 50,000 lines from over 20–25 countries and research institutions each year [4] (Fig. 8.2). The results from the international nurseries show a shift to higher frequencies of lines with resistance to SR races, since the screening activities were initiated in 2008. Similarly, close to 150,000 wheat land races and advanced durum wheat breeding lines and varieties have been evaluated in Debrezeit.

Reliable phenotypic data generated from these phenotyping platforms led to the characterization of over 35 SR genes/loci in collaboration with global partners (Matt Rouse CDL, unpublished data), Genomic prediction models for APR (Adult-Plant Resistance) showed promising results [23]. Release of over 17 varieties in Kenya and Ethiopia and more than 200 varieties released in several countries globally over the years is testament to the success of the impacts from the phenotyping platforms. CIMMYT-Kenya shuttle breeding has resulted in rapid cycling of over 2000 populations each year between Mexico and Kenya to evaluate and select lines in early generations against virulent SR. Candidates of the stage I (10,000 lines) and stage II (1500 lines) yield trials are also evaluated and the selected lines are included in international nurseries and trials and distributed to NARS partners.

8.4 International Research Networks in Mitigating the Threats of Emerging New Races-Early Detection, Forecasting and Prediction

In response to the resurgence of SR in eastern Africa and the threat of Ug99, and "sounding the alarm" by Dr. N. E. Borlaug in 2005, the international wheat research community led by Cornell university, established the BGRI (Borlaug Global Rust Initiative) to significantly reduce the vulnerability of wheat crop worldwide to three rusts diseases. Improved pathogen monitoring and surveillance activities greatly enhanced the tracking and spread of new and virulent variants of SR, YR and LR. Global cereal rust monitoring system (GCRMS) is an information platform that includes standardized protocols and methods for surveys, preliminary virulence testing, data, sample transmission and management at the field; national and global levels. Collected rust samples are sent under permit to several international specialist rust laboratories for pathotype analysis (GRRC-Denmark [YR+SR], CDL-Minnesota [SR+LR], AAFC- Canada [SR], and ICARDA-Turkey RCRRC- Izmir [SR+YR]). The GCRMS expanded substantially over the years and as of 2019, scientists from 40 countries are participating in wheat rust surveillance and over 44,000 geo-referenced survey records and 9000+ rust isolate records have been collected (see https://rusttracker.cimmyt.org/). For the first time, important *Pgt* race groups, e.g. the Ug99 group, have been successfully tracked in space and time. Other important new *Pgt* and *Pst* race groups that are spreading in Europe, the Middle East and Africa are also being monitored. Integrated data management is achieved through a centralized database (Wheat Rust Toolbox) managed by Aarhus University, Denmark and the tools and database are updated on a routine basis, hence delivering the most recent information in a timely manner. The Wheat Rust Toolbox includes a comprehensive user management system that permits controlled access to specific tools and functionality. Registered users have country-specific access to an on-line data entry system and a suite of country-specific data visualization options for their own data.

A series of reviews on current status of key SR races [4, 24, 25] have provided a recent and comprehensive overview of the status of the Ug99 race group, describing the rapid evolution of new races and its geographical expansion. Technology innovations are now enhancing the global rust monitoring system. Very recently Mobile And Real-time PLant disEase (MARPLE) diagnostics [26] has been developed by John Innes Center, UK and successfully deployed as a portable, genomics-based tool to identify individual strains of complex fungal plant pathogens. Advanced spore dispersal and meteorologically driven epidemiological models, developed by Cambridge University and UK Met Office, are now providing valuable new information on pathogen movements and the basis for near-real time, in-season rust early warning systems. An operational rust early warning system is now operational in Ethiopia and similar systems are being developed in Nepal and Bangladesh.

8.5 Types of Resistance, Strategies to Deploy Different Resistance Mechanisms to Attain Resistance Durability

Currently, over 220 rust resistance genes viz. 79 LR resistance genes, 82 YR resistance genes and 60 SR genes have been formally cataloged and designated of which majority of them confer race specific resistance and only a few genes confer slow rusting /partial adult plant resistance to the three rust diseases.

8.5.1 Race-Specific/Seedling Resistance

Race specific, or seedling resistance, also referred as qualitative resistance or all stage resistance is effective at all growth stages and belongs to the "R gene" class conferring NBS-LRR (Nucleotide Binding Site- Leucine Rich Repeat) domain. Some exceptions are known where R-genes are effective only in post-seedling or adult plant stages. R-genes may confer a major resistance effect/complete resistance expressed as varying degrees of hypersensitive response and are effective one or against few races of the pathogen. However, majority of the R-genes are intermediate and do not confer clean phenotype or adequate levels of resistance and some are influenced by temperature and light regimes. The ease of selecting these genes at both seedling and field stages has made it easier to incorporate such resistance in wheat breeding programs resulting in increased productivity (boom). However, deployment of single R-genes has often resulted in pathogen acquiring virulence post deployment as varieties in a short period leading to breakdown of resistance causing epidemics and severe yield losses (bust) cycles e.g. widespread virulence for *Yr9* and *Yr27*, virulence for SR gene *Sr31* and other important SR genes *Sr24*, *Srtmp* to the Ug99 race group and ineffectiveness of LR resistance genes in the United States. However, deployment of multiple R gene combinations often refereed as "pyramiding" can effectively enhance durability of resistance in an event of when one of the R gene breaks down other genes will continue to protect the variety and keep pathogen populations under check.

8.5.2 APR Genes Conferring Pleiotropic Effects

Race-nonspecific resistance often referred as adult plant resistance or partial resistance is effective against wider races of a pathogen species. APR is generally quantitative, exhibiting incomplete resistance that is usually expressed at later stages of plant development unlike race-specific resistance that is expressed at both seedling and adult plant stage. These genes help slow the disease progress through increased latency period, reduced infection frequency, reduced pustule size resulting in lower spore production. Several of the APR genes confer seedling susceptibility and usually produce medium to large compatible pustules at low frequency without hypersensitive response and expression of resistance is observed when the plants reach flag leaf or boot leaf stage. The phenotypic effect of such genes is relatively minor to moderate, however, additive effects of multiple APR genes (4–5) in combinations can result in very high levels of resistance [24]. The problem of "boom and bust" cycles prompted wheat breeders to embrace an alternate approach combining slow rusting or partial resistance to enhance resistance durability. Johnson and Law [27] defined durable resistance as "resistance that remained effective after widespread deployment over a considerable period of time". A general concept of a durable resistance source for cereal rusts is that it is minor in effect, polygenic, usually expressed at post-flowering/adult plant stage, non-race-specific and produce non-hypersensitive response to infection.

Noteworthy examples of durable resistance is the resistance to SR transferred from tetraploid emmer to North American bread wheat cultivars "Hope" and "H-44", and LR resistance in the South American wheat cultivar "Frontana". Since the early 1980s, significant progress has been achieved in understanding the genes involved with slow rusting and their efficient use in breeding [7]. Currently at CIMMYT, key slow rusting pleotropic genes such as *Lr34*. *Lr46*, *Lr67* and *Lr68* in combination with other minor effect genes continue to enhance durable resistance to the three rust diseases [4, 24].

Lr34 was first reported in cultivar "Frontana", and wheat cultivars containing *Lr34* are widely present and occupy more than 25 million ha in developing countries and is effective in reducing yield losses in epidemic years, and has been mapped on chromosome 7DS. This gene confers pleiotropic resistance effect (Table 8.1) on multiple diseases such as YR, SR, powdery mildew, barley-yellow dwarf virus and spot blotch (*Lr34/Yr18*, *Sr57*, *Pm38*, *Bdv1* and *Sb1*), respectively and is associated with a morphological marker leaf tip necrosis (LTN). *Lr34* was cloned and the gene

Table 8.1 Pleotropic APR genes used in CIMMYT wheat breeding program and linked markers

Genes	Reported linked markers	Marker type	Reference stock	Chromosomal location	Reference
Lr34/ Yr18/ Pm38/ Sr57	*wMAS000003, wMAS000004*	STS, SNP	Parula, Thatcher, Glenlea, Jupateco R, Opata, Bezostaya, Chinese Spring.	7DS	[29]
Lr46/ Yr29/ Pm39/ Sr58	*csLV46, csLV46G22*	CAPS	Pavon 76, Parula,	1BL	[30]
Lr67/ Yr46/ Pm46/ Sr55	*csSNP856*	SNP	RL6077	4DL	[31]
Sr2/Yr30	*csSr2, wMAS000005*	CAPS	Pavon76	3BS	[32]
Lr68	*cs7BLNLRR*	CAPS	Parula	7BL	[33]

encodes a full-size ATP-binding cassette (ABC) transporter [28] and gene-specific markers were developed which are widely used in marker assisted selection.

$Lr46$ was first described in 1998 in cultivar "Pavon 76", located on chromosome 1BL characterized by lower latency period [34], confers partial resistance to other diseases with corresponding designations $Yr29$, $Sr58$ and $Pm39$, respectively. $Lr46$ is also associated with LTN and is very common in both old and new wheat varieties including durum wheat.

The $Lr67$ gene was identified in the common wheat accession "PI250413" and transferred into "Thatcher" to produce the isoline "RL6077" (Thatcher*6/PI250413). $Lr67$ also shows pleotropic effect to SR and YR however, with *lower effect of* LR resistance than $Lr34$. Mapping studies mapped $Lr67/Yr46/Pm46$ on chromosome arm 4DL. Cloning elucidated that $Lr67$ gene encodes a hexose transporter [35].

$Lr68$ is another APR gene located on chromosome arm 7BL, conferring APR to LR identified in CIMMYT's wheat "Parula", known to carry $Lr34$ and $Lr46$ *and* likely to have originated from "Frontana" [33]. $Lr68$ showed a weaker effect than $Lr34$, $Lr46$ *and* $Lr67$ but combined effect of $Lr34$, $Lr46$ and $Lr68$ in Parula resulted in near immunity [33].

Stem rust gene $Sr2$ is one of the most important and widely used genes, conferring modest levels of resistance, and has been effective until date (over 100 years) even to the Ug99 and Digalu race groups of SR in East Africa. This gene was transferred from "Hope" and "H-44" into common cultivars and is derived from a tetraploid "Yaroslav" emmer and is located on chromosome 3BS. This gene was widely used by Dr. N. E. Borlaug when he initiated wheat breeding in 1944 in Mexico, which resulted in varieties such as "Yaqui 50" and several high yielding semi dwarf varieties that were deployed in different wheat programs [24]. The $Sr2$ gene shows pleiotropic effects with YR gene $Yr30$ that also confers moderate resistance. $Sr2$ gene is also associated with a morphological marker called pseudo-black chaff (PBC). Efforts to combine $Sr2$ with other minor effect genes to enhance SR resistance in breeding materials at CIMMYT has resulted in several resistant or moderately resistant varieties and recently lines combining $Sr2$ and Fhb1 have been developed [36]. Several new uncharacterized slow rusting genes, some potentially pleiotropic, have been identified in the recent years suggesting diversity for APR QTL and their potential in breeding.

Other adult plant resistance genes reported to confer partial or slow rusting include $Lr74$ $Lr75$, $Lr77$, and $Lr78$ for leaf rust, $Yr11$, $Yr12$, $Yr13$, $Yr14$, $Yr16$, $Yr36$, $Yr39$, $Yr52$, $Yr59$, $Yr62$, $Yr68$, $Yr71$, $Yr75$, $Yr77$, $Yr78$, $Yr79$, $Yr80$ and $Yr82$ [37] for yellow rust and more recently $Sr56$ identified in cultivar 'Arina' for stem rust [38].

8.6   Enhancing Resistance Durability Through Breeding Success, Setbacks and Lessons Learnt

Breeding for rust resistance has been a rigorous exercise owing to the continued evolution and selection of pathogen for new virulence to previously effective resistance genes largely through mutation or sexual recombination, or transboundary migration of races to new wheat production environments. In most developing countries, varieties with genetic resistance are preferred by farmers; therefore resistance is a required trait for release. Even though several race-specific resistance genes have been identified only a handful of genes are used actively in breeding as several genes are only effective in certain environments and majority are easily overcome in few years of deployment. Linkage drag associated with undesired genes transferred from secondary and tertiary gene pools or originating from unadapted genetic backgrounds remains a major constraint even with modern techniques to shorten e.g. translocations. One of the best approaches to utilize these race-specific resistance genes is through pyramiding, combinations of multiple effective genes in varieties. Molecular markers linked to some of the effective resistance genes have facilitated the selection for multiple resistance genes and releases of varieties that carry them. However, the lack of diagnostic markers to select genes in different genetic backgrounds leaves no option but use field-based selections under artificial epidemics, which continues to be the most common practice in several breeding programs.

Other approach is to utilize quantitative APR in breeding, although the individual effects of pleiotropic APR genes and other QTLs are small or moderate in their effect when present alone; near-immune levels of resistance have been achieved by combining 4–5 of these genes that often have additive effects. Incorporating such type of resistance has been found to enhance durability and significant progress was made for LR resistance, and more recently for resistance to Ug99 race group and stripe rust resistance in CIMMYT germplasm using a single back cross selected bulk scheme. Although breeding for APR resistance is cumbersome initially, additive effect of multiple minor APR genes enables combinations of high disease resistance, which can be simultaneously selected together with high yields with appropriate agronomic traits and the frequency of these genes can be increased within the breeding germplasm. Comparison of grain yield performance of 697 EYT lines (Stage II) 2018–2019 derived from Mexico Shuttle and Mexico Kenya Shuttle breeding schemes identified similar frequency of lines that combine high yield potential and stem rust resistance (Fig. 8.3) and significant progress has been achieved in combining yield potential and rust resistance in CIMMYT breeding lines.

One of the prerequisites for enhancing APR is the absence of epistatic race-specific resistance gene interactions in breeding materials, which enables selection of transgressive segregants with high levels of resistance under high disease pressure. The progress in breeding APR to the Ug99 race group was facilitated by extending shuttle breeding scheme to Kenya and were able to demonstrate success in achieving high levels of complex APR to rusts at CIMMYT.

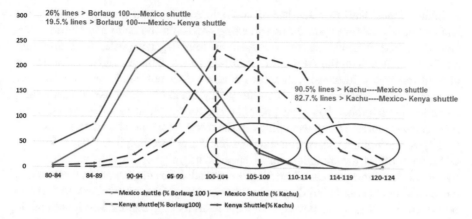

Fig. 8.3 Performance of grain yield of 697 EYT lines (Stage II) 2018–2019 derived from Mexico Shuttle and Mexico Kenya Shuttle breeding schemes

8.7 Integrating New Tools for Resistance Breeding Presents Opportunities for Wheat Improvement

The proven approach to enhance durability of genetic resistance is the deployment of combinations of multiple effective resistance genes. However, limitation to stack multiple genes is their segregation when parents possessing different genes are crossed and the need to grow large populations to identify multiple gene combinations and the need to have complementing diagnostic markers tagging the R-genes to ensure desired gene combinations are achieved. However, incomplete/moderate effect R-genes, race-nonspecific APR genes, or their combinations confers enhanced resistance levels due to additive effects, hence have been shown to be effectively selected in the field under high disease pressures [1, 24].

In the last two decades several rust resistance genes have been cloned viz. eleven SR resistance genes: *Sr13, Sr21, Sr22, Sr33, Sr35, Sr45, Sr46 Sr50, Sr55* (pleiotropic with *Lr67*), and *Sr57* (pleiotropic with *Lr34*) and more recently *Sr60* four LR resistance genes *Lr1, Lr10 Lr21*and *Lr22a* and six YR resistance genes *Yr5, Yr7, Yr10, Yr15, YrAS2388R* and *Yr36* (see Chap. 19). Also, in the last decade R gene enrichment sequencing (Ren-Seq), approaches have been widely used to clone resistance genes. Resistance genes from wild relatives can be introgressed to engineer broad-spectrum resistance in domesticated crop species using a combination of association genetics with R-gene enrichment sequencing (AgRenSeq) and a relatively new approach called MutRenSeq that combines chemical mutagenesis with exome capture and sequencing has been developed for rapid R-gene cloning [39]. Despite these advances, limited number of widely effective cloned genes and rapid evolution of new races with complex virulences and transcontinental migration reinforces a responsible strategy for their deployment.

The availability of multiple cloned resistance genes has opened the possibility of transforming wheat lines with a stack or cassette of multiple cloned effective resistance genes. This transgenic approach can help combine multiple resistance genes in a linkage block with one another on a single translocation thereby reducing the chances of segregation upon further breeding processes and up to five cloned genes can be stacked currently transgene cassette of four R-genes (*Sr22, Sr35, Sr45* and *Sr50*), combined with the APR gene *Sr55* [40] (https://2blades.org/). However, the current regulatory framework in most countries does not allow the cultivation of transgenic and cisgenic, wheat. If future policy decisions allow use of transgenic-cassettes this approach has great potential to develop wheat varieties with durable resistance. Genome editing technology in the recent years has shown great potential to surpass the bottlenecks of conventional resistance breeding (see Chap. 29). This technology offers the modification of specific target genes in elite varieties, thus bypassing the whole process of crossing. Recent advances in gene-editing technology can also offer avenues to building resistance durability. Genome editing was found to be effective in improving powdery mildew resistance by editing *Mlo* homologs in wheat to produce a triple knockout in hexaploid wheat [41]. As gene-editing technology develops, site-specific editing of alleles may become practical in the future.

8.8   Key Concepts

Geographical distribution of three rust diseases, impact, management strategies Rust resistance, Race specific genes, Adult plant resistance genes, breeding technologies and new molecular tools in the current era to enhance resistance durability.

8.9   Conclusions

The three rust diseases continue to be a significant challenge in several wheat production environments. Major threat is due to the extreme damage these diseases can cause to susceptible varieties. Severe localized epidemics have been reported in the last two decades largely due to the lack of resistance diversity in host and constantly evolving and migrating rust races that can pose a significant risk wherein breeding for these new incursions or newly evolved races could be a recent undertaking. Genetic resistance through deployment of both race specific genes and APR though quite widely used in breeding programs, effective combinations of partially effective pleotropic race- nonspecific genes such as *Sr2, Lr34, Lr46, Lr67*, and *Lr68* have been found to confer durable resistance in CIMMYT germplasm. Deployment of both APR genes with combinations of multiple race specific genes can be a better strategy to enhance resistance durability. Cloning of rust genes in the last decade and development of gene-specific DNA markers can facilitate pyramiding strategies into desired wheat backgrounds with a possibility to possible to transform wheat lines with a cassette of multiple cloned resistance genes. Significant progress in the area of global rust research including monitoring and surveillance, establishment of phenotyping platforms to facilitate the testing of global wheat germplasm and the identification and characterization of new sources of race-specific and APR genes, all of this has led to the development and rapid deployment of rust resistant cultivars target countries. CIMMYT breeding will continue to provide improved high-yielding wheat germplasm carrying high to adequate rust resistance to global wheat partners, mitigating the potential threat of these transboundary rust diseases. Even though fungicides are effective in controlling rusts and are widely used in developed countries, lack of availability at the right time and resources are still a limitation for small holder farmers and emerging concerns of new race groups requiring multiple applications (reduced fungicide efficacy) to avert losses highlights the importance of breeding and deployment of resistant germplasm to curtail epidemics.

Acknowledgements We greatly acknowledge the support of partnering institutions and financial support particularly from the DGGW Project funded by the Bill and Melinda Gates Foundation and the UK Department for International Development, USAID, USDA-ARS, AFFC-Canada, GRDC-Australia and all National partners of the International wheat improvement network.

References

1. Savary S, Willocquet L, Pethybridge SJ et al (2019) The global burden of pathogens and pests on major food crops. Nat Ecol Evol 3:430–439. https://doi.org/10.1038/s41559-018-0793-y
2. Huerta-Espino J, Singh R, Crespo-Herrera LA et al (2020) Adult plant slow rusting genes confer high levels of resistance to rusts in bread wheat cultivars from Mexico. Front Plant Sci 11:1. https://doi.org/10.3389/fpls.2020.00824
3. Kislev ME (1982) Stem rust of wheat 3300 years old found in Israel. Science 216:993–994. https://doi.org/10.1126/science.216.4549.993
4. Bhavani S, Hodson DP, Huerta-Espino J et al (2019) Progress in breeding for resistance to Ug99 and other races of the stem rust fungus in CIMMYT wheat germplasm. Front Agric Sci Eng 6:210–224. https://doi.org/10.15302/J-FASE-2019268
5. Singh RP (2006) Current status, likely migration and strategies to mitigate the threat to wheat production from race Ug99 (TTKS) of stem rust pathogen. CAB Rev Perspect Agri Vet Sci Nutr Nat Resour 1:177. https://doi.org/10.1079/PAVSNNR20061054
6. Wanyera R, Kinyua MG, Jin Y, Singh RP (2006) The spread of stem rust caused by Puccinia graminis f. sp. tritici, with virulence on Sr31 in wheat in Eastern Africa. Plant Dis 90:113–113. https://doi.org/10.1094/pd-90-0113a
7. Singh RP, Huerta-Espino J, Bhavani S et al (2011) Race non-specific resistance to rust diseases in CIMMYT spring wheats. Euphytica 179:175–186. https://doi.org/10.1007/s10681-010-0322-9
8. Newcomb M, Olivera Firpo PD, Rouse MN et al (2016) Kenyan isolates of Puccinia graminis f. sp. tritici from 2008 to 2014: virulence to SrTmp in the Ug99 race group and implications for breeding programs. Phytopathology 106:729–736. https://doi.org/10.1094/PHYTO-12-15-0337-R
9. Singh RP, Hodson DP, Huerta-Espino J et al (2008) Will stem rust destroy the world's wheat crop? Adv Agron 98:271–309. https://doi.org/10.1016/S0065-2113(08)00205-8
10. Olivera Firpo PD, Newcomb M, Szabo LJ et al (2015) Phenotypic and genotypic characterization of race TKTTF of Puccinia graminis f. sp. tritici that caused a wheat stem rust epidemic in southern Ethiopia in 2013–14. Phytopathology 105:917–928. https://doi.org/10.1094/PHYTO-11-14-0302-FI
11. Admassu B, Lind V, Friedt W, Ordon F (2009) Virulence analysis of puccinia graminis f. sp. tritici populations in Ethiopia with special consideration of Ug99. Plant Pathol 58:362–369. https://doi.org/10.1111/j.1365-3059.2008.01976.x

12. Olivera Firpo PD, Newcomb M, Flath K et al (2017) Characterization of *Puccinia graminis* f. sp. *tritici* isolates derived from an unusual wheat stem rust outbreak in Germany in 2013. Plant Pathol 66:1258–1266. https://doi.org/10.1111/ppa.12674

13. Jin Y, Singh RP (2006) Resistance in U.S. wheat to recent eastern African isolates of Puccinia graminis f. sp. tritici with virulence to resistance gene Sr31. Plant Dis 90:476–480. https://doi.org/10.1094/PD-90-0476

14. Patpour M, Justesen AF, Tecle AW et al (2020) First report of Race TTRTF of wheat stem rust (Puccinia graminis f. sp. tritici) in Eritrea. Plant Dis 104:973

15. Ali S, Gladieux P, Leconte M et al (2014) Origin, migration routes and worldwide population genetic structure of the wheat yellow rust pathogen Puccinia striiformis f.sp. tritici. PLoS Pathog 10:e1003903. https://doi.org/10.1371/journal.ppat.1003903

16. Beddow JM, Pardey PG, Chai Y et al (2015) Research investment implications of shifts in the global geography of wheat stripe rust. Nat Plants 1:15132. https://doi.org/10.1038/nplants.2015.132

17. Hovmøller MS, Walter S, Bayles RA et al (2016) Replacement of the European wheat yellow rust population by new races from the centre of diversity in the near-Himalayan region. Plant Pathol 65:402–411. https://doi.org/10.1111/ppa.12433

18. Ali S, Rodriguez-Algaba J, Thach T et al (2017) Yellow rust epidemics worldwide were caused by pathogen races from divergent genetic lineages. Front Plant Sci 8. https://doi.org/10.3389/fpls.2017.01057

19. Roelfs AP, Singh RP, Saari EE (1992) Rust diseases of wheat: concepts and methods of disease management. CIMMYT, Mexico

20. Kolmer JA, Acevedo MA (2016) Genetically divergent types of the Wheat Leaf Fungus *Puccinia triticina* in Ethiopia, a Center of Tetraploid Wheat Diversity. Phytopathology 106:380–385. https://doi.org/10.1094/PHYTO-10-15-0247-R

21. Kolmer JA, Mirza JI, Imtiaz M, Shah SJA (2017) Genetic differentiation of the Wheat Leaf Rust Fungus *Puccinia triticina* in Pakistan and genetic relationship to other worldwide populations. Phytopathology 107:786–790. https://doi.org/10.1094/PHYTO-10-16-0388-R

22. Ordoñez ME, Kolmer JA (2007) Virulence phenotypes of a worldwide collection of Puccinia triticina from durum wheat. Phytopathology 97:344–351. https://doi.org/10.1094/PHYTO-97-3-0344

23. Rutkoski JE, Poland JA, Singh RP et al (2014) Genomic selection for quantitative adult plant stem rust resistance in wheat. Plant Geno 7. https://doi.org/10.3835/plantgenome2014.02.0006

24. Singh RP, Hodson DP, Jin Y et al (2015) Emergence and spread of new races of Wheat Stem Rust Fungus: continued threat to food security and prospects of genetic control. Phytopathology 105:872–884. https://doi.org/10.1094/PHYTO-01-15-0030-FI

25. Singh RP, Herrera-Foessel S, Huerta-Espino J et al (2014) Progress towards genetics and breeding for minor genes based resistance to Ug99 and other rusts in CIMMYT high-yielding spring wheat. J Integr Agric 13:255–261

26. Radhakrishnan GV, Cook NM, Bueno-Sancho V et al (2019) MARPLE, a point-of-care, strain-level disease diagnostics and surveillance tool for complex fungal pathogens. https://doi.org/10.1186/s12915-019-0684-y

27. Johnson R, Law CN (1975) Genetic control of durable resistance to yellow rust (Puccinia striiformis) in the wheat cultivar Hybride de Bersée. Ann Appl Biol 81:385–391. https://doi.org/10.1111/j.1744-7348.1975.tb01654.x

28. Krattinger SG, Lagudah ES, Spielmeyer W et al (2009) A putative ABC transporter confers durable resistance to multiple fungal pathogens in wheat. Science 323:1360–1363. https://doi.org/10.1126/science.1166453

29. Lagudah ES, Krattinger SG, Herrera-Foessel S et al (2009) Gene-specific markers for the wheat gene Lr34/Yr18/Pm38 which confers resistance to multiple fungal pathogens. Theor Appl Genet 119:889–898. https://doi.org/10.1007/s00122-009-1097-z

30. Kolmer JA, Lagudah ES, Lillemo M et al (2015) The *Lr46* gene conditions partial adult-plant resistance to stripe rust, stem rust, and powdery mildew in thatcher wheat. Crop Sci 55:2557–2565. https://doi.org/10.2135/cropsci2015.02.0082

31. Forrest K, Pujol V, Bulli P et al (2014) Development of a SNP marker assay for the Lr67 gene of wheat using a genotyping by sequencing approach. Mol Breed 34:2109–2118. https://doi.org/10.1007/s11032-014-0166-4

32. Mago R, Spielmeyer W, Lawrence GJ et al (2002) Identification and mapping of molecular markers linked to rust resistance genes located on chromosome 1RS of rye using wheat-rye translocation lines. Theor Appl Genet 104:1317–1324. https://doi.org/10.1007/s00122-002-0879-3

33. Herrera-Foessel SA, Singh RP, Huerta-Espino J et al (2012) Lr68: a new gene conferring slow rusting resistance to leaf rust in wheat. Theor Appl Genet 124:1475–1486. https://doi.org/10.1007/s00122-012-1802-1

34. Martínez F, Niks RE, Singh RP, Rubiales D (2001) Characterization of Lr46, a gene conferring partial resistance to wheat leaf rust. In: Hereditas. Blackwell Publishing Ltd., pp 111–114

35. Moore JW, Herrera-Foessel S, Lan C et al (2015) A recently evolved hexose transporter variant confers resistance to multiple pathogens in wheat. Nat Genet 47:1494–1498. https://doi.org/10.1038/ng.3439

36. He X, Brar GS, Bonnett D et al (2020) Disease resistance evaluation of elite cimmyt wheat lines containing the coupled Fhb1 and Sr2 genes. Plant Dis 104:2369–2376. https://doi.org/10.1094/PDIS-02-20-0369-RE

37. Pakeerathan K, Bariana H, Qureshi N et al (2019) Identification of a new source of stripe rust resistance Yr82 in wheat. Theor Appl Genet 132:3169–3176. https://doi.org/10.1007/s00122-019-03416-y

38. Bansal U, Bariana H, Wong D et al (2014) Molecular mapping of an adult plant stem rust resistance gene Sr56 in winter wheat cultivar Arina. Theor Appl Genet 127:1441–1448. https://doi.org/10.1007/s00122-014-2311-1

39. Steuernagel B, Witek K, Jones JDG, Wulff BBH (2017) MutRenSeq: a method for rapid cloning of plant disease resistance genes. In: Methods in molecular biology. Humana Press Inc., pp 215–229

40. Luo M, Xie L, Chakraborty S et al (2021) A five-transgene cassette confers broad-spectrum resistance to a fungal rust pathogen in wheat. Nat Biotechnol 39:561–566. https://doi.org/10.1038/s41587-020-00770-x

41. Wang Y, Cheng X, Shan Q et al (2014) Simultaneous editing of three homoeoalleles in hexaploid bread wheat confers heritable resistance to powdery mildew. Nat Biotechnol 32:947–951. https://doi.org/10.1038/nbt.2969

Chapter 9
Globally Important Non-rust Diseases of Wheat

Xinyao He, Navin C. Gahtyari, Chandan Roy, Abdelfattah A. Dababat, Gurcharn Singh Brar, and Pawan Kumar Singh

Abstract While the three rusts are the most predominant wheat diseases in the global scale, various other diseases dominate in different geographical regions. In this chapter, some major non-rust diseases of wheat with global and/or regional economic importance are addressed, including three spike diseases (Fusarium head blight, wheat blast, and Karnal bunt), four leaf spotting diseases (tan spot, Septoria nodorum blotch, spot blotch, and Septoria tritici blotch), and several root diseases.

Keywords Head blight diseases · Leaf spotting diseases · Root diseases

9.1 Learning Objectives

- To learn the major epidemic regions, causal agent(s), epidemiology, management, genetics, resistance breeding etc. of each disease.

X. He · P. K. Singh (✉)
International Maize and Wheat Improvement Center (CIMMYT), Texcoco, Mexico
e-mail: x.he@cgiar.org; pk.singh@cgiar.org

N. C. Gahtyari
ICAR-Vivekananda Parvatiya Krishi Anusandhan Sansthan, Almora, Uttarakhand, India
e-mail: navin.gahtyari@icar.gov.in

C. Roy
Department of Plant Breeding and Genetics, Bihar Agricultural University,
Bhagalpur, Bihar, India

A. A. Dababat
International Maize and Wheat Improvement Center (CIMMYT), Ankara, Turkey
e-mail: a.dababat@cgiar.org

G. S. Brar
Faculty of Land and Food Systems, The University of British Columbia,
Vancouver, BC, Canada
e-mail: gurcharn.brar@ubc.ca

© The Author(s) 2022
M. P. Reynolds, H.-J. Braun (eds.), *Wheat Improvement*,
https://doi.org/10.1007/978-3-030-90673-3_9

9.2 Introduction

Wheat production is challenged by a range of diseases, rusts and non-rusts, causing on average 10–28% of yield losses globally according to a recent estimation [1]. The diseases can cause infection on all parts of the wheat plant (Fig. 9.1) and are strongly influenced by environmental conditions and disease management strategies. In the Sects. 9.3, 9.4 and 9.5, several major wheat diseases are presented according to their infection sites, i.e., spike, leaf, and root, and the most important information of each disease is summarized.

Fig. 9.1 Disease symptoms for (1) Fusarium head blight, (2) wheat blast, (3) tan spot, (4) spot blotch, (5) Septoria tritici blotch, and (6) cereal cyst nematode

9.3   Spike Diseases

9.3.1   Fusarium Head Blight

Fusarium head blight (FHB) is one of the most devastating diseases of wheat globally, with major epidemic regions in North America, Europe, East Asia, and the Southern Cone of South America. Many species in the genus *Fusarium* cause FHB, but it is *F. graminearum* species complex that has global importance and has been found in all major epidemic regions. The disease is favoured by warm and humid environment around anthesis, leading to yield reduction and quality deterioration. More importantly, the disease produces a range of mycotoxins, particularly deoxynivalenol (DON, or vomitoxin), which are toxic to humans and animals, raising a serious concern to food and feed safety. In the USA, losses attributable to FHB in wheat and barley between 1993 and 2001 were estimated at $7.67 billion. In China, the epidemic has increased significantly in the last two decades, affecting on average 5.3 Mha and reached 9.9 Mha in the 2012 great epidemic [2]. Yield reductions can reach up to 70% in Europe and South America [3].

FHB resistance is a typical quantitative trait, conditioned by numerous genes of minor effects. Several types of resistance have been proposed, represented by resistance to initial infection (Type I), resistance to disease spread within spike tissues (Type II), resistance to toxin accumulation (Type III), resistance to kernel infection (Type IV), and resistance to yield loss (Type V) [3]. Numerous sources of resistance were reported in literature; but only a few have been successfully utilized in breeding programs, such as 'Sumai 3', 'Wuhan 1', 'Frontana' etc. [3]. FHB resistance genes/QTL (Quantitative trait loci) have been mapped on all the 21 wheat chromosomes, though, only seven QTL have been formally designated as Mendelized genes, of which only *Fhb1*, *Fhb2*, *Fhb4*, and *Fhb5* are from common wheat, whereas *Fhb3*, *Fhb6*, and *Fhb7* are from wild wheat relatives [4]. So far, only *Fhb1* and *Fhb7* have been cloned, and their functional markers have been developed for marker-assisted selection (MAS).

Generally, two breeding strategies for FHB resistance could be utilized, i.e., exploitation of native resistance and introduction of exotic resistance. There is no strong FHB resistance available in the current CIMMYT gene pool; though, some moderately resistant lines have been identified and a few QTL with major effects have been mapped. Among those lines are 'Shanghai3/Catbird', 'Mayoor', 'Soru#1', 'IAS20*5/H567.71' etc. Apart from a major QTL on 2DL, others are either of low frequencies or of minor effects, but higher level of resistance can still be achieved via accumulating those QTL in elite breeding lines, similar to rust resistance breeding [5]. The limitation of using native resistance is, however, a lack of QTL/gene with strong Type II resistance, which could be compensated via introduction of exotic FHB resistance genes, like *Fhb1* and *Fhb7*. The former is the most well-known FHB resistance gene and has been extensively utilized in China, USA, and Canada; however, its resistance allele is tightly linked with the susceptibility allele of the stem rust gene *Sr2*, limiting its application in the CIMMYT wheat breeding.

To address this problem, several recombinant lines with both *Fhb1* and *Sr2* were introduced from Australia and included in various crosses with elite CIMMYT breeding lines [6].

Since no immunity to FHB has been found in wheat and high level of FHB resistance is difficult to achieve, other disease management strategies are also important in wheat production regions where FHB is a limiting factor. Removal of crop residue and rotation with non-host crops are helpful in reducing inoculum concentration. It is well known that maize-wheat rotation greatly increases the risk of FHB and thus should be avoided; otherwise, integrated disease management including deep tillage, fungicide application, and growing FHB resistant or moderately resistant cultivars are recommended.

9.3.2   Wheat Blast

Wheat Blast (WB) caused by the ascomycetes fungus *Magnaporthe oryzae* pathotype *triticum* (MoT) is one of the devastating diseases in warm and humid growing region. It can infect all the aerial parts of wheat, but completely or partially bleached spike is the typical symptom. WB is a new disease and was initially identified in the Parana state of Brazil in 1985; afterwards, its rapid widespread to the neighbouring states in Brazil and other countries of South America raised serious concerns. The first WB outbreak outside South America was reported in Bangladesh in 2016, raising a major concern on wheat production in South Asia (SA), as nearly 17% of the wheat growing areas in SA are vulnerable to WB. More recently, occurrence of WB has been reported from Zambia which can be a major threat for wheat production and trade in Africa [7]. Under favourable temperatures of 25–30 °C and high humidity, the disease can cause high yield loss ranging from 10% to 100% depending upon the level of infection.

The long-distance spread of the pathogen occurs through infected commercial grains, followed by the air transmission; therefore, grain treatment (chemical or irradiation) can effectively manage the primary inoculum load. For field WB management, foliar fungicides' application such as demethylation inhibitors (DMI), quinone outside inhibitor (QoI), succinate dehydrogenase inhibitors (SDHI) are suggested to be used in combination/rotation so as to reduce the fungal resistance against the fungicides especially QoI [8]. Various agronomic practices *viz.* optimizing planting dates, weed management, crop rotation with non-hosts, and avoid excessive nitrogen application are reported to be effective in WB control. However, these should be used in combination with genetic resistance to achieve a better management.

Regarding host resistance, the 2NS/2AS translocation has been widely acknowledged as a stable and effective resistance source, although virulent isolates have emerged recently in South America. The translocation was introduced from *Ae. ventricosa* and has been widely utilized in wheat breeding due to rust resistance genes (*Yr17, Lr37, Sr38*), as well as resistance genes for nematodes (*Cre5, Rkn3*) and WB. The 2NS/2AS translocation is an excellent example for the potential from crossing with wild relatives of wheat, for more examples refer to Chaps. 16, 17 and 18. Most well-known WB resistant lines have the 2NS/2AS translocation, e.g. 'Milan' and 'Borlaug #100' in the CIMMYT germplasm, 'Sausal CIAT', 'CD 116', 'Caninde #1' in South America, 'BARI Gom 33' in Bangladesh, 'HD2967', and 'DBW189' in India [9]. Recent genetic studies involving diverse wheat germplasm identified only one stable QTL on 2NS/2AS, whereas the remaining QTL were of small effects and were detected in only some environments (Singh et al. unpublished data). This highlights the importance of identification of new WB resistance genes for breeding use, which could alleviate the selection pressure that is being applied to 2NS virulent isolates, to prolong the lifespan of 2NS varieties.

A few resistance genes have been reported to have major effects at seedling (leaf resistance) but not at adult-plant (spike resistance) stages, among which, *Rmg2, Rmg3, Rmg7, Rmg8,* and *RmgGR119* are effective against MoT, whereas *Rmg1, Rmg4, Rmg5, Rmg6,* and *RmgTd(t)* are effective against non-MoT species. It is important to mention that *Rmg2, Rmg3,* and *Rmg7* have been overcome by new MoT isolates, whereas *Rmg8* and *RmgGR119* exhibited effective resistance in greenhouse but need to be validated in large scale field trials [9].

Early WB resistance breeding in South America depended heavily on natural infection, which was sporadic and unpredictable, with great variation in disease pressure. As for countries being threat by WB but still do not have the disease (like India), or those have WB but do not have the screening capacity (like Zambia), the request for an international precision phenotyping platform (PPP) is very strong, where interested cooperators can evaluate their wheat lines for reaction to WB. In collaboration with its national partners, CIMMYT has established three WB PPPs, with one in Bangladesh (Jashore), and two in Bolivia (Quirusillas and Okinawa) to screen germplasm and advanced lines from across the globe. High quality phenotypic data have been produced from the three PPPs, which greatly facilitated the WB resistance breeding, germplasm screening, as well as genetic studies [9].

9.3.3 Karnal Bunt

Tilletia indica (syn. *Neovossia indica*) is a hemibiotrophic fungus which was first described to cause disease in the Indian city of Karnal, hence called 'Karnal bunt' (KB). Currently, the disease is distributed in parts of Asia (India, Nepal, Pakistan, Iraq, Iran, Afghanistan), Africa (South Africa), and the Americas (USA, Mexico, Brazil). Though the estimated yield losses in KB affected regions are minimal (below 1%), it is an important disease from international trade perspective, where many member countries of WTO have zero tolerance quarantine laws. KB significantly deteriorates the wheat quality in terms of reduced vitamins, amino acids, weakened dough, and loss in flour recovery, ultimately affecting the human consumption negatively [10].

The conducive conditions for disease development are high humidity with cool temperature (<20 °C) favoring teliospore germination. Infected spikes disperse teliospores that become inoculum for the next season, and the teliospores are reported to remain viable for up to five years in soil under natural conditions, indicating the spatial and temporal dispersal capability of the disease. Boot emergence to anthesis is the optimum stage for a germinated teliospore to infect, however, an infection can happen as late as at late dough stage [11]. Treating seed with Chlorothalonil or mixture of carboxin & thiram and foliar spray with propiconazole, triadimefon and carbendazim are the suggested chemical control measures. The natural populations of *T. indica* have high genetic diversity owing to the sexual recombination leading to high diversity for virulency of KB strains as well as diversity in the wheat genotypes for resistant/susceptible reaction against the disease.

In the early days of KB resistance breeding at CIMMYT, important genetic stocks used were 'Aldan/IAS58' from Brazil, 'Shanghai-7' from China, and native CIMMYT lines 'Roek//Maya/Nac', 'Star', 'Vee#7/Bow' and 'Weaver'. To date, screening programs have resulted in the identification of numerous resistant sources for bread wheat and durum wheat from various countries as reviewed in Bishnoi et al. [10]. Additional resistant sources have been identified in primary to tertiary gene pools of wheat including *T. urartu* (AA) and *Ae. tauschii* (DD). Durum and triticale are generally more resistant than bread wheat.

Genetic resistance against KB is governed by polygenes with quantitative inheritance, although gene-for-gene interaction may exist to some extent. Many genes with small additive effects acting in an additive and epistatic mode impart KB resistance. Stacking additive genes along with an eye for significant epistatic gene interactions can enhance levels of KB resistance. In QTL mapping studies, as expected, majority of the identified QTL had minor effects, and only a few major QTL have been identified on chromosomes 4B, 5B, and 6B, where the one on 4B associated with SSR marker $Xgwm538$ had the largest effect (R^2 of 25%). A GWAS study on 339 accessions from Afghanistan led to the identification of a consistent QTL on chromosome 2BL along with some other novel genomic regions [12].

9.4 Leaf Spotting Diseases

9.4.1 Tan Spot

Tan spot (TS) is caused by the necrotrophic fungus *Pyrenophora tritici-repentis* (Died.) Drechs. The disease frequently appears in the warm and humid growing regions of bread and durum wheat, especially in Canada, Australia, USA, and South Africa. Yield and quality losses are common under high disease pressure. Reduced or no-till approaches to prevent soil erosion and water management are important reasons for increased disease pressure and TS infections can therefore be a challenge in using conservation agriculture practices. Another major reason that corresponds with increased pathogen virulence is the acquisition of a host-selective toxin (HST) PtrToxA by *P. tritici-repentis* from *Stagonospora nodorum* via horizontal gene transfer, which overcame the resistance of most cultivars carrying *Tsn1* gene. So far, three HSTs have been identified from *P. tritici-repentis*, acting as pathogen virulence factors in the TS pathosystem. Based on type of lesion (chlorosis or necrosis) and HSTs produced, *P. tritici-repentis* is classified into eight races using six differential genotypes (Table 9.1).

Pyrenophora tritici-repentis is a necrotroph and follows inverse gene-for-gene relationship where recognition of host sensitivity gene by pathogen produced HST results in a compatible (susceptible) interaction. This is opposite to Flor's classical gene-for-gene model in biotrophic diseases such as mildews and rusts, where host resistance gene is recognized by pathogen avirulence (*Avr*) gene, leading to an incompatible (resistant) reaction. High level of resistance has been found in several wheat genotypes although immunity is not reported [13]. Host resistance in wheat against TS can be qualitative or quantitative and some of the most well-characterized genes are *Tsn1* (interacts with PtrToxA), *Tsc2* (interacts with PtrToxB), and *Tsc1* (interacts with PtrToxC). *Tsn1* is the only cloned TS resistance gene, which is located on chromosome 5BL and a dominate functional marker *Xfcp623* is used for MAS [14]. *Tsc1* is located on chromosome 1A and *Tsc2* on 2BS, for which flanking markers are available for MAS. In addition to these three major genes, a recent meta-QTL study identified 19 QTL/loci for resistance to TS which can be utilized in wheat breeding programs [15].

Resistance breakdown is a major concern in R-genes conferring resistance to biotrophic pathogens as the pathogen *Avr* genes mutate rapidly. In case of TS

Table 9.1 Reaction of eight characterized races of *Pyrenophora tritici-repentis* on bread and durum wheat differential lines. R and S indicates resistant and susceptible response, respectively

Race	Associated toxins	Reaction of differential genotypes					
		Glenlea	6B662	6B365	Salamouni	Coulter	4B1149
1	PtrToxA, PtrToxC	S (necrosis)	R	S (chlorosis)	R	S (necrosis)	R
2	PtrToxA	S (necrosis)	R	R	R	S (necrosis)	R
3	PtrToxC	R	R	S (chlorosis)	R	S (necrosis)	R
4	None	R	R	R	R	R	R
5	PtrToxB	R	S (chlorosis)	R	R	S (necrosis)	R
6	PtrToxB, PtrToxC	R	S (chlorosis)	S (chlorosis)	R	S (necrosis)	R
7	PtrToxA, PtrToxB	S (necrosis)	S (chlorosis)	R	R	S (necrosis)	R
8	PtrToxA, PtrToxB, PtrToxC	S (necrosis)	S (chlorosis)	S (chlorosis)	R	S (necrosis)	R

resistance, if sensitivity genes are knocked-out or mutated, the pathogen cannot evolve as rapidly as biotrophs, so the resistance is more durable. Additionally, the fungus is saprophytic in nature and selection pressure on the pathogen would not be as high as in mildews or rusts. Molecular markers associated with major loci conferring susceptibility or resistance are very useful to select for TS resistant cultivars. Stacking of multiple QTL (including race non-specific) for TS resistance is an important and desirable strategy to manage the disease [15].

9.4.2 Septoria Nodorum Blotch

Stagonospora nodorum, a filamentous ascomycetes fungus, causes wheat leaf and glume blotch and affects wheat yield and quality in the warm and humid areas particularly in Australia, USA, parts of Europe and southern Brazil. Short incubation period enables the pathogen for multiple infection cycles within a season. The fungus can reproduce through asexual conidia and frequent sexual reproduction due to availability of both mating types (MAT1-1 and MAT1-2) that makes sexual reproduction possible.

Stagonospora nodorum produces multiple HSTs, of which 15 have been identified so far. The HSTs (e.g., SnToxA) interact with the corresponding host sensitivity genes (e.g., *Tsn1*) in an 'inverse gene-for-gene' manner that causes infection in the host, just as in TS. So far, nine necrotrophic effector (NE) and sensitivity gene interactions *viz.* SnToxA-*Tsn1*, SnTox1-*Snn1*, SnTox2-*Snn2*, SnTox3-*Snn3-B1*, SnTox3-*Snn3-D1*, SnTox4-*Snn4*, SnTox5-*Snn5*, SnTox6-*Snn6*, and SnTox7-*Snn7* have been identified in wheat. Three important NE genes in the pathogen *viz. SnToxA, SnTox1, SnTox3* and one important host sensitivity gene in wheat *viz. Tsn1* have been cloned which has helped in the extensive study of three important interactions *viz.* SnToxA-*Tsn1*, SnTox1-*Snn1* and SnTox3-*Snn3-B1* for better understanding the molecular basis of Septoria nodorum blotch (SNB) [16]. *Tsn1* was identified on chromosome 5BL [14], whereas both *Snn1* and *Snn3-B1* were mapped on 5BS [17]. Negative selection of host sensitivity genes during the breeding program would accelerate the breeding progress of resistant varieties.

An integrated disease management strategy including cultural practices, fungicides application, and use of resistant varieties is most effective in managing SNB. Infected seed and straw serve as the primary source of inoculum; therefore, seed treatment, crop rotation, and residue management reduce the chances of an epidemic in the disease-prone areas. SNB infection causes the greatest yield losses at the adult plant stage, for which resistance screening should be emphasized [18]. Genetic analysis revealed both qualitative and quantitative nature of resistance; but the latter dominates in field resistance against SNB [16]. Quantitative resistance is reported to have low to moderate heritability, thus high selection intensity should be kept to obtain higher genetic gain for SNB resistance. QTL associated with SNB resistance have been identified on multiple wheat chromosomes [18], yet few have been utilized in breeding.

9.4.3 Spot Blotch

Spot blotch (SB) caused by *Bipolaris sorokiana* (telemorph *Cochliobolus sativus*) is a destructive disease of wheat in the warm and humid growing regions, especially South Asia, Latin America, and Southern Africa. The pathogen causes average yield loss of 15–20%; but yield loss of up to 87% has been detected on the susceptible varieties [19]. The pathogen can infect all parts of the wheat plant, but leaf infection is the most typical, where infection starts from the older leaves and then progresses upward towards the younger leaves. High temperature (18–32 °C) and humidity (>90%) favours the disease establishment.

Identification of resistance sources through screening of national and international germplasm stocks was initiated in early 1980s and initial success was accomplished by replacing most susceptible varieties with the resistant lines in Brazil. Several resistant lines such as Saar, M 3, Yangmai 6, BH 1146, Shanghai 4, Ning 8201 including synthetic derivatives like 'Chirya 1', 'Chirya 3', 'SYN1' were identified as potential donors. Leaf tip necrosis (*Ltn+*) is associated with moderate resistance to SB, allowing breeders to use it as a phenotypic marker during selection. No host immunity has been reported for SB, and genetic studies on field SB resistance revealed a quantitative nature of inheritance [20].

To date, four major QTL (*Sb1-Sb4*) conferring SB resistance have been mapped. *Sb1* was mapped on chromosome 7DS, co-located with the cloned leaf rust resistance gene *Lr34* having pleiotropic effects on yellow rust (*Yr18*), stem rust (*Sr57*), powdery mildew (*Pm38*) and leaf tip necrosis (*Ltn+*). *Sb2* was identified on chromosome 5BL, *Sb3* on 3BS, and *Sb4* on 4BL [21]. These QTL can be used to develop new varieties or transferred into popular susceptible varieties through marker-assisted back cross (MABC) programme. Apart from these four *Sb* genes, *Tsn1* on 5BL has been shown to have major effects against *B. sorokiniana* isolates with *ToxA* [22]. Such *ToxA+* isolates have been identified in the *B. sorokiniana* populations of Australia, USA, India, and Mexico [23], implying that removing *Tsn1* from popular wheat varieties enhances resistance not only to TS and SNB, but also to SB. Contribution of QTL with minor effects is also significant in reducing SB severity, and such QTL have been mapped on chromosomes 1A, 1B, 1D, 2B, 2D, 3A, 3B, 4A, 4B, 4D, 5A, 5B, 6A, 7A, 7D in bi-parental and GWAS mapping studies [19].

9.4.4 Septoria Tritici Blotch

Septoria tritici blotch (STB) is caused by the fungal species *Zymoseptoria tritici* (teleo. *Mycosphaerella graminicola*). The pathogen is heterothallic with two mating types that have frequent sexual reproduction, resulting in a high level of genetic variation and an accelerated evolution and diversification of the fungal pathogen.

This in turn leads to problems like break down of host resistance and fungal resistance to fungicide. Losses to STB can range from 30% to 50% during severe epidemics, but typically are much lower. Epidemics are most severe in areas with extended periods of cool and wet weather, particularly North America (USA, Canada, Mexico), East Africa (Ethiopia, Kenya), South America (Brazil, Chile, Uruguay, Argentina) and the most damage occurs in Europe and CWANA (Central and West Asia and North Africa) region [24].

Host resistance to STB can be both qualitative and quantitative, but there is no clear difference between them since the gene-for-gene interaction in the wheat-STB pathosystem does not confer complete resistance. So far, 22 resistance genes have been designated, of which 21 were identified from hexaploid wheat, i.e. *Stb1* through *Stb19*, *StbSm3* and *StbWW*, and only one gene, *TmStb1*, has been found in *T. monococcum* [25]. So far, *Stb6* and *Stb16q* are the only two STB resistance genes that have been cloned, and their respective functional markers have been developed for MAS [26]. A total of 89 genomic regions carrying QTL or meta-QTL have been identified on all but 5D chromosomes, as summarized by Brown et al. [25].

The breeding effort for STB resistance began in 1970s in CIMMYT, using resistance sources from Brazil, Russia, Argentina, and China [27]. Nowadays, CIMMYT materials, represented by the International Septoria Observation Nurseries (ISEPTON), exhibit very good STB resistance under Mexican environments due to the consistent selection against the local *Z. tritici* strains. However, their performance in other countries varies greatly, due to different *Z. tritici* populations, although promising lines can still be identified. A vivid example is the resistance of durum wheat, which is nearly immune in Mexico but becomes highly infected in North Africa. Recent genetic studies on the STB resistance mechanism for CIMMYT lines revealed a nature of quantitative inheritance, with multiple minor QTL and limited major QTL (Singh et al., unpublished). This minor gene-based resistance mode is preferred as it likely confers durable resistance, as evidenced in resistance to many wheat diseases represented by rusts [5].

Plant height (PH) and days to heading (DH) are often negatively associated with STB resistance/escape, i.e., tall and late lines tend to have low STB. The association between short stature and high STB infection was a major issue that hampered the promotion of semi-dwarf wheat varieties in STB affected areas, especially in North Africa where STB is a priority biotic constrain. Efforts have been made to break such association, which resulted in the identification of intermediate maturing, high yielding semi-dwarf lines with high STB resistance [27]. It is noteworthy that such association exists in many abovementioned wheat diseases, like FHB, SNB, SB and TS. Such association is contributed mostly by disease escape, although tight linkage between resistance QTL and PH/DH associated genes and pleiotropic effects of the latter genes could be involved.

9.5   Root Diseases

Soil borne pathogens (SBPs) include the *Heterodera* species, cereal cyst nematode (CCN), *Pratylenchus* species, root lesion nematode (RLN) and many additional fungal species. Among the later are Take-all (GGT, *Gaeumannomyces graminis var. tritici*), *Pythium* spp, *Rhizoctonia solani*, Crown rot (CR, *Fusarium* spp), and common root rot (CRR, *Bipolaris sorokiniana*) (Table 9.2). These pathogens are favoured by different soil, cropping system and climate [28], and are found wherever cereal-based farming systems dominate. SBPs attack the roots of cereal crops resulting in a high yield loss and reduced grain quality. The damage caused by these pathogens is more visible in fields where drought and monoculture practices dominate. Rain-fed wheat under sustainable agriculture production, especially those grown under arid and semi-arid conditions, is being impacted by climate change due to hotter and drier soils. Under the harsh climatic condition characterized by low precipitation and high temperature, yield losses can exceed 50%. However, the available reports regarding wheat grain yield losses do not accurately portray the magnitude of economic losses at the regional or national levels, since those reports have been mostly linked to research plots located in infested areas of fields i.e., sick plots [29]. Further complications arise from reports initially attributed to yield reduction by *H. avenae* that are now identified as *H. filipjevi*, *H. latipons*, *H. australis*, or *H. sturhani* [30].

The pathogens have a wide host range and can survive in the soil/organic residue for many years, therefore crop rotation plays a paramount role in reducing their damaging impact. Root rot symptoms are difficult to identify clearly but generally are characterized by discolouration of roots, coleoptiles and stem bases of the infected seedling. Root rot fungi also may attack the upper parts of plants which may result in foliage lesions, head and seedling blight (Table 9.2).

Take-all (*G. graminis*) is the dominant root disease favoured by the moist and cool conditions in winter season followed by the moisture stress during anthesis. Fungicide application and rotation with non-host crops are effective options to control the disease [28]. *Pythium* is a pathogen having a wide host range causing root rot and seedling damping off. *Pythium* infects root system via root tips and root hairs and can also penetrate the embryo of germinated seed, leading to symptoms like stunting and yellowing of leaf tissue. Infected roots are stunted, and light brown-yellow colouration is seen near the tips. *Rhizoctonia* can prune off the root and limit water and nutrient absorption which ultimately leads to crop damage. It survives in the top of the soil (0–10 cm) on organic matter [31]. *Fusarium* spp. especially *F. culmorum* and *F. pseudograminearum* cause root diseases, including foot rot, root rot, and crown rot. Crown rot encompasses symptoms on the lower part of the wheat plant, and diseased plants are characterized by fungal colonization on the wheat stems, crown and root tissues leading to a honey-brown discolouration of the leaf sheaths and lower stem, and necrosis of the crown region. *Bipolaris* spp. especially *B. sorokiniana* cause common root rot of wheat worldwide, which produces a brown to black discolouration of the subcrown internode.

Table 9.2 Basic characteristics of the root rot diseases

Disease/ causal agent	Causal agent	Symptoms	Hosts	Survival
Take-all (GGT)	*Gaeumannomyces graminis var. tritici*	Patches, blackening of roots, plant are easy to pull from the soil	Wheat, barley, rye, oat, grasses	Grass, stubble
Pythium root rot	*Pythium* spp.	Patches yellow to brown root system	Wheat, barley, triticale, oats, grasses	Resting spores
Rhizoctonia bare patch	*Rhizoctonia solani*	Stunting of plants, seedling rots, roots stunted with spear point	Wheat, barley, triticale, grasses	Plant residue, hyphal fragments
Crown Rot (CR)	*F. pseudograminearum, F. culmorum*	Scattered plants, browning of stem base, crown, white heads, pinched no grain, pink lower nodes	Wheat, barley, triticale, grasses	Volunteer grass, stubble residue
Common root rot (CRR)	*B. sorokiniana*	Patches Dark brown discolouration on subcrown internode	Cereals, grasses	Spores in soil, stubble residue
Cereal cyst nematode (CCN)	*H. avenae, H. filipjevi, H. latipons*	Patches, stunted yellow plants, multiple short, branched roots, cysts visible on roots in spring	Wheat, barley, oat, triticale, and grasses	Eggs, cysts
Root-lesion nematode (RLN)	*Pratylenchus* spp.	Patches, chlorosis of lower leaves, stunting, fewer tillers, and delayed plant growth	Wheat, grasses	Eggs, nematodes

Three major species belong to CCN, *viz. Heterodera avenae, H. latipons*, and *H. filipjevi*, and the first is the most widely distributed CCN around the globe. Wheat producing regions with temperate climatic conditions in Asia, Africa, North and South America, Europe and the Mediterranean are typically CCN occurrence zones [29]. The *Pratylenchus* species, especially *P. thornei, P. crenatus, P. neglectus* and *P. penetrans*, are widely distributed pathogens for RLN [32]. CCN is monocyclic as it completes only one cycle per season while RLN is polycyclic due to a higher multiplication rate of three to five generations per year. RLN causes stunted and poorly tillered plants. The badly damaged roots are thin and poorly branched with short and knotted laterals. Above ground CCN symptoms can be identified easily through patches and stunted plants. Below-ground symptoms are white females on roots (immature cyst) which can be seen with naked eyes in spring time (Fig. 9.1) [32].

Identifying which root rot pathogen is present in the field by classical and/or molecular tools is the most important point to tackle the disease (Table 9.2).

Managing these diseases in the modern farming system is a difficult task due to their hidden nature compared to leaf diseases. A variety of management strategies have been studied to control root rots [28]. Better understanding of the pathogen biology is the first step to apply the best management strategy for targeted root rot disease. Sowing healthy and high-quality seeds at the correct depth and sowing time with adequate levels of nitrogen are main agronomy practices. As these pathogens have a wide range of host crop, rotation with non-host crops may help to reduce inoculum level in the soil [31]. If there is a registered fungicide, its seed treatment may support stand establishment. 'Green bridge' must be broken off, since the volunteer plants or weeds helps the fungi/nematode to survive during offseason [28, 32].

Using resistant crops of high yielding potential combined with good agronomy is the most efficient and economical way to improve the productivity of the crop and manage root rot diseases, especially in dryland areas. Tolerant varieties are also effective in reducing the yield losses; however, they may conduce inoculum build-up/increase in the soil. Wheat and its wild relatives have been screened for resistance against SBPs, and several *Cre* genes (*Cre1* to *Cre9, CreX, CreY*) against CCN have been identified, which are reported to follow gene-for-gene hypothesis. International collaborative efforts, *viz.* distribution and utilization of CIMMYT's International root disease resistance nurseries in the respective national breeding programs, is important to achieve desired resistance in locally adapted wheat varieties [32]. Other current and future research will address the use of endophytic microorganisms and other cultural practices to the yield losses incurred by SBPs. There is currently insufficient breeding for resistance to SBPs due to a lack of expertise and recognition of SBPs as a factor limiting wheat production potential, inappropriate breeding strategies, slow screening processes, and increased research funding is required for a more holistic approach to plant health management [30]. In conclusion, nematologists, breeders and agronomists need to draw a good strategy and work together to find solution to the complex issues facing agricultural production and use multidisciplinary approaches to move forward in ensuring food security for all.

9.6 Key Concepts

Host resistance is widely acknowledged as an economic and environment-friendly approach to manage wheat diseases, for which quantitative resistance is preferred over qualitative resistance due to the long-term durability of the former. For diseases where host resistance is less effective, alternative management tools like fungicide application and cultural practices should be utilized to obtain a satisfactory disease control.

9.7   Conclusions

For all wheat diseases, varietal resistance is an indispensable component in disease management, because it is cost-effective, environmentally friendly, and compatible with other management strategies, which is especially valuable to resource-poor farmers in developing countries who often have no access to fungicides. For developed countries, the increasing demand on organic production and the stricter regulation on fungicide application also call for varietal resistance. Therefore, host resistance becomes the focus of CIMMYT's breeding work. Quantitative loci should be preferred over qualitative genes in breeding to prolong the life span of the released resistant varieties, and when disease pressure is high, other management tools especially fungicide and agronomic management (rotation, plant density and sowing time etc.) should be combined with varietal resistance to obtain a reasonable control of the diseases. Wheat relatives have made great contribution to resistance against various diseases mentioned in this chapter, e.g., the 2NS/2AS translocation for resistance to WB, *Fhb7* for FHB, *Stb16q* for STB, etc. More efforts are needed to exploit and identify novel resistance genes from such materials, and some additional relevant information is available in Chaps. 16, 17 and 18.

References

1. Savary S, Willocquet L, Pethybridge SJ, Esker P, McRoberts N, Nelson A (2019) The global burden of pathogens and pests on major food crops. Nat Ecol Evol 3:430–439. https://doi.org/10.1038/s41559-018-0793-y
2. Zhu Z, Hao Y, Mergoum M, Bai G, Humphreys G, Cloutier S, Xia X, He Z (2019) Breeding wheat for resistance to Fusarium head blight in the Global North: China, USA, and Canada. Crop J 7:730–738
3. Buerstmayr H, Adam G, Lemmens M (2012) Resistance to head blight caused by Fusarium spp. in wheat. In: Sharma I (ed) Disease resistance in wheat. CABI, Wallingford/Cambridge, MA, pp 236–276
4. Bai GH, Su ZQ, Cai J (2018) Wheat resistance to Fusarium head blight. Can J Plant Pathol 40:336–346
5. Singh RP, Singh PK, Rutkoski J, Hodson DP, He X, Jørgensen LN, Hovmøller MS, Huerta-Espino J (2016) Disease impact on wheat yield potential and prospects of genetic control. Annu Rev Phytopathol 54:303–322. https://doi.org/10.1146/annurev-phyto-080615-095835
6. He X, Brar GS, Bonnett D, Dreisigacker S, Hyles J, Spielmeyer W, Bhavani S, Singh RP, Singh PK (2020) Disease resistance evaluation of elite CIMMYT wheat lines containing the coupled Fhb1 and Sr2 genes. Plant Dis 104:2369–2376
7. Tembo B, Mulenga RM, Sichilima S, M'Siska KK, Mwale M, Chikoti PC, Singh PK, He X, Pedley KF, Peterson GL, Singh RP, Braun HJ (2020) Detection and characterization of fungus (*Magnaporthe oryzae* pathotype *Triticum*) causing wheat blast disease on rain-fed grown wheat (Triticum aestivum L.) in Zambia. PLoS One 15:e0238724
8. Cruz CD, Valent B (2017) Wheat blast disease: danger on the move. Trop Plant Pathol 42:210–222
9. He X, Gupta V, Bainsla NK, Chawade A, Singh PK (2020) Breeding for wheat blast resistance. In: Kashyap PL, Singh GP, Kumar S (eds) Wheat blast. CRC Press, Boca Raton, pp 163–174

10. Bishnoi SK, He X, Phuke RM, Kashyap PL, Alakonya A, Chhokar V, Singh RP, Singh PK (2020) Karnal bunt: a re-emerging old foe of wheat. Front Plant Sci 11:7
11. Carris LM, Castlebury LA, Goates BJ (2006) Nonsystemic bunt fungi – Tilletia indica and T. horrida: a review of history, systematics, and biology. Annu Rev Phytopathol 44:113–133
12. Gupta V, He X, Kumar N, Fuentes-Davila G, Sharma RK, Dreisigacker S, Juliana P, Ataei N, Singh PK (2019) Genome wide association study of Karnal bunt resistance in a wheat germplasm collection from Afghanistan. Int J Mol Sci 20:3124
13. Singh PK, Duveiller E, Singh RP (2012) Resistance breeding for tan spot (Pyrenophora tritici-repentis) of wheat. In: Sharma I (ed) Disease resistance in wheat. CABI, Wallingford, pp 136–150
14. Faris JD, Zhang Z, Lu H, Lu S, Reddy L, Cloutier S, Fellers JP, Meinhardt SW, Rasmussen JB, Xu SS, Oliver RP, Simons KJ, Friesen TL (2010) A unique wheat disease resistance-like gene governs effector-triggered susceptibility to necrotrophic pathogens. Proc Natl Acad Sci U S A 107:13544–13549
15. Liu Y, Salsman E, Wang R, Galagedara N, Zhang Q, Fiedler JD, Liu Z, Xu S, Faris JD, Li X (2020) Meta-QTL analysis of tan spot resistance in wheat. Theor Appl Genet 133:2363–2375
16. Ruud AK, Windju S, Belova T, Friesen TL, Lillemo M (2017) Mapping of SnTox3–Snn3 as a major determinant of field susceptibility to Septoria nodorum leaf blotch in the SHA3/CBRD × Naxos population. Theor Appl Genet 130:1361–1374
17. Shi G, Zhang Z, Friesen TL, Raats D, Fahima T, Brueggeman RS, Lu S, Trick HN, Liu Z, Chao W (2016) The hijacking of a receptor kinase-driven pathway by a wheat fungal pathogen leads to disease. Sci Adv 2:e1600822
18. Francki MG (2013) Improving Stagonospora nodorum resistance in wheat: a review. Crop Sci 53:355–365
19. Gupta PK, Chand R, Vasistha NK, Pandey SP, Kumar U, Mishra VK, Joshi AK (2018) Spot blotch disease of wheat: the current status of research on genetics and breeding. Plant Pathol 67:508–531
20. Duveiller EM, Sharma RC (2009) Genetic improvement and crop management strategies to minimize yield losses in warm non-traditional wheat growing areas due to spot blotch pathogen Cochliobolus sativus. Phytopathol 157:521–534
21. Zhang P, Guo G, Wu Q, Chen Y, Xie J, Lu P, Li B, Dong L, Li M, Wang R, Yuan C, Zhang H, Zhu K, Li W, Liu Z (2020) Identification and fine mapping of spot blotch (Bipolaris sorokiniana) resistance gene Sb4 in wheat. Theor Appl Genet 133:2451–2459
22. Navathe S, Yadav PS, Chand R, Mishra VK, Vasistha NK, Meher PK, Joshi AK, Gupta PK (2020) ToxA Tsn1 interaction for spot blotch susceptibility in {I}ndian wheat: an example of inverse gene-for-gene relationship. Plant Dis 104:71–81
23. Wu L, He X, Lozano N, Zhang X, Singh PK (2021) ToxA, a significant virulence factor involved in wheat spot blotch disease, exists in the Mexican population of Bipolaris sorokiniana. Trop Plant Pathol 46:201–206
24. Goodwin SB (2012) Resistance in wheat to Septoria diseases caused by Mycosphaerella graminicola (Septoria tritici) and Phaeosphaeria (Stagonospora) nodorum. In: Sharma I (ed) Disease resistance in wheat. CABI, Wallingford/Cambridge, MA, pp 151–159
25. Brown J, Chartrain L, Lasserre-Zuber P, Saintenac C (2015) Genetics of resistance to Zymoseptoria tritici and applications to wheat breeding. Fungal Genet Biol 79:33–41
26. Saintenac C, Cambon F, Aouini L, Verstappen E, Smt G, Poucet T, Marande W, Berges H, Xu S, Jaouannet M, Favery B, Alassimone J, Sanchez-Vallet A, Faris J, Kema G, Robert O, Langin T (2021) A wheat cysteine-rich receptor-like kinase confers broad-spectrum resistance against Septoria tritici blotch. Nat Commun 12:433
27. Dubin H, Rajaram S (1996) Breeding disease-resistant wheats for tropical highlands and lowlands. Annu Rev Phytopathol 34:503–526
28. Cook RJ (2001) Management of wheat and barley root diseases in modern farming systems. Australas Plant Pathol 30:119–126

29. Smiley RW, Dababat AA, Iqbal S, Mgk J, Maafi ZT, Peng D, Subbotin SA, Waeyenberge L (2017) Cereal cyst nematodes: a complex and destructive group of heterodera species. Plant Dis 101:1692–1720
30. Dababat AA, Fourie H (2018) Nematode parasites of cereals. In: Coyne D, Hallmann J, Timper P, Sikora RA (eds) Plant parasitic nematodes in subtropical and tropical agriculture. CAB International, Boston, pp 163–221. https://doi.org/10.1079/9781786391247.0163
31. Cook RJ, Schillinger WF, Christensen NW (2002) Rhizoctonia root rot and take-all of wheat in diverse direct-seed spring cropping systems. Can J Plant Pathol 24:349–358
32. Dababat AA, Imren M, Erginbas-Orakci G, Ashrafi S, Yavuzaslanoglu E, Toktay H, Pariyar SR, Elekcioglu HI, Morgounov A, Mekete T (2015) The importance and management strategies of cereal cyst nematodes, Heterodera spp in Turkey. Euphytica 202:173–188

Chapter 10
Abiotic Stresses

Richard M. Trethowan

Abstract Abiotic stresses, such as drought and high temperature, significantly limit wheat yield globally and the intensity and frequency of these stresses are projected to increase in most wheat growing areas. Wheat breeders have incrementally improved the tolerance of cultivars to these stresses through empirical selection in the environment, however new phenotyping and genetic technologies and strategies can significantly improve rates of genetic gain. The integration of new tools and knowledge in the plant breeding process, including better breeding targets, improved choice of genetic diversity, more efficient phenotyping methods and strategy and optimized integration of genetic technologies in the context of several commonly used wheat breeding strategies is discussed. New knowledge and tools that improve the efficiency and speed of wheat improvement can be integrated within the scaffold of most wheat breeding strategies without significant increase in cost.

Keywords Drought · High-temperatures · Wheat breeding · Physiological traits · Physiological breeding

10.1 Learning Objectives

- Drought and heat stress are common constraints across most wheat growing regions.
- New phenotyping and genetic technologies and knowledge can be efficiently integrated in current wheat breeding strategies.
- Physiological trait breeding is effective in improving wheat adaptation to stress.
- Accurate breeding targets, relevant genetic diversity, efficient population screening methods and innovative whole wheat breeding program strategies are essential for sustained success.

R. M. Trethowan (✉)
Plant Breeding Institute, School of Life and Environmental Sciences, The University of Sydney, Sydney, Australia
e-mail: richard.trethowan@sydney.edu.au

M. P. Reynolds, H.-J. Braun (eds.), *Wheat Improvement*,
https://doi.org/10.1007/978-3-030-90673-3_10

10.2 Introduction

Abiotic stresses significantly limit wheat production globally and the extent and intensity of yield losses are increasing with climate change. Rainfall is declining and the distribution changing in many environments and the impacts will be more acute in rainfed production systems. Current yield losses in wheat are primarily a consequence of abiotic rather than biotic factors [1]; this was not always the case, but a consequence of the steady improvement of disease resistance over the past 100 years [2]. However, wheat breeders have also incrementally improved crop adaptation to stress. This was largely achieved by targeted use of diversity and extensive testing in the environment under prevailing stresses. Thus, empirical selection has improved adaptation to abiotic stresses across the world's wheat growing areas, despite the genetic complexity and low heritability of these traits compared to disease resistances.

Climate modeling indicates that instability will increase in the major wheat producing areas of the world [3]. However, some regions will suffer more than others, including Australia, North Africa and large parts of North and South America. Expected losses in wheat production due to drought and heat stress, exacerbated by climate change, for key wheat growing regions are outlined below.

10.2.1 Australia

Australian wheat productivity will be limited by climate change. The Agricultural Production Systems Modulator (APSIM) was used to estimate changes in wheat productivity and response to high temperature for the period 1985–2017 [4]. The production environment had become more variable over the period, and heat stress was found to reduce grain weight more than grain number. Of the yield losses estimated, 26% were associated with heat and the remainder with drought stress. Wheat breeders need to target both stresses as a priority.

10.2.2 North America

The impact of climate change on North America will be mixed. Climatic changes between 1981 and 2015 have led to higher rainfall and longer growth periods and this was positively associated with grain yield [5]. New winter wheat cultivars with higher yield potential and improved disease resistance are required to meet this shift. However, spring wheat was subjected to increased temperature stress in the critical June period, thus requiring some heat tolerance at anthesis. Wheat breeders need to target higher yield potential and improved heat tolerance at anthesis.

10.2.3   Europe

Increasing temperatures are projected to reduce wheat yield in Europe. Semenov and Shewry [6] simulated various climate scenarios and predicted that high temperatures, particularly at flowering, would limit wheat yield more than drought. They reasoned that lower summer rainfall would be offset by earlier maturation thus crops would escape the impact of drought. They concluded that wheat breeders should target the improvement of heat tolerance at anthesis as a priority.

10.2.4   Russia and Ukraine

Like North America, the impacts of climate change will be mixed. The most productive zones of Russia are likely to experience yield losses from reduced precipitation and heat waves during vegetative development [7]. However, milder and drier winters and warmer spring periods in northern production zones are likely to see increases in productivity. In Ukraine, modeling suggests moderate climate change will have little impact on wheat yield. Nevertheless, under high emissions scenarios and higher levels of warming, yield is expected to decrease by more than 11% [8]. Wheat breeders should target improved heat tolerance at all stages of development.

10.2.5   India

Climate change and increasing temperature will and have already reduced wheat yield in India [9]. Wheat yield is estimated to be 13% higher than it would have been without irrigation trends since 1970 [10]. Irrigation dampens the effect of high temperature and irrigated wheat has just 25% of the sensitivity of rainfed wheat. However, yield gains have slowed due to warming. These authors found that irrigation will have little impact on future warming as opportunity to expand the system is limited. Wheat breeders need to target both high temperature tolerance and better water use efficiency as a priority.

10.2.6   China

Climate change will limit the productivity of wheat in China. Under the most severe climate change scenarios, wheat yield in China is projected to decline by 9.4% by 2050, which represents the largest yield reduction of all Chinese crops [11]. This 50-year study of Chinese climate data concluded that terminal heat stress was more severe in cooler regions. They concluded that the vegetative period had changed little in these cooler areas, but temperatures post heading had increased significantly thus reducing yield. Development of cultivars with improved terminal heat tolerance should be a priority as much of the wheat production in these regions is irrigated, thus negating the impacts of drought stress.

While climate change has already impacted wheat production in many environments, empirical selection has mitigated the impact of climate change on yield. Thus, rates of genetic gain have plateaued, rather than declined, in many regions. However, rates of genetic gain are not constant over time and fluctuate depending on access to new technologies, such as the introduction of dwarfing genes in the 1960s and 1970s, climatic changes or biophysical yield limitations which have limited recent gains in various regions [12]. However, wheat breeders have access to better technology than ever before, and it can be expected that optimized use of technology will further lessen the impacts of drought and high temperatures. Historically, we can already document changes in wheat morphology and physiology [13]. Yield improvements were associated with shorter vegetative and longer grain filling periods, more grain per unit area, shorter plant stature, wider leaves and higher harvest indices. Modern varieties tend to be earlier maturing, more N use efficient and translocate more assimilate to the developing grain. These changes have come about by coupling empirical selection in the target environment, with access to new diversity, improved understanding of physiological limitations and more recently, better understanding of the genetic control of traits. Nevertheless, optimal integration of technologies remains a significant challenge to wheat breeding and this is discussed further in Chaps. 5 and 6.

10.3   Breeding for Improved Adaptation to Water-Limited and Heat Stressed Environments

Wheat breeders have many tools available and technology has advanced rapidly in recent years. Molecular markers for high value traits are routinely used in most programs, genomic selection forms part of many strategies, proximal and remote sensing have extended beyond the physiologist's experiments and is routinely used by some programs and many are considering ways to exploit gene editing effectively. However, no technology ensures high value varieties are delivered to farmers and strategy and the choices that breeders make are vital to success. Wheat breeding is needs driven. Breeding targets must be well defined and relevant to both producers and marketers. The breeder's choice of technology will reflect the available diversity, heritability of phenotypic screens and availability of markers for high value traits and other genomic strategies that improve rates of genetic gain.

This section will follow the breeder's decision-making process in the context of improving rates of genetic gain for heat and drought tolerance.

10.3.1   Relevant Breeding Targets

Most farmers are forthcoming in describing varietal limitations to wheat breeders. In fact, the farmer's wish list can sometimes be extensive and bear little relationship to the available genetic diversity. Nevertheless, most breeders are aware of production constraints and wheat market requirements. Many production constraints can be solved agronomically and the influence of genetics is so limited that they should not be selection targets. The effectiveness of rotation in controlling take-all (*Gaeumannomyces graminis* var. *tritici*) is one such example. Other traits, such as crown rot resistance in wheat, are managed by the interaction of genetics with management practices, like non-host rotation and interrow sowing [14] (see Chap. 9). Other traits with high heritability, such as rust resistance (see Chap. 8), are clear targets for genetic selection. However, the picture is less clear regarding abiotic stress tolerances as a combination of optimized management and targeted traits for specific environments is almost always the goal. Thus, a genetic ideotype that assumes optimized agronomic management for specific environmental conditions, and reflects the most probable or frequent environment type, would help the breeding process [15]. Definition of wheat breeding target environments is discussed in more detail in Chap. 3. Ideotypes for drought stress [16] and heat stress [17] have been developed. These ideotypes are general in nature and the traits identified may not be effective in all environments. For example, under Australian conditions, soluble stem carbohydrates improve drought response in northern Australia but not in southern areas [18]. Knowledge of how these traits interact with the environment is crucial. To explore this further, a national, field-based managed environment facility (MEF) was established at three locations representing the key wheat growing regions of Australia and the effectiveness of traits assessed [19]. The network, where all confounding effects, such as soil heterogeneity and moisture, were minimized, was effective in assigning trait values by region. Similarly, if conservation agriculture is used by farmers to reduce the loss of soil moisture, then more vigorous genotypes that emerge from depth and carry resistance to stubble borne pathogens would be required. Thus, the ideotype must reflect the most frequent genotype x management practice x environment interaction. Building this ideotype will then depend on available genetic diversity; there is no point in including traits for which no diversity exists, and trait heritability, which reflects the accuracy of phenotypic screens and/or availability of molecular markers and their degree of linkage. An example is an ideotype constructed for the specific conditions of northwestern NSW is reproduced in Fig. 10.1 [15].

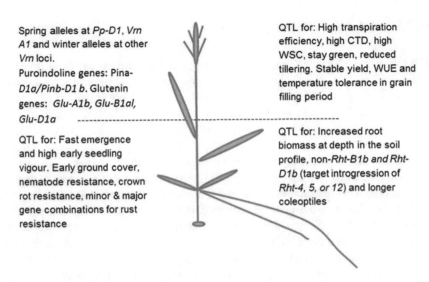

Spring alleles at *Pp-D1*, *Vrn A1* and winter alleles at other *Vrn* loci.
Puroindoline genes: Pina-*D1a/Pinb-D1 b*. Glutenin genes: *Glu-A1b*, *Glu-B1al*, *Glu-D1a*

QTL for: Fast emergence and high early seedling vigour. Early ground cover, nematode resistance, crown rot resistance, minor & major gene combinations for rust resistance

QTL for: High transpiration efficiency, high CTD, high WSC, stay green, reduced tillering. Stable yield, WUE and temperature tolerance in grain filling period

QTL for: Increased root biomass at depth in the soil profile, non-*Rht-B1b* and *Rht-D1b* (target introgression of *Rht-4, 5, or 12*) and longer coleoptiles

Fig. 10.1 An ideotype for drought stress tolerance in northwestern NSW, Australia. (Reprinted with permission from Ref. [15])

10.3.2   Meaningful Genetic Diversity

Genetic diversity should always be assessed and accessed from the adapted or primary wheat gene pool first. Species in the primary gene pool have completely homologous genomes with common wheat (AA, DD, AABB or AABBDD). Within the primary gene pool, the pathway to market is much shorter if the diversity is available in already adapted materials (AABB, AABBDD). The decision to access such diversity is a function of crossability with adapted wheat, either hexaploid or tetraploid, and the value of the trait. Much has been written about the value and use of synthetic wheat to improve the stress tolerance of wheat [20]. Primary synthetics, generated by crossing tetraploid wheat (such as *Triticum turgidum* cv *durum, T. dicoccum* or *T. dicoccoides*) with *Aegilops tauschii*, the donor of the D genome with subsequent embryo rescue and chromosome doubling, have been crossed to adapted wheat and cultivars released to farmers [21]. The D genome contributed by *Ae. tauschii* is more diverse than that in common wheat and this diversity has provided new alleles linked to stress adaptation. Wheat A and B genome diversity can also be introduced through direct crossing of wild tetraploid and adapted hexaploid wheat [17]. The wild tetraploid diversity, once introduced to hexaploid wheat, has been linked to improved drought [22] and high temperature [17] adaptation.

However, sometimes the diversity required for high value traits is not available in the primary gene pool. Only then is an exploration of the secondary gene pool; those materials with partial homology to the common wheat genome, warranted. Such materials include species such as *T. timopheevi* (AAGG) and *Aegilops speltoides* (BB) with one genome common to hexaploid wheat. One such example is the translocation of a segment of *Aegilops speltoides* in wheat linked to a more profuse root system and enhanced drought tolerance [23]. As a last resort, diversity for high value traits can be sourced from the tertiary gene pool where no homology with the common wheat genome exists. Historically, this diversity took a long time to introduce and was often associated with yield penalties caused by linkage drag. Examples include rye (RR) and *Thinopyrum elongatum* (EE). However, new genomic tools have made it easier to target and exploit tertiary diversity in wheat, and it is expected that the tertiary gene pool will be increasingly exploited to improve both the biotic and abiotic stress tolerance of tetraploid and hexaploid wheat. Chapters 16, 17 and 18 detail the conservation, characterization and use of genetic resources.

10.3.3   To Phenotype or Not?

Phenotyping is expensive and often comprises the greatest cost in any breeding program. Historically, parents are selected and crossed based on genetic and phenotypic information and availability of high value traits with high heritability that are amenable to high throughput screening. These have traditionally included traits such as disease resistance, plant height and phenology. However, marker assisted selection has broadened the suite of traits assessed in the early generations in recent years and grain quality, disease, phenology and even some major QTL linked to abiotic stress response have been used to truncate populations. If robust and tightly linked markers for high value traits exist, then it is not necessary to phenotype beyond parents and their fixed line progeny, thus reducing costs. It is assumed of course that the phenotypes used to identify the marker-trait linkages are accurate, repeatable and relevant. The same applies to the calculation of genomic estimated breeding values (GEBVs); it is assumed that the training population size is optimized, the phenotype accurate and the relationship between the training population and the breeding materials relatively close.

Nevertheless, some high value traits are difficult to phenotype and the available genetic information insufficient to justify a genomic approach alone. These include drought and heat tolerance as most observed QTL are of small effect and the influence of environment on QTL expression significant. Remote and proximal sensing are becoming increasingly valuable sources of information on genotype physiological responses to stress. Thermal infrared sun-induced fluorescence combined with solar-reflective hyperspectral remote sensing are considered state-of-the-art applications for assessing plant stress responses but require satellite access. However, proximal sensing using unmanned aerial vehicles (UAVs) or ground based phenomobiles, is now widely used by some breeding programs to capture real time thermal and spectral reflectance data on large numbers of genotypes. These data can be collected over time and responses with the highest heritability used to drive phenotypic selection and inform genomic prediction models. Kyratzis et al. [24] used an UAV to assess drought stress response in durum wheat and concluded that green NDVI (Normalized Difference Vegetation Index) was effective in discriminating genotypes. Nevertheless, the greatest limitation for the plant breeder is often the data processing to produce plot means and standard deviations that can be used for timely selection. The challenges of high-throughput phenotyping are discussed in Chap. 27.

Regardless of the technology used to capture field-based data, the information will have little value if field screening is confounded by heterogeneity or the season is not representative of the most common environment type. Pot-based screening in glasshouses or phenomics facilities can be effective in controlling environmental fluctuations for traits with high heritability. However, when applied to stresses such as heat and drought, the results rarely correlate with field responses thus limiting the utility of such data to the plant breeder [25]. While the field environment is subject to uncontrollable variation, the impact of confounding factors such as soil heterogeneity or season rainfall can be minimized.

As mentioned earlier, one such example is the MEF in Australia [19]. Materials are sown in a carefully managed crop sequences designed to limit wheat root diseases through rotation with non-host alternative crops, and soil heterogeneity is carefully assessed before sowing using an EM38 to detect differences in soil moisture and texture. The most homogenous areas are then selected for drought evaluation using an irrigation treatment split. Rainfall and soil moisture are assessed so that an environment type can be estimated, and this informs genotype responses. Once high value materials are identified, they are subsequently evaluated using rainshelters to control seasonal moisture and confirm drought responses in the field.

A slightly different approach can be used to evaluate genotype response to high temperature [26]. Here dates of sowing are used to screen thousands of genotypes for high temperature response at anthesis and grain filling. However, abnormal biomass development from a truncated vegetative period in late sown materials could influence estimations of grain number and seed weight under stress. To counter this, materials selected from delayed sowing are subsequently sown at an optimal time and portable field-based heat chambers used to apply a heat shock at anthesis for several days (Fig. 10.2). Those materials that maintain seed number and weight are then selected for final confirmation under controlled greenhouse conditions. This three-tiered phenotyping system thus overcomes the lack of relationship between glasshouse and field screening by inverting the process to initially screen in the field, followed by increasing levels of phenotyping precision on smaller numbers of lines. A more detailed discussion of heat stress methods and traits can be found in Chap. 22.

10.3.4 Physiological Wheat Breeding

The term 'physiological breeding' was first coined by Reynolds et al. [27] and refers to crossing parents carrying complementary traits with subsequent progeny screening under stress. Parental materials are selected based on the suite of traits required in a breeding program, including yield potential, yield under stress, seed weight, grain quality and disease resistance. However, all materials are subsequently

Fig. 10.2 Heat chambers with attached air conditioning units deployed in the field at Narrabri, NSW, Australia

assessed for the physiological traits deemed effective in the target environment. For example, these might comprise the traits in the ideotype in Fig. 10.1, if the target environment is northwestern NSW. However, some traits are more easily assessed than others and their amenability for high-throughput field-based phenotyping will determine if they are assessed on parents only or used to truncate segregating materials during progeny selection. Trethowan [15] categorized many physiological traits into those associated with emergence and establishment, early growth, pre-flowering and post-flowering. Traits such as osmotic adjustment can be assessed in the pre and post-flowering periods; but is difficult and time consuming to measure. This trait would therefore only be assessed on parents and again on fixed line progeny expressing drought tolerance in multi-environment testing. In contrast, canopy temperature depression; a trait assessable at the same developmental stages, is easily measured and can be used to select segregating materials, either using handheld sensors or remote or proximal sensing.

The concept of physiological breeding has been successfully applied in wheat breeding [27]. Materials developed using physiological crossing for drought tolerance at the International Maize and Wheat Improvement Centre (CIMMYT), were subsequently deployed in south Asia and found to be tolerant to drought [28]. Lines developed by crossing complementary physiological traits had on average, higher yield, superior grain weights and cooler canopies. New drought tolerant wheat cultivars were subsequently released to farmers in Pakistan from materials developed at CIMMYT using physiological breeding, including Barani-2017 and Kohat-2017. Targeted physiological trait introgression was successfully used to develop the

Australian wheat cultivar Drysdale [29]. Here carbon isotope discrimination, which is negatively correlated with transpiration efficiency, was backcrossed into an elite background and the transpiration efficient cultivar, Drysdale, was released in southern Australia in 2002 and the cultivar Rees for northern regions the following year. The Drysdale and Rees examples show that a single targeted physiological trait can have tangible benefits in a dry environment. This is discussed further in Chap. 23.

Wheat breeding and the enhancement of farmer profitability is more than just targeting stress adaptive traits. Most farmers in marginal environments tend to make most of their income in the better years. Hence, physiological traits that do not limit yield potential and have a higher heritability than yield alone would have high value in wheat improvement. For example, Pozo et al. [30] found that chlorophyll content was positively associated with yield under optimal and drought conditions, whereas carbon isotope discrimination was associated under optimal conditions only. In their study water soluble stem carbohydrates assessed at anthesis were not associated with yield at any level of moisture. Such findings help tailor trait selection for given production conditions.

10.3.5 Integration of Genomic Technologies in a Broader Physiological Breeding Strategy

Molecular markers linked to physiological traits that are deemed effective in the target environment and do not limit yield in the better years, significantly improve the effectiveness and cost of physiological breeding. Many studies have reported QTL linked to physiological traits with varying degrees of accuracy [31]. Unfortunately, many physiological traits are difficult to measure and have relatively low heritability, including stomatal conductance and photosynthetic rate, making marker development difficult and offering marginal value to the plant breeder [32]. Many QTLs are also cross specific and their expression fades when transferred to different backgrounds. Chen et al. [33] found that morphological traits such as plant height and peduncle length had high heritability while most physiological traits, including photosynthetic and transpiration rates, intercellular CO_2 concentration and stomatal conductance were low. Nevertheless, they concluded that five traits, including yield per plant, plant height, peduncle length, spike length and transpiration rate explained more than 90% of the variation in genotype response to drought.

Genomic selection, once the realm of the animal breeder/geneticist, is now integrated into many wheat breeding programs. Nevertheless, success depends on the accuracy, depth and relevance of the training population phenotype as much as the relatedness to the breeding population. Genomic estimated breeding values can be calculated for stress response based on a weighted index of associated traits. If makers linked to specific physiological traits are known, then they can be targeted in progeny selection following genotyping so that both GEBV and known QTL or gene profiles are optimized. A more detailed analysis of genomic selection can be found in Chaps. 6 and 32.

10.4 Examples of Integrating Physiological Breeding in Wheat Improvement Programs

To examine the practicalities of integrating physiological trait breeding in the wheat breeding process, three examples that compass commonly adopted breeding methods are presented. These methods include modified pedigree, selected bulk and a genomic strategy (Fig. 10.3). Pedigree breeding was not considered as so few programs use a strict pedigree breeding scheme due to the significant resources required. However, before crossing begins, it is necessary to accurately define the target environment (see Chap. 3 for more detail), the most likely probability of stress occurrence based on historical evidence and the suite of traits to be targeted [15]. Northwestern NSW in Australia will be used as an example; however, the principle can be applied to any environment. Details of genomic selection, including the available models and their applications, are provided in Chap. 6 and the general principle only, in the context of a wider physiological breeding strategy, will be discussed.

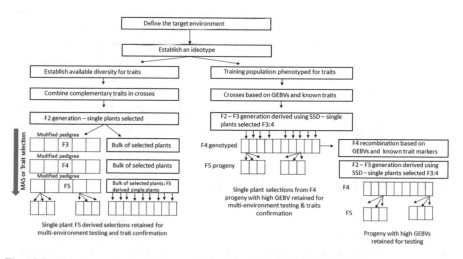

Fig. 10.3 Three strategies that integrate physiological breeding and selection

10.4.1 Defining the Environment in Northwestern NSW

The environment in northwestern NSW is characterized by summer dominant rainfall and extensive vertosol soils with high water holding capacity. The region lies between 26–30° latitude south. In season drought and heat stress are common, particularly from anthesis onwards [15]. Heat stress, defined as temperatures in excess of 35 °C for short periods of time, is common [26]. Stem, leaf and stripe rust, crown rot and root lesion nematode are major biotic constraints. The region produces high quality, high protein wheat that attracts a premium price. The optimum sowing time, based on simulation modeling, to minimize the risk of temperature extremes lies between 6 – 20th May [34].

10.4.2 Establishing an Ideotype for Northwestern NSW

Phenology is a primary driver of yield and matching phenology to the environment is critical to minimizing the impacts of stress. These responses are controlled by three loci each, PpdD1, Ppd2 and Ppd3 and VrnA1, VrnB1 and VrnD1 for photoperiod and vernalization responses, respectively. Daylength and vernalization insensitivity are controlled by dominant alleles at these loci and at this latitude, dominant alleles at PpD1 and VrnA1 with recessive alleles at the remaining vrn loci optimize the flowering window [15]. Plants should be semi-dwarf in stature to avoid lodging and if possible, height should be controlled by gibberellic acid sensitive dwarfing genes that do not significantly reduce coleoptile length and hence emergence and establishment [35]. Rapid early growth and ground cover will assist crop establishment in standing stubble as conservation agriculture is widely practiced in the region. In the region, stay-green, associated with deeper roots that extract soil moisture from depth, is an important character [36]. Genotypes with high water-soluble stem carbohydrates are also high yielding [18]. Genotypes with cooler canopies use soil moisture more effectively and continue to photosynthesize as temperature and moisture stress increases [16]. However, it is not clear whether high levels of transpiration efficiency will be beneficial given the trade off with high yield under less-limiting conditions in better years. Pollen fertility under heat stress and maintenance of grain weight under both heat and drought are important characteristics. Resistances to the rust diseases, root lesion nematodes and crown rot are required.

10.4.3 Breeding Method – Modified Pedigree

Once the ideotype has been developed and the trait selection determined, it is necessary to screen breeding materials, introductions and new diversity for the suite of traits required. Assessment may be phenotypic and/or genetic if tightly linked or perfect trait markers are available. At this stage, managed field environments, augmented by controlled environment testing can be used to establish trait profiles. Materials are then combined in backcrosses (when the source or non-recurrent parent is unadapted or carries deleterious characteristics), two or three-way crosses to combine physiological traits and other important diversity. A large F2 population is sown and single plants selected based on highly heritable and economically important traits such as rust resistance, plant height and maturity. If co-dominant physiological trait markers are available, they can also be used to drive F2 single plant selection.

Under the modified pedigree scheme, each selected plant becomes an F2:3 plot. Simple to measure physiological tools, such as canopy temperature depression or NDVI can be assessed manually or using proximal/remote sensing. A bulk of spikes taken from selected plants from each plot is then advanced to an F2:4 plot and the process repeated until individual plants are retained from the F5 generation to form the new fined line (F5:6). Marker assisted selection can augment this process as required. The near homozygous materials are then multiplied and evaluated across the target environment and physiological trait combinations confirmed in the best performing materials.

10.4.4   Breeding Method – Selected Bulk

The process of parental selection and crossing is identical to the modified pedigree or pedigree system up to the F2:3 generation. Individual plants selected from the F2 generation are bulked and not maintained as individual plots. Marker assisted selection for linked physiological traits on F2 plants can still be performed if required before bulking. Individual plant selections are bulked each generation and once the required level of homozygosity is reached, usually by F5, individual plants are selected, and these F5:6 selections become the new fixed lines. These enter multi-environment testing and confirmation of physiological trait expression as per the modified pedigree method. There are many fewer but much larger plots in the selected bulk method between F3 and F5 compared to modified pedigree. Nevertheless, these populations can still be assessed using easy to measure traits, such as canopy temperature depression or NDVI. Higher numbers of plant can be selected from plots with better agronomic type and superior physiological trait values, thus favorably skewing gene frequency.

10.4.5   Breeding Method – Genomic Selection

This breeding approach varies from the previous two strategies once the target ideotype has been determined. This section will focus on integrating physiological traits in a broader genomic breeding scheme. A training population representing the diversity required to assemble the ideotype, preferably in adapted backgrounds, of more than 2000 individuals is assembled. If physiological traits are found in unadapted materials, it is better to first derive lines carrying the trait in better agronomic backgrounds using backcrossing, otherwise the training population phenotype will be compromised by morphological and phenological extremes. In general, if plant height and phenology fall within a relatively narrow range, then the population phenotype is deemed comparable [27]. The training population phenotype must be largely field-based and should extend over time and space. The more accurate the phenotype, including physiological traits, the better the GEBVs upon which crossing and selection decisions will be based. Traits that correlate with yield under stress can be integrated using a weighted index based on heritability and GEBVs subsequently calculated using all the available information.

Crosses would then be made among genetically distant lines with high GEBVs that include, where possible, known marker-physiological trait associations. To optimize linkage disequilibrium, materials should be genotyped and recombined in crosses by the F4. At this stage a reasonable degree of homozygosity has been reached and the materials can be advanced from F2 – F3 rapidly using single seed decent or other methods of population advance that maintain gene frequency. Single plant selections taken at F3 would be genotyped and the F3:4 grown in plots. F4:5 progeny selections would be retained for multi-environment testing and trait validation from those F4 plots with high GEBVs. The F3:4 materials with highest GEBVs and greatest genetic distance would then be recombined in crosses and the process begun again without phenotyping. Following at least two breeding cycles based on genotype alone, the derived materials would be evaluated in multi-environment trials and phenotyped to confirm combined physiological traits. Superior materials would then cycle back to the training population along with a continuous flow of new alleles.

10.5   Key Concepts and Conclusions

Drought and heat stress tolerance will become increasingly important wheat breeding objectives in most wheat growing regions, increasing investment in these stresses and the opportunities for collaboration within and across regions. Most wheat breeding programs use a handful of breeding methodologies or modifications of these methods to derive new cultivars for farmers. The integration of new wheat breeding tools and knowledge does not entail a complete restructuring of breeding programs as such changes do have significant economic consequences. Instead, new technologies and knowledge can be integrated effectively with current commonly used breeding methods. These technologies are simply efficiencies that advance the overall goal of delivering better cultivars faster.

Physiological breeding is one such strategy that is easily integrated and entails better characterization of parents for physiological traits relevant to the target environment, implementation of an appropriate selection strategy that may entail high-throughput phenotyping, molecular markers or empirical selection under stress, followed by extensive evaluation of fixed lines under the stress and across multiple environments within the target region.

However, physiological breeding should become an obsolete term, as these traits are simply part of the suite of traits accessible to the plant breeder interested in improving crop adaptation to stress. Decisions to use these traits in crossing and selection will depend, as always, on heritability, ease of assessment and importance to farmers and industry. However, as Reynolds and others have shown [28, 37], at a minimum, they can be incorporated at crossing and confirmed following empirical selection in the target environment. Proximal and remote sensing are also changing the method (and scale) of assessment of physiological traits and these data can be used to truncate populations and favorably skew gene frequency. Physiological characterization can be easily incorporated into genomic selection strategies including the calculation of GEBVs based on weighted trait values.

References

1. Abhinandan K, Skori L, Stanic M, Hickerson NMN, Jamshed M, Samuel MA (2018) Abiotic stress signaling in wheat – an inclusive overview of hormonal interactions during abiotic stress responses in wheat. Front Plant Sci 9. https://doi.org/10.3389/fpls.2018.00734
2. Figueroa M, Hammond-Kosack KE, Solomon PS (2018) A review of wheat diseases—a field perspective. Mol Plant Pathol 19:1523–1536. https://doi.org/10.1111/mpp.12618
3. Toreti A, Cronie O, Zampieri M (2019) Concurrent climate extremes in the key wheat producing regions of the world. Sci Rep 9:5493. https://doi.org/10.1038/s41598-019-41932-5
4. Ababaei B, Chenu K (2020) Heat shocks increasingly impede grain filling but have little effect on grain setting across the Australian wheatbelt. Agric For Meteorol 284. https://doi.org/10.1016/j.agrformet.2019.107889
5. Morgounov A, Sonder K, Abugalieva A, Bhadauria V, Cuthbert RD, Shamanin V, Zelenskiy Y, DePauw R (2018) Effect of climate change on spring wheat yields in North America and Eurasia in 1981–2015 and implications for breeding. PLoS One 13:e0204932
6. Semenov M, Shewry P (2010) Modelling predicts that heat stress and not drought will limit wheat yield in Europe. Nat Proc 5. https://doi.org/10.1038/npre.2010.4335.1
7. Belyaeva M, Bokusheva R (2018) Will climate change benefit or hurt Russian grain production? A statistical evidence from a panel approach. Clim Change 149:205–217. https://doi.org/10.1007/s10584-018-2221-3
8. Müller D, Jungandreas A, Koch FJ, Schierhorn F (2016) Impact of climate change on wheat production in Ukraine. 2016 German-Ukrainian Agricultural Policy Dialogue. Institute for Economic Research and Policy Consulting
9. Gupta R, Somanathan E, Dey S (2017) Global warming and local air pollution have reduced wheat yields in India. Clim Chang 140:593–604
10. Zaveri EB, Lobell D (2019) The role of irrigation in changing wheat yields and heat sensitivity in India. Nat Commun 10:4144. https://doi.org/10.1038/s41467-019-12183-9
11. Xie W, Huang J, Wang J, Cui Q, Robertson R, Chen KZ (2020) Climate change impacts on China's agriculture: the responses from market and trade. China Econ Rev 62:101256. https://doi.org/10.1016/j.chieco.2018.11.007
12. Grassini P, Eskridge KM, Cassman KG (2013) Distinguishing between yield advances and yield plateaus in historical crop production trends. Nat Commun 4:2918. https://doi.org/10.1038/ncomms3918

13. Maeoka RE, Sadras VO, Ciampitti IA, Diaz DR, Fritz AK, Miedaner RP, Korzun V (2012) Marker-assisted selection for disease resistance in wheat and barley breeding. Phytopathology 102:560–566. https://doi.org/10.1094/PHYTO-05-11-0157
14. Rahman M, Davies P, Bansal U, Pasam R, Hayden M, Trethowan R (2020) Marker-assisted recurrent selection improves the crown rot resistance of bread wheat. Mol Breed 40:28. https://doi.org/10.1007/s11032-020-1105-1
15. Trethowan RM (2014) Defining a genetic ideotype for crop improvement. In: Fleury D, Whitford R (eds) Methods in molecular biology: crop breeding. Humana Press, New York, pp 1–20
16. Reynolds MP, Trethowan RM (2007) Physiological interventions in breeding for adaptation to abiotic stress. In: Spiertz JHJ, Struik PC, Van LHH (eds) Scale and complexity in plant systems research, gene-plant-crop relations. Springer, Dordrecht, pp 129–146
17. Ullah S, Bramley H, Mahmood T, Trethowan R (2019) A strategy of ideotype development for heat tolerant wheat. J Agron Crop Sci 206:229–241. https://doi.org/10.1111/jac.12378
18. Dreccer MF, van Herwaarden AF, Chapman SC (2009) Grain number and grain weight in wheat lines contrasting for stem water soluble carbohydrate concentration. Field Crop Res 112:43–54. https://doi.org/10.1016/j.fcr.2009.02.006
19. Rebetzke Greg J, Karine C, Ben B, Carina M, Deery Dave M, Rattey Allan R, Dion B, Barrett-Lennard Ed G, Mayer Jorge E (2012) A multisite managed environment facility for targeted trait and germplasm phenotyping. Funct Plant Biol 40:1–13
20. Trethowan RM, Mujeeb-Kazi A (2008) Novel germplasm resources for improving environmental stress tolerances of hexaploid wheat. Crop Sci 48:1255–1265
21. Aili L, Li D, Yang W, Kishii M, Mao L (2018) Synthetic hexaploid wheat: yesterday, today, and tomorrow. Engineering 4:552–558
22. Ma'arup R, Trethowan RM, Ahmed NU, Bramley H, Sharp PJ (2019) Emmer wheat (Triticum dicoccon Schrank) improves water use efficiency and yield of hexaploid bread wheat. Plant Sci 295. ISSN 0168-9452, https://doi.org/10.1016/j.plantsci.2019.110212
23. Djanaguiraman M, Prasad PVV, Kumari J, Sehgal S, Friebe B, Djalovic I, Chen Y, Siddique K, Gill B (2019) Alien chromosome segment from Aegilops speltoides and Dasypyrum villosum increases drought tolerance in wheat via profuse and deep root system. BMC Plant Biol 19:242. https://doi.org/10.1186/s12870-019-1833-8
24. Kyratzis AC, Skarlatos DP, Menexes GC, Vamvakousis VF, Katsiotis A (2017) Assessment of Vegetation Indices derived by UAV imagery for durum wheat phenotyping under a water limited and heat stressed mediterranean environment. Front Plant Sci 8:1114
25. Sallam A, Alqudah AM, Mfa D, Baenziger PS, Börner A (2019) Drought stress tolerance in wheat and barley: advances in physiology, breeding and genetics research. Int J Mol Sci 20:137. https://doi.org/10.3390/ijms20133137
26. Thistlethwaite Rebecca J, Tan DKY, Bokshi A, Ullah S, Trethowan RM (2020) A phenotyping strategy for evaluating the high-temperature tolerance of wheat. Field Crop Res 255:107905. https://doi.org/10.1016/j.fcr.2020.107905
27. Reynolds MP, Manes Y, Izanloo A, Langridge P (2009) Phenotyping approaches for physiological breeding and gene discovery in wheat. Ann Appl Biol 155:309–320. https://doi.org/10.1111/j.1744-7348.2009.00351.x
28. Pask A, Joshi AK, Manes Y, Sharma I, Chatrath R, Singh GP, Sohu VS, Mavi GS, Sakuru VSP, Kalappanavar IK, Mishra VK, Arun B, Muhajid MY, Hussain M, Gautam NR, Barma NCD, Hakim A, Hoppitt W, Trethowan R, Reynolds MP (2014) A wheat phenotyping network to incorporate physiological traits for climate change in South Asia. Field Crop Res 168:156–167. https://doi.org/10.1016/j.fcr.2014.07.004
29. Richards R (2004) Physiological traits used in the breeding of new cultivars for water-scarce environments. In: Proceedings of the 4th International Crop Science Congress. Brisbane, Australia
30. del Pozo A, Yáñez A, Matus IA, Tapia G, Castillo D, Sanchez-Jardón L, Araus JL (2016) Physiological traits associated with wheat yield potential and performance under water-tress in a mediterranean environment. Front Plant Sci 7:987

31. Khadka K, Earl HJ, Raizada MN, NavabiFront A (2020) A physio-morphological trait-based approach for breeding drought tolerant wheat. Front Plant Sci 11:715. https://doi.org/10.3389/fpls.2020.00715. PMID: 32582249; PMCID: PMC7286286

32. Rebetzke Greg J, Rattey Allan R, Farquhar Graham D, Richards Richard A, Condon Anthony (Tony) G (2012) Genomic regions for canopy temperature and their genetic association with stomatal conductance and grain yield in wheat. Funct Plant Biol 40:14–33

33. Chen X, Min D, Yasir TA, Hu Y-H (2012) Evaluation of 14 morphological, yield-related and physiological traits as indicators of drought tolerance in Chinese winter bread wheat revealed by analysis of the membership function value of drought tolerance (MFVD). Field Crop Res 137:195–201. https://doi.org/10.1016/j.fcr.2012.09.008

34. Luo Q, Trethowan R, Tan DKY (2018) Managing the risk of extreme climate events in australian major wheat production systems. Int J Biometeorol 62:1685–1694. https://doi.org/10.1007/s00484-018-1568-5

35. Rebetzke GJ, Richards RA (2000) Gibberellic acid-sensitive dwarfing genes reduce plant height to increase kernel number and grain yield of wheat. Crop Pasture Sci 51:235–246

36. Christopher JT, Manschadi AM, Hammer GL, Borrell AK (2008) Developmental and physiological traits associated with high yield and stay-green phenotype in wheat. Aust J Agric Res 59:354–364. https://doi.org/10.1071/AR07193

37. Reynolds M, Langridge P (2016) Physiological breeding. Curr Opin Plant Biol 31:162–171

Chapter 11
Wheat Quality

Carlos Guzmán, Maria Itria Ibba, Juan B. Álvarez, Mike Sissons, and Craig Morris

Abstract Wheat quality is a complex concept whose importance lies in determining the ability of each segment of the post-harvest processing and marketing industries to minimize cost while maximizing profit. Wheat quality is also a highly subjective concept that could be defined differently by the various stakeholders in the wheat value chain. It is usually subdivided into milling, processing, end-use and nutritional quality. Of these subcomponents, end-use quality, the ability of a wheat variety to produce a specific food according to the consumers preferences is probably the most important. Wheat is used to make hundreds of different products worldwide, each one with specific grain quality requirements. In this chapter are explained the main traits that define end-use quality (grain hardness, gluten, color and starch) and that need to be modulated to obtain the desired product properties. The genetic control as well as the environmental effects on those traits are also presented. Finally, breeding and selection strategies to genetically improve end-use quality for the most important wheat products globally (bread, noodles, cookies, and pasta) are presented in brief.

Keywords Grain quality · Flour · Dough viscoelastic properties · Bread-making · Pasta-making · Glutenins · Molecular markers

C. Guzmán (✉) · J. B. Álvarez
Departamento de Genética, Escuela Técnica Superior de Ingeniería Agronómica y de Montes, Universidad de Córdoba, Córdoba, Spain
e-mail: carlos.guzman@uco.es; jb.alvarez@uco.es

M. I. Ibba
International Maize and Wheat Improvement Center (CIMMYT), Texcoco, Mexico
e-mail: m.ibba@cgiar.org

M. Sissons
NSW Department of Primary Industries, Tamworth Agricultural Institute, Calala, Australia
e-mail: mike.sissons@dpi.nsw.gov.au

C. Morris
USDA-ARS Western Wheat Quality Laboratory, Pullman, WA, USA
e-mail: craig.morris@usda.gov

© The Author(s) 2022
M. P. Reynolds, H.-J. Braun (eds.), *Wheat Improvement*,
https://doi.org/10.1007/978-3-030-90673-3_11

11.1   Learning Objectives

- To understand what wheat quality is and how to integrate it into breeding programs.

11.2   Introduction – What Is Wheat Quality?

George Bernard Shaw wrote: *'Take care to get what you like or you will be forced to like what you get'*. This aphorism probably summarizes the clearest vision of quality in the context of wheat improvement. The concept of wheat quality can be simple (edible versus inedible) or very complex (adaptation to explicit or implicit consumer demands). Although many methodologies have been designed to measure wheat quality, in reality most of them have been used to assess whether or not the grain of a cultivar can be adapted to a specific end-use. A modern wheat cultivar could be considered of high quality for the manufacture of standard bakery products; however, if we use this flour to make traditional products, our appreciation of this cultivar could be very different.

In the agri-food industry, many stakeholders are involved in the wheat value chain, from the farmer to the consumer. This means that the term quality can have different meanings depending on each of these stakeholders. For the farmer, a high-quality wheat cultivar might be the one that requires the lowest inputs, gives the highest grain yield and the grain can be sold at the highest price in the market. However, the miller will classify the cultivars according to the performance of the grain to produce flour (in a broad sense), along with the energy requirements for obtaining it. Finally, the baker will discriminate these materials for their utilization in each baking product.

At the opposite ends of the value chain, we find two very different stakeholders. The consumer defines quality using subjective parameters that are often difficult to analyze. At the other extreme, we find the wheat breeder who must work with objective data to design new wheat cultivars. These cultivars may be appreciated by the farmer, desired by the miller and valued by the baker, and we must also add the hope that they are to the taste of the consumer. However, the possibility of a uniform response from all of them is clearly unlikely. Probably, for a given cultivar, these various perspectives can range from positive to negative. Consequently, the Manichean vision between good and bad is clearly a mistake here. Once the desired product is chosen, the materials with high quality will be those that best perform for this product.

In this context, wheat grain components play an important role, together with their physico-chemical properties, in defining grain quality characteristics. There are three main components of wheat grain: proteins (7–18%), lipids (1.5–2%) and carbohydrates (60–75%), and other minor components such as vitamins and minerals. Proteins and carbohydrates, especially starch and arabinoxylans (the main component of wheat grain fiber), have notable influence on three grain characteristics

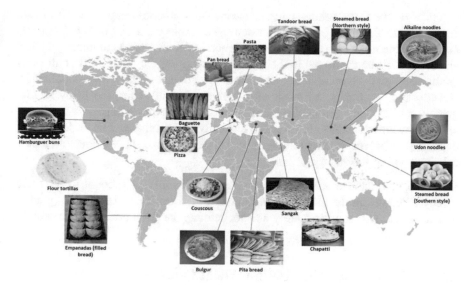

Fig. 11.1 Wheat products popular worldwide

closely linked to the technical wheat qualities required for diverse wheat products. These are the gluten viscoelastic properties, starch properties and grain hardness or texture, which are associated with milling, processing and end-use quality. There are other complementary parameters that sometimes have great importance such as flour or semolina yellow color.

11.3 Importance of Wheat Quality – Why We Need to Breed for It

Wheat, in contrast to other cereals, produces the greatest variety of consumer foods (Fig. 11.1). Each has unique attributes, which are often subtle in nature. The goal of delivering improved germplasm, i.e. 'breeding for quality', is to produce a genetic 'package' – a cultivar that possesses the greatest number of favorable alleles for grain and milling quality, processing and food manufacturing while importantly, aiming to meet the highest grade to obtain the best price for the grower. Most desirably, quality means the ability for each segment of the post-harvest processing and marketing industries to minimize cost while maximizing profit. This concept can be illustrated with a few examples: The 'correct' kernel hardness facilitates efficient milling that produces a flour of the 'correct' particle size and starch damage. The 'correct' glutenin profile produces doughs with the 'correct' mixing and rheological properties, and consumer traits such as product size and texture, and on and on.

Why are the cereal chemist and quality laboratory so integral to delivering improved wheat germplasm? Essentially it comes down to the fact that wheat cultivars do not last forever. Pests surmount resistances (see Chaps. 8, 9 and 19), farming practices evolve as do weather patterns creating new abiotic stresses (see Chap. 10), and the goal of attaining ever higher and more stable grain yields (see Chap. 21) necessitate the need to make crosses in the quest of seeking and combining 'better' alleles. As will be discussed in the Sect. 11.4, many of the main traits controlling quality are well characterized, and some are 'fixed' in breeding populations. Nevertheless, quality is the result of a large number of genes, too many to adequately select for by genotyping germplasm. In the never ending quest for better alleles, unwanted quality alleles will necessarily be introduced. For these reasons, delivering improved wheat germplasm will always involve some degree of empirical phenotyping for quality.

- **Exercise:** what is understood by *wheat quality* in your region/country? Are there any mechanisms to classify wheat grain based on its grain quality (grades, classes, etc.)? Is it grain quality a factor defining the grain price in the market?

11.4 Main Traits That Define Wheat Quality

11.4.1 Grain Hardness

Grain texture or hardness is the consequence of the degree of adhesion between the starch granules and the surrounding protein matrix inside the wheat endosperm. This trait has been used to classify wheat since antiquity, being the fundamental basis of differentiating the world trade of wheat grain. According to this trait, wheat is classified as very hard, hard or soft. Furthermore, this character is closely linked with the botanical classification of wheat: tetraploid wheat (subspecies of *Triticum turgidum* including durum wheat) exhibits very hard texture, whereas the *T. aestivum* group (hexaploid wheat including bread wheat) exhibits a texture that varies from hard to soft.

Grain hardness or texture is the single most important trait that determines end-use and technological utilization. It affects several parameters related to wheat milling: flour yield, energy requirement, particle size distribution of the flour and semolina, and percentage of starch damage (which strongly affects the dough water absorption linked with end-use quality). Due to differences in hardness, hard common wheat is used for bread-making while soft common wheat is preferred for cookies and pastries; very hard durum wheat is preferred for pasta. The very hard texture of durum wheat is associated with low flour yield and greater amounts of damaged starch in the flour. Durum flour is used in several Mediterranean regions to make traditional breads, which are usually denser than common wheat breads and have a more compact crumb texture. Durum wheat grain is primarily milled into semolina (larger particle size than flour) which is used to make pasta (made by extruding stiff semolina dough) or for couscous (made by agglomeration of semolina).

11.4.2 Gluten

In most cases, grain protein content varies between 7% and 18%. Of this protein, a large part (around 80%) is comprised of the proteins that form gluten. Gluten is the continuous protein viscoelastic network that develops when wheat flour is mechanically mixed with water. This protein network imparts to the wheat dough its unique properties which allow it to be processed into a wide range of products such as breads, noodles, pasta, cakes and biscuits. To give an example, in bread making gluten confers to the dough its viscoelasticity which allows the entrapment of carbon dioxide released by the yeast during leavening, whereas in pasta production it gives the necessary cohesiveness to extrude the dough and to form the desired shape. Products such as noodles, flatbreads and some cookies that need a sheeting procedure in their manufacture require flours with good extensibility to perform well in these processes.

The large complex polymer known as gluten is primarily comprised of two type of proteins: the monomeric gliadins (single-chain polypeptides), and the polymeric glutenins (multiple polypeptide chains linked by disulfide bonds), which can be separated based on their solubility in aqueous alcohols and acid solutions and alkali, respectively. Gliadins, which account for around 60% of the gluten, are classified into ω-gliadins, α/β-gliadins and γ-gliadins and contribute mainly to the viscosity and extensibility by working as plasticizers of the dough. Glutenins, which are subdivided into high molecular weight glutenins (HMW-Gs) and low molecular weight glutenins (LMW-Gs), are aggregating proteins with cysteine groups at the end and in the middle of the protein sequence. These cysteines enable intermolecular disulfide bonds, creating a large range in molecular weight. Glutenins are more responsible for the cohesive and elastic properties of the dough. All these gluten components show tremendous variation in the wheat germplasm pool leading to different gluten structures with contrasting properties and impact on dough physical and physico-chemical properties. These dough properties are also highly modulated by the protein or gluten content of the flour. Each type of these gluten networks with specific properties is more suitable to produce a specific type of wheat product. For all common wheat products certain levels of dough extensibility are necessary whereas dough strength requirements vary depending on the product: pan bread, strong gluten; hearth and flat breads and noodles, medium to strong gluten; and cookies and cakes, weak gluten. High quality pasta is made with durum with a high level of strong and tenacious gluten.

11.4.3 Color

Flour color plays a significant role in the end-use quality of wheat, particularly for Asian noodles, steamed bread and pasta since it affects consumer acceptance, market value and human nutrition (see Chap. 12). Color has two essential components:

inherent capacity to produce pigments (for example, presence of carotenoids or liberated flavonoids for alkaline noodles), and the capacity to not degrade those colors during processing. Desirable yellow color for 'white' salted noodles may range from very low (no yellow is preferred) to creamy yellow. For alkaline noodles, the pH-induced yellow color is appreciated and quite often, higher is better. For durum semolina, a high yellow color is desired. The color of the grain and the end-products derived, depends on genetic, environmental and processing factors. Genes coding for enzymes involved in pigment accumulation and degradation affect color. The main pigments are carotenoids (yellow pigment) and anthocyanins (responsible for blue to red grain), both are important for their aesthetic role and have been shown to benefit ocular health. Modern durum varieties and bread wheats have higher and lower yellow pigment, respectively, than older wheat varieties due to breeding selection. In durum wheat grains, the major carotenoid is the xanthophyll lutein, mostly located in the endosperm and consequently found in the flour/semolina. However, during flour processing, carotenoid degradation can occur by oxidases such as the lipoxygenases (LOX) and the polyphenol oxidase enzymes that can generate brown polymers that can mask the yellow color of pasta or make noodles appear dull. Fortunately, high yellow pigment levels work against LOX activity. Keeping the bran level in flour to a minimum is a good way to reduce these oxidative enzyme levels. Individual pigments can be measured using HPLC. Colorimetric methods such as NIR, extraction of yellow pigments or light reflectance using a Minolta CR-300 Chroma Meter of flour/semolina to measure lightness, red-green and yellow-blue chromaticity (CIE 1986) coordinates are fast and non-destructive.

11.4.4 Starch

Starch is the main component of wheat grain representing about 70% of the total dry matter, and is comprised of two polymers based on D-glucose residues: one linear formed by α-(1,4) residues (amylose) that represents 22–35% and the other (amylopectin) with α-(1,4) residues ramified each 20–30 residues by α-(1,6) linkages representing 65–78%. These polymers are synthesized in the amyloplast by two different synthetic routes; the search for cultivars with modifications of the enzymes involved in starch synthesis has been key to the generation of novel starches with special properties due to changes in the amylose/amylopectin ratio. The relationship between both polymers can affect the physical and chemical properties of starch (gelatinization, pasting and gelation), and consequently the quality of the end-products.

Starch properties greatly influence food products made from wheat flour or semolina, especially Asian noodles where low amylose content is desirable to obtain the desired texture. In addition to the quality of noodles, starch is associated with the shelf life of pre-cooked products and the nutritional value: a higher amylose content is associated with higher resistant starch content (functioning as fiber), which is associated with health (low glycemic index and better gut health), although with lower end-use quality too.

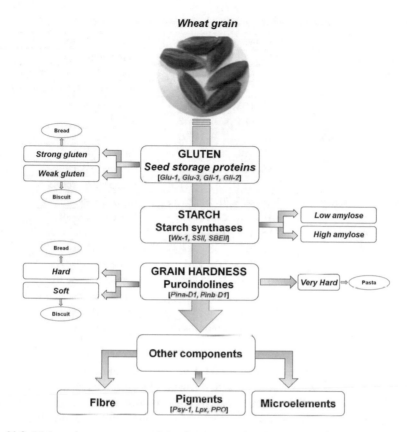

Fig. 11.2 Main grain components, traits and genes associated to wheat quality

11.5 Genetic Control of the Quality Traits and Environmental Effects

Genetic improvement is at the basis of crop breeding. For this reason, knowing the heritability of each quality trait, its genetic basis and how much of their variation is influenced by different environmental factors, is of fundamental importance for an effective improvement of wheat quality. Among the elements that influence wheat quality, grain hardness, gluten quality, flour color and starch properties have been the most studied. For this reason, extensive information is available on the genetic (Fig. 11.2 and Table 11.1) and environmental factors affecting their variation.

Specifically, variation in grain hardness is mainly determined by the *Puroindoline a* (*Pina-D1*) and *Puroindoline b* (*Pinb-D1*) genes, located at the *Hardness* locus, on the short arm of chromosome 5D. When the wild-type form of the two *Pin* genes is present (alleles *Pina-D1a* and *Pinb-D1a*), wheat kernels exhibit a soft texture. In contrast, when either of the two genes is mutated, wheat kernels exhibit a hard texture. Due to the lack of the D genome and hence the two *Pin-D1* genes, durum

Table 11.1 Genes associated with major influences on wheat quality traits

Trait	Chromosomes	Locus/gene	Protein/enzyme
Grain hardness	5DS	*Hardness*	Puroindoline a, b
Gluten quality	1AS, 1BS, 1DS	*Glu3*	Low-molecular-weight glutenins
	1AL, 1BL, 1DL	*Glu1*	High-molecular-weight glutenins
	1AS, 1BS, 1DS	*Gli1*	γ and ω-gliadins
	6AL, 6BL, 6DL	*Gli2*	α/β-gliadins
Yellow pigment accumulation	7AL, 7BL, 7DL	*Psy1*	Phytoene synthase
	4AL, 4BL, 4DL	*Pds1*	Phytoene desaturase
	2AS, 2BS, 2DS	*Zds1*	ζ-carotene desaturase
	3A, 3B, 3D	*ε-LCY*	Lycopene ε-cyclase
Yellow pigment degradation	4AS, 4BS, 4DS	*Lox1.1*	Lipoxygenase
Flour discoloration	2AL, 2BL, 2DL	*Ppo1*	Polyphenol oxidase
Starch functionality	7AS, 4AL, 7DS	*Wx1*	Granule bound starch synthase I
	7AS, 7BS, 7DS	*Ss1*	Starch synthase I
	7AS, 7BS, 7DS	*Ss2*	Starch synthase IIa
	1AS, 1BS, 1DS	*Ss3*	Starch synthase III
	7AL, 7BL, 7DL	*Sbe1*	Starch branching enzyme I
	2AL, 2BL, 2DL	*SbeIIa*	Starch branching enzyme IIa
	2AL, 2BL, 2DL	*SbeIIb*	Starch branching enzyme IIb

wheat kernels exhibit an extremely hard texture. Additional minor variation in kernel hardness among wheat varieties with the same *Pin* profile have also been identified. This variation could be determined by both environmental and genetic factors affecting, among the others, grain protein and moisture content, grain vitreousness and morphology, and pentosan quantity and quality.

Differently, moderate to high heritability has been observed for gluten quality with, on average, 60% of its variation being explained by differences in the genotype. Most of this variation is related to differences in the combination of the gluten-forming proteins, with the HMW-Gs (*Glu-1* loci, long arm of the group 1 chromosomes) and the LMW-Gs (*Glu-3* loci, short arm of the group 1 chromosomes) being typically the major determinants of these differences. Specifically, variation in the HMW-Gs has been shown to explain from 20% to 30% of the variation in gluten strength in common wheat and, among the *Glu-1* loci, the *Glu-D1* locus has typically a greater effect on gluten quality, followed by the *Glu-B1* and the *Glu-A1* loci, respectively. Wide allelic variation has been detected at each *Glu-1* locus and alleles associated with specific gluten characteristics have been identified. The effect of the LMW-Gs on gluten quality is different in common and durum wheat. In common wheat, variation in the LMW-Gs has typically a lower impact on gluten properties compared to the HMW-Gs, accounting for 10–20% of the observed variation. Differently, in durum wheat, the effect of the LMW-Gs on gluten quality is greater compared to the HMW-Gs. In both cases, alleles associated with variation in gluten strength have been identified. Besides the genetic factors, several studies have shown that the environment plays a significant role in determining gluten quality, influencing from 3% to 50% of its variation. Depending on the environment, the content and ratio of the gluten-forming proteins change greatly, thus affecting both the rheological and end-use quality. For example, drought stress is typically associated with an increase in grain protein content and in the gluten polymeric fraction which results in an increase in gluten strength and tenacity. In contrast, wheat lines grown under heat stress typically have a greater protein content but lower glutenin/gliadin ratio, resulting in a weaker and more extensible gluten. However, the response depends on when the heat stress occurs during grain development and its severity and duration.

Similar to kernel hardness, flour or semolina color typically exhibits high heritability. Indeed, ~90% of the variation observed in flour or semolina yellowness depends on the genotype. Even though all the genes and their relative allelic variants involved in the modulation of flour yellowness have not been identified, variation in the *Phytoene synthase I* (*PsyI*) genes have been associated in both common and durum wheat with major changes in flour and semolina carotenoid content, typically explaining >20% of the observed phenotypic variation. Additional smaller variation of this trait is influenced by the environment, which could affect both the expression level of the different enzymes involved in the synthesis of the yellow pigments, both the concentration of the pigments in the grain (the smaller the grain, the higher the concentration).

Major changes in flour color may result from the activity of specific enzymes, which are also highly genetically controlled. For example, degradation of the yellow color is mainly determined by the activity of LOX. Genes encoding this enzyme have been mapped and the alleles *Lox-B1.1c* and *TaLox-B1b* in durum and common wheat, respectively, have been associated with drastic reductions in LOX quantity and activity. Similarly, genes encoding polyphenol oxidase (PPO), which is associated with the undesirable discoloration of some wheat products have been identified and mapped. Among them, variation in the *Ppo-A1* and *Ppo-D1* genes have been associated with major variations in the activity of this enzyme with the alleles *Ppo-A1b* and *Ppo-D1a* being associated with lower PPO activity.

Like gluten quality, starch pasting properties exhibit moderate to high heritability and differences in the genotype have been shown to consistently explain more than 30% of the observed phenotypic variation. Mutations in key genes involved in the starch biosynthetic pathway have been associated with significant changes in starch physical properties. Specifically, mutations in the gene involved in amylose synthesis (Granule-bound starch synthase, GBSS, or waxy protein; *Wx-1* loci) have been associated with the synthesis of starch with either a higher proportion of amylopectin or with the complete absence of amylose (waxy starch). Similarly, mutations in the genes involved in the synthesis of amylopectin, like the Starch synthase (SS) or the Starch branching enzyme (SBE) genes, led to the synthesis of starch with greater amylose content (resistant starch). However, up to ~60% of the observed starch physical properties are also influenced by environmental conditions and variation in other grain components. Biotic and abiotic stresses during plant growth are associated with changes in the starch properties. For example, lodging is often associated with increased alpha amylase activity, which leads to more rapidly degraded starch in flour during mixing and fermentation causing different problems in the end-use quality of the products.

11.6 Breeding for Quality

11.6.1 Integrating Quality in the Breeding Process

Wheat breeding programs measure a range of plant and grain characteristics to improve grain yield, abiotic and biotic resistance, adaptation and grain quality suitable for markets (see Chaps. 5, 6 and 7). Integrating quality into the breeding process, although different between programs, has common features. Breeding programs make many crosses between parents possessing a value-added trait(s), which results in the creation of large populations to evaluate in the generation cycle. Therefore high-throughput, small-scale tests that allow discrimination between acceptable and unacceptable/borderline samples can help reduce the size of the material carried forward. Later generations (replicated field trials) produce more grain of fewer samples, which is amenable for conducting more time consuming and accurate tests.

11.6.2 Bread

Bread is probably one of the most universal foods and there is a huge diversity of types worldwide (pan, hearth, flat, steam breads, etc.). There are differences in specific grain quality requirements, processing conditions, and end-product properties for each type of bread (Fig. 11.3). All breads are made from viscoelastic and cohesive doughs prepared from refined or whole-meal flour. They are mostly produced from hard common wheat flour, but durum flour or semolina is also used in some areas to make bread. Due to the huge diversity of breads and contrasting consumer preferences, it is difficult to define what makes a good bread but in most of the cases, bread quality is related to the crust and crumb properties, color and other organoleptic and more subjective properties such as texture, aroma and taste.

In breeding, emphasis has been put in improving those traits related with the volume, texture and color of bread. Bread ('loaf') volume is a crucial trait for pan bread and certain types of hearth breads and depends highly on dough strength and extensibility. During the fermentation stage in bread-making, doughs with sufficient gluten strength will have cells with the capacity to retain the gases without collapsing. If the same dough has also high extensibility, those cells will enlarge giving the bread the desired large volume. Those same dough or gluten characteristics are also important to obtain a uniform, fine and silky crumb, which is desirable for pan breads (less important in hearth breads or flat breads). Crumb with light white color

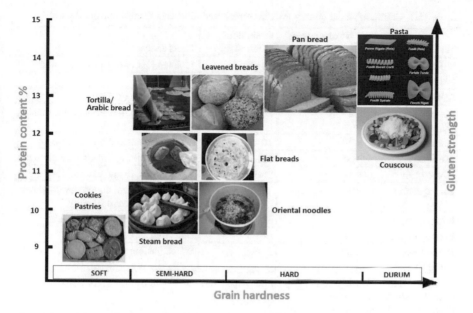

Fig. 11.3   General grain quality characteristics of wheat products

is a characteristic desired by most consumers, and thus 'whiteness' has been targeted by breeding programs, selecting germplasm with reduced or almost null amount of pigments.

Phenotyping for bread-making quality is not easy. For bread and any other wheat product, manufacturing that product (with laboratory scale methods or in full scale) should be the critical and ultimate test to define the suitability of a wheat cultivar to produce that product. Several small-scale bread-making protocols are available and are used routinely in wheat quality labs depending on capability. However, it is not always possible to perform such tests due to insufficient grain, high cost and time required to do the analysis, etc. Consequently, tests to evaluate traits related to or predictive of bread-making are usually applied (Table 11.2). Flour color is measured with a Minolta color meter or similar instrument.

11.6.3   Noodles

Here we delineate *noodles* from pasta. Although both may resemble strands of dough prepared from wheat flour, which are then boiled prior to consumption, noodles are most commonly prepared from common wheat by sheeting and cutting, whereas pasta is extruded and is made from durum wheat semolina. Most noodles are simple in composition: flour, water and salt. The key first difference among noodle types is, What kind of salt? Two approaches are encountered: normal table salt, sodium chloride at 1–5% on a flour weight basis (flour representing 100%), the

Table 11.2 Common wheat quality tests/machines used globally to determine quality traits

Test	N° of samples[a] per day	Grain/flour required (g)	Traits analyzed
NIRS	150–300	20–40	Moisture, hardness, protein, color
PPO Activity[b]	60	0.1	PPO activity
SDS-Sedimentation	100	0.5–2	Overall gluten quality
Solvent Retention Capacity	25[c]	20	Damaged starch, overall gluten quality, arabinoxylans, gliadin
Glutomatic	30	10	Gluten content and gluten strength
Mixograph	35	10, 35	Optimum mixing time, gluten strength
Alveograph	14	250	Gluten strength and extensibility
Farinograph	7	10, 50, 300	Water absorption, dough development time, softening and stability
Extensograph	12	300	Dough extensibility and strength
Falling number	70	6–7.5	Detecting sprouting damage
Rapid Visco Analyzer	28	3–4	Starch pasting viscosities

[a]Number of samples analyzed per day by one experienced technician working for eight hours
[b]L-DOPA whole kernel assay
[c]Performed with four solvents

second is a mixture of alkaline salts termed kansui. Kansui is often equal amounts of potassium and sodium carbonate (e.g. 0.5% each), less frequently sodium hydroxide. The use of kansui lends its name to the second type of noodle based on formula, 'alkaline noodles'. In addition to kansui, alkaline noodles will often have 1–1.5% NaCl.

After the basic formulation, processing dictates the next delineation. Classifications include fresh, dried, boiled (usually parboiled) and frozen. A unique style of noodle that has grown tremendously in popularity is the 'instant noodle'. Instant noodles, as the name implies, are quick-cooking due to the fact that they are essentially already cooked. Processing involves steaming and (usually) frying. For the consumer, 'cooking' is really simply rehydrating. Raw fresh noodles are termed Chinese raw noodles and Japanese Udon noodles, both are styles of 'white salted' noodles. Similarly, raw alkaline noodles may be 'Cantonese' in Southeast Asia or 'Chukamen' in Japan. White salted noodles are often dried to extend shelf life. Parboiled alkaline noodles are consumed throughout Southeast Asia ('hokkien') and Taiwan ('wet noodles').

Although the variety of noodle types and processing techniques is great, the fundamental basis for quality lies with the flour itself. From a consumer standpoint, most of the concern is with color and texture. Desirable yellow color for 'white' salted noodles may range from very low (white is preferred) to creamy yellow. For alkaline noodles, the pH-induced yellow color is appreciated and quite often, higher is better. Discoloration is primarily the result of PPO, but not entirely. For screening germplasm, color is conveniently measured on flour or raw noodle sheets using a Minolta color meter or similar instrument. Resting raw noodle sheets for 24 h at room temperature can be used to determine undesirable darkening, ΔL^*. A highly efficient system of screening germplasm for PPO activity uses L-DOPA (L-3,4-dihydroxyphenylalanine) as a substrate on five intact kernels.

After appearance, texture is next in importance. Texture is a complex trait to measure, but descriptive adjectives include firmness, springiness, stickiness, and gumminess. The surface character of the noodle (smoothness) is also important. Texture is assessed using either trained sensory panelists or instrumental approaches, for example the TA-XTPlus C. The primary genetic determinants of texture are starch and glutenin composition. Starch composition is relatively simple, in that either a 'normal' ratio of amylose to amylopectin is preferred as it conveys a firmer texture ('bite'), or a reduced amylose, 'partial waxy' genotype is preferred. Partial waxy wheats are produced by selecting a null allele at one of the $Wx-1$ genes, usually on chromosome 4A ($Wx-B1$). Partial waxy germplasm can be selected using DNA markers or empirically using the Flour Swelling Volume test or pasting viscometers such as the RapidVisco Analyzer or MicroAmylograph. Partial waxy varieties are preferred for Udon noodles. The role of glutenins is more complex, but can be viewed from the standpoint of dough rheology or simply the assessment of texture using sensory or instrumental analysis. The big caveat on texture is the role that protein content plays, mostly independent of glutenin haplotype.

11.6.4   Cookies

A large proportion of soft wheats are used to make cookies and cakes. The foremost genetic consideration from a quality standpoint is *soft kernel texture*, which is conditioned by the puroindoline genes/proteins.

Although a number of flour analyses can be performed to predict consumer end-product quality, such as the Solvent Retention Capacity tests, quite often laboratory bake tests are employed. Two common tests involve baking 'sugar-snap' cookies, which represent low moisture soft wheat products, and cakes, which represent high moisture, batter-based products. At the USDA Western Wheat Quality Lab, Japanese sponge cakes provide objective information for selecting superior soft wheat germplasm. As noted above, the variety of wheat foods is too numerous to characterize individually. Thus, these two 'model systems' provide sufficient prediction of consumer products to guide breeding programs.

11.6.5   Pasta

Durum wheat breeders consider a range of quality specifications before releasing a new variety (Table 11.3) but only measure a few in the early stages of the breeding cycle due to resource limitations caused by large numbers of samples to be

Table 11.3 Quality traits of different durum wheat samples

Traits/sample	A	B	C	D	E	F
Protein (%)	11.0	13.0	15.5	12.5	10.0	13.0
Test weight (kg/hl)	79.0	79.0	78.0	82.0	72.5	78.5
Falling number	389	200	600	650	720	500
Screenings	3.2	4.1	8.1	3.2	1.9	2.5

The acceptable samples are D and F with A having low protein (required >12%), B low FN (required >250 s), C high screenings and E low TW (required >76 kg/hL) and protein

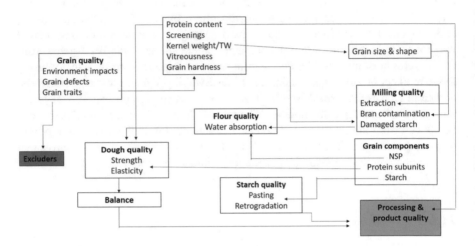

Fig. 11.4 Interactions defining pasta-making quality

evaluated. Key measures are grain protein and weight, screenings, some indicator of dough strength and color of wholemeal or semolina (Table 11.3). To understand the interactions leading to pasta quality refer to Fig. 11.4. Grain quality can be affected by (1) environmental impacts on the grain that negatively affect processing and the breeder avoids testing such grain: blackpoint (10% limit), fungal staining, frosted, white grain, heat and insect damaged grain; (2) grain defects: test weight <76 kg/hl impacts milling yield; falling number <250 s impacts pasta appearance and cooking loss; screenings >5% reduces milling yield; low vitreosity kernels tends to produce more flour on milling and creates poor pasta strength, such defects can result in exclusion from testing by the breeder; (3) grain traits: key traits to ensure good pasta quality are grain protein >12–13%, highly vitreous (>70%), hard, large and sound grain (thousand kernel weight >35 g; test weight >76 kg/hl) and acceptable glutenin allelic composition which impacts gluten strength. Generally, if these minimum standards are met, such grain when milled produces particles with the correct size distribution, with minimal bran in good yield and when mixed with water, creates dough that absorbs water uniformly. This dough when extruded or sheeted makes pasta with a good gluten matrix surrounding the starch granules which ensures good texture after cooking. Of course, all these measures depend on the genotype being tested and the interaction with the environment.

High-throughput tests that are inexpensive are desired by cereal chemists in a breeding program (Table 11.2). Near infrared spectroscopy (NIRS) is probably the best known instrument using either manufacturer supplied in-built or in-house developed calibrations. This technology allows non-destructive assessment of grain samples (35–150 g) at about 0.2–2 min/sample and can be automated. NIRS predictions for protein, moisture, wet gluten, test weight, yellow pigment, hardness and ash are being used by breeders. A more recent tool is image analysis to measure grain vitreousness, semolina speck counts and blackpoint but has yet to find wide application in breeding programs. Color assessment is performed rapidly on whole-meal or semolina using a colorimeter to measure yellowness (b*). Most small scale tests useful to a breeding program tend to focus on measures of dough quality (SDS sedimentation, mixograph, gluten index, glutopeak) requiring 1–10 g. Instrumental and cooking tests to evaluate dried pasta do not require large amounts of sample and there are standardized international methods available. There is also no standard method to prepare laboratory scale pasta.

- **Exercise:** choose one important wheat product of your region and identify what are the main quality traits that define its end-use quality and what grain components affect it. Assess how local breeding programs integrate quality into their breeding schemes to ensure high end-use quality of this product.

11.6.6 Molecular Markers Useful to Select for the Above-Mentioned Traits

As reported in the previous sections, several high-throughput, small-scale and highly repeatable tests have been developed in order to accurately and efficiently define the quality of a specific wheat line. However, even if phenotypic characterization will always be needed due to the inherent complexity of quality traits, the use of molecular markers (here intended as PCR-based molecular markers) and other genotyping tools could greatly improve the efficiency and speed of wheat quality selection.

Up to now, several molecular markers targeting the genes associated with major quality traits such as kernel hardness, gluten quality, flour or semolina color and discoloration, and starch quality are available and are routinely used by most wheat breeding programs. However, in most of the cases, the available molecular markers are only targeting a subset of all the genes contributing to a specific quality trait and are often discriminative for only few of the alleles detected for each gene. For this reason, in the context of wheat quality, molecular markers should be preferably used to introgress or detect the presence of specific allelic variants associated with a trait of interest, rather than to predict the overall quality profile of a specific wheat line.

In contrast, genomic selection (GS) has arisen as a promising tool for the prediction of wheat quality. Using genomic selection, most of the wheat quality traits could be predicted with an accuracy ranging from ~60% for traits like gluten strength, to ~40% for traits more highly influenced by the environment such as protein content and dough extensibility. In contrast to single-locus molecular markers, genomic selection can capture the genetic complexity of the different quality traits at once, thus making the selection process more efficient and accurate. However, it is important to take into consideration that the accuracy of GS is highly affected, among the others, by the size of the training population, its relationship with the testing population and the quality of the phenotypic data. For these reasons, the application of GS to wheat quality prediction is likely to be restricted to those breeding programs that have the necessary resources to develop reliable prediction models.

11.7   Key Concepts

Grain quality is a complex and diverse concept that is mainly defined by the end-product. There is no wheat with bad or good quality; there is wheat with the correct quality to elaborate a given product or there is wheat with undesirable quality to make another product(s).

11.8   Conclusions

Grain quality is important as it defines the end-use of wheat and contributes to maximize profit across the wheat value chain. It adds value to the rest of breeding activities as it is a key set of characteristics for the trading and commercialization of the grain. Grain quality should be an integral part of the breeding process and considered within the variety development process. This is a highly feasible objective due to the knowledge acquired about the genetic control of several quality traits, which is in overall high, making genetic improvement approaches possible. Grain yield and quality are not confronted and can be obtained at the same time if the right breeding and selection strategies are implemented.

Further Reading

1. Battenfield SD, Guzmán C, Gaynor RC, Singh RP, Peña RJ, Dreisigaker S, Fritz AK, Poland JA (2016) Genomic selection for processing and end-use quality traits in the CIMMYT spring bread wheat breeding program. The Plant Genome 9:1–12
2. Clarke JM, DeAmbrogio E, Hare RA, Roumet P (2012) Genetics and breeding durum wheat. In: Sissons M, Abecassis J, Marchylo B, Carcea M (eds) Durum wheat chemistry and technology, 2nd edn. AACC International Press, pp 15–36

3. Colasuonno P, Marcotuli I, Blanco A, Maccaferri M, Condorelli GE, Tubeorsa R, Parada R, Costa de Camargo A, Schwember A, Gadaleta A (2019) Carotenoid pigment content in durum wheat (*Triticum turgidum* L. var *durum*): an overview of quantitative trait loci and candidate genes. Front Plant Sci 10:1–18
4. Ficco DBM, Mastrangelo AM, Trono D, Borrelli GM, De Vita P, Fares C, Beleggia R, Platani C, Papa R (2014) The colours of durum wheat: a review. Crop Pasture Sci 65:1–15
5. Guzman C, Alvarez JB (2016) Wheat waxy proteins: polymorphism, molecular characterization and effects on starch properties. Theor Appl Genet 129:1–16
6. Haile JK, N'Diaye A, Clarke F, Clarke J, Knox R, Rutkoski J, Bassi FM, Pozniak CJ (2018) Genomic selection for grain yield and quality traits in durum wheat. Mol Breed 38:75
7. He XY, He ZH, Zhang LP, Sun DJ, Morris CF, Fuerst EP, Xia XC (2007) Allelic variation of *polyphenol oxidase* (*PPO*) genes located on chromosomes 2A and 2D and development of functional markers for the *PPO* genes in common wheat. Theor Appl Genet 115:47–58
8. Hernández-Espinosa N, Mondal S, Autrique E, Gonzalez-Santoyo H, Crossa J, Huerta-Espino J, Singh RP, Guzmán C (2018) Milling, processing and end-use quality traits of CIMMYT spring bread wheat germplasm under drought and heat stress. Field Crops Res 215:104–112
9. Igrejas G, Ikeda T, Guzman C (eds) (2020) Wheat quality for improving processing and human health. Springer, Cham
10. Pasha I, Anjum FM, Morris CF (2010) Grain hardness: a major determinant of wheat quality. Food Sci Technol Int 16:0511–0512
11. Peña-Bautista RJ, Hernandez-Espinosa N, Jones JM, Guzmán C, Braun HJ (2017) Wheat-based foods: their global and regional importance in the food supply, nutrition, and health. Cereals Foods World 62:231–249
12. Rasheed A, Wen W, Gao F, Zhai S, Jin H, Liu J, Guo Q, Zhang Y, Dreisigacker S, Xia X, He Z (2016) Development and validation of KASP assays for genes underpinning key economic traits in bread wheat. Theor Appl Genet 129:1843–1860
13. Shewry PR, Halford NG, Lafiandra D (2003) Genetics of wheat gluten proteins. Adv Genet 49:111–184
14. Shevkani K, Singh N, Bajaj R, Kaur A (2017) Wheat starch production, structure, functionality and applications – a review. Int J Food Sci Technol 52:38–58
15. Sissons M, Abecassis J, Cubadda R, Marchylo B (2012) Methods used to assess and predict quality of durum wheat, semolina and pasta. In: Sissons M, Abecassis J, Marchylo B, Carcea M (eds) Durum wheat chemistry and technology, 2nd edn. AACC International Press, pp 213–234
16. Wrigley C, Batey I, Skylas D, Sharp P (eds) (2006) Gliadin and Glutenin: the unique balance of wheat quality. AACCI Press, St. Paul

Chapter 12
Nutritionally Enhanced Wheat for Food and Nutrition Security

Velu Govindan, Kristina D. Michaux, and Wolfgang H. Pfeiffer

Abstract The current and future trends in population growth and consumption patterns continue to increase the demand for wheat. Wheat is a major source and an ideal vehicle for delivering increased quantities of zinc (Zn), iron (Fe) and other valuable bioactive compounds to population groups who consume wheat as a staple food. To address nutritious traits in crop improvement, breeding feasibility must be assessed and nutrient targets defined based on their health impact. Novel alleles for grain Zn and Fe in competitive, profitable, Zn enriched wheat varieties have been accomplished using conventional breeding techniques and have been released in South Asia and Latin America, providing between 20% and 40% more Zn than local commercial varieties and benefitting more than four million consumers. Future challenges include accelerating and maintaining parallel rates of genetic gain for productivity and Zn traits and reversing the trend of declining nutrients in wheat that has been exacerbated by climate change. Application of modern empirical and analytical technologies and methods in wheat breeding will help to expedite genetic progress, shorten time-to-market, and achieve mainstreaming objectives. In exploiting synergies from genetic and agronomic options, agronomic biofortification can contribute to achieving higher Zn concentrations, stabilize Zn trait expression, and increase other grain minerals, such as selenium or iodine. Increasing Fe bioavailability in future breeding and research with other nutrients and bioactive compounds is warranted to further increase the nutritious value of wheat. Crop profiles must assure value propositions for all actors across the supply chain and consider processors requirements in product development.

Keywords Nutritional quality · Biofortification · Micronutrients · Zinc · Yield gains · Genomic selection

V. Govindan (✉)
International Maize and Wheat Improvement Center (CIMMYT), Texcoco, Mexico
e-mail: velu@cgiar.org

K. D. Michaux · W. H. Pfeiffer
HarvestPlus, IFPRI, Washington, DC, USA
e-mail: k.michaux@cgiar.org; w.pfeiffer@cgiar.org

12.1 Learning Objectives

- Understanding product development prospects for nutritional quality traits in wheat breeding and the knowledge to develop a roadmap for application.
- Mainstreaming nutritional quality traits in wheat breeding and novel approaches.
- Parental selection, crossing strategies, speed breeding and selection strategies.
- Integration of genomic selection and population improvement approaches for simultaneous gains for grain yield and Zn concentration.
- Product deployment and value chain development for biofortified wheat.

12.2 Introduction

12.2.1 Improving Nutrition of Crops for Human Health

In the past, the agriculture sector focused on producing enough food for a rapidly growing global population, in large part by increasing the production of widely consumed, acceptable, cheap, high calorie staple food crops, including wheat [1]. Yet, food insecurity and malnutrition remain a challenge across the lifespan and may worsen as result of a still rising global population, climate change, and the recent COVID-19 pandemic. Forty-seven million children under five years are too thin for their age, chronic undernutrition affects 144 million children (<5 years), 462 million adults are underweight, and an estimated two billion people are affected by one or more micronutrient deficiencies [2]. Vitamin A, iron, zinc, and iodine deficiencies are the most common worldwide (Table 12.1). Even more alarming are

Table 12.1 Global prevalence rates for key micronutrients and consequences of deficiency

Micronutrient	Consequences of deficiency	Estimated global prevalence rates for key population groups
Vitamin A	Impaired vision; night blindness; increased risk of morbidity and mortality	29% of children (<5 year) and 15% of pregnant women are vitamin A deficient
Iron	Leading cause of anemia; impaired physical and cognitive performance; increased risk of neonatal and maternal mortality	42% of children (<5year) and 29% of women are anemic, with between 30% and 50% of all anemia cases attributed to iron deficiency
Zinc	Stunted growth and development; impaired immune function; loss of appetite; cell damage; increased risk of T2DM and CVD	17% of the global population is at risk of inadequate zinc intake
Iodine	Impaired thyroid function; goiter; abnormal growth and developmental, including irreversible brain damage.	29% of school-aged children (6–12 year) have inadequate iodine intake

With data from [3]

the rates of overweight and obesity and diet related non-communicable diseases (NCDs) that are increasing at a devastating pace, particularly in low- and middle-income countries (LMICs). Nearly two billion adults and 38 million children (<5 years) are overweight or obese and NCDs are the cause of more than 70% of global deaths every year [2]. The number of overweight and obese people is expected to increase by an additional 40% by 2025, if the rate of increase remains unchanged [2].

In recent years there has been a shift in resources and political will that has encouraged researchers across a range of disciplines, including agriculture, public health and nutrition, to address not only the quantity of staple food crops but also the quality (i.e., potential nutritional impact) [1]. One of the biggest advancements was the nutritional enrichment of crops called "biofortification", using agriculture as vehicle for a public health intervention. Biofortification is the use of conventional plant breeding, bioengineering techniques, and agronomic practices to improve the nutritional value of staple food crops [4]

Wheat is an ideal crop for biofortification and conventionally bred high zinc (Zn) wheat is one of its success stories. Wheat ranks second in global cereal production and is a valuable source of nutrients and other non-nutritious bioactive compounds (see Chap. 24). For example, compared to other staple cereal grains, such as maize and rice, wheat is relatively high in protein content (average protein content in wheat cultivars is between 10% and 15% per dry weight) and it supplies an average of 20% of daily global protein intake for humans [5]. However, while wheat has a wealth of genetic resources available in its secondary and tertiary gene pools [6], the levels of micronutrients, such as Zn, in commonly consumed, commercial varieties are not high enough to meet daily requirements of people in countries where wheat constitutes the main source of calories. Thus, biofortified whole grain wheat can supply essential micronutrients to vulnerable at-risk populations with inadequate intake, particularly young children and women of reproductive age [4].

12.2.2 Importance of a Whole Grain Diet

For optimal nutrition and health, global dietary guidelines recommend consumption of whole wheat grain [7]. Whole grain wheat includes the bran, germ and endosperm, whereas in refined grain, only the endosperm remains after milling [8]. Nutritionally superior, whole grain wheat provides energy, protein, some fats, vitamins, minerals, dietary fiber, and bioactive compounds [8]. If the bran and germ are removed during processing, the refined flour provides mostly energy in the form of carbohydrates (~80%), and very little other nutrients, dietary fiber, or bioactive compounds.

Dietary fiber and bioactive components provide important health benefits to humans. Dietary fiber is vital for maintaining gut health through its fermentation in the large intestine, which produces short chain fatty acids (SCFA) and increases the abundance of beneficial gut bacteria [9]. Higher intakes of whole grains and dietary fiber are causally associated with overall better health, including a lower risk of metabolic syndrome, reductions in overweight and obesity, and reduced risk of mortality and in the incidence of a wide range of diet related NCDs (9).

Bioactive compounds are heterogeneous group of molecules found in small quantities in wheat, and include tocopherols, carotenoids, certain minerals, and phenolic compounds [10] (Table 12.2). While most of these compounds do not provide any nutritional value, they have strong antioxidant properties and provide protection against inflammation and oxidative stress [11]. Although the mechanisms of action are not fully understood, in vitro, animal, and epidemiological studies suggest that consumption of bioactive compounds may help to reduce risk factors and the incidence of cardiovascular disease (CVD), type 2 diabetes mellitus (T2DM), certain cancers, and age-related eye disease [10–15].

12.2.3   Significance of Processing, Retention and Bioavailability on Nutritional Impact of Wheat

A number of factors affect the nutrition and health impact of biofortified wheat. Nutrients and non-nutritious bioactive compounds are not distributed equally throughout the edible portion of the plant and the concentration of these compounds in the grain is under genetic control and affected by growing conditions. Most of the B-vitamins and carotenoids are found in the germ fraction; more than 80% of tocopherol is found in the pericarp, testa, and aleurone; and most of the minerals and phenolic compounds have been reported in the aleurone and bran layers [8, 16] (Fig. 12.1). Hence, it is important to consider the extent and type of milling when determining the potential health benefits of biofortified wheat.

A study of the retention of Zn in biofortified and non-biofortified conventional wheat varieties in different grain milling fractions and flours of various extraction rates showed that biofortified and conventional wheat flour milled at a very low 60% extraction level contained only 14% of the Zn concentration compared to whole grain wheat [17]. In contrast, when milled at 80% extraction, the biofortified wheat flour contained 13 ppm more Zn than the non-biofortified variety [17]. Furthermore, two absorption studies with foods made from biofortified and non-biofortified wheat flour showed that Zn absorption was significantly increased with biofortification (by up to 40%), regardless of the extraction level [18, 19], thereby helping to meet dietary Zn requirements without changing food sources.

Additionally, not all compounds that are present in the plant can be absorbed; for most compounds to produce a physiological effect, they need to be bioavailable (i.e., absorbed in sufficient quantity and transported into the bloodstream). Absorption of nutrients and non-nutritious bioactive compounds are affected by the food matrix; the amount and form of the compound; dietary promoters (e.g., dietary fat and protein) and inhibitors (e.g., phytic acid); and processing conditions [8].

Table 12.2 Potential health benefits of nutritious and non-nutritious bioactive compounds in whole grain wheat

Bioactive compounds	Example(s)	Known/purported health benefits
B-vitamins	Thiamine (B1), riboflavin (B2), niacin (B3), pyridoxine (B6), folate (B9)	Essential group of water-soluble vitamins required for energy production, one carbon metabolism, DNA synthesis/repair, red blood cell production. Antioxidant properties; associated with decreased risk of cataracts; migraine prevention; intakes associated with improved risk factors for T2DM (e.g., increase intestinal absorption of sugar and reduced risk of hyperglycemia) and stroke (e.g., may lower blood pressure and help prevent dyslipidemia and atherosclerosis); may have anti-carcinogenic activity
Vitamin E	Tocopherols, tocotrienols	Essential vitamin found in food as eight fat-soluble isomers; antioxidant properties; anti-inflammatory agent; intake associated with lower risk of cancer, CVD, T2DM and obesity (e.g., lowers blood cholesterol levels and improves insulin sensitivity)
Minerals	Zinc, iron, selenium, magnesium, manganese	Essential micronutrients required for proper immune function and physical growth; antioxidant properties; role in prevention of age-related eye diseases; role in prevention of chronic diseases, including T2DM, CVD and cancer (e.g., zinc may improve glycemic control in diabetic patients, higher magnesium intake is associated with lower risk of hypertension and metabolic syndrome, selenium intake is associate with lower risk of cancer and cancer-related mortality)
Choline and compounds	Choline, betaine	Essential nutrient and methyl donor; required for very low-density lipoprotein assemble and secretion; neuroprotective effect (i.e., essential for normal fetal brain growth and development); intake associated with lower risk of fatty liver disease, optic neuropathies (e.g., glaucoma), neural tube disorders, preeclampsia, and inflammation in asthma patients
Phenolic acids	Ferulic acid	Intake associated with reduced risk of T2DM (e.g., may help improve insulin sensitivity) and certain cancers (e.g., colon, prostate, and breast cancer) by improving insulin sensitivity
Phenolic lipids	Alkylresorcinols	Plays a role in blood clotting and may help reduce risk factors for obesity and CVD (e.g., inhibiting triglyceride synthesis)
Lignans (phytoestrogens)	Secoisolariciresinol diglycoside	May play a role in preventing CVD and breast cancer

(continued)

Table 12.2 (continued)

Bioactive compounds	Example(s)	Known/purported health benefits
Phytosterols	Sisterol	Intake is associated with reduced risk factors for CVD and T2DM (e.g., may help lower blood cholesterol levels and circulating cholesterol, blood lipids, blood pressure, and blood sugars)
Non-provitamin A carotenoids	Lutein, zeaxanthin	May reduce risk of macular degeneration and cataracts
Phytic acid	myoinositol hexaphosphate	Intake associated with protection against kidney stones and reduced risk of some cancers; may help lower blood lipids

Sources: [10–15]

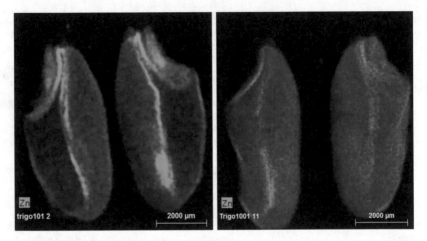

Fig. 12.1 Localization of zinc in wheat grain with μXRF (left high zinc wheat' Zinc-shakti', right CIMMYT control variety 'Baj')

12.3 Crop Improvement for Nutritional Quality

12.3.1 Setting Breeding Target Levels

Target levels were set to achieve a measurable impact on health for the primary target population: women of reproductive age and children [4]. For wheat, a target increment of +12 ppm for Zn need to be added to country or region-specific baselines to achieve a required contribution to the Estimated Average Requirement for Zn from the biofortified wheat (Fig. 12.2). While a global baseline can be assumed at 25 ppm and are derived from commercial varieties, country specific baselines may vary widely. Target increments are adjusted for per capita intake, bioavailability, and retention losses during processing, storage and cooking [20]. Hence, target increments can be achieved by breeding for higher micronutrient concentration and

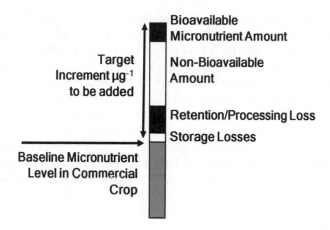

Fig. 12.2 Setting breeding target levels for grain Zn in wheat

by increasing the bioavailability and retention of micronutrients in the final product. In addition, it is projected that climate change will drastically affect the mineral and protein content in wheat. Gradual increases to target levels may be required, but our understanding is limited. Thus, there is an urgent need to quantify the impact of climate change on the nutritional quality of wheat.

12.3.2   Genetic Diversity for Nutritional Quality Traits

Large-scale screening of the genetic diversity spectrum in gene banks and collections at the International Maize and Wheat Improvement Center (CIMMYT) and partner institutions revealed genetic variation for minerals in primitive wheats, wild relatives, and landraces that would surpass targets, but insufficient variation in commercial varieties and breeding program materials. However, achieving Zn varieties with 40 ppm or higher Zn requires using unadapted genetic resources and pre-breeding materials. Though there is a large variation for Zn and Fe in wheat progenitor species, care must be taken in analysis as grain shriveling, which can be also caused be abiotic or biotic stress, may lead to a concentration effect. Grain nutrient content rather than concentration should be determined [4]. Contaminant Fe and Zn from soil/dust or threshing equipment can be assessed by measuring index elements such as Al, which are abundant in nature but absent in plants, or by measuring Ti and Cr to estimate contamination from metal parts. Breeding programs require fast, accurate, and inexpensive methods for screening large numbers of breeding lines and germplasm for nutritional traits. Energy dispersive X-ray fluorescence (EDXRF) has been investigated as a reliable alternative to ICP-based methods for high-throughput analysis of Zn and Fe in wheat.

Wheat grown in Central and East Asia reports grain Zn concentrations ranging from 14 to 35 ppm in 150 bread wheat lines grown in China, to 20–39 ppm in 66 advanced wheat genotypes grown in Kazakhstan, Kyrgyzstan and Tajikistan. Grain Zn concentration is even more variable among wheat grown in South Asia and Mexico: 29–40 ppm of grain Zn has been reported among 40 CIMMYT bread wheat lines [21]; 30–98 ppm among 518 accessions of *T. dicoccoides*; 27–53 ppm among 93 advanced lines developed by the crossing of Mexican landraces, *Triticum dicoccoides* and *Ae. Tauschii,* with durum and common wheat genotypes [22]; and 25–60 ppm in 185 recombinant inbred lines derived from crossing *T. spelta* and *T. aestivum* grown under field conditions. Grain Fe concentration usually varies by 1.2-fold in tetraploid cultivars, 1.8-fold in hexaploids cultivars, and 2.9-fold in diploid cultivars. Additionally, a set of Indian and Turkish landraces showed large variation in macrominerals[1] (1.7–6.4 fold), certain microminerals[2] (3.6–5.6 fold) and protein (2.2 fold) [23].

12.3.3   Targeted Breeding Approach

The HarvestPlus crop development alliance with CIMMYT and other national public and private sector partners and centers of excellence focused on improving grain Zn and Fe concentrations in South Asia, where Zn and Fe deficiency are widespread [2]. To date, Zn concentrations between 50% and 100% of the target increment have been achieved in competitive first and second wave varieties in India, Pakistan, Bangladesh and Nepal [24]. In developing these varieties, Zn density from synthetic wheat lines derived from wild wheat relatives, *Aegilops tauschii* (D genome donor of wheat), *Triticum spelta* and wild *T. dicoccon* were combined with yield and farmer preferred trait packages using elite breeding lines and major local varieties in crosses. In addition, due to the positive correlation between Zn and Fe content, Fe increased by approximately 0.5 ppm annually via correlated selection response. A similar increase was observed for other micronutrients positively associated with Zn, such as Mn and Mg [24].

In targeted breeding, each year 400–500 simple crosses are made between elite high/moderate Zn lines, elite high Zn lines, and best lines with normal Zn. Segregating populations from these crosses are advanced and selected for agronomic and disease resistance. F5/F6 lines retained are selected for grain characteristics and grain Zn and Fe concentration. High Zn F5/F6 lines are advanced to stage 1 replicated yield trials in Zn-homogenized fields at CIMMYT, Mexico, which show good grain Zn prediction in South Asia and other target population of environments (TPEs). Grain Zn and Fe is determined for lines with equal or higher yields

---

[1]Macrominerals: calcium (Ca), potassium (K), Magnesium (Mg), sodium (Na), phosphorus (P), and sulphur (S).

[2]Microminerals: Zinc (Zn), Iron (Fe), Copper (Cu), Manganese (Mn).

compared to checks following analyses for end-use processing quality. Varieties for South Asia must have resistance to Ug99 and yellow rust and lines are also simultaneously phenotyped for resistance during two seasons at Njoro, Kenya. The selected superior lines are then distributed to national agricultural research system (NARS) partners in South Asia and other TPEs in international nurseries which serve as germplasm distribution and investigative tools [21].

12.3.4   Genetic Architecture and Association of Nutritional Quality Traits in Wheat

Zn concentration on a global scale varies widely due to a variation of variable and permanent environmental factors, in particular edaphic conditions as soil Zn deficiency is widespread. Ranges of 20–31 ppm in grain Zn have been found [25] and genotypes showed consistent ranking across sites and years for grain Zn concentrations [26, 27].

There is a positive significant relationship with Zn concentration and N, Ca, Cu, Fe, K, Mg, Mn, Mo, P, S and Se concentrations in wheat grain (Cu et al. 2020). Grain Zn concentration is more strongly correlated with grain Cu, Fe, Mn, P and S ($r = 0.61, 0.46, 0.43, 0.53, 0.46$, respectively) than with Ca, K, Mg, Mo and Sr ($r = 0.36, 0.20, 0.33, 0.18, 0.15$, respectively) [26]. Given the positive correlation of grain Zn with these elements, increases in grain Zn concentration may also help to increase the grain Ca, Cu, Fe, K, Mg, Mn, Mo, P, and S concentrations. Co-localization of grain Zn and Fe QTLs on chromosome 2B has been reported [30]. Similarly, a slight increase in S concentration reportedly increased the grain Zn concentration, possibly due to the increase in the amino acid methionine that further increased levels of phytosiderophores and nicotinamide, which are involved in uptake and translocation of Zn (Fig. 12.3).

With respect to the relationship between grain Zn concentration and grain yield, the evidence is conflicting. The most recent data suggest there is no direct trade-off between increased grain Zn concentration and yield in wheat once Zn density is established in high yielding elite donors, which is in contrast to previous studies. These reported differences may be due to linkage drag in early biofortification materials from using unadapted high Zn parents, as a negative correlation of grain Zn and grain yield was observed in unadopted wild sources. Identifying the genomic regions that regulate the accumulation of Zn and Fe in grain without any confounding effects on yield or pleiotropic Zn and Fe QTL with seed size or other yield components would allow breeders to develop high yielding biofortified cultivars.

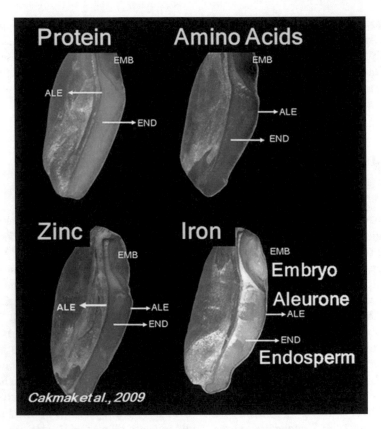

Fig. 12.3 Co-localization of grain Zn and Fe with protein and amino acids

12.3.5 Genetic Control of Nutritional Quality Traits

An enabling knowledge base regarding the genetic control of nutritional traits is crucial for breeding effectiveness. Genetic and QTL mapping studies at CIMMYT revealed that small-to-intermediate-effect QTL of additive effects govern the inheritance of grain Zn and Fe. Several studies also identified promising large-effect QTL regions for increased grain Zn on chromosomes 1B, 2B, 3A, 4B, 5B, 6B and 7B and some QTL regions have a pleiotropic effect for grain Fe (Hao et al. 2014; Velu et al. 2016; Cu et al. 2020). Moreover, 2B and 4B QTL has a pleiotropic effect for increased thousand-kernel weight (TKW), suggesting that a simultaneous improvement of grain Zn and consumer preferred seed size is possible. Four QTLs have been identified for combination in adequate genetic backgrounds in forward breeding (Hao et al. 2014).

Although several QTL of moderate effect on grain Zn have been found in different germplasm sources, the genetic control of the trait appears polygenic. Multiple years of phenotyping results and several studies at CIMMYT show a relationship

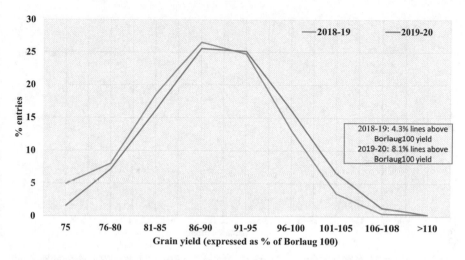

Fig. 12.4 Grain yield trends of wheat lines derived from two cohorts of Zn breeding pipeline evaluated in stage 1 replicated (3 reps) yield trials at Ciudad Obregón 2018–2019 and 2019–2020

between the two traits, while a moderately high heritability for Zn and Fe suggest that grain yield and grain Zn are independently inherited. The variance components from CIMMYT's Ciudad Obregón site in Mexico showed genotypic (main) effects attributed to a larger share of total variation for grain Zn (61%) than the environment (39%), whereas multi-site analyses of an association genetics panel across locations in India showed 27% variation attributed to genotypic effects, 30% variation explained by the genotype-by-environment (GxE) interaction, and 43% by environment and error variance [27].

Recent yield data from the stage 1 yield trials in Ciudad Obregón showed about 1% average yield gain per year was achieved over the past two years while enhancing grain Zn concentration by +1–2 ppm annually (Figs. 12.4 and 12.5), suggesting the feasibility of combining high yield with high Zn concentration. Moreover, the lack of association between grain yield and grain Zn will support their simultaneous genetic gain as realized in CIMMYT's current breeding scheme [24].

12.3.6   Agronomic Biofortification

By exploiting synergies from genetic and agronomic options, agronomic biofortification can help achieve higher Zn concentrations, stabilize Zn trait expression caused by spatial, temporal, and systems environmental fluctuations, and increase other grain minerals, such as selenium or iodine, to nutritionally important levels (see Chap. 24). Zinc has moderate phloem mobility, hence foliar application or a combination of soil and foliar application, markedly increases grain Zn content. Furthermore, grain Zn concentration is severely affected by the availability of a

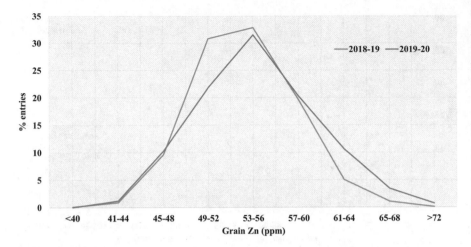

Fig. 12.5 Grain Zn concentration of wheat lines derived from two cohorts of Zn breeding pipeline evaluated in stage 1 replicated (3 reps) yield trials at Ciudad Obregón during 2018–2019 and 2019–2020

physiological pool of Zn in vegetative tissues, hence foliar application substantially increases Zn in the wheat endosperm. Soil Zn application is less effective in increasing grain Zn concentration because of poor Zn mobility and its rapid adsorption in alkaline calcareous soils. Field trial data from India and Pakistan revealed that improved stand establishment for biofortified wheat increased biomass and yield; the better ground cover can reduce evapo-transpirative moisture losses especially under rainfed production.

12.3.7 Mainstreaming Nutritional Quality Traits in Wheat Breeding and Novel Approaches

Maintaining traits of genetic gains for grain yield along with increased grain Zn concentration, thereby closing the yield gap between non-biofortified and biofortified lines (currently 4–6%), poses a major future challenge. For adoption by farmers, the performance of Zn enhanced lines/varieties must be at least on par with current non-biofortified varieties, in the absence of price premiums for Zn wheat. Adding Zn as a core trait in breeding and converting progenitors requires accelerating the breeding cycle, expanding operational scale, phenotyping bread wheat breeding lines for Zn, and then applying the latest technologies in phenotyping, genotyping, molecular-assisted and genomic selection (GS) methods (see Chap. 7). To maximize reach, mainstreaming must be expanded to public and private national program partners and the enabling infrastructure built. This includes optimizing selection environments for Zn to overcome the heterogeneity problem by using soil Zn application in key TPEs.

Capturing favorable and simultaneous additive effects in yield and Zn improvement requires selection indices with weights for yield and Zn, as well as considering heritability and genetic variance estimates in target locations to guide the development of inter-population recurrent selection schemes for crossing well-defined parental lines. Correlated selection response will lead increases in Fe and other positively correlated elements. While the medium high heritability for Zn increases gains in selection, both yield and Zn content are polygenic traits, and an increased breeding effort and new approaches are required to increase allele frequencies and capitalize on transgressive segregation. In order to achieve faster genetic gains, CIMMYT is generating genotypic data for high Zn wheat lines and training populations specific for biofortification breeding established. Prediction models developed using novel statistical genetic models (e.g., gBLUP -Genomic best linear unbiased prediction-), which incorporate available genomic and phenomic information, are validated and utilized in the rapid breeding pipeline to select potential parents and progenies with high breeding values for Zn and grain yield. So far, genomic predictions for Zn and Fe are moderately high ($r = 0.4–0.6$) across locations in Mexico and India using the association mapping panel from the biofortification program [32]. Therefore, GS models for these traits could also be used to select parents.

12.3.8 Speed Breeding

Modern breeding techniques that use cutting-edge genomics and accelerated breeding cycles are to be exploited to accelerate achieving linear progress, at 1–2% genetic gains per cycle for grain Zn and yield (see Chap. 30). Generally, genetic gain is determined by Eq. 12.1.

Breeder's equation:

$$\Delta g = \frac{i * r * \sigma g}{L} \tag{12.1}$$

where:

Δg = Genetic gain
i = selection intensity
r = selection accuracy
σg = genetic variance
L = breeding cycle length

Targeted genetic gain can be achieved at a faster rate by shortening the breeding cycle or generation time rather than selection intensity and heritability, which are highly trait and environment dependent. Shortening the breeding cycle or generation time can be achieved by adopting speed breeding [34]. This is different from shuttle breeding, which was pioneered by N.E. Borlaug in the late 1940s. Shuttle breeding was developed to advance two generations per year and half development

Rapid Cycle Recurrent Selection (RCRS)

Fig. 12.6 Proposed RCRS breeding scheme with a two-year breeding cycle

Fig. 12.7 Happy wheat plants in a speed breeding facility at CIMMYT, Toluca, Mexico

time, as well as add yield stability, adaptation range, and photoperiod insensitivity by selecting in contrasting environments. Speed breeding allows three to four generations per year, for example, at CIMMYT's Toluca facility.

Further, speed breeding allows faster recombination of elite lines through genomic estimated breeding values. CIMMYT uses a rapid-cycle recurrent selection (RCRS) population improvement approach with a two-year breeding cycle (Fig. 12.6); wheat plants are advanced in greenhouses using a speed breeding green house facility (Fig. 12.7).

12.3.9   Population Improvement

Population improvement increases the allele frequency of positive alleles for grain yield and grain Zn using recurrent selection. In a closed population development system, each cohort (cycle) of materials should originate from crosses between parents from previously evaluated and selected materials.

12.3.10   Genomic Selection

The application of GS in wheat breeding is enabled by the availability of high-throughput molecular markers, which cover the entire genome and facilitate trait value prediction. Experimental studies based on multi-environment wheat trials demonstrated that genomic selection models accurately predict genetic values of complex traits, such as grain yield, grain nutritional quality, or stress adaptation under different conditions [35]. However, prediction accuracy values for grain yield varies widely across different cohorts of materials and at different TPEs and studies that consider GxE interactions are still under development. Nevertheless, retrospective GS analysis showed promising results, supporting its application in breeding.

12.4   Product Development and Dissemination

The aim of the breeding scheme is to extract products from the population improvement scheme to deliver improved wheat lines to NARS partners. CIMMYT focuses on population improvement, introducing new alleles through trait/diversity introgression onto elite parents containing positive alleles that are either absent, or present only in low frequencies in the core germplasm, in forward-breeding/population improvement activities. Introgressing new germplasm needs to be based on yield and Zn values and on a robust selection index.

12.4.1   Adoption and Commercialization of Biofortified Wheat

Traditional breeding focuses on improving traits of known economic value and developing product profiles for existing markets. Traits are targeted for selection based on whether they can provide better crop and/or utilization options to farmers. In general, these traits are related to productivity, biotic/abiotic stress, and end-use quality. Biofortification additionally enhances traits whose value is measured in health and nutrition outcomes. Improvements in nutrition and health status from consuming biofortified wheat occur gradually. Enhanced traits, such as Zn or Fe, are invisible and do not affect sensory characteristics; therefore, crop profiles must assure value propositions for all actors across the value chain and consider processors requirements in product development (see Chap. 11) (Fig. 12.8).

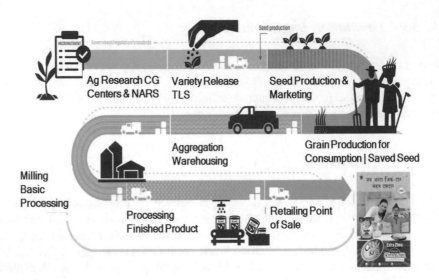

Fig. 12.8 Value chain for biofortified wheat

Producer and consumer insight research is critical for farmer adoption and consumer acceptance. Research and concept testing in India revealed that more than 90% of farmers and rural and urban consumers would definitely or likely grow and consume Zn wheat, once they aware of the health benefits. Overall, awareness of the health benefits from consuming biofortified wheat, in particular minerals, is low and awareness campaigns, demand creation, and promotion materials are important in developing sustainable markets for nutritionally enhanced biofortified crops. Supply chain development is of equal importance, particularly in countries that lack segregation. Identity preservation, quality control, and traceability must be guaranteed when developing procurement systems for timely volume supply for private and institutional buyers. In this context, standards at the grain level were identified as a missing enabler of the wheat value chain making it difficult for buyers to specifically procure high Zn grain. HarvestPlus partnered with the British Standards Institute to create an international Publicly Available Standard (PAS) for Zn with planned publication and use in June 2021; development of a PAS for Fe has also commenced. The PAS are harmonized with breeding targets of the HarvestPlus/CIMMYT crop development alliance and helps facilitate including micronutrients as value added traits and considered in release.

The impact pathway for commercializing biofortified wheat is outlined in Fig. 12.9. The primary beneficiaries of biofortification are smallholder farm households in LMICs who rely on staple crops for caloric and nutrient intakes and often lack reliable access to diversified diets, fortified foods or supplements. In addition, partnerships along the value chain are crucial to catalyzing nutritious, biofortified food systems worldwide that deliver adequate micronutrients through regular diets. To achieve a global reach, it is critical to market and advocate for biofortification at all levels, including developing a regulatory framework for biofortified crops,

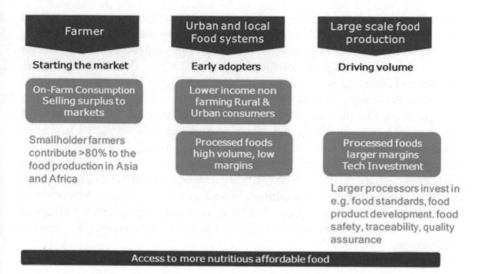

Fig. 12.9 Impact pathway for commercializing biofortified wheat

mainstreaming micronutrients in crop development and in food systems, and incorporating biofortification into public and private policies, programs, and investments.

12.5 Key Concepts

The chapter provides the prospects of wheat improvement for nutritional quality traits, guiding through steps in practical application. Specifically, on the mainstreaming of grain zinc in CIMMYT wheat germplasm using novel approaches of applying modern genomics and quantitative genetics strategies of population improvement coupled with rapid recycling of parents based on true breeding value (TBV) to accelerate rate of genetic gain for grain yield and grain zinc concentration. In addition, product deployment and value chain development for biofortified wheat also discussed.

12.6 Conclusions and Future Perspectives

Crop improvement for nutritional quality is essential to improving public health in developing and developed market economies. Biofortified wheat varieties that provide between 20% and 40% more Zn than local commercial varieties have been released since 2016 in South Asia and are grown by roughly one million households, benefitting more than four million consumers. Through mainstreaming grain Zn, CIMMYT's wheat breeding program will achieve more than 75–80% of elite lines with enhanced Zn (and Fe) within the next ten years. This will be realized using modern genomics and speed breeding techniques to reduce breeding cycle time and accelerate rates of genetic gains for both high yield and increased grain Zn concentration. Eventually, higher frequency of elite wheat lines with high yield, high Zn, and other agronomic traits will become available to NARS partners who can then select and promote biofortified wheat varieties to farmers and consumers in target countries, thus helping to improve global nutrition and food security. Research with other nutrients and bioactive compounds is warranted to further increase the nutritious value of wheat. In exploiting synergies from genetic and agronomic options, agronomic biofortification can contribute to achieving higher mineral concentrations and stabilize Zn trait expression caused by spatial, temporal, and systems environmental fluctuations. Since Zn or Fe content is a quantitatively inherited invisible trait, crop profiles must assure value propositions for all actors across the supply chain and consider processors requirements in product development.

References

1. Poole N, Donovan J, Erenstein O (2020) Agri-nutrition research: revisiting the contribution of maize and wheat to human nutrition and health. Food Policy. https://doi.org/10.1016/j.foodpol.2020.101976
2. FAO, IFAD, UNICEF, WFP, WHO (2020) The state of food security and nutrition in the world 2020. Transforming food systems for affordable healthy diets. Rome
3. World Health Organization (2020) Micronutrient database. In: The vitamin and mineral nutrition information system. https://www.who.int/vmnis/database/en/
4. Pfeiffer WH, McClafferty B (2007) HarvestPlus: breeding crops for better nutrition. Crop Sci 7:S88–S105. https://doi.org/10.2135/cropsci2007.09.0020IPBS
5. Shiferaw B, Smale M, Braun HJ, Duveiller E, Reynolds M, Muricho G (2013) Crops that feed the world 10. Past successes and future challenges to the role played by wheat in global food security. Food Secur 5:291–317. https://doi.org/10.1007/s12571-013-0263-y
6. Velu G, Ortiz-Monasterio I, Cakmak I, Hao Y, Singh RP (2014) Biofortification strategies to increase grain zinc and iron concentrations in wheat. J Cereal Sci 59:365–372
7. Reynolds A, Mann J, Cummings J, Winter N, Mete E, Te Morenga L (2019) Carbohydrate quality and human health: a series of systematic reviews and meta-analyses. Lancet 393:434–445. https://doi.org/10.1016/S0140-6736(18)31809-9
8. Luthria DL, Lu Y, John KMM (2015) Bioactive phytochemicals in wheat: Extraction, analysis, processing, and functional properties. J Funct Foods 18:910–925. https://doi.org/10.1016/j.jff.2015.01.001
9. Barber TM, Kabisch S, Pfeiffer AFH, Weickert MO (2020) The health benefits of dietary fibre. Nutrients 12:1–17
10. Gani A, Wani SM, Masoodi FA, Hameed G (2012) Whole-grain cereal bioactive compounds and their health benefits: a review. J Food Process Technol 03. https://doi.org/10.4172/2157-7110.1000146
11. Saini P, Kumar N, Kumar S, Mwaurah PW, Panghal A, Attkan AK, Singh VK, Garg MK, Singh V (2020) Bioactive compounds, nutritional benefits and food applications of colored wheat: a comprehensive review. Crit Rev Food Sci Nutr 61:3197–3210
12. Bruins MJ, Van Dael P, Eggersdorfer M (2019) The role of nutrients in reducing the risk for noncommunicable diseases during aging. Nutrients 11. https://doi.org/10.3390/nu11010085

13. Bjørklund G, Dadar M, Pivina L, Doşa MD, Semenova Y, Aaseth J (2019) The role of zinc and copper in insulin resistance and diabetes mellitus. Curr Med Chem 27:6643–6657. https://doi.org/10.2174/0929867326666190902122155

14. Al-Ishaq RK, Overy AJ, Büsselberg D (2020) Phytochemicals and gastrointestinal cancer: cellular mechanisms and effects to change cancer progression. Biomolecules 10. https://doi.org/10.3390/biom10010105

15. Abdel-Aal ESM, Akhtar H, Zaheer K, Ali R (2013) Dietary sources of lutein and zeaxanthin carotenoids and their role in eye health. Nutrients 5:1169–1185. https://doi.org/10.3390/nu5041169

16. Li J, Liu J, Wen W, Zhang P, Wan Y, Xia X, Zhang Y, He Z (2018) Genome-wide association mapping of vitamins B1 and B2 in common wheat. Crop J 6:263–270. https://doi.org/10.1016/j.cj.2017.08.002

17. Hussain S, Maqsood MA, Rengel Z, Aziz T, Abid M (2013) Estimated zinc bioavailability in milling fractions of biofortified wheat grains and in flours of different extraction rates. Int J Agric Biol 15:921–926

18. Rosado JL, Hambidge KM, Miller LV, Garcia OP, Westcott J, Gonzalez K, Conde J, Hotz C, Pfeiffer W, Ortiz-Monasterio I, Krebs NF (2009) The quantity of zinc absorbed from wheat in adult women is enhanced by biofortification. J Nutr 139:1920–1925. https://doi.org/10.3945/jn.109.107755

19. Signorell C, Zimmermann MB, Cakmak I, Wegmüller R, Zeder C, Hurrell R, Aciksoz SB, Boy E, Tay F, Frossard E, Moretti D (2019) Zinc absorption from agronomically biofortified wheat is similar to post-harvest fortified wheat and is a substantial source of bioavailable zinc in humans. J Nutr 149:840–846. https://doi.org/10.1093/jn/nxy328

20. Van Der Straeten D, Bhullar NK, De Steur H, Gruissem W, MacKenzie D, Pfeiffer W, Qaim M, Slamet-Loedin I, Strobbe S, Tohme J, Trijatmiko KR, Vanderschuren H, Van Montagu M, Zhang C, Bouis H (2020) Multiplying the efficiency and impact of biofortification through metabolic engineering. Nat Commun 11:1–10. https://doi.org/10.1038/s41467-020-19020-4

21. Velu G, Singh RP, Huerta-Espino J, Peña RJ, Arun B, Mahendru-Singh A, Mujahid MY, Sohu VS, Mavi GS, Crossa J, Alvarado G, Joshi AK, Pfeiffer WH (2012) Performance of biofortified spring wheat genotypes in target environments for grain zinc and iron concentrations. Field Crop Res 137:261–267. https://doi.org/10.1016/j.fcr.2012.07.018

22. Guzmán C, Medina-Larqué AS, Velu G, González-Santoyo H, Singh RP, Huerta-Espino J, Ortiz-Monasterio I, Peña RJ (2014) Use of wheat genetic resources to develop biofortified wheat with enhanced grain zinc and iron concentrations and desirable processing quality. J Cereal Sci 60:617–622. https://doi.org/10.1016/j.jcs.2014.07.006

23. Bhati KK, Aggarwal S, Sharma S, Mantri S, Singh SP, Bhalla S, Kaur J, Tiwari S, Roy JK, Tuli R, Pandey AK (2014) Differential expression of structural genes for the late phase of phytic acid biosynthesis in developing seeds of wheat (Triticum aestivum L.). Plant Sci 224:74–85. https://doi.org/10.1016/j.plantsci.2014.04.009

24. Velu G, Crespo Herrera L, Guzman C, Huerta J, Payne T, Singh RP (2019) Assessing genetic diversity to breed competitive biofortified wheat with enhanced grain Zn and Fe concentrations. Front Plant Sci 9. https://doi.org/10.3389/fpls.2018.01971

25. Li BY, Zhou DM, Cang L, Zhang HL, Fan XH, Qin SW (2007) Soil micronutrient availability to crops as affected by long-term inorganic and organic fertilizer applications. Soil Tillage Res 96:166–173. https://doi.org/10.1016/j.still.2007.05.005

26. Khokhar JS, Sareen S, Tyagi BS, Singh G, Wilson L, King IP, Young SD, Broadley MR (2018) Variation in grain Zn concentration, and the grain ionome, in field-grown Indian wheat. PLoS One 13. https://doi.org/10.1371/journal.pone.0192026

27. Velu G, Singh RP, Joshi AK (2020) A decade of progress on genetic enhancement of grain zinc and iron in CIMMYT wheat germplasm. In: Wheat and barley grain biofortification. Elsevier, pp 129–138

28. Zhao FJ, Su YH, Dunham SJ, Rakszegi M, Bedo Z, McGrath SP, Shewry PR (2009) Variation in mineral micronutrient concentrations in grain of wheat lines of diverse origin. J Cereal Sci 49:290–295. https://doi.org/10.1016/j.jcs.2008.11.007
29. Cu ST, Guild G, Nicolson A, Velu G, Singh R, Stangoulis J (2020) Genetic dissection of zinc, iron, copper, manganese and phosphorus in wheat (Triticum aestivum L.) grain and rachis at two developmental stages. Plant Sci 291. https://doi.org/10.1016/j.plantsci.2019.110338
30. Hao Y, Velu G, Peña RJ, Singh S, Singh RP (2014) Genetic loci associated with high grain zinc concentration and pleiotropic effect on kernel weight in wheat (Triticum aestivum L.). Mol Breed 34:1893–1902. https://doi.org/10.1007/s11032-014-0147-7
31. Srinivasa J, Arun B, Mishra VK, Singh GP, Velu G, Babu R, Vasistha NK, Joshi AK (2014) Zinc and iron concentration QTL mapped in a Triticum spelta × T. Aestivum cross. Theor Appl Genet 127:1643–1651. https://doi.org/10.1007/s00122-014-2327-6
32. Velu G, Crossa J, Singh RP, Hao Y, Dreisigacker S, Perez-Rodriguez P, Joshi AK, Chatrath R, Gupta V, Balasubramaniam A, Tiwari C, Mishra VK, Sohu VS, Mavi GS (2016) Genomic prediction for grain zinc and iron concentrations in spring wheat. Theor Appl Genet 129:1595–1605. https://doi.org/10.1007/s00122-016-2726-y
33. Velu G, Singh RP, Crespo-Herrera L, Juliana P, Dreisigacker S, Valluru R, Stangoulis J, Sohu VS, Mavi GS, Mishra VK, Balasubramaniam A, Chatrath R, Gupta V, Singh GP, Joshi AK (2018) Genetic dissection of grain zinc concentration in spring wheat for main-streaming biofortification in CIMMYT wheat breeding. Sci Rep 8. https://doi.org/10.1038/s41598-018-31951-z
34. Voss-Fels KP, Stahl A, Hickey LT (2019) Q&A: Modern crop breeding for future food security. BMC Biol. 17:1–7
35. Juliana P, Poland J, Huerta-Espino J, Shrestha S, Crossa J, Crespo-Herrera L, Toledo FH, Govindan V, Mondal S, Kumar U, Bhavani S, Singh PK, Randhawa MS, He X, Guzman C, Dreisigacker S, Rouse MN, Jin Y, Pérez-Rodríguez P, Montesinos-López OA, Singh D, Mokhlesur Rahman M, Marza F, Singh RP (2019) Improving grain yield, stress resilience and quality of bread wheat using large-scale genomics. Nat Genet 51:1530–1539. https://doi.org/10.1038/s41588-019-0496-6

Chapter 13
Experimental Design for Plant Improvement

Ky L. Mathews and José Crossa

Abstract Sound experimental design underpins successful plant improvement research. Robust experimental designs respect fundamental principles including replication, randomization and blocking, and avoid bias and pseudo-replication. Classical experimental designs seek to mitigate the effects of spatial variability with resolvable block plot structures. Recent developments in experimental design theory and software enable optimal model-based designs tailored to the experimental purpose. Optimal model-based designs anticipate the analytical model and incorporate information previously used only in the analysis. New technologies, such as genomics, rapid cycle breeding and high-throughput phenotyping, require flexible designs solutions which optimize resources whilst upholding fundamental design principles. This chapter describes experimental design principles in the context of classical designs and introduces the burgeoning field of model-based design in the context of plant improvement science.

Keywords Model-based · Classical design · Linear mixed model

The original version of the chapter has been revised. A correction to this chapter can be found at
https://doi.org/10.1007/978-3-030-90673-3_34

K. L. Mathews (✉)
University of Wollongong, Wollongong, NSW, Australia
e-mail: kmathews@uow.edu.au

J. Crossa
International Maize and Wheat Improvement Center (CIMMYT), Texcoco, Mexico
e-mail: j.crossa@cgiar.org

13.1   Learning Objectives

* Understand fundamental experimental design concepts.
* Describe the structural differences between classical designs.
* Understand the purpose of model-based design and how it can enhance plant improvement experiments.

13.2   Introduction

Good experimental design underpins wheat improvement research, whether it is conducted in the field, glasshouse or laboratory. Experimental design theory has developed over the last two decades from the classical designs described in texts like Cochran and Cox [1] to optimal model-based designs introduced in Martin [2] and extended in Butler et al. [3] and Cullis et al. [4]. However, the fundamental design principles of replication, randomisation and controlling for heterogeneity promoted by Fisher [5] remain the same (Sect. 13.3).

Typical plant improvement experiments (PIEs) evaluate treatments in replicated experiments which follow one of a few classical experimental design structures (Sect. 13.4). Examples of treatments include genetic entities such as lines, hybrids or varieties in breeding trials; agronomic factors such as fertilizer or irrigation amounts and pathotypes in disease rating trials. Classical designs primarily differ in the way they control for expected heterogeneity in the experiment, that is, their plot, or block structure (Sect. 13.5). These structures are rigid with respect to the number of treatments and/or replication per treatment and can constrain research outcomes.

In contrast, model-based designs are flexible and directly link to the data analysis. They are enabled by the development of statistical modelling technology and advances in computational power. Hence, it is now possible to design experiments which optimize resource use and improve treatment prediction accuracy. Importantly, classical designs can be generated in the model-based design paradigm, as demonstrated in Sect. 13.5.

It is important to understand that the new technologies of high throughput phenotyping, genomic selection and rapid cycle breeding are as dependent on robust experimental design as older breeding technologies. The success of these technologies will depend on cohesive multi-disciplinary teams which include biometricians. This chapter aims to provide researchers with a good understanding of experimental design concepts and a taste of what is possible in the model-based design paradigm. As such, it is a resource for basic knowledge and a springboard to other resources for out-of-scope topics.

13.3   Fundamental Design Concepts

The following terms and concepts form the basis for understanding classical and model-based designs in a plant improvement context. These definitions follow two recommended texts: [6, 7].

13.3.1 Definitions

Experimental Purpose is the aim of the experiment. Examples include, selecting breeding lines for variety release and testing the hypothesis that two pesticides are equally effective in controlling aphids.

Experimental Unit is the smallest unit to which a treatment is applied. For example, in a yield trial a field plot is the experimental unit as a variety (the treatment) is allocated to an entire plot. In an agronomy trial where a herbicide treatment is applied along a row (containing 10 field plots, say) then the experimental unit for herbicide is a row.

Treatment Factors are factors of interest imposed by the researcher, each treatment factor describes what can be applied to an experimental unit. The treatment structure is a meaningful way to divide up the set of treatments.

Observational Unit is the smallest unit on which a response (trait) is measured. It is often called a *plot* but it may not reflect an actual field plot. For example, the observational unit (*plot*) could represent a tiller or grain sample, sampled from within a field plot. In yield trials, the observational unit (*plot*) is a physical field plot (i.e. the intersection of a row and column in a field layout) and yield is measured on the whole plot. The term *field plot* is used for clarity.

Plot Factors are the non-treatment factors whose structure describes the observational units (*plots*).

Design Function describes how the treatments are allocated to plots. The process of randomization which determines this allocation takes many forms and considers the logistical constraints of the experiment and the experimental purpose.

13.3.2 Replication

A *replicate* is a copy of a treatment, such that the number of replicates of a treatment is the number of experimental units to which a treatment is applied [7].

   A common question from researchers is "how many replicates do I need?" The ability to detect a statistically significant difference between treatments, or power, depends on the underlying population variance (σ^2) and the sample size (replication, n). The formula for the variance of the sample mean is σ^2/n. Theoretically, it is clear that increasing n should decrease the variance of the sample mean thereby increasing the power of the experiment but this is not always the case (see [6, 7] for further details).

13.3.3   Randomization

Randomization is the process of allocating treatments to experimental units. Randomization minimizes bias in the experiment ensuring representative sampling of each treatment. Bailey [6] describes four types of bias, each illustrated here with a plant improvement example:

Systematic: allocating the varieties 1, 2, …, 20 to plots 1, 2, …, 20, i.e., variety 1 to plot 1, variety 2 to plot 2 etc. in the first replicate for all trials in a multi-environment trial series.

Selection: compositing the grain samples from the varieties with lower plot yields but not those with higher plot yields.

Accidental: measuring a grain quality trait on varieties which reach maturity before others.

Cheating: allocating an irrigation treatment to a lower lying part of the field than a non-irrigation treatment.

13.3.4   Blocking: Controlling for Variability

Biologically, individual experimental units (e.g., field plots) vary from one another prior to the application of treatments. Common sources of variability are fertility and moisture gradients in the field; lighting and air conditioning in glasshouse and processing equipment such as mills in laboratories. If this variability is ignored in the design (or analysis) then the measurement error (residual variation) can be inflated which results in less accurate comparisons between the treatments of interest (see Chap. 15).

   Blocking the experimental units into groups that are considered to be homogeneous attempts to control known (or anticipated) local variation [5], thereby reducing the residual variance and increasing the precision (power) of the experiment. Complete blocks contain an experimental unit for each treatment, incomplete blocks do not. Spatial variability and experimental logistics determine block size, shape and orientation.

13.3.5   Pseudo-Replication

Pseudo-(or false) replication is when multiple measurements are taken from an experimental unit. Pseudo-replication frequently occurs when treatments are allocated to big blocks. For example, two trials of a double-haploid population are conducted to assess drought tolerance, one in an irrigated block and the other in a non-irrigated block. There is replication of the breeding lines within each trial (block) but there is no replication of the irrigation treatments and is thus not 'real' [6, 7].

13.3.6   Orthogonality and Balance

Orthogonality and balance describe the structure of an experiment [7]. Two factors are orthogonal if they can be evaluated independently of each other, i.e. their estimated effects are the same irrespective of the presence (or not) of the other factor in the model [7]. A balanced design (e.g., RCBD) has equal precision on all treatment comparisons [7].

Non-orthogonal designs are possible for situations where resources are limited. In non-orthogonal designs treatment factors are deliberately not equally replicated or deliberately confounded with other factors, such as blocks. Identifying which factors in a design are orthogonal (or not) enables appropriate inference about the key factors of interest. Non-orthogonality can occur between any two factors (treatment or plot) in an experiment. For example, in a randomized complete block design (RCBD, Sect. 13.4.2) the treatment and plot factors are orthogonal but if there is a missing data point then they are not.

Balanced incomplete block designs are an example where there is non-orthogonality between the block and treatment factors but they are balanced because each pair of treatments occurs equally often within the same blocks [7].

13.3.7   Resolvability

A design is resolvable if its blocks (complete or incomplete) can be grouped into sets such that each treatment occurs exactly once in each set, e.g., a RCBD is resolvable (Sect. 13.4.2).

Resolvability ensures orthogonality between treatment and block factors. It is not necessary for an optimal design. However, near-optimal designs are achieved when near-resolvability is attained.

13.3.8   Optimality Criterion

An optimal design is selected based on a pre-determined criterion. Two common criteria, $A-$ and $D-$optimality, seek to minimize a variance. A minimizes the average pair-wise variance of treatment *differences* whilst D minimizes the variance of treatment means. A is common in plant improvement experiments where the treatment comparisons are of equal interest [2, 3, 8, 9]. A lower value indicates greater optimality.

13.3.9   Model Notation

We use the notation of Wilkinson and Rogers [10] to describe the relationship between factors in treatment and plot structures. Let A and B be two factors, where their structure can be independent (A + B, main effects), interacting (A:B), crossed (A*B, a factorial) or nested (A/B, where B is nested with A). The latter two expand such that,

$$A * B = A + B + A : B, \text{ and}$$

$$A / B = A + A : B$$

See Piepho et al. [11] and Welham et al. [7] for further details. Note that the interaction operand " : " is not consistent across statistical packages.

13.4   Classical Designs

In this section we describe classical designs commonly used in plant improvement experiments. The number of treatment factors, their levels and structure, together with management practices and logistics influence the plot structure and subsequent experimental design. These designs differ primarily in their plot structures, whereas their treatment structures are often similar. The Design Tableau approach of Smith and Cullis [12] helps define the treatment and *plot factors*, the design function and the resulting treatment and *plot structures*. Common treatment structures are single factor, factorial and nested (Sect. 13.4.1). For each design we describe the fundamental principles, the plot structure and assume a single factor treatment structure unless otherwise stated.

To assist with reading the text this font is used for treatment and plot factors. The notation for defining factor levels follows John and Williams [13] such that there are v treatments, r replicates, s blocks and k plots within blocks.

13.4.1   Treatment Structures

The treatment underpins the experimental purpose and informs the experimental hypothesis. The treatment structure describes the relationship between all treatment factors and their allocation to experimental units. Three common treatment structures in PIEs are:

Single factor: PIEs often aim to evaluate genetic material for selection or commercialization. The genetic material can be breeding lines, hybrids or varieties, which we call Variety, for simplicity. Variety is the treatment factor and the treatment structure is simply: Variety.

Factorial: A factorial treatment structure is possible with two or more factors. A full factorial experiment is when all combinations of all treatment factor levels are evaluated. Partial factorial treatment structures are possible [1]. Agronomy experiments frequently employ factorial treatment structures. For example, an experiment to identify optimal seeding (Seeding) and nitrogen rates (Nitrogen) employs a factorial treatment structure written, using the crossed notation (Sect. 13.3.9), as:

$$Seeding * Nitrogen - Seeding + Nitrogen + Seeding : Nitrogen.$$

Factorial treatment structures have the following advantages over a series of experiments with single treatment factors:

1. the presence of between treatment factor interactions can be tested;
2. the interaction effects are non-zero then the optimal combination of treatments can be identified;
3. there is higher replication for the individual treatment factors.

Nested: Nested treatment structures are hierarchical, often due to biology. For example, selecting breeding lines often occurs within families and the treatment structure is written as:

$$Family / Line = Family + Family : Line.$$

13.4.2 Plot Structures

The plot structure describes the relationship between all plot factors (e.g. blocks, columns, rows, machines) and fully defines the observational units. The design function links the treatment and plot structures. Any of the following designs can have any of the treatment structures described in Sect. 13.4.1.

13.4.2.1 Randomized Complete Block Designs (RCBDs)

RCBDs have the following characteristics: all experimental units (e.g., field plots) within a *block* are considered homogeneous, i.e. similar in all respects that affect plant growth; each block contains a *complete* set of treatments so that blocks are resolvable for treatments; within a block the treatments are *randomly* allocated to the experimental units. The plot structure is

$$Block / Plot = Block + Block : Plot,$$

where Block:Plot defines the observational units and represents the residuals (errors). The treatment structure can be any of those described in Sect. 13.4.1.

Blocks are orthogonal to treatments so that the difference between treatments is independent of blocks. Usually, these experiments have a small number of treatments and the block size is not large. RCBDs are not recommended for PIEs with more than 10 treatments because within block homogeneity cannot be assured.

13.4.2.2 Alpha-Lattice Designs

The aim of the alpha-design algorithm, introduced by Patterson and Williams [14], is to generate resolvable incomplete block designs for 'any number of varieties v and block size k such that v is a multiplier of k'. This design function determines how v treatments are allocated to k plots within s blocks within r replicates whilst minimizing the concurrence of treatment pairs within a block. An alpha (0,1)-lattice design has zero or one treatment pair concurrences in a block.

Alpha-lattice designs are suitable whenever the number of treatments, v, is a multiple of the block size, k and are easily adapted when it is not. A rule of thumb is to choose a block size which is equal to or slightly smaller than the square root of the number of treatments, i.e., $k = \sqrt{v}$.

Figure 13.1 presents an alpha-lattice design for $v = 30$ varieties with $r = 2$ replicates, $s = 6$ blocks within each replicate and $k = 5$ plots within a block arranged as 4 rows by 15 columns. The plot structure for this design is:

Replicate / Block / Plot = Replicate + Replicate : Block + Replicate : Block : Plot,

where Replicate:Block:Plot defines the observational units and represents the residuals (errors).

The treatment structure contains a single factor, Variety.

13.4.2.3 Row-Column Designs

Heterogeneity between rows and between columns in PIEs is well known [15, 16]. Row-column designs block in both row and column directions to minimize the effect of spatial heterogeneity. They usually employ incomplete blocks – blocks that do not contain all treatments – and are resolvable when rows and/or columns are grouped together to create single replicate blocks. Piepho et al [17] provide a concise review of these designs.

Figure 13.2 presents a row-column design for $v = 18$ varieties with $r = 3$ replicates arranged as 9 rows by 6 columns. Each row and column is an incomplete block. The design is resolvable in both the row and column directions with 3 row-blocks (RowBlock) and 3 column-blocks (ColBlock). Varieties (treatments) are allocated to field plots such that there is one replicate in each row- and

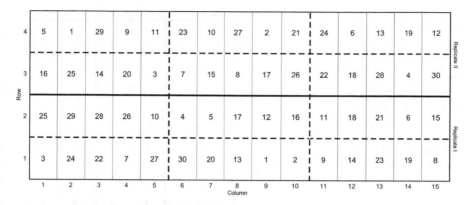

Fig. 13.1 An alpha-lattice design for $v = 30$ varieties and $r = 2$ replicates, $s = 6$ blocks within each replicate and $k = 5$ plots within a block arranged as 4 rows by 15 columns, where the replicates are in the row direction. Replicates are delineated by the horizontal bold line and blocks by the dashed lines

column-block. The plot structure for a row-column design depends on the direction of any resolvable blocks in the design and will contain row and column terms.

The plot structure for the design presented in Fig. 13.2 is:

$$\text{RowBlock / Row} + \text{ColBlock / Column} + \text{Plot}$$
$$= \text{RowBlock} + \text{RowBlock : Row} + \text{ColBlock} + \text{ColBlock : Column} + \text{Plot},$$

where Plot is described by RowBlock:Row:ColBlock:Column, defines the observational units and represents the residuals (errors).

The treatment structure contains a single factor, **Variety**.

13.4.2.4 Latinized Designs

Layouts with evenly distributed treatments are desirable to minimize the event of treatment pairs occurring together and conforms with the concept of blocking to minimize residual variation. The importance of balance and evenness depends on the intended analysis model and a researcher may forego these characteristics in some situations (see Sect. 13.5).

Latinized designs extend the concept of Latin Squares (see [6, 7]) where each treatment occurs exactly once in each row and each column. The popular mind-puzzle *Sudoku* is an example of a latinized row-column design. The design in Fig. 13.2 is a resolvable, latinized row-column design. No variety is in the same row or column more than once.

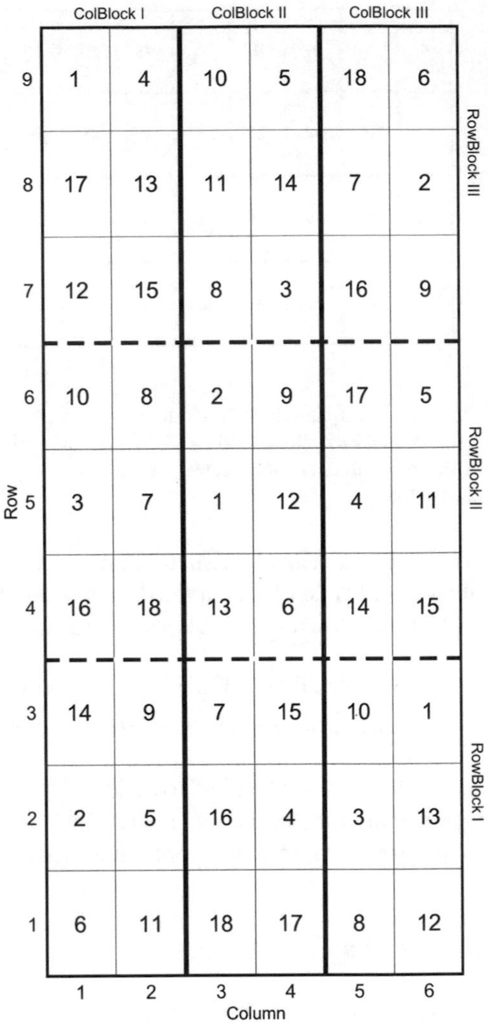

Fig. 13.2 A row-column design for $v = 18$ varieties and $r = 3$ replicates with $s = 3$ blocks of size $k = 6$ plots arranged as 9 rows by 6 columns. Rows and columns are incomplete blocks. Dashed horizontal and bold vertical lines delineate the row (**RowBlock**) and column replicates (**ColBlock**), respectively

	ColBlock I		ColBlock II		ColBlock III		
9	1	4	10	5	18	6	RowBlock III
8	17	13	11	14	7	2	
7	12	15	8	3	16	9	
6	10	8	2	9	17	5	RowBlock II
5	3	7	1	12	4	11	
4	16	18	13	6	14	15	
3	14	9	7	15	10	1	RowBlock I
2	2	5	16	4	3	13	
1	6	11	18	17	8	12	
	1	2	3	4	5	6	

Row (vertical axis) / Column (horizontal axis)

13.4.2.5 Split Plot Designs

Split plot designs are utilized for factorial treatment structures (Sect. 13.4.1) where one factor is applied to main plots and a second factor is applied to sub- (or split) plots. They are advocated in the following scenarios:

1. There is a factorial treatment structure and the levels of one factor must be applied to large plots (e.g., irrigation, tillage, herbicide application) for practical purposes.
2. There is a factorial treatment structure, but the aim of the experiment is to investigate the treatment factor allocated to the sub-plots and its interaction with the

main plot treatment factor; usually because the differences between the levels of the main plot treatment factor are known (e.g., irrigation).
3. A long-term experiment is in progress with treatments applied to the main plots. Another treatment which can be allocated to sub-plots within the main plots is of interest.

Figure 13.3 presents a split-plot experiment for evaluating the effect of nitrogen levels (0, 50, 100 kg/ha) on $v = 20$ varieties with $r = 2$ replicates and treatment structure,

$$\text{Nitrogen} * \text{Variety} = \text{Nitrogen} + \text{Variety} + \text{Nitrogen} : \text{Variety}.$$

The layout of 6 rows by 20 columns is equally divided in the row and column directions into 6 main plots (MainPlot) (Fig. 13.3). The two replicates (Block) contain three MainPlots each, and each MainPlot contains 20 sub-plots (Plot). The plot structure is:

$$\text{Block} / \text{MainPlot} / \text{Plot}$$
$$= \text{Block} + \text{Block} : \text{MainPlot} + \text{Block} : \text{MainPlot} : \text{Plot},$$

where Block:MainPlot:Plot defines the observational units and represents the residuals (errors).

Note, that factors which accommodate spatial variability, such as row and column, are not included. We will extend this example in Sect. 13.5 to illustrate how

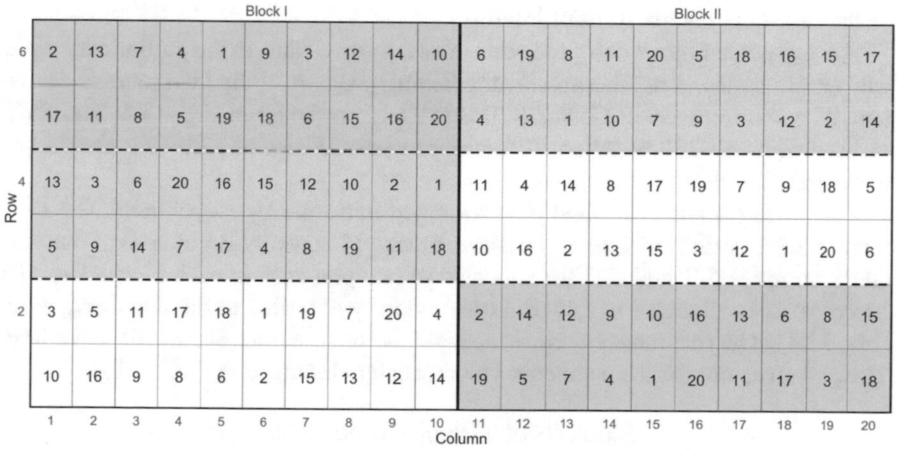

	Block I										Block II									
6	2	13	7	4	1	9	3	12	14	10	6	19	8	11	20	5	18	16	15	17
	17	11	8	5	19	18	6	15	16	20	4	13	1	10	7	9	3	12	2	14
4	13	3	6	20	16	15	12	10	2	1	11	4	14	8	17	19	7	9	18	5
	5	9	14	7	17	4	8	19	11	18	10	16	2	13	15	3	12	1	20	6
2	3	5	11	17	18	1	19	7	20	4	2	14	12	9	10	16	13	6	8	15
	10	16	9	8	6	2	15	13	12	14	19	5	7	4	1	20	11	17	3	18
	1	2	3	4	5	6	7	8	9	10	11	12	13	14	15	16	17	18	19	20

Row / Column

Fig. 13.3 Split plot design for $v = 20$ varieties and 3 nitrogen treatments (0, 50, 100 kg/ha) in $r = b = 2$ replicates (blocks), arranged in 6 rows by 20 columns. The bold vertical line delineates between blocks. The dashed lines delineate the main plots within blocks. The nitrogen treatments are allocated to the main plots within blocks. The varieties are allocated to the plots within main plots

model-based design can assist to minimize the effects of the expected spatial variability.

Strip-plots designs are a variation of a split-plot design. They are used when two treatment factors need to be applied to large areas, e.g., investigating the response to micronutrient combinations. Suppose there are two treatment factors (A with *a* levels and B with *b* levels), instead of randomizing the B within A as in a split-plot, both factors are arranged in strips across the replicates. The experimental area is divided into horizontal and vertical strips (rows and columns). Each level of factor A is allocated to all the plots in a row, and the levels of B are allocated to all plots in a column. This design provides high precision on the interaction between treatments at the expense of the main effects [1].

13.4.2.6 Augmented Designs

Augmented designs are widely used in the design of early stage variety trials. Early stage variety trials have large treatment numbers (hundreds to thousands of lines) with minimal seed availability for replication within and across environments. Augmented designs contain a combination of replicated and unreplicated treatments [18]. The replicated treatments (a set of check varieties, say) are allocated to a classical plot structure which accounts for spatial heterogeneity and the unreplicated treatments (usually the treatments of interest, the set of breeding lines, say) *augment* the replicated design. Each unreplicated treatment is allocated to one (incomplete) block only while each replicated treatment appears in each block at least once. The systematic repetition of the replicated treatments enables estimation of the block effects and residual (error) variance resulting in more precise estimates of the treatment comparisons of interest.

An augmented block design for one of twenty-five trials in the preliminary yield trial (PYT) series of the durum wheat breeding program at the International Maize and Wheat Improvement Center (CIMMYT) is presented in Fig. 13.4. The PYT series evaluates 4200 breeding lines grouped into 25 sets of 120. Two checks are evaluated in each trial.

Each trial contains 128 field plots arranged in 8 rows by 16 columns. The augmented trial is divided into equal sized blocks (4 rows by 8 columns). The two check varieties (C1 and C2) are allocated to one plot each in each block. The 120 breeding lines allocated to this trial are randomly allocated to the remaining plots (Fig. 13.4). The treatment structure, a single factor structure (Section 0), is Variety. The plot structure for the augmented block design described in Fig. 13.4 is:

$$\text{Block} / \text{Plot} = \text{Block} + \text{Block} : \text{Plot},$$

where Block:Plot defines the observational units and represents the residuals (errors). Note this plot structure is dependent on the experimental design of the replicated checks and is specific to this example.

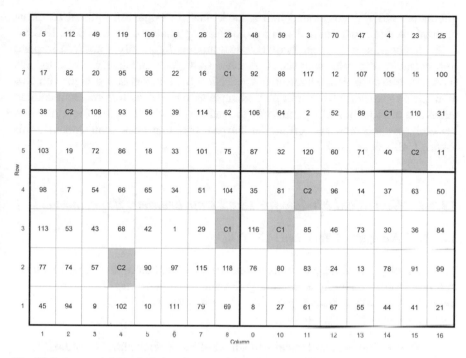

Fig. 13.4 Augmented block design for one trial from the CIMMYT durum breeding program preliminary yield trial series. There are 120 breeding lines (labelled 1–120) and 2 check varieties (C1 and C2). The bold black lines delineate the blocks

Traditionally, each trial in a series is analyzed separately. However, this compromises the selection decisions as the spatial variability within and between trials is large. It is advisable to analyze all trials together and model the spatial variation appropriately, following Gilmour et al. [15], for example. We describe model-based partially replicated trials, which extend these augmented grid designs in Sect. 13.5.

13.5   Model-Based Designs

Classical designs can constrain comparative experiments resulting in sub-optimal and costly outcomes [6, 16]. A model-based approach can generate classical designs whilst accommodating less structured design specifications such that the design is based on the intended analytical model [2–4]. Model-based designs uphold the fundamental design concepts described in Sect. 13.3 and can generate and enhance the classical designs described in Sect. 13.4. They can include terms for anticipated peripheral effects such as those induced by trial management practices along row and columns. Furthermore, correlated structures for the treatment and/or residual effects are easily incorporated into a model-based design. An optimal (or near-optimal) design is determined using pre-defined optimality criterion (Sect. 13.3.8).

In this section we review two statistical models frequently employed in plant improvement research: analysis of variance (ANOVA) and linear mixed models (LMMs). Next we demonstrate the application of model-based design with two examples: extension of the split plot example to include random row and column terms and introduction of a partially replicated design which models the correlation between residuals and the correlation between breeding lines simultaneously following Cullis et al. [4, 9].

13.5.1 Statistical Models for Plant Improvement Experiments

13.5.1.1 Analysis of Variance (ANOVA)

The method of ANOVA partitions experimental observations into their treatment and plot factors, enabling a test of significance to be performed for the difference between treatment means. For example, each observation from a RCBD experiment (Sect. 13.4.2) can be written as:

observation = overall mean + treatment effect + block effect + residual.

This partitioning is summarized in an ANOVA table (see Welham et al. [7] for details).

The principle of least-squares, employed in ANOVA, seeks to minimize the residual sum of squares thus obtaining the best estimate of σ^2. The residuals ($\equiv$ errors) are assumed to be independently, identically, normally distributed with mean zero and variance σ^2. The treatment and block factors are fixed effects and have no distribution.

13.5.1.2 Linear Mixed Model

A LMM modifies the linear model of ANOVA to allow terms to be fitted as *random* or *fixed*, hence *mixed*. Each random term is assumed to be independent with effects sampled from a normal distribution with a common variance, called the variance component. The residual maximum likelihood (REML) method provides unbiased estimates of the variance components [19] in a LMM. It is the method implemented in LMM software such as ASReml-R [20], REML in GenStat [21] and PROC MIXED in SAS.

Identifying which terms to fit as fixed or random is non-trivial [11, 16, 22]. A sensible starting point is the randomization-based model where all plot structure factors are fitted as random and all treatment factors are fitted as fixed [6, 7]. Smith and Cullis [22] have developed an instructive tool, Design Tableau, to identify the LMM best suited to the design and analysis of an experiment.

LMMs have some significant advantages over ANOVA models. They accommodate non-orthogonality and imbalance arising from missing data or complex experimental designs. When terms are modelled with a variance, (i.e., fitted as random) recovery of inter-block information and appropriate modelling of effects representing different sources of variation (e.g., blocks, rows and/or column) are enabled. There are three characteristics of PIEs which can be accommodated in a LMM: extraneous and residual (plot-to-plot) variability, and complex variance structures between treatments.

Extraneous variation arises from management practices and is modelled by fitting row and/or column effects as random and estimating their variance components. If extraneous variation is expected, then it can be included in a model-based design.

Accurate estimation of the plot-to-plot variability (residual variation) is achieved via spatial modelling, such as the two-dimensional separable auto-regressive models of order 1 (known as AR1⊗AR1) of Gilmour et al. [15]. Spatial models assume that spatial dependence exists between plots, i.e., plots close together are more similar than plots further apart. Accommodating this dependence between field plots in the design [2, 9] is logical and particularly important in trials with minimal replication (see Sect. 13.5.2.2).

The treatments (breeding lines, varieties or hybrids) in PIEs are often related. Pedigree information, in the form of a numerator relationship matrix, $\mathbf{A}$, captures the genetic similarity between treatments. Inclusion of the $\mathbf{A}$ matrix in the analysis enables estimation of additive and non-additive effects [23–25]. Using a model-based design approach it is possible to include the pedigree information using $\mathbf{A}$ in the design process [3, 4]. Alternatively, if marker data are available then the kinship or genomic relationship matrix can replace the $\mathbf{A}$ matrix.

13.5.2 Examples

These designs were generated using the R library 'od', a freely available optimal design software [26].

13.5.2.1 Accounting for Extraneous Variation

Consider the split plot design experiment (Sect. 13.4.2.5) within the model-based design paradigm of "design how it would be modelled". The LMM for the analysis of this experiment, using randomization-based theory, would include the plot structure terms Block and Block:MainPlot as random effects, i.e. assign a variance to each of them, σ^2_{Block} and $\sigma^2_{BlockMainPlot}$, say. The term, Block:MainPlot:Plot defines the observational units, i.e. the residuals, which are assumed to be normally distributed with mean zero and variance σ^2. In addition, extraneous variation introduced by management practices conducted across rows and column is accounted for by including random row and column effects with variances σ^2_{Row} and σ^2_{Column}, respectively [15]. Thus, the plot structure for this experiment is now:

Block + Block : MainPlot + Block : MainPlot : Plot + Row + Column,

where Block:MainPlot:Plot defines the observational units and represents the residuals (errors).

The layout presented in Fig. 13.3 was generated using this model. The resulting design is latinized (Sect. 13.4.2.4) with respect to rows such that each variety occurs exactly once in each row and no variety is allocated to a column more than once. The A-optimality criterion increased slightly from 0.362 for the classical design to 0.377 for the model-based design. This is considered acceptable.

13.5.2.2 Partially Replicated Designs

Partially replicated (p-rep) designs are model-based designs which were introduced as an alternative to augmented grid designs (Sect. 13.4.2.6) for early stage variety trials [9]. The key principle is to replace the replicated check lines in an augmented grid design with test lines. This increases the response to selection due to an increased replication of the lines under selection. The theoretical development underpinning this design is described in Cullis et al. [9] and extended in Cullis et al. [4] to include the use of pedigree information.

A yield evaluation trial is planned for 504 breeding lines (Varieties), but the field layout is limited to 24 columns by 26 rows, 624 plots. A p-rep trial is designed where 384 varieties are allocated to one field plot and 120 allocated to two field plots, a p-rep of 24%. The trial is blocked in the row direction (Fig. 13.5). Extraneous variation in both the column and row directions is known to exist due to irrigation infrastructure and management practices. Thus, the plot factors are Block with 2 levels, Row with 26 levels, Column with 24 levels and Plots, described by Row:Column, with 624 levels.

The plot structure is:

Block + Row + Column + Row : Column

Starting values for the variance components of the peripheral random effects, Block, Row and Column were estimated from the previous year's dataset. The term Row:Column specifies the observational units and represents the residuals. An even spread of replicated treatments was achieved using a separable spatial model with an auto-correlation model fitted in the row direction only (written AR1$\otimes$I). Thus, extraneous and spatial variation is captured in this model.

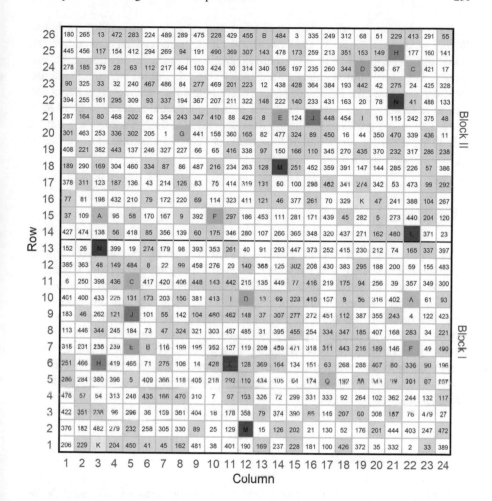

Fig. 13.5 Partially replicated design for $v = 504$ varieties in 624 plots, arranged in 26 rows by 24 columns. The bold horizontal line delineates the Blocks. Colors represent different check lines. The gray shaded plots are those allocated with 2 replicate varieties

The treatment structure is:

Variety

where Variety is fitted as a random effect and partitioned into an additive component with variance σ_a^2 and additive variance-covariance matrix **A** (Sect. 13.5.1.2) and a non-additive component with variance σ_e^2. Starting values for these variances were estimated from the previous year's dataset.

The resulting design (Fig. 13.5) is resolvable with respect to replicated varieties and blocks, thus it is also near-orthogonal. The inclusion of the pedigree information, modelling of extraneous and residual variation ensures reasonable balance of varieties (treatments) across the layout and is the anticipated analysis model.

Early generation variety trials are often evaluated in multi-environment trial (MET) series (see Chap. 3). Cullis et al. [9] states, '*p*-rep designs are particularly suited to this setting [MET] since there is potential to balance test line replication across trials'. Near-optimal designs are achieved by aiming for resolvability across locations and can take family, or pedigree, structures into account. It is not necessary to have equal numbers of lines, nor even equal partial-replication at all locations. This is a significant advantage over the classical design approach given the gain in accuracy for prediction of the genetic effects and selection that is achieved by using model-based design methods [4].

13.6 Summary

Plant improvement datasets are costly and time consuming to collect. It is crucial then, that the best statistical methods (design and analysis) be employed to ensure that the return on investment is optimized. The fundamental design principles of replication, randomization and blocking need to be understood and upheld in classical and model-based designs. Classical designs provide a rigorous, systematic structure and are important in plant improvement research. Model-based designs are flexible and tailored to the experimental purpose and constraints. Model-based design theory allows an easing of some design concepts, such as orthogonality and resolvability, whilst maintaining optimality for the experimental purpose and intended analysis.

13.7 Key Concepts

- Replication, randomisation and blocking are fundamental experimental design concepts required for rigorous plant improvement experiments
- Understanding and minimizing bias and pseudo-replication in experimental designs enhances plant improvement research outcomes
- Classical designs primarily differ in their plot structures – which is somewhat driven by their treatment structures
- Model-based design theory use the anticipated statistical model to generate the design
- Classical experimental designs can be generated within the model-based paradigm
- Model-based designs enhance plant improvement outcomes by optimising resources and flexibility accommodating logistical constraints whilst improving prediction accuracy for the treatment effects under evaluation.

13.8   Review Questions

1. Figure 13.6 presents a field layout for 2 replicates of 24 varieties arranged as 6 rows by 8 columns. Use it to answer the following questions: (a) Where are the blocks in this design?; (b) Describe the incomplete and complete blocks?; (c) Are the blocks resolvable? Why?; (d) Are treatments orthogonal to blocks?; (e) Describe an alternate layout and why this one may have been selected.
2. Describe the difference between a crossed and nested treatment structure. Provide an example of each found in plant improvement experiments.
3. A yield evaluation trial for 30 varieties with 3 replicates is planned for a field layout with 15 rows and 6 columns. There is a known fertility gradient in the row direction whilst all trial management practices take place across columns. (a) What are the treatment factor(s)? What is the treatment structure?; (b) What type of designs could be employed? What are their plot factors and structures?; (c) Another possible layout for this design was 30 rows by 3 columns. What design principles were considered in determining the final layout?
4. An experiment is conducted to investigate variety response to frost events. In order to maximize the opportunity for a variety to experience frost four trials were sown at different times, two weeks apart. Each trial contained two replicates. Discuss why variety differences across the four trials cannot be attributed to time of sowing only.
5. An early generation yield trial is planned for a location where sowing and harvesting operations occur along columns in a serpentine pattern. Describe what

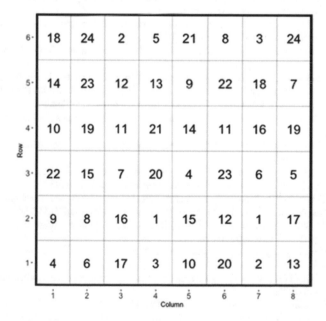

Fig. 13.6 Field layout for $v = 24$ varieties and $r = 2$ replicates arranged as 6 rows by 8 columns

sources of variation could occur and how the trial could be designed optimally to minimize this variation.

Acknowledgements The authors would like to thank our colleagues who reviewed this manuscript.

References

1. Cochran WG, Cox GM (1957) Experimental designs. Willey, New York
2. Martin RJ (1986) On the design of experiments under spatial correlation. Biometrika 73:247–277
3. Butler DG, Smith AB, Cullis BR (2014) On the design of field experiments with correlated treatment effects. J Agric Biol Environ Stat 19:539–555. https://doi.org/10.1007/s13253-014-0191-0
4. Cullis BR, Smith AB, Cocks NA, Butler DG (2020) The design of early stage plant breeding trials using genetic relatedness. J Agric Environ Biol Stat 25:553–578
5. Fisher RA (1935) The design of experiments. Oliver and Boyd, Edinburgh
6. Bailey RA (2008) Design of comparative experiments. In: Series in statistical and probabilistic mathematics. Cambridge University Press, Cambridge. https://doi.org/10.1017/CBO9780511611483
7. Welham SJ, Gezan SA, Clark SJ, Mead A (2015) Statistical methods in biology: design and analysis of experiments and regression. CRC Press, Boca Raton
8. Bueno Filho J, Gilmour S (2003) Planning incomplete block experiments when treatments are genetically related. Biometrics 59:420–1541. https://doi.org/10.1111/.00044
9. Cullis BR, Smith AB, Coombes NE (2006) On the design of early generation variety trials with correlated data. J Agric Biol Environ Stat 11:381
10. Wilkinson GN, Rogers CE (1973) Symbolic description of factorial models for analysis of variance. J R Stat Soc Ser C (Applied Stat) 22:392–399
11. Piepho HP, Büchse A, Emrich K (2003) A Hitchhiker's guide to mixed models for randomized experiments. J Agron Crop Sci 189:310–322
12. Smith A, Cullis B (2018) Design tableau: an aid to specifying the linear mixed model for a comparative experiment. In: National Institute for Applied Statistics Research Australia Working Paper Series. https://niasra.uow.edu.au/workingpapers/index.html
13. John J, Williams ER (1995) Cyclic and computer generated designs, 2nd edn. Chapman and Hall, London
14. Patterson HD, Williams ER (1976) A new class of resolvable incomplete block designs. Biometrika 63:83–92
15. Gilmour AR, Cullis BR, Verbyla AP (1997) Accounting for natural and extraneous variation in the analysis of field experiments. J Agric Environ Biol Stat 2:269–293
16. Casler MD (2015) Fundamentals of experimental design: guidelines for designing successful experiments. Agron J 107:692–705. https://doi.org/10.2134/agronj2013.0114
17. Piepho HP, Williams ER, Michel V (2015) Beyond latin squares: a brief tour of row-column designs. Agron J 107:2263–2270
18. Federer W (1961) Augmented designs with one-way elimination of heterogeneity. Biometrics 17:447–473
19. Patterson HD, Thompson R (1971) Recovery of interblock information when block sizes are unequal. Biometrika 58:545–554
20. Butler DG, Cullis BR, Gilmour AR, Gogel BJ, Thompson R (2018) ASReml-R Reference Manual Version 4. VSN International Ltd. Hemel Hempstead, UK. http://www.vsni.co.uk/

21. VSN International (2019) Genstat for windows, 20th edn. VSN International, Hemel Hempstead
22. Smith AB, Cullis BR, Thompson R (2005) The analysis of crop cultivar breeding and evaluation trials: an overview of current mixed model approaches. J Agric Sci 143:449–462. https://doi.org/10.1017/S0021859605005587
23. Oakey H, Verbyla A, Pitchford W, Cullis BR, Kuchel H (2006) Joint modelling of additive and non-additive genetic line effects in single field trials. Theor Appl Genet 113:809–819
24. Crossa J, Burgueño J, Cornelius PL, McLaren G, Trethowan R, Krishnamachari A (2006) Modeling genotype × environment interaction using additive genetic covariances of relatives for predicting breeding values of wheat genotypes. Crop Sci 46:1722–1733
25. Burgueño J, Crossa J, Cornelius PL, McLaren G, Trethowan R, Krishnamachari A (2007) Modeling additive × environment and additive × additive × environment using genetic covariances of relatives of wheat genotypes. Crop Sci 47:311–320
26. Butler DG, Cullis BR (2019) od: generate optimal experimental designs. www.mmade.org

Chapter 14
Seed Systems to Support Rapid Adoption of Improved Varieties in Wheat

Arun Kumar Joshi and Hans-Joachim Braun

Abstract New varieties of crops are developed to provide farmers seeds of cultivars that are acquainted with specific environmental or management conditions to realize best yield and quality. Seed is the carrier of genetic potential for the performance of a crop, hence is considered the most vital input in agriculture. Wheat being self-pollinated, it is not necessary to buy seed every year as in case of hybrids. Seeds are multiplied through an informal or formal approach. In most developing countries, informal wheat seed sector is dominant. Seed production follows well defined steps wherein a particular class of seed is grown to deliver another class of seed to the farmer. In general, there are four classes of seeds in wheat – nucleus, breeder, foundation and certified, although in some cases registered seed is also produced. The strength of the seed sector varies across countries – strong in developed countries but moderate to weak in the Global South. In most countries seed production and its marketing is regulated and both public and private sectors are involved. In counties with a not so strong seed sector, a fast track approach for varietal release and seed dissemination has been advocated to meet the challenges of climate change and transboundary diseases.

Keywords Improved seed · Seed system · Pre-release · Policy · Training · Participatory

14.1 Learning Objectives

- Improved seed is key for food and nutritional security.
- Seed system is within breeding process.
- Fast track release and dissemination is important.
- Policy change and capacity building are required.

A. K. Joshi (✉)
International Maize and Wheat Improvement Center (CIMMYT), New Delhi, India

Borlaug Institute for South Asia (BISA), Ludhiana, Punjab, India
e-mail: a.k.joshi@cgiar.org

H.-J. Braun
International Maize and Wheat Improvement Center (CIMMYT), Texcoco, Mexico

14.2 Introduction: Need for Efficient Wheat Seed System and Issues That Affect Its Functioning

High yielding varieties with better quality and nutrition, that are adapted to a particular environment, increase the choice of healthy and nutritious food while generating a viable income for farmers [1] (For history of wheat breeding, see Chap. 2). In a self-pollinated crop like wheat, traditionally it takes about 15–20 crop cycles starting from crossing, testing and then to finally release a variety. Combining fast track breeding systems with fast track testing and release can bring this down in spring wheat to 6 (most extreme) to 10 crop cycles (more realistically). For a long time, shuttle breeding has been used at CIMMYT for spring wheat in which two breeding cycles are managed in a year, to reduce time to develop a variety in the breeding process and improve adaptation of breeding lines (For details, see Chap. 7).

New varieties, if not delivered to farmers on time may become susceptible to new races of pathogens and may not remain useful [2, 3]. Like breeding, seed dissemination in farmers fields may also take considerable time [4, 5]. The speed of seed dissemination is much faster in developed countries where there is abundance of private companies compared to the developing world [3]. There are several reasons of slow rate of variety turnover. A major challenge is the requirement of high seeding rate, 80–120 kg in spring wheat and about 180 kg in winter wheat. This demands a huge amount of seed to be produced and marketed which makes it quite challenging compared to other major cereal crops like maize and rice which require low (about 1/5th to 1/10th) seed rate. Since it is not necessary to purchase seed every year for a self-pollinated crop like wheat, it has a weaker economic business model compared to hybrid crops. Another factor is the low degree of commercialization in developing countries where majority of farmers are small holders [6].

Most countries require that as a pre-requisite for a variety to be admitted to the national list, it must meet the criteria if DUS (Distinctiveness from other varieties, Uniformity and Stability). DUS determines whether a newly bred variety is distinct (D) compared to existing varieties within the same species, whether the characteristics are expressed uniformly (U) and that these characteristics are stable (S), and do not change over subsequent generations [7]. For agricultural plant varieties incl. wheat, value for cultivation and use is often also required. A variety is considered to have value for cultivation and use if its qualities taken as a whole offer a clear improvement for cultivation, for use of the harvest or use of products derived from the harvest compared to comparable listed varieties. The value features of a variety are determined by properties shown in cultivation testing and laboratory testing relating to cultivation, resistance, yield, quality and it must be distinctly different from other varieties [8].

A seed production program can not compromise on the three issues: (1) the nature of improved variety, (2) seed purity (both physical and genetic) and, (3) seed germination. In addition, the seed must be made available in required quantities before the optimum sowing date at a reasonable cost. Thus, seed production involves

Fig. 14.1 Seed-to-Seed cycle in seed production

all those stages that fall between the procurement of right class of seed for initiating seed production and the distribution of the next class of seed to the grower. In other words, seed production is achieved through a "Seed-to-Seed" cycle (Fig. 14.1) and hence, is different from commercial crop production, where locus of production is the grain for consumption or the market.

14.3 Importance of Quality Seed in Modern Agriculture

Seed is considered the most cost efficient means of increasing agricultural production since it carries genetic potential to express best yields in a given environment. Agronomic interventions and policy decisions also play a key role. However, efficacy of other agricultural inputs in enhancing productivity and production, such as fertilizers, pesticides and irrigation is largely determined by the quality of seed. Food security is therefore dependent upon the seed security of farming communities [9]. Not only improved seed is necessary to realize yield potential, but promotion of varieties with different resistance genes enhance genetic diversity in farmers' fields and reduce the risk of pandemics.

The value of quality seed in obtaining optimum production and productivity has been proven in numerous studies of different crops including wheat [10]. Good seed is also important to keep farmers in good confidence and living without fear from losses due to diseases and abiotic stresses. During sudden emergence of a new virulence or a disease, seed of resistant varieties work just like a vaccine and save farmers from crop failure (Dr. Alison Bentley, personal communication, April 14, 2021).

14.4 Systems of Deed Dissemination

Since wheat is sown in large areas with high seeding rates, most seed is produced close to seed markets where wheat is grown. In other crops having low seed rate, seed can be imported from different regions, but not in wheat due to high transportation costs. In general, there exist two broad systems of seed dissemination and adoption: formal (organized) and informal (unorganized) [11].

14.4.1   Formal and Informal Seed Dissemination

The formal seed system involves an organized way of seed business in which improved varieties are developed and seed of a particular class is produced and marketed. In contrast, in the informal system, farmers themselves produce and disseminate from their own harvest following farmer to farmer exchange. The informal system is mostly observed in the Global South, where farmer to farmer seed dissemination is well known in countries like Afghanistan, Bangladesh, Ethiopia, Nepal, Pakistan and in some parts of India [3, 12]. The formal sector dominates in high income countries. However, both formal and informal seed systems may exist in developing countries. Among high income countries, the public wheat sector is strong in North America, while Australia, West and Central Europe and in the Southern cone of Latin America the formal private sector prevails.

In view of significant informal seed sector, countries like Nepal implemented participatory research program as an official way of varietal selection and seed dissemination. Much earlier, participatory research [13, 14] with farmers was recognized as a way of varietal selection and promoting seed dissemination. The approach of participatory varietal selection (PVS), in which new pipeline varieties are tested in farmers fields, has inherent ability to fast track the dissemination of seeds of new varieties since pipeline varieties enter seed production on the day 1 of their introduction to farmers fields which happens much before their official release [15, 16].

14.4.2   Seed System in Developed Countries and UPOV

The global private spending on agricultural R&D (excluding R&D by food industries) has increased from about $5.1billion in 1990 to $15.6billion by 2014 [17]. In wheat and small grains, the total spending on R&D in 2014 was $1 billion [17]. Since farmers can save and reproduce their own wheat seed, seed royalties and consequent return of investments used to be limited. This limitation was recognized in developed countries and attempts were made to strengthen property rights of plant breeder over time. One such effort was the International Union for the

Protection of Plant Varieties (UPOV) which is an international agreement that attempts to create a common approach to plant breeders rights [18]. UPOV was established by the International Convention for the Protection of New Varieties of Plants that was adopted in Paris in 1961 and revised in 1972, 1978 and 1991. UPOV's mission is to provide and promote an effective system of plant variety protection, with the aim of encouraging the development of new varieties of plants, for the benefit of society [18].

There are currently 75 members of UPOV, 57 countries have legislation to implement the UPOV 1991 convention. The implementation of UPOV 1991 in different countries was done in different years. For example, USA (1994), Germany (1998), UK (1999), Australia (2001), European Union (2005), France (2012) and Canada (2015). Major wheat producing countries that are not member of UPOV are India, Pakistan, Nepal, Iran, Afghanistan, Kazakhstan and Ethiopia. A well-functioning royalty collection system is in Australia. The interesting component of the Australian system is that royalties are collected when commercial grain is delivered to the elevator. Breeding companies have therefore no incentive to invest in large seed production programs. They give seed of their varieties to seed producing farmers who make their money on selling seed. Thus seed farmers become the seed promoters. The Australian model is suitable for countries where most wheat is exported i.e., grain delivered to an elevator where variety is determined and royalty payments can be calculated. It could be a model for wheat exporting countries like Canada, Kazakhstan, Ukraine and Argentina.

14.4.3 Pre-release Seed Multiplication

Pre-release seed multiplication means multiplication of the seed of a variety before its formal release. This assures sufficient seed is available for large scale multiplication. Due to production costs, in many developing countries, the amount of breeder seed available at the time of varietal release is about one ton only. This amount is too small to allow rapid dissemination. Moreover, as soon as farmers learn about release of a new variety there is demand, but no seed is available. This leads to multiple consequences, such as dissatisfaction among farmers, slow rate of adoption and sometimes may lead to black marketing of seed or sale of spurious seed. The benefits of pre-release seed multiplication, as explained in Sect. 14.5, are of immense value during outbreak of a new disease like Ug99 [4] or wheat blast [19].

14.5 Type of Varieties in Wheat and Classes of Quality Seed

Seed production methodology is influenced by the type of the variety being pursued. A brief concept is explained below.

14.5.1 Land Race, Pure Line Varieties and Hybrid Varieties

14.5.1.1 Land Race

Cultivars that are under cultivation in farmers fields for a long time beyond the records of organized breeding are termed land races. They are grown sparsely and are generally observed in far flung areas where new alternatives are not available. Land races are mostly heterogeneous and hence their seed production is done by bulking seeds of similar looking plants.

14.5.1.2 Pure Line Varieties

Pure line varieties are the advanced generations of the progenies of single plants selected from a heterogeneous population generated through an organized crossing and breeding program. Therefore, pure line varieties are highly homozygous and homogeneous and do not change their genetic makeup from generation to generation except though outcrossing and/or mutations. Therefore seed production needs simple steps with focus on purity through use of pure source seed, rouging, avoiding seed mixture etc.

14.5.1.3 Hybrid Varieties

A hybrid variety is the first generation seed derived from cross between genetically unrelated parents. Their seed production involves development and crossing of generally two parents (inbreds/pure lines). Therefore, maintenance of inbreds/pure lines is essential. Hybrid seed production program involves use of male sterility or chemical hybridizing agents. Seed of hybrid varieties when grown more than once, show reduced expression of yield and other traits due to inbreeding depression. Hence their seed need to be purchased every year.

As of today, most released wheat varieties are pure lines and hybrid wheat is globally insignificant. Producing hybrid seed is much more complex than pure line seed. This chapter focuses on seed production of pure line varieties.

14.5.2 Classes of Improved Seed

Seeds of high yielding varieties, that are genetically and physically pure and carry high germination (%) are called Improved Seed. There are different classes of seeds recognized in different countries (Table 14.1). However, for easy understanding, there are four recognized classes: Nucleus seed (NS), Breeder Seed (BS), Foundation Seed (FS) and Certified Seed (CS) (Fig. 14.2). The last two classes of seed fall under

Table 14.1 Wheat seed classes nomenclature in some countries

Class of seed	S. Asia (India, Nepal, Bangladesh, Pakistan)	OECD	AOSCA	Ethiopia	Egypt
First seed available	Nucleus seed	Nucleus seed	Nucleus seed	Nucleus seed	Nucleus seed
1st generation supplied by plant breeders	Breeder	Breeder	Breeder	Breeder	Breeder
2nd generation	Foundation	Pre-basic	Foundation	Pre-basic	Foundation
3rd generation	Certified	Basic	Registered	Basic	Registered
4th generation		Certified 1	Certified	Certified 1	Certified
5th generation		Certified 2	–	Certified 2	–

Modified with permission from Ref. [2]

OECD includes European countries; AOSCA is Association of Official Seed Certifying Agencies. AOSCA's membership includes Seed Certifying Agencies across the US, and Global membership including Canada, Argentina, Brazil, Chile, Australia, New Zealand, and South Africa

For all classes, stage I and stage II are permitted in most countries when quality seed is in short supply

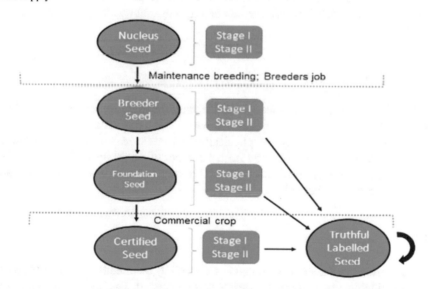

Fig. 14.2 Classes of improved seed and their general status during seed production

certified seed category and are generally certified by a government designated certification agency. In some countries, another class of seed (Registered Seed) is designated between foundation and certified seed class. Each class of seed may be reproduced again from same type of seed and are classified as stage I and Stage II. For instance Breeder seed stage I and Breeder seed stage II. The stage II class is allowed (from stage I of the same class) only under emergency situations when a particular class of seed is in extremely short supply. There is another class of seed

Fig. 14.3 Nucleus seed production in wheat by progeny rows of single spikes at BISA Ludhiana, India

called Truthful Labelled (TL) seed. This seed can be produced from any class of seed and is not certified by any certification agency, but quality assurance is given by the producer/company.

14.5.2.1 Nucleus Seed

Nucleus Seed (NS) is the first seed available when a variety is produced. It is generally produced by the breeding institution which owns the variety. It is considered 100% pure – both genetically and physically. Nucleus seed is produced by growing progeny rows using seeds collected from a number of single spikes (Fig. 14.3). Off type single rows are removed and only rows having characteristic traits of the variety are harvested and bulked.

14.5.2.2 Breeder Seed

BS is the progeny of nucleus seed and is produced by the breeder/sponsored breeder or the original institute where the variety was developed. The BS crop is monitored by an Inspection Team comprising of plant breeders and officials designated by government.

14.5.2.3 Foundation Seed

FS is the progeny of BS and is used to produce the next class of seed i.e., certified seed. This seed is not used for commercial cultivation. Foundation seed is also certified by a certifying agency for minimum seed standards. Production of FS is done under the control of government designated agencies (Seed Certification Agency).

14.5.2.4   Certified Seed

CS, the last category of seed, is the progeny of foundation seed. Certified seed is given to farmers for commercial cultivation. Production of CS is also done under the control of a government designated agencies (Seed Certification Agency).

14.5.2.5   Truthful Labelled Seed

As per Seeds Act prevailing in various countries, *certification is voluntary* but *labelling is compulsory*. It means presence of proper labelling in seed bags is necessary. As a result, there is another class of seed called Truthfully Labelled (TL) seed where there is no involvement of Certification Agency, but labelling is done. TL seed can be produced from any class of seed.

14.6   How to Judge the Quality of Seed

Seed quality is judged by seed testing. Seed testing refers to an evaluation of seed quality parameters to ensure that the seed conforms to the 'Minimum Seed Standards'. Seed testing involves tests that are meant to verify the following three parameters: (1) Physical and genetic purity, (2) moisture content and (3) seed germination. The information obtained from seed testing is printed on the seed tags attached to the seed bags or packets.

Seed testing as a science and standard procedure has developed during the nineteenth century. The first seed-testing laboratory was established in 1869 in Saxony, Germany while in 1876 in USA and in 1961 in India. Presently, seed testing laboratories are present in almost all countries and are regulated by both government and private sectors. The procedure for seed testing is based on guidelines from International Seed Testing Association [7] and other publications of the host country issued in support of the standard procedure. The standard procedure for seed testing involve, seed sampling, purity analysis, germination test, moisture test and test for seed health.

14.7   Steps Involved in Seed Production and Minimum Seed Standards

A seed production program must ensure attainment of the defined genetic constitution of the aggregate of seed being produced.

14.7.1 Steps in Seed Production

Quality seed of wheat is produced in a step wise fashion. Each class of seed is produced under strict supervision and must meet minimum seed quality standards [2]. In general, seed multiplication of a variety involves the following five steps: (1) Procurement of a class of seed, (2) Reporting to monitoring/certification agency, (3) Seed production in the field, (4) Seed processing, (5) Delivering seed to market (Fig. 14.4).

14.7.2 Minimum Seed Standards

Each class of seed must conform to certain level of quality standards, termed minimum seed standards. These standards mainly reflect two things – the performance of the seed crop in the field and the characteristics of the seed being made available to the consumers. Seed standards are very stringent in case of the nucleus and breeder seed compared to the foundation and certified. The major parameters that define minimum seed standards are based on field and seed standards.

14.7.2.1 Field Standards

Isolation distance is considered the most important field standard in a seed production program. The isolation distance includes distance of the seed production field from the fields of other varieties of wheat and from same variety not conforming to the purity standards. It also includes distance of the seed field from the field carrying infection of certain air borne diseases such as loose smut. In most countries,

Fig. 14.4 Steps in quality seed production of wheat

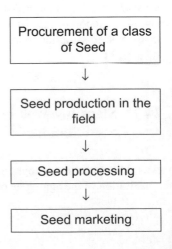

the isolation distance in case of pure line varieties is only 3m, however, if there is a loose smut infected field, the isolation distance shall be 150 m. Outcrossing in wheat, for most varieties, is between 0% and 1%, though there are varieties which show, depending on the year, outcrossing rates of up to 9% [20].

14.7.2.2 Seed Standards

Among seed standards, genetic purity is of utmost importance. However, appropriate germination and physical purity are equally important. A genetically pure seed without proper germination will lead to crop failure, hence is of little value. Physical impurities may also impair crop performance by leading to lesser plant population or by causing unwanted infestation by weeds. Likewise, moisture percent below a threshold helps in maintaining seed life and vigor during transportation or storage. For wheat, the minimum germination for certified seed is 85%, while moisture should be <12%. (Tables 14.2 and 14.3).

14.8 Need for Rapid Seed Dissemination and Challenges to Support Rapid Adoption of New Varieties

A major bottle neck in realizing the impact of improved wheat varieties in many countries is the long time gap between identification of a variety through yield trials and the time the variety is available for cultivation in farmers' fields. This takes usually from 5 to 8 years, since seed multiplication usually starts only when a variety is registered or released. Initiating seed multiplication at the time a variety enters registration trials can shorten the period by three years, but more importantly will provide farmers three years earlier protection from air borne diseases like rusts (For details on wheat rusts, see Chap. 8). Thus there are solid justifications for the need of rapid seed dissemination through pre-release seed multiplication. However, pre-release seed multiplication is often not done, since there is the risk that a line will be dropped from registration trials due to poor performance, which converts expensive breeder seed into commercial grain. This means, pre-release seed multiplication can be economically risky for a seed producer and therefore private companies will only do pre-release seed production when they are sure that there will be significant demand for the new variety. If the public sector produces seed by this approach and shares with private sector, the risk will be carried by Government institutions. This approach of using government institutions for pre-release seed multiplication was successfully implemented for the fast track release and dissemination of U99 resistant wheat varieties in six countries of Asia and Africa [4].

Table 14.2 Minimum seed certification standards for foundation (F) and certified (C) seed of pure line varieties in wheat and other cereals as applicable in India and South Asia

No.	Parameter – need for all parameter units	Wheat F	Wheat C	Paddy F	Paddy C	Barley F	Barley C	Triticale F	Triticale C
	General requirements								
1.	Distance to fields of other varieties (M)	3.0	3.0	3.0	3.0	3.0	3.0	3.0	3.0
2.	Distance to fields of the same variety not conforming to varietal purity requirements for certification (M)	3.0	3.0	3.0	3.0	3.0	3.0	3.0	3.0
3.	Distance to field of varieties with infection of disease in excess of 0.10% and 0.50% in Foundation and Certified seed respectively. (M)	150	150	–	–	150	150	150	150
4.	Off types %	0.05	0.20	0.05	0.20	0.05	0.20	0.05	0.20
5.	Inseparable other crop plants	0.01	0.05	0.01	0.02	0.010	0.050	0.01	0.05
6.	Plants affected by seed borne disease	0.10	0.50	–	–	0.10[e]	0.50[e]	0.10[f]	0.50[f]
7.	Objectionable weed plant (Max.)	–	–	0.01[c]	0.02[c]	–	–	–	–
8.	Plant effect by ergot disease	–	–	–	–	–	–	0.02[g]	0.04[g]
9.	Pure seed % (Mini.)	98.0	98.0	98.0	98.0	98.0	98.0	98.0	98.0
10.	Inert matter % (Max.)	2.0	2.0	2.0	2.0	2.0	2.0	2.0	2.0
11.	No other crop seeds/Kg (Max.)	10.0	20.0	10.0	20.0	10.0	20.0	10.0	20.0
12.	Germination % (Mini.)	85.0	85.0	80.0	80.0	85.0	85.0	85.0	85.0
13.	Moisture % (Max.)	12.0	12.0	13.0	13.0	12.0	12.0	12.0	12.0
14.	For-vapor-proof containers % (Max.)	8.0	8.0	8.0	8.0	8.0	8.0	8.0	8.0
15.	No Total weed seeds/kg (Max.)	10.0	20.0	10.0	20.0	10.0	20.0	10.0	20.0
16.	Other distinguishable varieties/ Kg (Max.)	–	–	10.0	20.0	10.0	20.0	–	–
17.	No Objectionable weed seeds/kg (Max.)	2.0[a]	5.0[a]	2.0[c]	5.0[c]	–	–	2.0[h]	5.0[h]
18.	Seed infested % (Mini.)	None[b]	None[b]	0.10[d]	0.50[d]	–	–	0.05[i]	0.25[i]
19.	Husk less seeds % (Max.)	–	–	2.0	2.0	–	–	–	–

[a]*Hirankhuri* (Convolvulus arvensis L.), Gulli danda (*Phalaris minor Retz*)
[b]Nematode galls of Ear-cockle (*Anguina tritici Milne.*), *Tundu* (*Corynebacterium michiganense*)
[c]Wild rice (*Oryza sativa* L. var. fatua Prain (Syn. O. Sativa L.f. spontanea Rosch)
[d]Seed infested by paddy bunt (*Neovossia horroda* (Tak)
[e]Loose smut (*Ustilago nuda*)
[f]Loose smut disease (*Ustilago tritici*)
[g]Ergot disease
[h]Wild morning glory (Hirankhuri, Gulli danda)
[i]Karnal bunt (KB – *Neovossia* indica); There is zero tolerance for KB in KB free countries

Table 14.3 Minimum seed certification standards for foundation (F) and certified (C) seed of hybrid varieties in wheat and other cereals as applicable in India

No.	Parameter	Wheat		Paddy		Barley	
		F	C	F	C	F	C
	General requirements						
1.	Fields of other varieties including commercial hybrid of the same variety	200	100	200	100	200	100
2.	Fields of the same hybrid (code designation) not conforming to varietal purity requirements for certification.	200	100	200	100	200	100
3.	Field of same varieties with infection of loose smut disease in excess of 0.10% and 0.50% in Foundation and Certified seed respectively.	200	150	–	–	200	150
4.	Off types in seed parent	0.05	0.20	0.05	0.20	0.05	0.50
5.	Off types in pollinator	0.05	0.20	0.05	0.20	0.05	0.50
6.	Pollen shedding ear heads in seed parent	0.05	0.10	0.05	0.10	0.05	1.00
7.	Inseparable other crop plants	0.01	0.05	–	–	0.01	
8.	Plants affected by disease	0.10	0.50			0.10[f]	0.50
9.	Pure seed % (mini.)	98.0	98.0	98.0	98.0	98.0	98.0
10.	Inert matter % (maxi.)	2.0	2.0	2.0	2.0	2.0	2.00
11.	Other crop seeds/kg (max1)	10/kg	20.0	10.0	20.0	10.0	None
12.	Germination % (mini.)	85.0	85.0	80.0	80.0	85.0	85.0
13.	Moisture % (maxi.)	12.0	12.0	13.0	13.0	12.0	12.0
14.	For-vapor-proof containers %(maxi.)	8.0	8.0	8.0	8.0	8.0	8.0
15.	Other distinguishable varieties/kg (maxi.)	–	–	10.0	20.0	10.0	20.0
16.	Total Weed seed/kg (maxi.)	10.0	20.0	10.0[d]	20.0[d]	10.0	10g
17.	Objectionable weed seed/kg (maxi.)	2.0[b]	5.0[b]	2.0	5.0	–	None
18.	Seed infested (%)	None[c]	None[c]	0.10[e]	0.50[e]	–	None
19	Seed infested by Karnal bunt[a]	0.05	0.025	–	–	–	–
20.	Husk less seeds % (maxi.)	–	–	2.0	2.0	–	2.0

[a]Karnal bunt (KB – *Neovossia indica*); There is zero tolerance for KB in KB free countries
[b]*Hirankhuri* (*Convolvulus arvensis L.*), Gulli danda (*Phalaris minor Retz*)
[c]Seed infested with Nematode galls of ear- cockle
[d]Wild rice (*Oryza sativa* L. var. fatua Prain (Syn. O. Sativa L.f. spontanea Rosch).
[e]Paddy bunt (*neovossia horrida*)
[f]Loose smut (*Ustilago nuda*)

14.9   Case Studies of Rapid Seed Dissemination

Development of seed of high yielding, stress tolerant varieties that can adapt to unfavorable climatic conditions and have capacity to thwart the hazard posed by a range of pests and diseases are at the forefront of the agriculture industry of different countries [21]. Some examples of rapid variety release and seed dissemination are described in Sects. 14.9.1 and 14.9.2.

14.9.1   Thwarting the Threat of Stem Rust Race UG99

In the beginning of this century it was found that stem rust race caused by *Puccinia graminis* f. sp. *tritici*, in particular race Ug99 and its derivatives [22] were virulent on about 90% wheat varieties of the world [23] (See Chap. 8). The consultative group centers (CIMMYT and ICARDA) and BGRI, in collaboration with national research centers from countries under threat, developed high yielding Ug99 resistant varieties and disseminated rapidly in the most threatened areas [4]. Rapid seed multiplication and dissemination of Ug99 resistant varieties was initiated in Nepal, Bangladesh, Afghanistan, Pakistan, Egypt, Ethiopia, Iran and India [4]. In Nepal [4] and Bangladesh [15], PVS was used aggressively. The objective was to ensure that Ug99 resistant varieties must occupy about 5% of the area sown to wheat in each country to ensure sufficient seed to displace current popular varieties. Approaches used for rapid multiplication and distribution included pre-release seed multiplication, while resistant varieties were released in a fast track manner [4].

14.9.2   Case of Wheat Blast in South Asia

The emergence of wheat blast caused by Magnaporthe oryzae pathotype triticum (MoT) in year 2016 in Bangladesh [19, 24], its first occurrence outside Latin America, raised alarm bells in whole of South Asia [25, 26] (See Chap. 9). Wheat blast is exemplary for the benefits of global testing and consequent data sharing for entries in International Screening Nurseries and Yield Trials, which are distributed by CIMMYT worldwide. When wheat blast was detected in Bangladesh, data from Bolivia allowed to identify lines that were resistant and had good agronomic performance in Bangladesh. These lines were tested in Bangladesh under local environment and management system in the crop season 2016–2017. Side by side a pre-release seed multiplication was also initiated. Bangladesh Agricultural Research Institute (BARI) released wheat blast resistant variety (BARI Gom 33) in 2017 [19] (Fig. 14.5). The same approach was used for testing lines for other countries of South Asia (India, Nepal and Pakistan) where wheat blast represents a potential threat for future wheat production. In 2020, Borlaug 100 was released in Bangladesh (as WMRI 3) as well as in Nepal (as NL 1307). So far, more than two dozen wheat blast resistant varieties have been released in India (Dr. G.P. Singh, Director ICAR-IIWBR, Karnal, personal communication, June 24, 2021).

14.10   Future Need of Rapid Seed Dissemination

Due to increasing demand for wheat and the emergence of new challenges, the need of rapid breeding and seed dissemination appears to be of a higher necessity in future [3]. There are considerable number of challenges that may become more serious and unpredictable just like COVID-19 (Coronavirus disease) in case of human

Fig. 14.5 BARI Gom 33, a Zinc-enriched, blast resistant variety released in Bangladesh

beings. Some of the current and future issues that need rapid seed dissemination are: biotic stresses (wheat blast, new virulences of wheat rusts, aphid), abiotic stresses (high temperature and reduced availability of water) and biofortified varieties to address malnutrition (For details on biotic and abiotic stresses, see Chaps. 8, 9, 10 and 12).

Recently, Atlin et al. [3] proposed seven steps that should be taken by different countries to speed up varietal turnover: (i) Quick identification of new promising varieties supported by reliable data; (ii) A robust demonstration in place for promoting these varieties; (iii) De-certification of obsolete varieties to ensure promotion of new ones; (iv) Withdrawal of seed subsidies of old varieties; (v) No support of funding for the production of seed of obsolete varieties; (vi) Setting targets for the average varietal age in seed production and in farmers' fields; (vii) Establishing a simple variety release processes to encourage private sector.

14.11   Policy Changes by Countries to Ensure Rapid Seed Dissemination

The importance of improved seed or food security and environmental sustainability places high importance to the policies through which a strong seed system can be built and sustained. Countries having similar socio-economic and infrastructural facilities may share common policies and learn from one another [27].

Seed policy changes can play effective role in supporting entire agriculture operations of a country. For example, policy change in Turkey in the mid-1980s that allowed foreign investment led to major change in their seed sector. The private sector became strong in Turkey by another policy change in which government pays 60% of price difference between certified and commercial grain back to a farmer against submission of an invoice that farmer had purchased certified seed. With this introduction, the seed sector in Turkey completely changed and there are now many private companies producing seed and variety replacement rate has increased considerably. The success of this policy is dependent upon subsidies.

Countries like India also introduced a new seed policy in 1988 which allowed farmers to obtain best planting materials available [28]. Another policy change (2003) in India promoted participatory seed production in farmers' fields. A recent policy change among South and Southeast Asian countries occurred between 2013 and 2018. Three agreements were made:

(i) Between Government of Bangladesh and India in 2013 for rice by which it was agreed that they can release varieties of one country in to another.
(ii) The agreement between Bangladesh and India on rice was extended to Nepal in 2014.
(iii) In 2017, the agreement extended to other cereals, pulses, oil seeds, vegetables and fiber crops, was made among five countries (India, Nepal, Bangladesh, Sri Lanka and Cambodia) with the provision that variety released in a country can be released in any other by using the data of the country of release.

14.12 Seed System Is Within the Breeding Process – Conservation and Sustainable Use of Crop Genetic Resources

Seed is an inherent part of the entire breeding process which involves the entire range of activities involved in the conservation, pre-breeding, breeding or genetic improvement, testing, release and delivery to farmers through seed systems (Fig. 14.6). In informal seed systems, seed may be obtained through farmer to farmer exchange and the seed may come from a land race. But whether formal or informal, seed system falls within the breeding process since it depends on a number of varieties developed over a given period, which in long term is dependent upon conservation and sustainable use of crop genetic resources.

Formal seed systems are highly regulated. The seed of a variety is the part of scientific and technical network, both on its upstream and downstream. The upstream is the entire breeding process which starts on the day first cross is made to develop a variety. This cross becomes possible since there is collection of diverse genotypes having a range of genes for agronomic, stress tolerance and quality traits. Therefore investment in germplasm conservation and breeding research is necessary to ensure a strong seed system in place (For details on genetic resources, see Chap. 16). On the downstream, a seed of a variety will be of value if it is maintained as it is and is liked by farmers. This requires a standard way of producing high quality of seed with an efficient marketing system. To maintain purity of a released variety, fundamentals of maintenance breeding must be applied. Through genetic purity, productivity gains achieved are maintained and do not deteriorate over time. Maintenance research may not be profitable [29] since it will not lead to a measurable increase in production, but is important to realize actual genetic potential of a variety over a given period of time [30].

Fig. 14.6 Relationship of improved seed and the breeding process

14.13   Building Capacity in Seed Assurance in Developing Countries

The strengthening of the seed sector will require high class capacity building to have enough people trained in a country or region to support this system. A seed delivery system is like a value chain composed of interlinked components – from the development of well-adapted and nutritious varieties and their adoption by farmers, through the production and distribution, including sales, of quality seeds and planting materials, to on-farm utilization of recommended inputs by farmers [9]. The effective functioning of the value chain depends on the extent to which the stakeholders are able to put into practical use the relevant knowledge and skills required for producing quality seeds and planting materials [31].

To support capacity building in wheat seed system, almost all national research systems and international institutions like that in CGIAR (mainly CIMMYT, ICARDA), organize regular training programs. CIMMYT has been organizing a wheat improvement course since 1968 which includes components of maintenance breeding and seed production (CIMMYT, 2021). ICARDA's seed unit conducts courses focusing on seed production. Likewise, FAO assists member countries in carrying out a number of capacity building activities [9]. In addition to this, FAO has developed a Seeds Toolkit [31] to support capacity building of seed practitioners for the whole value chain of seed.

14.14 Key Concepts

Rapid seed system is necessary for rapid adoption of new varieties and to address new challenges. Strong seed systems are key to achieve food and nutritional security, environmental sustainability and carry immense potential to achieve the United Nations Millennium Development Goals. Large number of varieties are released across the world, but many do not reach small holder farmers who are in majority in the developing world. Whenever improved seeds have reached to farmers at large scale, as was the case during the Green Revolution, significant changes were observed. Many other examples showed the importance of rapid seed dissemination. This is more urgent today since problems are surmounting due to climate change, dwindling water resources, soil fatigue and a number of transboundary insect-pests and diseases. However, seed system is not so strong in most of the developing countries where wheat is grown. There is urgent need to invest in this system.

14.15 Conclusions

Rapid seed dissemination is key to adoption of new varieties and thus is crucial to whole plant breeding efforts and benefit to global agriculture.

   An efficient seed system is integrated with in an efficient breeding system. The key to good seed is a good plant breeding program including pre-breeding that generates valuable genetic stocks. A seed system can be formal or informal and involves well defined steps of seed production. The objective of a seed system is to ensure absence of crop failure in the fields of farmers, especially small holders and marginal. This could be possible when both variety and seed is of desired level and is produced and delivered when it is needed the most. Any delay in seed delivery will be wastage of the efforts put into breeding a variety. Rapid seed dissemination is key to adoption of new varieties and is crucial to whole plant breeding efforts and benefit to global agriculture.

References

 1. McEwan MA, Almekinders CJM, Andrade-Piedra JJL, Delaquis E, Garrett KA, Kumar L, Mayanja S, Omondi BA, Rajendran S, Thiele G (2021) "Breaking through the 40% adoption ceiling: mind the seed system gaps." A perspective on seed systems research for development in One CGIAR. Outlook Agric 50:5–12. https://doi.org/10.1177/0030727021989346
 2. van Gastel AJG, Bishaw Z, Gregg BR (2002) Wheat seed production. In: Curtis BC, Rajaram S, Gómez Macpherson H (eds) Bread wheat improvement and production. Food and Agriculture Organization of the United Nations, Rome
 3. Atlin GN, Cairns JE, Das B (2017) Rapid breeding and varietal replacement are critical to adaptation of cropping systems in the developing world to climate change. Glob Food Sec 12:31–37. https://doi.org/10.1016/j.gfs.2017.01.008

4. Joshi AK, Azab M, Mosaad M, Moselhy M, Osmanzai M, Gelalcha S, Bedada G, Bhatta MR, Hakim A, Malaker PK, Haque ME, Tiwari TP, Majid A, Kamali MRJ, Bishaw Z, Singh RP, Payne T, Braun HJ (2011) Delivering rust resistant wheat to farmers: a step towards increased food security. Euphytica 179. https://doi.org/10.1007/s10681-010-0314-9
5. Krishna VV, Spielman DJ, Veettil PC (2016) Exploring the supply and demand factors of varietal turnover in Indian wheat. J Agric Sci 154:258–272. https://doi.org/10.1017/S0021859615000155
6. Spielman DJ, Smale M (2017) Policy options to accelerate variety change among smallholder farmers in South Asia and Africa South of the Sahara. In: IFPRI Discussion Paper, 1666. http://ebrary.ifpri.org/cdm/ref/collection/p15738coll2/id/131364
7. International Seed Testing Association (2021) ISTA 2021 rules. https://www.seedtest.org/en/ista-rules-2019-_content%2D%2D-1%2D%2D3410.html
8. BSA (2021) Admission to the National List. https://www.bundessortenamt.de/bsa/en/variety-testing/national-listing
9. FAO (2020) Seeds. http://www.fao.org/seeds/en
10. Copeland L, McDonald M (2001) Principles of Seed Science and Technology, 4th edn. Kluwer Academic Publishers, Norwell
11. Joshi AK, Chand R, Chandola VK, Prasad LC, Arun B, Tripathi R, Ortiz Ferrara G (2005) Approaches to germplasm dissemination and adoption – reaching farmers in the Eastern Gangetic Plains. In: Proceedings of 7th international wheat symposium Nov 27–Dec 2. Mar de Plata, Argentina
12. Chatrath R, Mishra B, Ortiz Ferrara G, Singh SK, Joshi AK (2007) Challenges to wheat production in South Asia. Euphytica 157:447–456. https://doi.org/10.1007/s10681-007-9515-2
13. Witcombe JR, Joshi KD, Rana RB, Virk DS (2001) Increasing genetic diversity by participatory varietal selection in high potential production systems in Nepal and India. Euphytica 122:575–588. https://doi.org/10.1023/A:1017599307498
14. Witcombe JR, Joshi A, Goyal SN (2003) Participatory plant breeding in maize: a case study from Gujarat, India. Euphytica 130:413–422. https://doi.org/10.1023/A:1023036730919
15. Pandit D, Mandal M, Hakim M, Barma N, Tiwari TP, Joshi AK (2018) Farmers' preference and informal seed dissemination of first Ug99 tolerant wheat variety in Bangladesh. Czech J Genet Plant Breed 47:S160–S164
16. Joshi AK, Mishra VK, Sahu S (2017) Variety selection in wheat cultivation. In: Langridge P (ed) Achieving sustainable cultivation of wheat Volume 2, First Edit. Burleigh Dodds Science Publishing, London
17. Fuglie K (2016) The growing role of the private sector in agricultural research and development world-wide. Glob Food Sec 10:29–38. https://doi.org/10.1016/j.gfs.2016.07.005
18. UPOV (2021) Frequently asked questions. https://www.upov.int/about/en/faq.html
19. Mottaleb KA, Govindan V, Singh PK, Sonder K, He X, Singh RP, Joshi AK, Barma NCD, Kruseman G, Erenstein O (2019) Economic benefits of blast-resistant biofortified wheat in Bangladesh: the case of BARI Gom 33. Crop Prot 123:45–58. https://doi.org/10.1016/j.cropro.2019.05.013
20. Lawrie RG, Matus-Cádiz MA, Hucl P (2006) Estimating out-crossing rates in spring wheat cultivars using the contact method. Crop Sci 46:247–249. https://doi.org/10.2135/cropsci2005.04-0021
21. Wimalasekera R (2015) Role of seed quality in improving crop yields. In: Hakeem KR (ed) Crop production and global environmental issues. Springer, Cham, pp 153–168
22. Pretorius ZA, Singh RP, Wagoire WW, Payne TS (2000) Detection of virulence to wheat stem rust resistance gene Sr31 in Puccinia graminis. f. sp. tritici in Uganda. Plant Dis 84:203–203. https://doi.org/10.1094/PDIS.2000.84.2.203B
23. Singh RP, Hodson DP, Huerta-Espino J, Jin Y, Njau P, Wanyera R, Herrera-Foessel SA, Ward RW (2008) Will stem rust destroy the world's wheat crop? Adv Agron 98:271–309
24. Malaker PK, Barma NCD, Tiwari TP, Collis WJ, Duveiller E, Singh PK, Joshi AK, Singh RP, Braun H-J, Peterson GL, Pedley KF, Farman ML, Valent B (2016) First report of wheat blast

caused by Magnaporthe oryzae Pathotype triticum in Bangladesh. Plant Dis 100:2330. https://doi.org/10.1094/PDIS-05-16-0666-PDN

25. Chowdhury AK, Saharan MS, Aggrawal R, Malaker PK, Barma N, Tiwari TP, Duveiller E, Singh PK, Srivastava AK, Sonder K, Singh R, Braun H, Joshi AK (2017) Occurrence of wheat blast in Bangladesh and its implications for South Asian wheat production. Indian J Genet Plant Breed 77:1–9. https://doi.org/10.5958/0975-6906.2017.00001.3

26. Bishnoi S, Kumar S, Singh GP (2021) Wheat blast readiness of the Indian wheat sector. Curr Sci 120:262–263

27. Turner M (2001) The role of national seed policies in re-structuring the seed sector in CEEC, CIS and other Countries in Transition. Seed Policy Program Cent East Eur Countries, Commonw Indep States Other Ctries Transit 1–13

28. Joshi AK, Singh BD (2004) Principles of seed technology. Kalyani Publishers, New Delhi

29. Plucknett DL, Smith NJH (1986) Sustaining agricultural yields. Bioscience 36:40–45. https://doi.org/10.2307/1309796

30. Dubin HJ, Brennan JP (2009) Combating stem and leaf rust of wheat: historical perspective, impacts, and lessons learned. IFPRI Discussion Paper 00910

31. FAO (2018) Seed toolkit. Module 3: seed quality assurance

Chapter 15
Crop Management for Breeding Trials

Nora Honsdorf, Jelle Van Loon, Bram Govaerts, and Nele Verhulst

Abstract Appropriate agronomic management of breeding trials plays an important role in creating selection conditions that lead to clear expression of trait differences between genotypes. Good trial management reduces experimental error to a minimum and in this way facilitates the detection of the best genotypes. The field site should be representative for the target environment of the breeding program, including soil and climatic conditions, photoperiod, and pest and disease prevalence. Uniformity of a field site is important to provide similar growing conditions to all plants. Field variability is affected by natural and management factors and leads to variability in crop performance. Additionally, pest and disease incidence tend to concentrate in patches, introducing variability not necessarily related to the susceptibility of affected genotypes. Precise agronomic management of breeding trials can reduce natural field variability and can contribute to reduce variability of crop performance. Through specialized agronomic management, contrasting selection conditions can be created in the same experimental station. The use of adequate machinery like plot seeders and harvesters contributes to precise trial management and facilitates operation. Machine seeding assures even seeding depth and density. Plot combines can be equipped with grain cleaners, on-board weighing systems and sensors to measure grain humidity and weight, which can greatly facilitate data collection.

N. Honsdorf (✉)
Kiel University, Kiel, Germany

International Maize and Wheat Improvement Center (CIMMYT), Texcoco, Mexico
e-mail: honsdorf@pflanzenbau.uni-kiel.de

J. Van Loon · N. Verhulst
International Maize and Wheat Improvement Center (CIMMYT), Texcoco, Mexico
e-mail: J.VanLoon@cgiar.org; n.verhulst@cgiar.org

B. Govaerts
International Maize and Wheat Improvement Center (CIMMYT), Texcoco, Mexico

Cornell University, Ithaca, NY, USA
e-mail: b.govaerts@cgiar.org

M. P. Reynolds, H.-J. Braun (eds.), *Wheat Improvement*,
https://doi.org/10.1007/978-3-030-90673-3_15

257

Keywords Agronomic management · Mechanization · Test/selection/target environment · Soil heterogeneity · Experimental error

15.1 Learning Objectives

- To understand the experimental error of field trials and ways to reduce it.
- To understand the importance of agronomic management for creating appropriate selection environments.

15.2 Introduction

Field experimentation is an essential part of plant breeding programs. Appropriate agronomic management of breeding trials plays an important role in creating selection conditions that lead to clear expression of trait differences between genotypes. Good trial management reduces random variation (experimental error) between plots (smallest experimental unit) to a minimum and in this way facilitates the detection of the best genotypes.

During the different selection stages throughout the breeding process, different kinds of experimental layouts and experiment management are needed, including different types of machinery (Fig. 15.1). For example, the seeds obtained from an initial cross might be sown in short rows by hand, but later generations are sown mechanically in plots and might be tested under various environmental conditions and in several locations. Traits like plant height, maturity and disease resistance can be measured in very small plots, while realistic yield estimates require large plots. Through specialized agronomic management, contrasting selection conditions can be created in the same experimental station, for example optimum vs. low nutrient or optimum vs. reduced irrigation environments. Selection environments (SE) are usually created through a combination of natural site characteristics and modifications by agronomic management.

Although genetic and genomic data have been becoming more and more important in plant breeding, data obtained from field trials do not lose their significance. It is under varying field conditions where the plants must perform and ultimately those data can only be obtained from practical experiments. Techniques, like genomic selection, require accurate field experimentation, since data from field trials are used to predict performance of untested populations. The quality of the prediction also depends on the quality of the phenotypic input data.

The chapter is divided into two parts. The first part treats the selection of field sites, the creation of selection environments through agronomic management and mechanization of breeding trials. The second part deals with experimental error and ways to reduce its impact in field experimentation.

Fig. 15.1 Test plots of different sizes at CIMMYT's experiment station in Ciudad Obregon, Mexico. (Image by Lorena González/CIMMYT)

15.3 Selection and Management of Field Sites

The field site should be representative for the target environment of the breeding program [1]. This includes soil and climatic conditions, photoperiod, and pest and disease prevalence. Some conditions can be created artificially; disease pressure can be increased through artificial inoculation or generation of humid environments where pathogens thrive. In arid environments, different levels of drought stress can be mimicked through irrigation practices. Creation of environments the opposite way around is more challenging; mimicking drought in humid environments needs for example rainout shelters and keeping trials disease free in areas with high disease pressure would need large amounts of pesticides. Air temperature (and vernalizing cold) are impossible to change in field trials but photoperiod can be extended using low intensity lamps. The 'modifiability' of an environment should be considered if genotypes are to be tested under different conditions in the same site.

Uniformity of a field site is important to provide similar conditions for all genotypes grown in an experiment or nursery. It is affected by natural and management factors (field history). Natural factors include soil characteristics, but also landscape aspects, like hills, slopes, depressions, or rows of trees that cause shading in some areas of a field. Heterogeneous soil conditions can be caused by previous experiments, through different types of management. Fertilization or tillage experiments can lead to patchy soil conditions. An ideal field site is flat with homogeneous soil conditions and without shading. However, these ideal experimental sites are often not available, and some degree of heterogeneity is present [2]. It is important to know the variability of the experimental area, recognize possible impacts and take them into consideration for trial layout and analyses.

15.3.1 Agronomic Techniques for Creation of Selection Environments

Selection environments are the trials that breeders use to make selections for desired traits. The SE must be designed (or chosen) in a way that maximizes the power of prediction for performance in the target population of environments (TPE) (see Chap. 3). Careful trial management is necessary to create selection conditions that are similar to the ones found in farmers' fields where the new varieties eventually need to succeed. Certain aspects of SE can be created or influenced by agronomic management, those include for example water and nutrient availability and disease pressure. The 'Quick facts' Table 15.1 provides a summary of agronomic factors to be considered in breeding trials.

Drought is the most important abiotic stress worldwide. Drought often appears in conjunction with other abiotic stresses, like high temperatures and high radiation (see Chap. 10). While it is common to generate artificial drought conditions for breeding trials and agronomic experiments, accompanying factors are difficult to create artificially in field trials and therefore are usually neglected. A difficulty of drought stress experiments is that drought can appear in many different ways, at different growth stages and with different intensities. For example, drought can be caused by constant reduction of soil humidity throughout the season. Or a lack of water could appear even though regular rainfalls are present, but those are not intensive enough to meet crop demands. In the first case the stress level is rising over time, while in the second case water availability might stay low but constant. Sometimes strong rainfalls alternate with prolonged drought phases, exposing crops to extremely contrasting conditions within one cropping cycle. It is important to define the type of drought stress that is to be mimicked in a selection environment.

Arid regions are naturally the most appropriate places to conduct drought stress experiments. In arid regions, where irrigation is a prerequisite for crop cultivation, different amounts of irrigation water and types of irrigation can be used to create relevant drought stress environments. In humid regions, rainout shelters can be used to create drought experiments. Those shelters can be fixed or mobile. The latter version has the advantage that it only covers the crop during precipitation events and therefore allows the crop to be exposed to natural radiation and wind conditions during the rest of the time [3]. Lateral water flow can be a constraint in drought experiments with rainout shelters and soil humidity needs to be monitored throughout the experiment.

Nutrient availability can vary widely from field to field in the same target environment. Economic constraints that do not allow farmers to buy fertilizer or simply the lack of products to purchase are important reasons. Low input systems, like

Table 15.1 Quick facts 'Important agronomic factors to consider in breeding trials'

Factor	Impact/Relevance
Representativeness	Soil and climatic conditions, photoperiod, pest and disease prevalence and agronomic management of the field site should be representative for target environments.
Field history	Previous experiments can lead to patchy field conditions and may require a uniformity treatment to achieve homogeneous conditions.
Nutrient management	Optimum nutrient supply can even out heterogeneous soil conditions and is required for expression of maximum yield potential.
	For a low nutrient selection environment, the field plot needs to be depleted prior to start of selection, only target nutrient should be limited.
Weed control	Weeds should be controlled to avoid competition with crop.
Pest and disease control	Pest and diseases need to be controlled for maximum expression of yield potential.
	Favorable conditions for pests and diseases can be created if resistance or tolerance are part of the breeding targets.
Soil management	Tillage, direct seeding, removing or leaving crop residues and their combinations lead to different germination conditions. Soil management should be representative for the target environment.
Irrigation	Different types of irrigation, like drip, furrow and sprinkler irrigation differ in the way they make water available to plants and should be chosen according to the most common technique used by farmers in the target population of environments.
Crop rotation	Appropriate crop rotations improve nutrient status and reduce pathogen pressure in the field. To keep the field healthy, wheat should be rotated with non-cereal crops.
Lodging	Lodging can be reduced by timing of N application and irrigation, reducing plant stand, growing plants on beds and the application of growth regulators.

organic farming, require genotypes with high nutrient use efficiency under low nutrient conditions as well. Ideally, efficient genotypes are also highly responsive when additional nutrients are available. Selection and evaluation under both low and high nutrient conditions allows researchers to identify genotypes that perform well under both conditions [4]. In order to create those environments, the soil nutrient status needs to be analyzed prior to trial establishment. This is especially important for the establishment of low nutrient selection environments. If nutrient levels are too high, a uniform crop needs to be grown without fertilizer addition to remove excess nutrients. If a particular nutrient, e.g. N, is to be removed it needs to be assured that all other nutrients are sufficiently available so that crop growth is only limited by the target nutrient. Soil nutrient status should be monitored regularly to assure the desired level is maintained.

Conservation agriculture (CA) is a form of agronomic management that consists of minimum tillage, maintaining crop residues on the soil surface and crop diversification. This type of management can improve soil health and water availability. Soil cover and higher top soil bulk density compared to conventional tillage conditions can be a constraint for early crop development. Certain diseases, e.g. Fusarium head blight and Yellow leaf spot can be favored by retention of crop residue, especially in monoculture or certain crop rotations (e.g. wheat-maize). Most breeding programs operate under conventional tillage conditions and the special conditions caused by CA are not included as SE. The area where wheat is grown under CA is expanding, also due to promotion by national and international organizations. Therefore, it is an important question whether CA requires varieties with different characteristics compared to those used in and developed under conventional tillage conditions. Characteristics discussed as especially beneficial are strong early vigor and disease resistance. CIMMYT's durum wheat breeding program conducted a parallel selection experiment under conventional tillage and CA conditions. Subsequently all genotypes from both selection streams were evaluated under both tillage regimes. For the case of CIMMYT's widely adapted durum wheat material no relevant difference between selection under conventional tillage or CA were detected [5]. These results indicate that for the conditions tested, there is no need for specialized breeding programs and selection can take place under CA without negative consequences on genotype performance under conventional conditions.

Selection environments with high disease pressure can be created through management suitable to create conditions where disease thrive. Humidity is an important factor that can be manipulated through sprinkler irrigation (Fig. 15.2). High plant stand densities and monotonous crop rotation that favor the development of plant diseases and pests are ways to create relevant conditions.

15.3.2 Mechanization for Breeding Trials

Breeding trials follow specific designs with numerous small plots arranged to grow a wide array of genotypes. Often a considerable amount of seeding and harvest operations in breeding programs is done by manual labor. Manual operations are highly labor-intensive, especially for sowing and harvest, and can result in more variable seeding depth and spacing due to human error. Therefore, specific experimental machinery has been developed that can handle precision plot sowing and harvest, reducing variability and speeding up operations.

15.3.2.1 Plot Seeders

Limited plot sizes and randomized plot designs require accurate seed metering that can respect the complex lay-out of small plots or rows of genotypes to be tested, often placed at short equidistant intervals or at varying densities. Small-seed plot seeders uniformly distribute a measured or counted quantity of seed per surface area unit (Fig. 15.3). Cone seeders are the most common type, as these can handle very small amounts of seeds and do not require large amounts of seed to be held in hoppers as is the case for conventional planters. During sowing, manually prepared seed

Fig. 15.2 Artificially created humid environment using sprinklers to create optimum conditions for fusarium screening at CIMMYT's headquarter El Batán, Mexico. (Image by Pawan Kumar Singh/CIMMYT)

packages or seeds preloaded in cartridge arrays are dropped on top of a cone-shaped plate and released or 'tripped' at the start of each plot or plant row (Fig. 15.4).

With seeds evenly distributed at the bottom of the cone, the latter revolves to deposit the seeds to the soil as the machine advances. Newer systems allow electronic calibration of cone revolution speed to match plot length, while older systems use a clutch in combination with a mechanical gear box. Depending on experimental layout, multiple seed tubes that deliver seed to the soil can be connected to a single cone seed meter to plant several plant rows simultaneously, or alternatively, multiple cone meters can be placed on a seeder. The use of multiple cone meters allows to sow different genotypes with each cone (for plots that are half the machinery width, two cone seed meters would be mounted – Fig. 15.3) or the same genotype (with plots width matching the machine's). If multiple seed tubes are connected to a single cone seed meter, it is necessary to include a motorized divider mechanism to evenly distribute the seed among the tubes. Next to seed meters and similar to regular planters, soil penetration and seeding depth can be configured using opening tines or disc coulters depending soil conditions, followed by pressing wheels for adequate soil-seed contact to promote germination. Precision sowers allow for seed singulation, i.e. sowing individual seeds separately along the row at a defined spacing. These are already available for commercial sowing of many crops and are becoming available for plot sowing, potentially playing a role in breeding in the future.

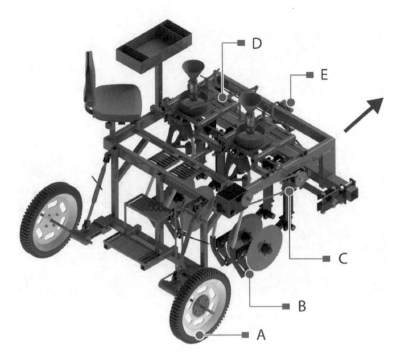

Fig. 15.3 Experimental plot seeder with two cone-shaped seed meters and mechanical drive train (marked blue, *C*) for seed distribution, with (*A*) traction wheel, (*B*) disc coulters (optional in case of crop residue) and seed delivery tynes, calibration mechanism with black gear box, and (*D*) operating lever to release seeds on cone-shaped metering device. This model is designed to be pulled by a tractor with three-point hitch (*E*), while self-propelled models also exist. Red arrow indicates direction of movement during operation

15.3.2.2 Machinery for Harvest

Similar to plot seeders, specialized farm machine manufacturers offer a variety of self-propelled experimental harvesters and mini-combines. Critical steps during harvest are the cutting of ears which, if done with inadequate equipment, can cause significant grain losses, and threshing where yield losses occur due to grain damage. The choice of mechanized harvesting for experimental plots depends on plot size, harvest volume and final objective. Despite their reduced size, the front-end of experimental harvesters remain largely similar to commercial grain harvester combines, consisting of a pickup reel, cutter bar and thresher drum. Plot combines can be equipped with grain cleaners, on-board weighing systems and sensors to measure grain humidity and weight, which can greatly facilitate data collection. Bagging options and crop dividers for continuously harvesting of experimental plots are all part of the possibilities and choices to consider. Adjusting harvesting speed and minimal machine vibration help reduce the amount of grain that falls outside the header.

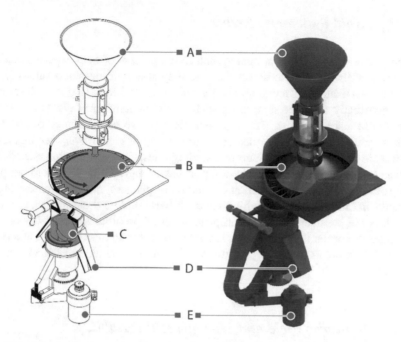

Fig. 15.4 Schematic representation of experimental cone seed meter system in open (left) and closed (right) position: (*A*) seed insertion funnel, (*B*) cone-shape seed distribution plate, (*C*) seed divider mechanism to distribute seeds evenly among tubes, (*D*) seed tubes entry, (*E*) electric motor that drives divider system. Red arrows indicate revolving movement of cone and divider mechanism during operation

15.4   The Experimental Error

Experimental error is a term used to describe variation that occurs among plots that receive the same treatment. Two types of error are distinguished – random and systematic. The random error is not a mistake due to poor trial management. Rather it describes the slight variation that exists even if all management is carried out in the most careful way. Sources of this variation are for example differences between plants of the same genotype, natural soil variability, topographic gradients, measurement inaccuracies, etc. Systematic errors follow a constant pattern. They usually occur due to incorrectly calibrated instruments or equipment. Experimental errors can never be eliminated completely but should be reduced as much as possible. Common sources of error are described below. Besides the reduction of experimental error, appropriate experimental designs, including appropriate blocking of replicated trials and statistical analyses are essential for its precise estimation (see Chap. 13).

15.4.1   Avoid Systematic Errors

Systematic errors can occur at many points in the process of field experimentation. They are not always easy to detect, because they affect all measured values in the same way. An example is a wrongly calibrated scales that adds 10 g to each sample. Or an incorrectly designed measuring stick where the number of plants is counted along 90 cm instead of 1 m. The experimenter can also be the source of error, by constantly making the same mistake, e.g. putting 40 g of seed in every seeding envelope instead of 50 g leading to incorrect seeding rates. Low germination rates due to poor seed storage conditions can be a factor that strongly influences yield per plot. Other sources of error can be wrong calibration of planters, mistakes numbering or arranging field entries and weighing errors at harvest. And there are many more! While it is not possible to have a complete list of all possible sources of error, it is even more important that the experimenter has possible sources in mind, and organizes and documents every step of the trial in order to be able to detect systematic errors.

15.4.2   Minimize Field and Management Variability

Variability in crop performance can be due to variability in the availability of soil and above-ground resources or field operations. Additionally, pest and disease incidence tend to concentrate in patches, introducing variability not necessarily related to the susceptibility of affected genotypes. Precise agronomic management of breeding trials can reduce natural field variability and can contribute to reduce variability of crop performance.

Variability in resources only results in spatial variability in crop development when the resource is limiting crop performance [6]. Shatar and Mcbratney [7] examined relationships between sorghum yield and soil properties in Australia and found that most of the measured soil properties varied spatially, but only a few were responsible for variation in yield. Along the field boundaries, changes in the amount of plant available water mostly caused variation in sorghum yield, while in the center of the field, soil held more water so that production reached a level at which the potassium content limited production. Machado et al. [8] reported a positive effect of soil NO_3-N on sorghum grain yield in a year when water was abundant, but a negative effect in a year when water was limited.

Within-field spatial variability can be the result of inherent variation in field conditions. However, agronomical practices also influence spatial within-field plant variability. Kravchenko et al. [9] found that in a zero-input treatment, overall variability (coefficient of variation) was significantly higher compared to treatments with low or conventional input. In semi-arid highlands in Mexico, conservation agriculture, i.e. zero tillage with residue retention and wheat-maize rotation, resulted in high soil health and uniform crop performance, while under zero tillage with residue removal, soil health and crop performance followed micro-topography with higher values where micro-topography was lower [6].

For breeding trials, researchers should use agronomic management that maximizes the uniformity in the distribution of resources and results in vigorous crops, like conservation agriculture, only inducing stresses that represent the chosen selection environment. Appropriate crop rotations help to eliminate variability due to previous trials and reduce the disease and weed seed burden.

Lodging introduces variability, because it tends to occur in patches related to micro-topography, wind- and rainfall patterns and a domino-effect where lodging-prone genotypes drag down neighboring plots (Fig. 15.5). Several agronomic management practices can minimize lodging. The most commonly used management factors to minimize lodging are reduced or delayed N fertilization and reduced seeding density [10]. Planting systems can decrease lodging, for example, bed planting with furrow irrigation had over 50% less lodging than flat planting with flood irrigation in Mexico [11]. Plant growth regulators can reduce lodging by decreasing plant height and increasing the physical strength of the basal part of the culm internode [12]. If management in SE is optimized to minimize lodging, while these management factors are different in areas where the materials will be used, materials should be screened under lodging-inducing conditions before they are released to screen out materials prone to lodging.

Fig. 15.5 Logged plots surrounded by standing plots at CIMMYT's experimental station in Ciudad Obregon, Mexico

Researchers should carefully design and monitor field operations to minimize errors. The protocol of each experiment should include a field management plan with an overview of options to manage common problems, like pests, diseases and weed pressure in the experimental field. This includes options of pesticides available in the area and preferably threshold values for pest incidence that require their use. Weed, pest and disease incidence and the technologies for their management tend to evolve rapidly, so it is important to collaborate with pathologists, agronomists, weed scientists to get the latest insights and updated technologies. When working at an experimental station, station managers can offer experience and insight and should be involved in discussion to stimulate continuous improvement of agronomic management of wheat breeding experiments.

During the growing season, researchers should prepare detailed instructions for all operations, adjusting field management to the development of the growing season, since weed pressure, disease and pest incidence are highly variable and dependent on weather conditions. Regular monitoring of the field is essential to ensure any problems are caught early and can be managed before they introduce variability that affects the experiment. Walk-throughs should be done at least twice a week.

It is important to keep detailed records of all field operations, to allow a good description of experimental conditions for reports and publications, to spot potential problems and to design and monitor improvements in management over time. When pesticides are used, active ingredients with different modes of action should be varied in time, to prevent the development of resistance and this can be monitored through these records. The records of field operations should include dates, products used (concentration of active ingredients, dose used), names of the persons executing field operations and preferably also time of day, since that affects effectivity of certain active ingredients. For certain types of operations, e.g., tillage operations, a more detailed description is necessary (e.g., including tillage depth, implements used), but these can be made once, using a brief description from then on (e.g., 2 passes of disking). Keeping a physical copy of the operations records in a visible place, can help make sure that records are always up-to-date and emphasize the importance of careful management and record keeping.

Variability in field operations that can cause variability in crop performance includes uneven applications of inputs like fertilizer, irrigation water or pesticides and errors in sowing, like clogged tubes. To prevent these errors, it is important to regularly revise and give maintenance to machinery and equipment, give detailed instructions on calibrations, make sure that field operators have a good understanding of machinery and equipment calibrations and their importance for the validity of the research and to check machinery calibrations before field operations. Again, frequent field monitoring is important to spot mistakes and, where possible, correct them before they affect the outcome of the experiment.

15.4.3 *Account for Soil Variability*

Field heterogeneity is caused by natural soil heterogeneity and topography, agronomic management and previous experiments – the field history. Some variation is present in all experimental sites. The degree of heterogeneity however varies. In an ideal site variation is low (i.e., does not affect yield in a significant way) and does not need to be considered in the trial design. However, often a considerable field variability is present, resulting for example in a productivity gradient. Highly variable field sites produce highly variable phenotypic data and can mask true genetic differences between genotypes. Therefore, it is important to assess field variability and account for it with experimental design and/or with spatial analyses [13].

It is important to know the field history. In some cases, significant variability is caused by previous experiments and/or agronomic management. Different types of tillage or fertilization experiments can lead to patchy distribution of soil conditions. In some cases, a uniformity treatment of the experimental area is necessary to create homogeneous conditions. In case of previous fertilizer experiments for example, it is advisable to cultivate crops without fertilizer to extract nutrients and achieve a homogeneous nutrient status in the experimental area before starting experiments or selections. Different types of soil management, like plowing, zero tillage, removing or leaving of plant residues also affect soil quality and uniformity treatment should be considered.

Natural soil heterogeneity can hardly be changed but can lead to very different growing conditions even at short distances. Soil texture or depth, for example can vary and influence nutrient and water availability, which in turn leads to different levels of productivity. Soil and yield maps help to identify similar areas and allow the experimenter to choose appropriate trial layouts.

Field heterogeneity can be measured by growing uniformity trials. The experimental area is sown with one variety and treated uniformly. For harvest the area is divided into many small plots (the smaller the more precise) and yield for each plot is determined separately. The result is a yield map that enables the identification of more and less productive areas.

Instead of manual harvest, uniformity can also be assessed in automated ways using sensors. Yield monitors mounted onto combines in combination with differentially-corrected global positioning system receivers enable the automated collection of georeferenced yield data and subsequent creation of high-resolution yield maps. This way of yield monitoring is a common application in precision agriculture and can be a valuable instrument in trial planning.

Remote sensing technologies enable rapid, non-destructive mapping of areas with high and low productivity within fields (see Chap. 27). Vegetation Indices (VI) based on multispectral remote sensing are a standard method for monitoring crop growth and can provide estimates for grain yield in wheat through correlation analyses [14]. One example of a widely used VI is the Normalized Difference Vegetation Index (NDVI), which uses red and infrared bands to estimate canopy growth. In

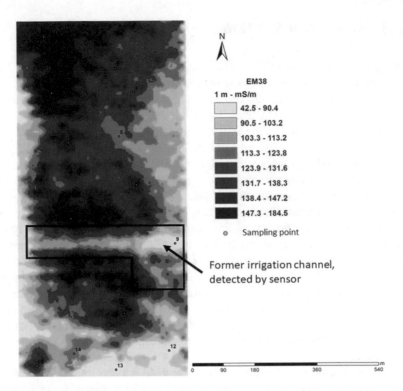

Fig. 15.6 Field map showing soil heterogeneity by differences in electrical conductivity measured with the conductivity meter EM-38. (Modified with permission from [16])

wheat this index is widely used to monitor crop growth throughout the growing cycle and to estimate grain yield.

Yield and productivity maps serve as a proxy for the assessment of soil heterogeneity. Proximal soil sensing is an established way to assess soil characteristics and create soil maps. An established parameter is apparent soil electrical conductivity (ECa). Georeferenced ECa measurements are used to create soil maps that enable estimates of heterogeneity (Fig. 15.6). ECa measurements are used to map variation of soil salinity, clay content, soil water content and organic matter [15].

15.4.4 Border Effects

In plant breeding programs and genetic studies large numbers of genotypes need to be tested in field trials. Due to resource constraints, usually, the area per genotype is reduced with growing number of genotypes tested. Growing conditions in very short rows containing only a few seeds are very different compared to commercial fields. In the latter all plants are of the same genotype. In test fields with small plots, difference in plant height and canopy architecture between adjacent plots can lead to confounding effects when rating plant performance. Especially for complex traits like yield, this poses challenges. Reduction of yield was observed with growing height of plants in adjacent plots [17]. When gaps are narrow, intergenotypic competition between neighbors is growing. The leaf angle can also contribute to inter-plot competition. Genotypes producing more horizontal leaves gain more space in competition across a gap path row. Such genotypes benefit especially from large gaps between plots, because they are able to spread horizontally below and above ground to obtain more gap resources. Ideally plots are large enough to allow the removal of border and harvest only the inner rows. In this way, confounding effects of adjacent plots are largely removed [18]. Most of the time, especially in early generations this is not feasible and the whole plot can be harvested for expedience. End-trimming to reduce the effect of the (unnatural) lateral space between plots is a compromise when harvest of inner rows only is not possible. It is important that no large gaps exist between plots and that the spacing between plots is always the same, to avoid introducing additional variability.

15.5 Summary

Appropriate agronomic management of plot trials (1) chooses the location or creates the general field test environment which favors the traits that the breeder wishes to measure, and (2) reduces experimental error to a minimum in the test. The latter assures the breeder that measured trait values of genotypes most closely reflect true values for the particular test environment. Field sites representative for the target cropping environment in terms of soil and climatic conditions, photoperiod, and pest and disease prevalence assure relevant selection conditions, but management to create environments differing from the natural one at any location can be useful (e.g., manipulating water supply). Maximum uniformity of growing conditions within any given test is a prerequisite to compare genotypes. Both natural (e.g., soil, topography, disease) and careless management can disrupt uniformity. Along with appropriate blocking of plots in replicated trials, careful agronomic management of breeding trials, including the setting of appropriate levels across all input levels can reduce variability of crop performance from natural causes. Spatially uniform management of input applications, and meticulous operation and checking of plot seeders and harvesters are essential for maximizing uniformity and accuracy. Usually plot border rows are harvested with the rest of the plot for yield, but overlooked is the bias created by both the extra (unnatural) space of the lateral path that border rows have and/or the competition between adjacent plots for the resources of this space. Even seeding depths and densities, and plot spacing and end-trimming are critical and can be assured by properly calibrated machinery. Plot combines equipped with grain cleaners, on-board weighing systems and sensors to measure grain humidity greatly facilitate data collection and accuracy, but must be regularly checked and calibrated.

15.6   Key Concepts

Good planning, knowledge of field site conditions and appropriate agronomic management reduce the experimental error, mimicking the target environment precisely is important to develop adapted varieties.

15.7   Review Questions

1. What is experimental error?
2. Why is agronomic management important in breeding trials?
3. How can agronomic management be used to reduce experimental error?
4. Why are field trials important when a wide variety of molecular genetic tools is available?
5. How can mechanization facilitate the implementation of breeding trials?

15.7.1   Review Question Answers

1. The experimental error is defined as the difference between a measured value of a quantity and its true value.
2. Good agronomic management creates selection conditions that lead to clear expression of trait differences between genotypes. It also reduces field variability which facilitates the detection of the best performing genotypes
3. Researchers should use agronomic management that maximizes the uniformity in the distribution of resources and results in vigorous crops. Researchers should carefully design and monitor field operations to minimize errors. Regular monitoring of the field is essential to ensure any problems with weeds, diseases or pests are caught early and can be managed before they introduce variability that affects the experiment. Variability in field operations like uneven applications of inputs should be prevented by regular revision and maintenance of machinery and equipment, providing detailed instructions and oversight for field operations.
4. Plants must perform under varying field conditions and complex traits like yield cannot be easily predicted. Ultimately, those data can only be obtained from practical experiments. Techniques, like genomic selection, require data from field experimentation, to predict performance of untested populations.
5. The use of adequate machinery like plot seeders and harvesters contributes to precise trial management and facilitates operation. Machine seeding assures even seeding depth and density. Plot combines can be equipped with grain cleaners, on-board weighing systems and sensors to measure grain humidity and weight, which can greatly facilitate data collection.

15.8 Conclusions

Agronomic management is very important in breeding programs to reduce the experimental error and create selection environments relevant for the target regions.

Acknowledgements This work was made possible by the generous support of the CGIAR Research Program on Wheat (WHEAT).

References

1. Hobbs PR, Sayre KD (2001) Managing experimental breeding trials. In: Reynolds MP, Ortiz-Monasterio JI, McNab A (eds) Application of physiology in wheat breeding. CIMMYT, Mexico City
2. Petersen RG (1994) Agricultural field experiments: design and analysis. Marcel Dekker, New York
3. Mwadzingeni L, Shimelis H, Dube E, Laing MD, Tsilo TJ (2016) Breeding wheat for drought tolerance: progress and technologies. J Integr Agric 15:935–943. https://doi.org/10.1016/S2095-3119(15)61102-9
4. Ortiz-Monasterio JI, Manske G, van Ginkel M (2012) Chapter 3. Nitrogen and phosphorous use efficiency. In: Reynolds MP, Pask AJD, Mullan DM (eds) Physiological breeding I: interdisciplinary approaches to improve crop adaptation. CIMMYT, Mexico City, p 174
5. Honsdorf N, Verhulst N, Crossa J, Vargas M, Govaerts B, Ammar K (2020) Durum wheat selection under zero tillage increases early vigor and is neutral to yield. Field Crop Res 248. https://doi.org/10.1016/j.fcr.2019.107675
6. Verhulst N, Govaerts B, Sayre KD, Deckers J, François IM, Dendooven L (2009) Using NDVI and soil quality analysis to assess influence of agronomic management on within-plot spatial variability and factors limiting production. Plant Soil 317:41–59. https://doi.org/10.1007/s11104-008-9787-x
7. Shatar TM, Mcbratney AB (1999) Empirical modeling of relationships between sorghum yield and soil properties. Precis Agric 1:249–276. https://doi.org/10.1023/A:1009968907612
8. Machado S, Bynum ED, Archer TL, Bordovsky J, Rosenow DT, Peterson C, Bronson K, Nesmith DM, Lascano RJ, Wilson LT, Segarra E (2002) Spatial and temporal variability of sorghum grain yield: influence of soil, water, pests, and diseases relationships. Precis Agric 3:389–406. https://doi.org/10.1023/A:1021597023005
9. Kravchenko AN, Robertson GP, Thelen KD, Harwood RR (2005) Management, topographical and weather effects on spatial variability of crop grain yields. Agron J 97:514–523. https://doi.org/10.2134/agronj2005.0514
10. Berry PM, Griffin JM, Sylvester-bradley R, Scott RK, Spink JH, Baker CJ, Clare RW (2000) Controlling plant form through husbandry to minimise lodging in wheat. Field Crop Res 67:59–81. https://doi.org/10.1016/S0378-4290(00)00084-8
11. Tripathi SC, Sayre KD, Kaul JN (2005) Planting systems on lodging behavior, yield components, and yield of irrigated spring bread wheat. Crop Sci 45:1448–1455. https://doi.org/10.2135/cropsci2003-714
12. Peng DL, Chen XG, Yin YP, Lu KL, Yang WB, Tang YH, Wang ZL (2014) Lodging resistance of winter wheat (Triticum aestivum L.): lignin accumulation and its related enzymes activities due to the application of paclobutrazol or gibberellin acid. Field Crop Res 157:1–7. https://doi.org/10.1016/j.fcr.2013.11.015

13. Prasanna BM, Araus JL, Crossa J, Cairns JE, Palacios N, Das B, Magorokosho C (2013) Chapter 13. High-throughput and precision phenotyping for cereal breeding programs. In: Cereal genomics II. Springer, p 438
14. Campos I, González-Gómez L, Villodre J, Calera M, Campoy J, Jiménez N, Plaza C, Sánchez-Prieto S, Calera A (2019) Mapping within-field variability in wheat yield and biomass using remote sensing vegetation indices. Precis Agric 20:214–236. https://doi.org/10.1007/s11119-018-9596-z
15. Corwin DL, Lesch SM (2005) Apparent soil electrical conductivity measurements in agriculture. Comput Electron Agric 46:11–43. https://doi.org/10.1016/j.compag.2004.10.005
16. Rodrigues JF, Ortiz-Monasterio I, Zarco-Tejeda PJ, Schulthess J, Gerard B (2015) High resolution remote and proximal sensing to assess low and high yield areas in a wheat field. In: Proceedings of the European Conference on Precision Agriculture, 10th conference. ECPA, Tel Aviv, Israel
17. Clarke FR, Baker RJ, DePauw RM (1998) Interplot interference distorts yield estimates in spring wheat. Crop Sci 38:62–66. https://doi.org/10.2135/cropsci1998.0011183X003800010011x
18. Rebetzke GJ, Fischer RTA (2013) Plot size matters: interference from intergenotypic competition in plant phenotyping studies. Funct Plant Biol 41:107–118

Part III
Translational Research to Incorporate
Novel Traits

Chapter 16
A Century of Cytogenetic and Genome Analysis: Impact on Wheat Crop Improvement

Bikram S. Gill

Abstract Beginning in the first decade of 1900, pioneering research in disease resistance and seed color inheritance established the scientific basis of Mendelian inheritance in wheat breeding. A series of breakthroughs in chromosome and genome analysis beginning in the 1920s and continuing into the twenty-first century have impacted wheat improvement. The application of meiotic chromosome pairing in the 1920s and plasmon analysis in the 1950s elucidated phylogeny of the Triticum-Aegilops complex of species and defined the wheat gene pools. The aneuploid stocks in the 1950s opened floodgates for chromosome and arm mapping of first phenotypic and later protein and DNA probes. The aneuploid stocks, coupled with advances in chromosome banding and in situ hybridization in the 1970s, allowed precise chromosome engineering of traits in wide hybrids. The deletion stocks in the 1990s were pivotal in mapping expressed genes to specific chromosome bins revealing structural and functional differentiation of chromosomes along their length and facilitating map-based cloning of genes. Advances in whole-genome sequencing, chromosome genomics, RH mapping and functional tools led to the assembly of reference sequence of Chinese Spring and multiple wheat genomes. Chromosome and genomic analysis must be integrated into wheat breeding and wide-hybridizaton pipeline for sustainable crop improvement.

Keywords Genome analyzer methods · Wheat phylogeny · Aneuploidy · Chromosome banding · in situ hybridization · Deletion stocks · Genome sequencing

B. S. Gill (✉)
Wheat Genetics Resource Center, Department of Plant Pathology, Kansas State University, Manhattan, KS, USA
e-mail: bsgill@ksu.edu

© The Author(s) 2022
M. P. Reynolds, H.-J. Braun (eds.), *Wheat Improvement*,
https://doi.org/10.1007/978-3-030-90673-3_16

16.1 Learning Objectives

- Become familiar with the history of wheat genetics, cytogenetics and genomics research, the scientists who did the work, the significance of their discoveries and how it impacted wheat genetics and breeding research.

16.2 Introduction

The author [1] had the pleasure of doing graduate work with Professor Charley Rick, who obtained his PhD with Karl Sax (pioneer wheat cytogeneticist) at Harvard University in 1940 and the same year began his career at UC Davis. I did postdoctoral research (1973–1975) with Dr. Ernie Sears (Father of Wheat Genetics), who obtained his PhD with EM East, Harvard University in 1936, and the same year began his research career with USDA at the University of Missouri. My co-supervisor at the University of Missouri was Professor Gordon Kimber, who trained at the famous Plant Breeding Institute at Cambridge in UK. As a founding director of Wheat Genetics Resource Center (1984–present), a position from which I retired in 2018, we conducted collaborative research with major wheat research groups in the US and worldwide, including CIMMYT and ICARDA [2]. Many of my academic pedigree and first and second generation scientists are active in crop research. From this vantage point, I want to highlight major breakthroughs over a century of wheat cytogenetic and genome analysis research and how it impacted crop improvement. Due to limitations on space and citations, for original citations, the reader may be referred to secondary citations in books [3–6] or review articles [7–11].

16.3 Validation of Mendel's Laws of Inheritance in Wheat Laid the Foundation for Scientific Breeding

Soon after the rediscovery of Mendel's laws of genetic inheritance in 1900, Biffen [12] reported that yellow rust resistance in a winter wheat cultivar was controlled by a single recessive gene that segregated in a ratio of 3:1. This was the first documented case of Mendelian inheritance for disease resistance in plants. However, other workers were unable to reproduce Biffen's results until Stakeman in 1914 [13] in Minnesota documented physiological races in the fungus with differing specificities to resistance genes in the host. These discoveries laid the foundation for breeding for disease resistance in wheat and other crops. Borlaug, who trained with Stakeman in Minnesota, will go on to work on a Rockefeller Foundation funded project in Mexico in the 1940s and usher in the Green Revolution to fight world hunger.

However, one unsolved problem remained: how do Mendel's law of discrete inheritance factors account for continuous, quantitative or blending inheritance?

Nilson-Ehle in 1909 [14] solved this riddle by an ingenious analysis of seed color inheritance in wheat where he observed ratios of 63:1, 15:1 and 3:1 red to white seeds in F2:3 families. Nilson-Ehle proposed a multifactorial hypothesis to explain red seed color inheritance; three seed color genes were segregating in some F3 families, which gave 63:1 ratio; two were segregating in others, which gave 15:1 ratio; and one gene was segregating in some that gave 3:1 ratio of red and white seeds. This led to the wide acceptance of Mendel's laws for all types of qualitative and quantitative genetic traits and the pioneering work in wheat laid the foundation for scientific breeding for crop improvement.

16.4   Genome Analyzer Method, Wheat Phylogeny and Gene Pools

By 1915, three cultivated wheat species had been described by Schulze in Germany (cited in [3], p. 5) and Flakesberger (cited in [7]) in Russia. In an episode worthy of a suspense movie, T. Minami of Hokkaido University in Japan, in the middle of the First World War, requested these wheat seed stocks from Flakesberger in autumn 1915. Minami probably got these seeds in spring 1916 as he wrote a letter of acknowledgement in May, 1916 [7]. In 1918, a young graduate student, Tetsu Sakamura, (cited in [7]) analyzed chromosome counts of these species and discovered chromosome numbers of 2n = 14, 2n = 28 and 2n = 42 and concluded that polyploidy played a major role in wheat species phylogeny.

Sakamura also produced F1 hybrids between diploid and tetraploid species, and between tetraploid and hexaploid species. A second graduate student, H. Kihara, in 1924 (cited in [7]) analyzed chromosome pairing in triploid and pentaploid hybrids. And as often happens in science, Sax in 1922 [15] independently discovered polyploidy in wheat and also reported on the chromosome pairing in triploid and pentaploid wheat hybrids (Fig. 16.1).

Kihara ([3], p. 14) designated the tetraploid wheat (*T. turgidum*) genome as AABB and the hexaploid wheat (*T. aestivum*) genome as AABBDD (D as a designation of the unique genome of Dinkel wheat) and, by inference, diploid wheat (*T. monococcum*) as AA. Kihara reported crucial observations on the breeding behavior of pentaploid hybrids; they were semisterile and most of the progeny had chromosome numbers either close to 2n = 28, 35 or 42. This meant that, although based on F1 plant meiotic pairing of 14″ + 7′, a range of gametes (chromosome ranging from 14, 15, 16 to 21) are expected but mainly gametes with n = 14 or n = 21 functioned. This led Kihara [3] to propose the concept of the genome ([3], p. 69) as a physiological entity necessary for cell function, which was 1x = 7 unique chromosomes for wheat as mainly gametes with n = 7 or multiples of 7 such as 14 or 21 were functional.

Kihara in 1930 (cited in [7]) called phylogenetic analysis based on meiotic pairing analysis the genome analyzer method and went on to elucidate phylogenetic

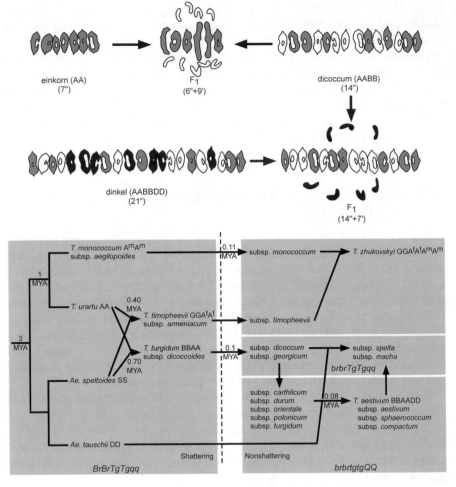

Fig. 16.1 First breakthrough in chromosome and genome analysis. Top panel: Homologous chromosomes pair during meiosis and this method, called the genome analyzer method, was used to elucidate chromosome and genome relationships among einkorn (*T. monococcum*), dicoccum (*T. turgidum*) and dinkel (*T. aestivum*) wheat species hybrids. At metaphase I (MI) of meiosis; einkorn, dicoccum and dinkel showed 7, 14 and 21 bivalents, respectively, indicating polyploidy driven speciation. The F1 hybrids between einkorn and dicoccum showed typically 3 rod and 3 ring bivalents and 9 univalents; we now know that chromosome 4A of polyploid wheat is highly rearranged and does not pair with 4A of diploid wheat. The F1 hybrids between dicoccum and dinkel wheat showed 14 bivalents and 7 univalents. The fact that chromosomes of these three species of wheat pair and recombine means that genes can be transferred from einkorn to dicoccum and dinkel, and from dicoccum to dinkel by interspecific hybridization and breeding. Figure modified with permission from [16]. Bottom panel: Current understanding of phylogeny and time line of wheat speciation [17], domestication and domestication genes (*Br/br* britille/nonbrittle rachis, *Tg/tg* tought/soft glume, *q/Q* speltoid/square spike)

relationships of the wheat and *Aegilops* species (summarized in Kihara 1951, cited in [7]). In 1937, the drug colchicine was found to induce polyploidy by artificial chromosome doubling. McFadden and Sears in 1944 (cited in [18] produced an amphiploid by colchicine chromosome doubling of an F1 hybrid between wild emmer and *Aegilops tauschii* (*syn Ae. squarrosa*). Mcfadden and Sears [18] found that F1 hybrids between the amphiploid and bread wheat were fertile and their chromosomes paired as 21 bivalents (21″) in meiosis. Kihara in 1944 ([3], p. 82) independently produced F1 hybrids between cultivated tetraploid wheat *T. persicum* and *Ae. tauschii* and found that they were naturally fertile; he called them synthesized wheats now referred as synthetic hexaploid wheats.

The seminal and independent discoveries of *Ae. tauschii* as the D-genome donor of bread wheat, and artificial synthesis of bread wheat at the height of Second World War laid a scientific basis for the exploitation of tetraploid wheat and *Ae. tauschii* for wheat improvement. The US occupation of Japan also provided an opportunity for USDA scientist SD Salmon to procure seed of the semidwarf wheat Norin 10 (*Rht1-B1*, *Rht2-D1*), and USDA scientist Vogel at Washington State began breeding short-statured wheats (see Chap. 2).

Tetraploid wheat (*T. turgidum*, 2n = 28, genomes AABB) and *Ae. tauschii* (2n = 14, genome DD), the latter belonging to a different genus, are considered as primary gene pool species of wheat. Although there are crossability and sterility barriers because of ploidy variation, the D-genome chromosomes of *Ae. tauschii* and bread wheat readily pair and recombine (Riley and Chapman 1960 cited in [19]) as do the A- and B-genome chromosomes of emmer and bread wheat. McFadden (cited in [2]) made the first crosses between emmer and bread wheat in 1915, a wide-crossing method he termed "radical breeding", and over the next 20 years bred the wheat variety 'Hope'. Among a suite of abiotic and biotic stress traits, Hope carried a durable stem rust resistance gene *Sr2*.

Kihara ([3], pp. 15, 73) noted that pairing between A^m-genome chromosomes of *T. monococcum* with the A genome of polyploid wheat was loose. Naranjo et al. 1987 (cited in [2]). discovered that chromosome 4A of polyploid wheat is highly rearranged and no longer pairs with 4A of diploid wheat. Lilienfeld and Kihara in 1934 (cited in [7], see also [3], p. 75) found that another tetraploid, *T. timopheevii*, had a genome composition of AAGG. Sax and Sax as early as in 1924 (cited by Linc et al. 1999 cited in [2]) reported that *Ae. cylindrica* had one genome in common with wheat, which was later identified as the D genome; many other polyploid species also carry D genome (Chap. 17). All these species that share partial chromosome homologies with bread wheat constitute the secondary gene pool of wheat. Doussinault et al. in 1983 (cited in [10]) transferred eyespot resistance (*Pch1*) from D-genome of *Ae. ventricosa* (DDM^vM^v) to chromosome 7D of wheat by homologous recombination. Later research by Barianna and McIntosh 1993, 1994 (cited in [10]) detected a cryptic transfer by spontaneous recombination involving homoeologus chromosomes $2M^v$ of *Ae ventricosa* and 2A of *Ae ventricosa* carrying resistance genes for rust (Lr37, Sr38, Yr17), powdery mildew, root knot nematode, wheat blast and T2A·2Mv may also boost wheat yield [20].

Kihara in 1924 (cited in [7]; see also [3], p. 14) also analyzed wheat X rye hybrids, and reported an almost complete lack of meiotic pairing between wheat and rye chromosomes; 28 univalents were observed in most cells thereby precluding genetic transfers by natural recombination. Such species constitute the tertiary gene pool of wheat. However, univalent chromosomes are prone to breakage at the centromere and spontaneous translocations involving wheat and alien chromosomes are not uncommon. Spontaneous 1B/1R substitutions and a T1BL·1RS translocation, where the long arm of chromosome 1B of wheat was translocated to the short arm of chromosome 1R of rye, was discovered in German wheat varieties by Kattermann in 1937 (cited in [16]). Wheats bred with the T1RS·1BL have a robust root system, high yield and resistance to all three rusts (*Lr26, Sr31, Yr9*), powdery mildew (*Pm8*) and some insects. This translocation was deployed with great success first, in Germany and Russia, and then worldwide from breeding efforts at CIMMYT. The *Sr31* provided worldwide resistance to stem rust until Ug99 race in Uganda in 1998.

The genome analyzer method not only elucidated phylogeny of the wheat-*Aegilops* complex (Fig. 16.1, and Figure 1 in [2]) but also defined the wheat gene pools, thereby laying the theoretical foundation for their exploitation in wheat improvement. Borlaug used McFadden's Hope, Vogel's reduced height germplasm and shuttle breeding in Mexico to develop short-statured and rust-resistant varieties that launched the Green Revolution in south and west Asia beginning in the late 1960s. CIMMYT breeders bred the world's highest yielding, second generation Green Revolution wheats based on T1B·1R. More recently, *Ae, tauschii*, either through direct hybridization [19] or synthetic hexaploids [21] has provided a major flux of new variation for wheat crop improvement.

16.5   Wheat Aneuploidy, Chromosome Mapping, and Comparative Genetics

While Kihara's genome analysis provided a rough road map of genomic and phylogenetic relationships of wheat and *Triticeae* species, it revealed very little about the genetic effects of individual chromosomes. In 1936, Sears began a long-term project on wheat polyploidy by producing a large number of amphiploids from his wide-hybridization experiments. Sears (see Sears and Miller cited in [22]) selected 'Chinese Spring', a wheat land race from China, because of its high crossability with rye and, by inference, with other wild species. Unexpectedly, in addition to authentic wheat/rye hybrid plants, he recovered two haploid wheat plants. Upon pollination of haploids, Sears recovered 11 plants that were aneuploid 2n-1 or 2n-2 (in contrast to ploidy variation of multiples of basic genome of 1x = 7). In the progeny of one monosomic, Sears recovered a nullisomic-3B plant (missing 3B chromosome pair) that was asynaptic and isolated 17 of the possible 21 monosomic/nullisomics. Nullisomic phenotyping was used to assign a number of traits to individual chromosomes, such as the red seed trait that Ehle analyzed in

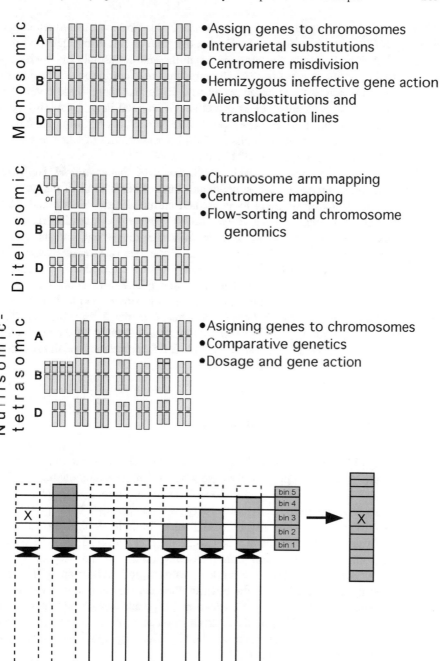

Monosomic
- Assign genes to chromosomes
- Intervarietal substitutions
- Centromere misdivision
- Hemizygous ineffective gene action
- Alien substitutions and translocation lines

Ditelosomic
- Chromosome arm mapping
- Centromere mapping
- Flow-sorting and chromosome genomics

Nullisomic-tetrasomic
- Asigning genes to chromosomes
- Comparative genetics
- Dosage and gene action

NT   DtS   DtL   delS1   delS2   delS3   delS4   bins   RH map

bin 5
bin 4
bin 3
bin 2
bin 1

Fig. 16.2 Second breakthrough in chromosome and genome analysis based on aneuploid stocks. Top panel. Sears isolated many types of aneuploid stocks for targeted mapping of genes to individual chromosomes or arms bypassing genetic complexities posed by polyploidy. Three most commonly used type of aneuploid stocks and their uses are shown; such stocks are available for the 21 chromosomes of Chinese Spring wheat. Bottom panel: The aneuploid stocks in combination with deletion stocks (see Fig. 16.5) and radiation hybrid (RH) mapping [24] provide a pipeline for targeted mapping of genes as shown for trait x

1909, awnedness, pubescence and speltoidy (q). In addition, he isolated telosomics (missing a chromosome arm), trisomics and tetrasomics and also elucidated their breeding behavior. Sears also isolated the first nullisomic-tetrasomic stock where he showed that a specific A-genome (2A) chromosome could compensate for a missing B-genome chromosome (2B) based on gametophytic and sporophytic compensation (for methodology details, see Friebe et al. 1993c cited in [2]), ushering comparative genetic analysis. Sears [23] report, "The Aneuploids of Common Wheat" on the isolation of cytogenetics stocks for the 21 chromosomes of wheat is considered the "Wheat Cytogenetics Bible" ([22]; Fig. 16.2).

In wheat breeding, one particular application was the aneuploid facilitated isolation of intercultivar wheat substitution lines that facilitated mapping of qualitative and quantitative traits to individual chromosomes (Morris and Sears 1967, cited in [22]). McFadden's Hope cultivar genome was partitioned into 21 individual chromosome substitution lines in Chinese Spring wheat. Loegering et al. in 1957 (cited in [22]) used this material to map Hope stem rust resistance gene $Sr2$ on chromosome 3B. Law [25] using substitution lines, constructed a linkage map of chromosome 7B for a number of qualitative and quantitative characters. Sears cytogenetic stocks were widely shared and ensued a worldwide explosion of wheat genetics research and the first "Wheat International Genetics Symposium" (IWGS) was organized in Winnipeg in 1958 to coordinate and review wheat research at 5-year intervals. The last IWGS that was held in 2018, replaced by the International Wheat Congress to be held at 2-year intervals.

16.6   Chromosome Manipulation

Sears aneuploidy research also laid the foundation for directed chromosome manipulation, which he appropriately described as "chromosome engineering", a term reserved for introgressing chromosome segments into a crop plant from different genomes of the secondary and tertiary gene pool species. These procedures are discussed in Chap. 18, see also Qi et al. [11]. O'Mara in 1940 [26] produced a set of rye chromosome additions in wheat using the first man-made crop 'triticale'. Since then, many alien addition lines involving dozens of species have been produced (WGRC website https://www.k-state.edu/wgrc/). Wheat aneuploids and alien additions can be used to produce wheat-alien chromosome translocations as first demonstrated by Sears in 1952 (cited in [22]), and several sets have been produced [27]. Sears in 1956 (cited in [22]) also pioneered irradiation as a method to transfer alien genes into wheat and radiation hybrid mapping played a major role in the genome assembly of wheat [24].

One of the most fundamental discoveries from aneuploidy research was the identification of a pairing homoeologous gene $Ph1$ on 5B (Okamoto 1957, Riley and Chapman 1958, cited in [28]), which controls diploid-like pairing and disomic inheritance in polyploid wheat. Mello-Sampayo in 1971 (cited in [29]) identified a second gene, $Ph2$ with an intermediate effect, on 3D and encodes a mismatch repair protein MSH7-3D that inhibits homoeologous recombination. A large number of suppressors and promotors of pairing have been identified on many wheat chromosomes and in wheat species hybrids [28]. Sears in 1977 (cited in [22]) used irradiation to isolate *ph1b,* a deficiency mutant of *Ph1*. Alien chromosome transfers into wheat by induced homoeologous pairing were first demonstrated for the transfer of yellow rust resistance from *Ae. comosa* (Riley et al. 1968, cited in [28]) and leaf rust resistance from *Agropyron* (Sears 1973, cited in [22]). The *ph1b*-induced, homoeologous pairing, coupled with modern chromosome identification and molecular marker tools, is now the method of choice in alien gene transfer [11] (see Chap. 18).

16.7 Plasmon Analysis, Wheat Phylogeny and Hybrid Wheat

Kihara (1951; cited by Tsunewaki in Chapter 16 in [4]) also is credited for initiating studies on the production of nuclear-cytoplasmic substitutions and plasmon diversity in the wheat-Aegilops complex. His student, T. Tsunewaki, SS Maan in USA and Panayotov in Bulgaria, had long-running projects on alloplasmic wheat (Maan, 1975, 1991; Panytov 1983, cited in the Chapter 16 by Tsunewaki in [4]). Kihara and Tsunewaki in 1962 (cited in Chapter 16 in [4]) reported the use of alien cytoplasm for producing haploids. Tsunewaki's group sequenced the mitochondrial and chloroplast (cp) genomes [30, 31] and demonstrated that *Ae. speltoides* contributed cytoplasmic genomes to both lineages of polyploid wheats (Chapter 16 in [4]). This has been validated by sequencing and haplotype analysis of cp genomes of a large number of diploid and polyploid *Triticum* and *Aegilops* species [17]. The analysis revealed that the older emmer lineage evolved 700,000 years ago compared to the timopheevii lineage that evolved 400,000 years ago (Fig. 16.1). One of the most important outcomes of plasmon analysis for wheat improvement was the discovery of a hybrid wheat production system based on Timopheevii cytoplasm (Wilson 1962, cited in [32]). Maan (cited in [32]) and his colleague Lucken at North Dakota led a major public sector effort in developing and freely sharing refined *Rf* gene stocks and improved A, B and R lines for a commercially viable hybrid wheat crop. Hybrid wheat received a further boost with the recent molecular cloning of fertility restoration genes *Rf1-1A* and *Rf3-1B* and sterility inducing mitochondrial orf279 transcript and molecular elucidation of their mode of action [32].

16.8 Protein Markers

In the mid-1960s, my fellow graduate students began using gel electrophoresis to study protein variation especially of isozymes and seed storage proteins, presumed to be direct products of genes based on the classic one gene-one protein hypothesis. Indeed, beginning with first results of aneuploid mapping of isozymes (Brewer et al.

1969 cited in [22]), especially Hart in USA and Gale and his group in the UK, identified a large set of isozyme homoeoloci that were conserved among wheat and alien
homoeologous chromosomes (Chapter 12 by Hart in [4]). Thus protein markers
rather than time consuming analysis of sporophytic and gametophytic compensation could be used to measure chromosome homoeologous relationships. The protein markers also found applications in wheat breeding for marker-assisted selection
for linked markers for disease resistance, such as eye spot resistance (McMillen
et al. 1986 cited in [10]), bread making quality (see Chap. 11) and many other traits.
Protein markers gave the first indications of patterns of native wheat species diversity and wheat phylogeny, including the birth place of bread wheat (Wang et al.
cited in [16]).

16.9   Molecular Cytogenetic Methods Provide Insights into
Chromosome Substructure and Rapid Analysis
of Alien Introgressions

Sears developed an exquisite cytogenetic system in wheat, yet nothing was known
about the structure of individual wheat chromosomes. All chromosome identification
was indirect, based on time-consuming meiotic pairing and aneuploidy analysis of
F1 plants. Beginning in the late 1960s, rapid identification of somatic chromosomes
in plants and animals was achieved with the discovery of Giemsa and fluorescence
staining techniques (see Gill and Kimber 1974a, b cited in [1]). Simple methods
were developed for DNA digestion, gel electrophoresis, cloning, labelling and
mapping in Southern blots and *in situ* on chromosomes on a glass slide. The first
experiments on wheat DNA analysis were initiated by Richard Flavell in the UK
and by Rudi Appels in Australia (relevant references cited in Chapter 23 by Dvorak
in [6]). We knew that the wheat genome was polyploid, but it was also large at 16
billion bp, and more than 80% was repetitive consisting of dispersed and tandemly
repeated arrays (Flavell et al. 1974 and Bennett and Smith 1976 cited in Chapter
23 in [6]; Li et al. 2004 cited in [1]).

While still a graduate student at Davis, I won a grant from DF Jones Research
Foundation to explore the application of new staining techniques for wheat
chromosome identification for which Ernie Sears offered laboratory facilities at
Missouri. Arriving in Missouri in the spring of 1973, Ernie found space for my work
in Kimber's laboratory, for Ernie did all his monumental work by himself in his
large office (shared with his wife and fellow geneticist Lottie Sears), where one
table was devoted to a small microscope and another with a sink for fixing wheat
spikes for cytology and, incidentally, brewing coffee! I hit pay dirt soon and, based
on distinctive patterns of heterochromatic bands, we cytogenetically identified the
seven chromosomes of rye (Gill and Kimber 1974a cited in [1]) and the 21
chromosomes of wheat (Gill and Kimber 1974b cited in [1]). A few years later, with
colleagues Friebe and Endo, we published detailed cytological maps and a
nomenclature system for the 21 chromosomes of wheat (Gill et al. 1991 cited in [1])
(Fig. 16.3).

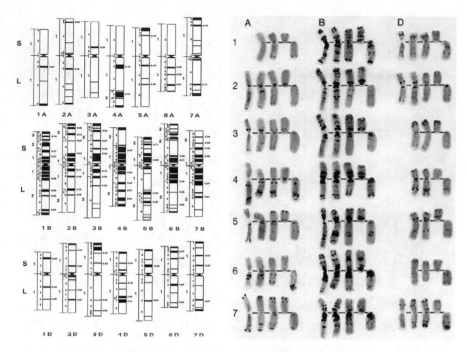

Fig. 16.3 Third breakthrough in chromosome and genome analysis based on the cytogenetic identification, and resolution and description of the substructure of heterochromatic (dark staining) and euchromatic (light staining) regions of the 21 chromosomes of wheat. (Modified with permission from Gill et al. 1991, cited in [1])

In 1984, Lane Rayburn, a postdoctoral fellow in my lab from Louisiana, travelled in his cowboy attire to Stanford University (birthplace of DNA cloning) to clone "dang wheat DNA" (Rayburn and Gill 1986 cited in Chapter 23 in [6]). Rayburn isolated a clone pAs1 for identification of the D-genome chromosomes of wheat (Rayburn and Gill 1987 cited in Chapter 23 in [6]) and also developed a rapid biotin-labelling method for mapping DNA sequences on chromosomes in situ (Rayburn and Gill 1985 cited in [1]). Scharweizer and Heslop Harrison in the UK developed methods for genomic in situ hybridization (GISH), where parental genomes could be distinguished in interspecific F1 hybrids (cited in [8]). Single-copy gene sequences also can be mapped by fluorescence in situ hybridization (FISH) to discern genetic homology [34]). Thus, armed with these tools, a cytogeneticist can establish a system for any unknown species (Fig. 16.4), cytogenetically identify individual chromosomes and also discern their genomic origin and follow chromatin transfer in wide hybrids [9].

Advances in wide hybridization techniques (Zenkteler and Nitzse 1984, Laurie and Bennett 1986 and 1988 including the discovery of wheat/maize system for haploid breeding, cited in [8]) and new cytogenetic tools were applied to the analysis of alien introgressions [2, 5, 8, 10]. In the 1950s, wheat streak mosaic virus (WSMV), vectored by the wheat curle mite, devastated the Great Plains wheat crop.

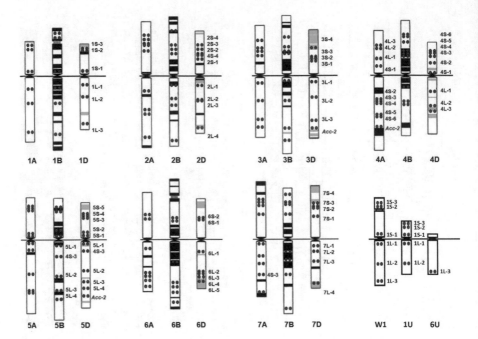

Fig. 16.4 Fourth breakthrough in chromosome and genome analysis based on fluorescence *in situ* hybridization (FISH) mapping of DNA sequences on chromosomes. FISH and unique gene probe sets (shown as red dots) allow rapid cytogenetic identification of wheat and alien chromosomes. Wheat group 1 probe set (W1) revealed a translocation between chromosomes 1 U and 6 U of *Ae umbellulata* (bottom right). (Modified with permission from [34])

The greenhouse where I began wheat genetics research in Kansas in 1979, was built by wheat growers to tackle this menace. Daryl Wells and his group at South Dakota threw everything into the alien gene transfer tool kit, including irradiation and crosses with high-pairing *Ae. speltoides*, to induce alien transfer and produced a number of lines immune to WSMV from Agrotricum/wheat crosses (Lay et al. 1971 and Wells et al. 1973, 1982; cited in [10, 33]). Among this material, using C-banding and in situ hybridization, Friebe et al. 1991a (cited in [33]) identified a compensating translocation T4DL·4Ai#2S, but this line also contained another translocation T7AS-7SS·7SL (5% of 7AS of wheat and 95% *Ae. speltoides* 7S) that was preferentially transmitted. It took us some effort to eliminate this unwelcome alien chromosome. The T4DL·4Ai#2S harboring *Wsm1*, and more recent recombinants using molecular cytogenetic and DNA marker tools [11], are impacting production agriculture for control of WSMV. As usually happens, the *Wsm1* recombinant also has a potent gene that provides resistance to all races of Ug99 (Yu Jin, personal communication, April 8, 2021, Manhattan, KS, USA).

In the southern Great Plains, EE Sebesta was using irradiation to transfer rye genes for greenbug (*Gb6*) and Hessian fly (*H25*) resistance to wheat. I remember visiting him in Oklahoma and he proudly showed me the irradiation gun he used to produce Amigo wheat, the donor of T1RS·1AL that does not have the adverse effect on breadmaking properties and has been widely used in production agriculture with great impact (Sebesta et al. 1995b cited in [10]). However, Sebesta was greatly devastated, for he had bred Amigo to control greenbug only to learn a new biotype had overcome the resistance. Our cytogenetic results (Lapitan et al. 1986 cited in [10]) showed that the Amigo translocation arose spontaneously by centric misdivision rather than by irradiation. Our colleague Jim Hatchett was screening another set of Sebesta's wheat-rye irradiation materials for Hessian fly resistance. We did not work on this material while Sebesta was alive. Our posthumous analysis (Mukai et al. 1993 cited in [10]) showed that Sebesta had accomplished a rare feat and inserted a tiny bit rye chromatin harboring Hessian fly resistance *H25* into a wheat chromosome and this picture made the cover of *Chromosoma*. I have always regretted that Sebesta was not able to appreciate the beauty of his creation during his lifetime!

One more story before I close this section. Bob McIntosh spent a mini-sabbatical in Kansas to work on mapping gene *Lr45* introgressed from rye that he was unable to map by monosomic analysis. Within a few weeks, Bob determined that *Lr45* was located on the translocation chromosome T2AS-2RS·2RL, consisting of a small chunk of wheat 2AS arm but half of rye 2RS arm and all of rye 2RL arm; too much alien chromatin to be useful for breeding (McIntosh 1995a, cited in [10]). Apparently, McIntosh was a victim of Murphy's Law, for he analyzed 19 of the 21 monosomic progenies that gave noncritical ratios, except the critical monosomic 2A cross that he failed to make!

16.10   Chromosome Physical and DNA Marker Linkage Maps Reveal Wheat Chromosome Structural and Functional Differentiation

I spent time at UC Riverside working with Giles Waines in 1976–1977, where Lennert Johnson had amassed one of the most well-documented wild wheat collections. In Kansas, we focused our efforts on exploiting this collection for wheat improvement. *Ae. tauschii* proved to be a rich source of genetic diversity resistance genes, and we developed a pipeline for direct introgression using wheat/*Ae. tauschii* crosses and backcrosses [2, 19]. For documenting gene novelty, monosomic methods of gene mapping were cumbersome (Gill et al. 1987 cited in [6]) and we soon, in parallel with molecular cytogenetics research, began exploring RFLP (restriction fragment length polymorphism) markers for genetic mapping and tagging of useful genes.

My student Kam-Morgan was the first in our group to explore, and feel the pain and pleasure, of RFLP mapping in wheat. Because more than 90% of wheat genome consists of repetitive DNA, catching a signal of hybridization probe of a single copy clone on a X-ray film is technically demanding. But worse, 90% of the time, Lauren found that her probes did not detect polymorphism, were uninformative and wasted effort! We shifted our strategy to mapping the D-genome of wheat using an *Ae. tasuchii* mapping population where 75% of the probes were polymorphic. Kam-Morgan et al. in 1989 (cited in [1]) reported the first rudimentary linkage map of 5D chromosome.

Graduate student Kulvinder Gill made the first robust linkage map of *Ae. tasuchii*, a wild crop relative that was proving to be a gold mine for wheat improvement, using an in-house *PstI*-digested clone library that targets transcribed genes (Gill et al. 1991 cited in [1]). He mapped a rust resistance gene 43 cM away from marker locus D14 at the tip of chromosome arm 1DS. Postdoctoral fellow Ed Lubbers (Lubbers et al. 1991 cited in [1]) used RFLPs to analyze the structure of Ae tauschii gene pool and more recent analysis has identified two major lineages of Ae. tauschii and the birthplace of bread wheat (Wang et al. cited in [16]). RFLPs are great for comparative mapping but, for plant breeding applications, alternative breeder-friendly markers and maps were developed and of these, microsatellite marker maps, Dart arrays and more recent SNP arrays are noteworthy (see Chap. 28) (Chapter 9 by Paux and Sourdille in [6]).

I spent my sabbatical leave Down Under in Rudi Appels lab in Australia in 1986–1987 to learn the basics of DNA cloning, mapping and sequencing. As usually happens, Rudi became interested in our *Ae. tauschii* introgression research, and recruited Evans Lagudah to lead a GRDC project. During one of the all-important tea breaks, Sir Otto Frankel showed me a wheat chromosome banding photograph from Endo vividly demonstrating a chromosome breaking effect of an alien chromosome. Endo had visited our lab in 1981 to hone his skills in chromosome banding techniques. I immediately contacted Endo and we began a US-Japan Collaborative project on the isolation of deletion stocks (Fig. 16.5).

We constructed the first-generation, deletion bin-based physical maps of molecular markers for the 21 chromosomes of wheat [35]. The data provided the first glimpse of structural and functional differentiation along the chromosome length. Recombination was suppressed around the centromeric regions and gene density was low; on the contrary, recombination and gene density was high towards the chromosome ends. The deletion stocks, together with Sears' aneuploid stocks, now could be used for targeted mapping of genes to small chromosome intervals (Fig. 16.2, bottom panel).

It was time of great molecular fervor during the 7th IWGS (1988) held in Cambridge, UK and some of us there under the leadership of Cal Qualset began discussions on the need for a coordinated international public effort for the molecular mapping of the wheat genome. The first meeting of the International Triticeae Mapping Initiative (ITMI) was held in California in 1989 and Ernie Sears attended to bless this new "wild west" of wheat research. An ITMI single-seed descent (SSD) molecular mapping population was based on a cross of Ernie's iconic genetic model variety Chinese Spring with the first SHW genotype produced by McFadden and Sears [18]. Besides coordinating mapping efforts of the seven wheat homoeologous groups by seven research laboratories around the world, an ITMI\–NSF-funded project was launched on deletion bin mapping of the expressed portion (cDNAs) of the wheat genome using a subset of deletion stocks (Qi et al. 2003 cited in [1]). The

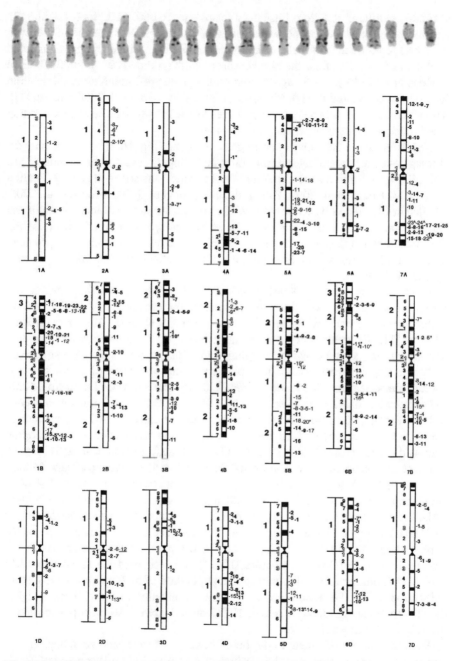

Fig. 16.5 Fifth breakthrough in chromosome and genome analysis based on deletion stocks for targeted mapping of genes to specific regions of chromosomes. Top panel shows normal chromosome 5A (left) and 23 5A-deletion chromosomes involving the long arm from the smallest to the largest deletion (left to right). These deletion breakpoints are listed on the ideogram of 5AL on the right. The *Q* gene was mapped to a tiny segment of overlapping distal deletions 7 and 23, which led to the cloning of *Q* gene (Simons et al. 2006 cited in [1]) and many other genes in wheat. The breakpoints of 436 deletions are depicted similarly on the ideogram of 21 chromosomes of wheat. (Modified with permission from [35])

second-generation, deletion bin-based maps of 16,000 EST loci for the 21 chromosomes of wheat (results were published in a special volume 168 of Genetics in 2004) confirmed the gene density/recombination frequency gradients and evolutionary novelty along the chromosome length (Akhunov et al. 2003a, b cited in Chapter 23 in [6]). All 64 agronomic gene tags mapped in the terminal deletion bins (Qi et. 2004 cited in [1]). My student Deven See (See et al. 2006 cited in [1]), who was a welder before he got late into science, used to say that Darwin's workshop was located at the ends of wheat chromosomes.

The deletion bin EST maps and targeted mapping (Fig. 16.2) paved the way for cloning genes for several agronomic traits, including disease resistance genes *Lr10* and *Lr21*, vernalization genes *Vrn1* and *Vrn2* and the domestication gene *Q* (Feuillet et al. 2003, Huang et al. 2003, Yan et al. 2003, Yan et al. 2004, Simons et al. 2006, all cited in Chapter 12 by Krattinger, Wicker and Keller in [6]). Reduced height and photoperiod genes were identified based on comparative mapping (relevant references cited in Chapters 17 and 20 in [6]). The cloned genes not only provided perfect markers for wheat breeding but also identified various alleles at each agronomic locus.

Even more important, cloned genes provide insights about their origin and evolution suggesting experimental approaches for creating new alleles, as we learned from our work with the *Lr21* gene (Huang et al. 2003, Huang et al. 2009 cited in [1]). Graduate student Li Huang developed a high-resolution mapping population and, after intensive mapping, found that D14 was the closest marker. Only one plant had the D14 allele of the resistant *Ae. tauschii* donor but was susceptible to leaf rust. Li made a cosmid library and sequenced a 40-kb cosmid clone and it had only one disease resistance-like gene that was identical in sequence to D14. Finally, discussing the results at one of the daily WGRC 'lunch munch' meetings following years of frustration, we decided to forget about the exceptional plant and use the cosmid clone harboring D14 in transformation. Harold Trick gave us transgenic plants in a few months and D14 positive plants were resistant. Marker D14 was *Lr21*! We sequenced the exceptional F2 plant (with the resistant D14 allele but susceptible) and found that it had suffered a gene conversion and had an 800-bp DNA insertion from the susceptible parent. Sequencing of *lr21* alleles, we identified an H1H1 haplotype in the spring wheat "Fielder" and an H2H2 haplotype in the winter wheat "Wichita"; intriguingly, *Lr21* had a hybrid haplotype of H1H2. We crossed Fielder (H1H1) and Wichita (H2H2) and recovered the resistance allele H1H2 from intragenic recombination in a population of 5876 plants (Huang et al. 2009 cited in [1]). The recombination associated mutation rate is 170 times higher than the spontaneous mutation rate of 10^{-6}; indeed, Darwin's workshop is located at the ends of chromosomes!

Building on Sears' aneuploidy based concept of comparative mapping and chromosome homoeologous relationships of wheat and alien species, Tanksley's famous "garden blots" extended the concept of homoeology to the grass pangenome (Ahn et al. 1993 and other relevant references cited in Sorrells et al. 2003 in Chapter 17 in [6]). Thus, all grass genome information can be leveraged for the improvement of grass crops.

16.11 Reference Wheat Genome Sequence

As we entered the twenty-first century, Arabidopsis was sequenced in 2000 and the sequencing of rice as a model for cereal crops was underway (relevant references in Chapter 24 in [6]). In wheat, we were doing tedious chromosome walking, dirty Southerns and getting "blot" fatigue! Watching our students working with stone-age tools, I and many other wheat workers were convinced that a reference sequence and investments in wheat were needed if wheat crop technology was to stay competitive with other crops. Following an exploratory wheat genome sequencing workshop at ITMI meetings in Winnipeg in 2002, Rudi Appels and I co-organized a USDA/NSF-funded workshop in Washington DC and made a strategic plan for wheat genome sequencing [36].

The key technology component of the new strategy for mitigating disadvantages posed by a large genome size and polyploidy was the exploitation of a "chromosome genomics" platform, pioneered by Dolezel's group in the Czech Republic (see Chapter 10 on chromosome genomics by Dolezel et al. in [6]) where they could fractionate single chromosomes and arms for sequencing or prepare DNA libraries for physical mapping. Wheat chromosomes were assigned to genome centers in 13 countries (http://www.wheatgenome.org). We had a double ditelosomic chromosome field planting in Aberdeen, ID, and sent seed material for chromosome fractionation to Dolezel's group and from there DNA or BAC libraries went to genome centers. We were unable to get US funding for wheat genome sequencing and the leadership shifted to INRA, France under the overall leadership of Kelley Eversole (see Chapter 24 in [6]). Instead, the NSF in US chose to fund sequencing of diploid *Ae. tauschii* led by Jan Dvorak at UC Davis. The shot-gun sequencing papers (unanchored contig sequences, limited value) were followed by the first reference (ordered and anchored to chromosome and genetic maps, high value) sequence of chromosome 3B and survey sequences of the 21 chromosomes of wheat (IWGSC 2014 cited in [37]).

I began the chapter by recounting Sakamura's discovery of wheat chromosome constitution and ploidy in 1918. One hundred years later, the wheat reference genome as well as the diploid D and A, the tetraploid AABB and ten elite wheat variety genome sequences have been deciphered providing information on agronomically important genomic regions (relevant references in [37, 38]). Wheat gene discovery platforms (see Chapter 13 in [6]) are driving the pace of gene discovery for precise gene tinkering using technologies such as CRISPER (see Chap. 29). Sequence-based analysis of genetic diversity, monitoring of genetic diversity during germplasm enhancement and MAS (see Chap. 28) and genomic selection (see Chap. 32) are poised to drive the efficiency and pace of genetic gain for wheat crop improvement. The applications of genomics information for conservation, management and utilization of wheat genetic resources are discussed elsewhere [16].

16.12   Key Concepts

The conceptual advances discussed in detail in the chapter relate to the definition of wheat gene pools defined by meiotic pairing analysis; aneuploidy facilitated genetic and comparative mapping based on gametophytic and sporophytic compensation; chromosomal structural and functional differentiation, chromosome engineering and gene novelty; wheat genome structure and function based homoeology, gene discovery and breeding; the concepts are briefly described due to space limitations and reader is highly encouraged to consult the original sources as cited through-out the chapter.

16.13   Conclusions

As the brief review shows, each genetic, chromosome and genomic advance facilitated the efficiency and productivity of wheat breeding. Now we are entering a new phase where one must be able to decipher the reference genomes of the parents and selected breeding lines and make selections based on masses of phenotypic and genomic data. In wide hybridization, each cross has an impact of an earthquake and one must use the concepts of homoeology to distinguish chaff from grains and cryptic transfers may be more important than the targeted transfer!

Acknowledgments   The author is grateful to John Raupp for excellent art work and drafting of the figures, and to John Raupp, Wanlong Li and Bernd Friebe with the editing of the manuscript and Bikram S Gill Chair for financial support.

References

1. Raupp WJ, Friebe B (2013) Bikram Gill: cytogeneticist and wheat man. In: Plant breeding reviews 37. Wiley, pp 1–34
2. Gill BS, Friebe B, Raupp WJ, Wilson DL, Cox TS, Sears RG, Brown-Guedira GL, Fritz AK (2006) Wheat genetics resource center: the first 25 years. Adv. Agron. 85, Academic, pp 73–136
3. Kihara H (1982) Wheat studies - retrospect and prospects, vol Volume 3. Kodansha Ltd., Tokyo
4. Jauhar P (1996) Methods of genome analysis in plants. CRC Press
5. Molnár-Láng M, Ceoloni C, Doležel J (2015) Alien introgression in wheat cytogenetics, molecular biology, and genomics. Springer
6. Feuillet C, Muehlbauer G (2009) Genetics and genomics of the triticeae. Springer, New York
7. Tsunewaki K (2016) Memoir on the origin of wheat stocks used by Prof. Tetsu Sakamura, on the centennial of his discovery of the correct chromosome number and polyploidy in wheat. Genes Genet Syst 91:41–46. https://doi.org/10.1266/ggs.15-00077
8. Jiang J, Friebe B, Gill BS (1993) Recent advances in alien gene transfer in wheat. Euphytica 73:199–212. https://doi.org/10.1007/BF00036700

9. Jiang J, Gill BS (1994) Nonisotopic in situ hybridization and plant genome mapping: the first 10 years. Genome 37:717–725. https://doi.org/10.1139/g94-102

10. Friebe B, Jiang J, Raupp WJ, McIntosh RA, Gill BS (1996) Characterization of wheat-alien translocations conferring resistance to diseases and pests: current status. Euphytica 91:59–87. https://doi.org/10.1007/BF00035277

11. Qi L, Friebe B, Zhang P, Gill BS (2007) Homoeologous recombination, chromosome engineering and crop improvement. Chromosom Res 15:3–19. https://doi.org/10.1007/s10577-006-1108-8

12. Biffen RH (1905) Mendel's laws of inheritance and wheat breeding. J Agric Sci 1:4–48. https://doi.org/10.1017/S0021859600000137

13. Stakman E (1914) A study of cereal rusts: physiological races. Minn Agric Expt Stat Bull 138

14. Nilson-Ehle H (1909) Einige ergebnisse von Kreungungen bei Hafer und weizen. Bot Not:257–294

15. Sax K (1922) Sterility in wheat hybrids. II. Chromosome behavior in partially sterile hybrids. Genetics 7:513–552

16. Gill B, Friebe B, Koo DH, Li W (2019) Crop species origins, the impact of domestication and the potential of wide hybridization for crop improvement. In: Zeigler R (ed) Sustaining global food security. CSIRO Publishing, Clayton, p 538

17. Gornicki P, Zhu H, Wang J, Challa GS, Zhang Z, Gill BS, Li W (2014) The chloroplast view of the evolution of polyploid wheat. New Phytol 204:704–714. https://doi.org/10.1111/nph.12931

18. McFadden ES, Sears ER (1946) The origin of Triticum spelta and its free-threshing hexaploid relatives. J Hered 37(81):107. https://doi.org/10.1093/oxfordjournals.jhered.a105590

19. Gill BS, Raupp WJ (1987) Direct genetic transfers from Aegilops squarrosa L. to hexaploid wheat1. Crop Sci 27. https://doi.org/10.2135/cropsci1987.0011183X002700030004x

20. Gao L, Koo D-H, Juliana P, Rife T, Singh D, Lemes da Silva C, Lux T, Dorn KM, Clinesmith M, Silva P, Wang X, Spannagl M, Monat C, Friebe B, Steuernagel B, Muehlbauer GJ, Walkowiak S, Pozniak C, Singh R, Stein N, Mascher M, Fritz A, Poland J (2021) The Aegilops ventricosa 2N(v)S segment in bread wheat: cytology, genomics and breeding. Theor Appl Genet 134:529–542. https://doi.org/10.1007/s00122-020-03712-y

21. Ogbonnaya FC, Abdalla O, Mujeeb-Kazi A, Kazi AG, Xu SS, Gosman N, Lagudah ES, Bonnett D, Sorrells ME, Tsujimoto H (2013) Synthetic hexaploids: harnessing species of the primary gene pool for wheat improvement. In: Plant breeding reviews. Wiley, pp 35–122

22. Riley R (1990) Ernie Sears: wheat cytogeneticist. In: 2nd International Wheat Genetics Symposium. College Ag, Univ Ext, Univ Missouri-Columbia, p 7

23. Sears ER (1954) The aneuploids of common wheat. Res Bull Mo Agr Exp Stn 572:59

24. Tiwari VK, Heesacker A, Riera-Lizarazu O, Gunn H, Wang S, Wang Y, Gu YQ, Paux E, Koo D-H, Kumar A, Luo M-C, Lazo G, Zemetra R, Akhunov E, Friebe B, Poland J, Gill BS, Kianian S, Leonard JM (2016) A whole-genome, radiation hybrid mapping resource of hexaploid wheat. Plant J 86:195–207. https://doi.org/10.1111/tpj.13153

25. Law CN (1967) The location of genetic factors controlling a number of quantitative characters in wheat. Genetics 56:445–461

26. O'mara JG (1940) Cytogenetic studies on Triticale. I. A method for determining the effects of individual Secale chromosomes on Triticum. Genetics 25:401–408

27. Liu W, Koo D-H, Friebe B, Gill BS (2016) A set of Triticum aestivum-Aegilops speltoides Robertsonian translocation lines. Theor Appl Genet 129:2359–2368. https://doi.org/10.1007/s00122-016-2774-3

28. Sears ER (1976) Genetic control of chromosome pairing in wheat. Annu Rev Genet 10:31–51. https://doi.org/10.1146/annurev.ge.10.120176.000335

29. Serra H, Svačina R, Baumann U, Whitford R, Sutton T, Bartoš J, Sourdille P (2021) Ph2 encodes the mismatch repair protein MSH7-3D that inhibits wheat homoeologous recombination. Nat Commun 12:803. https://doi.org/10.1038/s41467-021-21127-1

30. Ogihara Y, Isono K, Kojima T, Endo A, Hanaoka M, Shiina T, Terachi T, Utsugi S, Murata M, Mori N, Takumi S, Ikeo K, Gojobori T, Murai R, Murai K, Matsuoka Y, Ohnishi Y, Tajiri H, Tsunewaki K (2002) Structural features of a wheat plastome as revealed by complete sequencing of chloroplast DNA. Mol Gen Genomics 266:740–746. https://doi.org/10.1007/s00438-001-0606-9

31. Ogihara Y, Yamazaki Y, Murai K, Kanno A, Terachi T, Shiina T, Miyashita N, Nasuda S, Nakamura C, Mori N, Takumi S, Murata M, Futo S, Tsunewaki K (2005) Structural dynamics of cereal mitochondrial genomes as revealed by complete nucleotide sequencing of the wheat mitochondrial genome. Nucleic Acids Res 33:6235–6250. https://doi.org/10.1093/nar/gki925

32. Melonek J, Duarte J, Martin J, Beuf L, Murigneux A, Varenne P, Comadran J, Specel S, Levadoux S, Bernath-Levin K, Torney F, Pichon J-P, Perez P, Small I (2021) The genetic basis of cytoplasmic male sterility and fertility restoration in wheat. Nat Commun 12:1036. https://doi.org/10.1038/s41467-021-21225-0

33. Friebe B, Mukai Y, Dhaliwal HS, Martin TJ, Gill BS (1991) Identification of alien chromatin specifying resistance to wheat streak mosaic and greenbug in wheat germ plasm by C-banding and in situ hybridization. Theor Appl Genet 81:381–389. https://doi.org/10.1007/BF00228680

34. Danilova TV, Friebe B, Gill BS (2014) Development of a wheat single gene FISH map for analyzing homoeologous relationship and chromosomal rearrangements within the Triticeae. Theor Appl Genet 127:715–730. https://doi.org/10.1007/s00122-013-2253-z

35. Endo TR, Gill BS (1996) The deletion stocks of common wheat. J Hered 87:295–307. https://doi.org/10.1093/oxfordjournals.jhered.a023003

36. Gill BS, Appels R, Botha-Oberholster A-M, Buell CR, Bennetzen JL, Chalhoub B, Chumley F, Dvořák J, Iwanaga M, Keller B, Li W, McCombie WR, Ogihara Y, Quetier F, Sasaki T (2004) A workshop report on wheat genome sequencing: International Genome Research on Wheat Consortium. Genetics 168:1087–1096. https://doi.org/10.1534/genetics.104.034769

37. The International Wheat Genome Sequencing Consortium, Appels R, Eversole K, Stein N, Feuillet C, Keller B, Rogers J, Pozniak CJ, Choulet F, Distelfeld A, Poland J, Ronen G, Sharpe AG, Barad O, Baruch K, Keeble-Gagnère G, Mascher M, Ben-Zvi G, Josselin A-A, Himmelbach A, Balfourier F, Gutierrez-Gonzalez J, Hayden M, Koh C, Muehlbauer G, Pasam RK, Paux E, Rigault P, Tibbits J, Tiwari V, Spannagl M, Lang D, Gundlach H, Haberer G, Mayer KFX, Ormanbekova D, Prade V, Šimková H, Wicker T, Swarbreck D, Rimbert H, Felder M, Guilhot N, Kaithakottil G, Keilwagen J, Leroy P, Lux T, Twardziok S, Venturini L, Juhász A, Abrouk M, Fischer I, Uauy C, Borrill P, Ramirez-Gonzalez RH, Arnaud D, Chalabi S, Chalhoub B, Cory A, Datla R, Davey MW, Jacobs J, Robinson SJ, Steuernagel B, van Ex F, Wulff BBH, Benhamed M, Bendahmane A, Concia L, Latrasse D, Bartoš J, Bellec A, Berges H, Doležel J, Frenkel Z, Gill B, Korol A, Letellier T, Olsen O-A, Singh K, Valárik M, van der Vossen E, Vautrin S, Weining S, Fahima T, Glikson V, Raats D, Číhalíková J, Toegelová H, Vrána J, Sourdille P, Darrier B, Barabaschi D, Cattivelli L, Hernandez P, Galvez S, Budak H, Jones JDG, Witek K, Yu G, Small I, Melonek J, Zhou R, Belova T, Kanyuka K, King R, Nilsen K, Walkowiak S, Cuthbert R, Knox R, Wiebe K, Xiang D, Rohde A, Golds T, Čížková J, Akpinar BA, Biyiklioglu S, Gao L, N'Daiye A, Kubaláková M, Šafář J, Alfama F, Adam-Blondon A-F, Flores R, Guerche C, Loaec M, Quesneville H, Condie J, Ens J, Maclachlan R, Tan Y, Alberti A, Aury J-M, Barbe V, Couloux A, Cruaud C, Labadie K, Mangenot S, Wincker P, Kaur G, Luo M, Sehgal S, Chhuneja P, Gupta OP, Jindal S, Kaur P, Malik P, Sharma P, Yadav B, Singh NK, Khurana JP, Chaudhary C, Khurana P, Kumar V, Mahato A, Mathur S, Sevanthi A, Sharma N, Tomar RS, Holušová K, Plíhal O, Clark MD, Heavens D, Kettleborough G, Wright J, Balcárková B, Hu Y, Salina E, Ravin N, Skryabin K, Beletsky A, Kadnikov V, Mardanov A, Nesterov M, Rakitin A, Sergeeva E, Handa H, Kanamori H, Katagiri S, Kobayashi F, Nasuda S, Tanaka T, Wu J, Cattonaro F, Jiumeng M, Kugler K, Pfeifer M, Sandve S, Xun X, Zhan B, Batley J, Bayer PE, Edwards D, Hayashi S, Tulpová Z, Visendi P, Cui L, Du X, Feng K, Nie X, Tong W, Wang L (2018) Shifting the limits in wheat research and breeding using a fully annotated reference genome. Science 361:eaar7191. https://doi.org/10.1126/science.aar7191

38. Walkowiak S, Gao L, Monat C, Haberer G, Kassa MT, Brinton J, Ramirez-Gonzalez RH, Kolodziej MC, Delorean E, Thambugala D, Klymiuk V, Byrns B, Gundlach H, Bandi V, Siri JN, Nilsen K, Aquino C, Himmelbach A, Copetti D, Ban T, Venturini L, Bevan M, Clavijo B, Koo D-H, Ens J, Wiebe K, N'Diaye A, Fritz AK, Gutwin C, Fiebig A, Fosker C, Fu BX, Accinelli GG, Gardner KA, Fradgley N, Gutierrez-Gonzalez J, Halstead-Nussloch G, Hatakeyama M, Koh CS, Deek J, Costamagna AC, Fobert P, Heavens D, Kanamori H, Kawaura K, Kobayashi F, Krasileva K, Kuo T, McKenzie N, Murata K, Nabeka Y, Paape T, Padmarasu S, Percival-Alwyn L, Kagale S, Scholz U, Sese J, Juliana P, Singh R, Shimizu-Inatsugi R, Swarbreck D, Cockram J, Budak H, Tameshige T, Tanaka T, Tsuji H, Wright J, Wu J, Steuernagel B, Small I, Cloutier S, Keeble-Gagnère G, Muehlbauer G, Tibbets J, Nasuda S, Melonek J, Hucl PJ, Sharpe AG, Clark M, Legg E, Bharti A, Langridge P, Hall A, Uauy C, Mascher M, Krattinger SG, Handa H, Shimizu KK, Distelfeld A, Chalmers K, Keller B, Mayer KFX, Poland J, Stein N, McCartney CA, Spannagl M, Wicker T, Pozniak CJ (2020) Multiple wheat genomes reveal global variation in modern breeding. Nature 2020:1–7. https://doi.org/10.1038/s41586-020-2961-x

Chapter 17
Conserving Wheat Genetic Resources

Filippo Guzzon, Maraeva Gianella, Peter Giovannini, and Thomas S. Payne

Abstract Wheat genetic resources (WGR) are represented by wheat crop wild relatives (WCWR) and cultivated wheat varieties (landraces, old and modern cultivars). The conservation and accessibility of WGR are fundamental due to their: (1) importance for wheat breeding, (2) cultural value associated with traditional food products, (3) significance for biodiversity conservation, since some WCWR are endangered in their natural habitats. Two strategies are employed to conserve WGR: namely *in situ* and *ex situ* conservation. *In situ* conservation, i.e. the conservation of the diversity at the location where it is found, consists in genetic reserves for WCWR and on farm programs for landraces and old cultivars. *Ex situ* conservation of WGR consists in the storage of dry seeds at cold temperatures in germplasm banks. It is currently the most employed conservation strategy for WGR because it allows the long-term storage of many samples in relatively small spaces. Due to the great number of seed samples of WGR and associated passport data stored in genebanks, it is increasingly important for the management of *ex situ* collections to: (1) employ efficient database systems, (2) understand seed longevity of the seed accessions, (3) setup safety backups of the collections at external sites.

Keywords Germplasm banks · Genetic reserves · On farm conservation · Seed conservation · Seed viability · Wheat wild relatives

F. Guzzon (✉) · T. S. Payne
International Maize and Wheat Improvement Center (CIMMYT), Texcoco, Mexico
e-mail: f.guzzon@cgiar.org; t.payne@cgiar.org

M. Gianella
Department of Biology and Biotechnology "L. Spallanzani", University of Pavia, Pavia, Italy
e-mail: maraeva.gianella01@universitadipavia.it

P. Giovannini
Global Crop Diversity Trust, Bonn, Germany
e-mail: peter.giovannini@croptrust.org

© The Author(s) 2022
M. P. Reynolds, H.-J. Braun (eds.), *Wheat Improvement*,
https://doi.org/10.1007/978-3-030-90673-3_17

17.1 Learning Objectives

- To know the principal categories of wheat genetic resources,
- To know the principles of *in situ* conservation of wheat genetic resources,
- To know the principles of *ex situ* seed conservation of wheat genetic resources in germplasm banks.

17.2 Introduction – Plant Genetic Resources (PGR) and their Conservation

Wheat domestication occurred 9000 to 12,000 BCE, resulting in cereal crops within the genus *Triticum*, two of which are among the most widely grown crops worldwide, namely bread wheat (*T. aestivum* subsp. *aestivum*) and durum wheat (*T. turgidum* subsp. *durum*). Wheat genetic resources are represented by several domesticated and wild taxa.

Overall, plant genetic resources for food and agriculture (PGRFA) are defined as "any genetic material of plant origin of actual or potential value for food and agriculture" [1]. Genetic diversity is the foundation for crop improvement and is an insurance against unforeseen threats to agricultural production such as plant pathogens and climate changes [2].

Wheat genetic resources can be grouped in the following biological/agronomic categories:

- Cultivated wheats: wheat species were gathered by ancient societies, gradually resulting in the domestication of several wheat crop taxa. Cultivated materials consist of:

    - Landraces (or primitive cultivars): "dynamic populations of a cultivated plant that have historical origin, distinct identity and lacks formal crop improvement, as well as often being genetically diverse, locally adapted and associated with traditional farming systems" [3];
    - Old cultivars: sometimes known as obsolete cultivars, the term refers to cultivated varieties which have fallen into disuse;
    - Modern cultivated varieties (modern cultivars): agronomic varieties in current use and newly developed varieties;
    - Special stocks: such as advanced breeding lines (i.e. pre-released varieties developed by plant breeders), mapping populations, CRISPR-edited lines and cytogenetic stocks.

- Crop wild relatives (CWR): wild plant species that are genetically related to cultivated crops. CWR are not only the wild ancestors of the domesticated plant but also other more distantly related species.

Another category of PGR of significance are the neglected crops, also referred as underutilized or orphan crops: "crop species that have been ignored by science and development but are still being used in those areas where they are well adapted and competitive" [4]. An example is the einkorn (*Triticum monococcum* subsp. *monococcum*) currently cultivated by small-holder farmers in limited areas in Europe, Middle East and North Africa. In recent years, there is a renewed interest for einkorn, mainly due to its nutraceutical properties and adaptations to organic agriculture [5].

The aim of plant genetic resources conservation is to ensure that the maximum possible allelic genetic diversity, and therefore potential useful traits for breeding of a crop, is maintained and is available and accessible for utilization. Crop domestication and selection have favored preferred haplotypes and have reduced genetic diversity. The conservation of landraces and CWR is particularly important considering that in those plants is concentrated the bulk of genetic diversity and of potential useful traits within a crop genepool. The conservation of modern cultivar is also of great importance since breeders often wish to access "improved" or refined sources of PGR diversity. Conserving PGR is important not only in order to provide useful traits for crop improvement but also for cultural reasons, since many landraces and neglected crops are connected to local identities, especially through local foods and ceremonial products.

Two main strategies are employed for the conservation of PGR, namely *in situ* and *ex situ* conservation. *In situ* conservation, i.e. the conservation of the diversity in its natural habitat, means the designation, management and monitoring of a population at the location where it is currently found. On the other hand, the *ex situ* conservation, i.e. the conservation of a genetic resources outside its natural habitat, is intended as the sampling, transfer and storage of a sample of a population of a certain species away from the original location where it was collected. Several *ex situ* conservation strategies are employed for different crops e.g. *in vitro* storage, seed banking, field genebanks, DNA banks. Seed banking allows the storage of many seed accessions in relatively small spaces; seed collections are economically viable and can provide a good sample of the genetic diversity within the crop genepool, usually remaining viable for the long-term [6].

17.3 Wheat Genetic Resources (WGR)

17.3.1 Domesticated Wheats

Two species of wheat are widely cultivated, namely: the hexaploid *Triticum aestivum* (ABD genome) and the tetraploid *T. turgidum* (AB genome, Table 17.1). Both species include several subspecies (Table 17.1). As previously mentioned, einkorn (*Triticum monococcum* L. subsp. *monococcum*, A genome) is a locally cultivated, diploid wheat.

Table 17.1 Domesticated wheats. The more common domesticated subspecies of *T. aestivum* and *T. turgidum* are also presented

Taxonomic name	Common English Name	Genome(s)	Accessions conserved *ex situ*[a]
Triticum monococcum L. subsp. *monococcum*	Einkorn	A	6971
Triticum monococcum L. subsp. *sinskajae* (Filat. & Kurkiev) Valdés & H. Scholz	Naked einkorn	A	23
Triticum turgidum L.	Rivet wheat	AB	179,701
Triticum turgidum L. subsp. *dicoccon* Schrank (Thell.)	Emmer	AB	8793
Triticum turgidum L. subsp. *durum* (Desf.) van Slageren	Durum wheat	AB	149,485
Triticum turgidum L. subsp. *carthlicum* (Nevski) Á. Löve & D. Löve	Persian wheat	AB	1382
Triticum turgidum L. subsp. *polonicum* (L.) Thell.	Polish wheat	AB	766
Triticum turgidum L. subsp. *turanicum* (Jakubz.) Á. Löve & D. Löve	Khorasan wheat	AB	461
Triticum turgidum L. subsp. *turgidum*	Poulard wheat	AB	7171
Triticum timopheevii (Zhuk.) Zhuk. subsp. *timopheevii*	Chelta Zanduri	AG	189
Triticum aestivum L.		ABD	511,130
Triticum aestivum L. subsp. *aestivum*	Bread wheat	ABD	243,634
Triticum aestivum subsp. *compactum* (Host) Mac Key	Club wheat	ABD	1921
Triticum aestivum subsp. *macha* (Dekapr. & Menabde) Mac Key	Macha wheat	ABD	374
Triticum aestivum L. subsp. *spelta (L.)* Thell.	Spelt	ABD	7070
Triticum aestivum subsp. *sphaerococcum* (Percival) Mac Key	Indian wheat	ABD	684
Triticum zhukovskyi Menabde & Eritzjan	Zhukovsky's wheat	AAG	71

[a]Accessions conserved *ex situ* estimated using data from [7], FAO-WIEWS, USDA GRIN and data provided directly by CIMMYT. The number of accessions of *T. aestivum* and *T. turgidum* includes also the accessions of the different subspecies

Two additional species of wheat were cultivated in western Georgia but are probably currently extinct under cultivation and conserved only in germplasm banks: *T. timophevii* subsp. *timopheevii* (Chelta Zanduri or Timopheevi wheat, tetraploid, AG) and *T. zhukovskyi* (Zhukovsky's wheat, hexaploid, AGG, Table 17.1). The Zhukovsky's wheat was described in the 1960s growing in a restricted area of western Georgia. This hexaploid wheat is an allopolyploid, spontaneous hybrid between Timopheevi wheat (*T. timopheevii*) and einkorn (*T. monococcum*). Zhukovsky's wheat and the two parental species used to be cultivated together in a complex of domesticated wheats named *zanduri*.

Wheat landrace cultivation was endemic throughout the Mediterranean Basin, Europe, Near East, Ethiopia, Caucasus, China and Southern Asia, since time immemorial. Wheat landraces were subsequently diffused to Australia, South Africa and the Americas. For example, the Creole wheats descendant of Spanish wheats imported from the sixteenth century were cultivated in Mexico for four centuries by small-scale farmers. In many areas of the world those landraces were replaced since the twentieth century by modern, improved varieties.

Formal wheat breeding started in the eighteenth century, eventually resulting in a plethora of old and modern cultivars. Noteworthy examples of old cultivars of bread wheat are: 'Sherriff's Squarehead', selected in the end of the nineteenth century in Great Britain, 'Ardito' and 'Mentana' selected in Italy in the first decades of twentieth century, 'Marquis' selected in Canada at the beginning of twentieth century, the semi-dwarf cultivar 'Norin 10' selected in Japan in 1935 and the cultivar 'Bezostaya 1' selected in Russia in the1950s. Several old cultivars of durum wheat also exist, e.g. the renowned 'Senatore Cappelli' released in Italy in 1915. Today, many old cultivars figure in the pedigree of modern wheat varieties and are therefore of great priority for conservation (see Chap. 2 for a history of wheat breeding).

17.3.2 Wheat Crop Wild Relatives (WCWR)

A crop "genepool concept" was defined by Harlan and De Wet [8] based on formal taxonomy and genetic relatedness, determined by the crossing ability between related species. Three main categories are considered: Primary Gene Pool (GP-1) comprising the domesticated crop and its closed wild forms with which the crop can cross producing fertile hybrids; Secondary Gene Pool (GP-2) which includes less closely related species, from which gene flow, even if difficult, is still possible using conventional breeding techniques; Tertiary Gene Pool (GP-3) which includes species from which gene transfer to the crop is impossible without the use of "rather extreme or radical measures". The gene pool levels here presented are based on: "The Harlan and de Wet Crop Wild Relative Inventory" (see: https://www.cwrdiversity.org/checklist/). An additional gene pool level classification system is historically used in wheat based on chromosome pairing and recombination (see Sect. 16.4).

The primary gene pool (GP-1, Fig. 17.1) of wheat comprises, beside the aforementioned domesticated wheats (Table 17.1), also the four wild species of the genus *Triticum* (*sensu* van Slageren 1994 [9]) included in Table 17.2.

GP-2 includes 22 species of the genus *Aegilops* and *Amblyopyrum muticum* (Table 17.3, Figs. 17.1 and 17.2). The geographic center of diversity, the areas where the most *Aegilops* grows in sympatry, is the Fertile Crescent, Turkey, the southern Caucasus, as well as the shores of the Aegean Sea. Spontaneous crosses between *Aegilops* species and cultivated wheats have been observed in several areas of the natural distribution of *Aegilops*. Those hybrids are classified in the genus x *Aegilotriticum* and are mostly sterile.

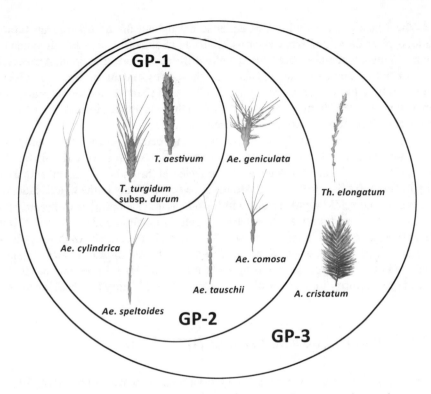

Fig. 17.1 Schematic representation of the genepool of wheat, only some species are shown

Table 17.2 Wild wheats of the genus *Triticum*

Taxonomic name	GP-1 ancestor of	Native to	Genome(s)	Accessions conserved *ex situ*[a]
Triticum monococcum L. subsp. *aegilopoides* (Link) Thell.	Einkorn	Near East, Western Asia, southern Balkans	A	5816
T. timopheevii (Zhuk.) Zhuk. subsp. *armeniacum* (Jakubz.) van Slageren	Timopheevi	Near East, southern Caucasus	AG	1849
T. turgidum L. subsp. *dicoccoides* (Körn. ex Asch. & Graebn.) Thell.	Emmer & tetraploid wheats	Near East	AB	11,535
T. urartu Tumanjan ex Gandilyan	Tetraploid wheats	Near East, southern Caucasus	A	2274

[a]Accessions conserved *ex situ* estimated using data from Genesys PGR, FAO-WIEWS, USDA GRIN and data provided directly by CIMMYT

Nevertheless, hybridization events between *Aegilops* and *Triticum* species were indeed involved in the process of evolution and domestication of tetraploid and

Table 17.3 The species of *Aegilops*, organized in the different sections in which is divided the genus, and *Amblyopyrum*. Data on the genome, ploidy and natural distribution are also provided

Section	Species name	Genome(s)	Ploidy	Distribution	Accessions conserved *ex situ*[a]
Aegilops	*Ae. umbellulata* Zhuk.	U	Diploid	Turkey, Fertile Crescent, Caucasus, Iran	794
	Ae. biuncialis Vis.	UM	Tetraploid	Mediterranean Basin, Fertile crescent, Caucasus, Russia, Ukraine	2505
	Ae. columnaris Zhuk.	UM	Tetraploid	Turkey, Crete, Fertile Crescent, Iran	509
	Ae. geniculata Roth	MU	Tetraploid	Mediterranean Basin, Caucasus, Turkey, Crimea	3218
	Ae. kotschyi Boiss.	SU	Tetraploid	Middle East, North Africa, Arabia, Central Asia	613
	Ae. neglecta Req. ex Bertol.	UM/UMN	Tetra/ Hexaploid	Mediterranean Basin, Crimea, Middle East, Turkmenistan	1818
	Ae. peregrina (Hack.) Maire & Weiller	SU	Tetraploid	Middle East, Greece, North Africa, Arabia	1642
	Ae. triuncialis L.	UC	Tetraploid	Mediterraean Basin, Crimea, Caucasus, Central Asia	6647
Comopyrum	*Ae. comosa* Sm.	M	Diploid	Southern Balkans, Cyprus, Turkey	423
	Ae. uniaristata Vis.	N	Diploid	Croatia, Greece, Albania, Italy, Turkey	79
Cylindropyron	*Ae. caudata* L.	C	Diploid	Aegean, Turkey, Fertile Crescent	701
	Ae. cylindrica Host	DC	Tetraploid	Eastern Europe, Middle East, Caucasus, Central Asia	3893

(continued)

Table 17.3 (continued)

Section	Species name	Genome(s)	Ploidy	Distribution	Accessions conserved *ex situ*[a]
Sitopsis	*Ae. bicornis* (Forssk.) Jaub. & Spach	S[b]	Diploid	Cyprus, North Africa, Middle East	505
	Ae. longissima Schweinf. & Muschl.	S[l]	Diploid	Egypt, Israel/ Palestine, Jordan	1779
	Ae. sharonensis Eig	S[sh]	Diploid	Israel/Palestine, Lebanon	2546
	Ae. searsii Feldman & Kislev ex K. Hammer	S[s]	Diploid	Israel/Palestine, Syria, Jordan, and Lebanon	519
	Ae. speltoides Tausch	S	Diploid	Fertile crescent, Turkey, Southeastern Europe	3369
Vertebrata	*Ae. tauschii* Coss.	D	Diploid	Caspian seashores, Caucasus, Central Asia, China	7186
	Ae. crassa Boiss.	DM/DDM	Tetra/ Hexaploid	Middle East, Central Asia	608
	Ae. vavilovii (Zhuk.) Chennav.	DMS	Hexaploid	Middle East	345
	Ae. ventricosa Tausch	DN	Tetraploid	Mediterranean Basin, North Africa	486
	Ae. juvenalis (Thell.) Eig	DMU	Hexaploid	Central Asia, Azerbaijan, Fertile Crescent	132
Genus *Amblyopyrum*	*Amblyopyrum muticum* (Boiss.) Eig	T	Diploid	Turkey, Armenia	181

[a]Accessions conserved *ex situ* estimated using data from Genesys PGR, FAO-WIEWS, USDA GRIN and data provided directly by CIMMYT

hexaploid wheats. The wild tetraploid wheats (i.e. *T. turgidum* subsp. *dicoccoides* and *T. timopheevi* subsp. *armeniacum*) resulted from hybridization events that occurred a few hundred thousand years ago between *T. urartu* and an unknown species of the genus *Aegilops*, probably similar to the only existing outcrossing species of this genus, *Ae. speltoides*. Hexaploid wheats belonging to *T. aestivum* do not have a single wild progenitor. This crop arose from hybridization events that occurred probably 8000 BCE in the coastal areas of the Caspian Sea, between the domesticated *T. turgidum* susbsp. *dicoccon* and the wild species *Ae. tauschii* (Fig. 17.3).

Wild species of *Triticum* and *Aegilops* have significantly contributed to wheat improvement, especially in terms of biotic resistances, as well as for grain yield and

Fig. 17.2 Examples of
Wheat Crop Wild Relatives
(WCWR): (**a**) *T. turgidum*
subsp. *dicoccoides* at
CIMMYT screenhouse
(Texcoco, Mexico); (**b**) *Ae.
biuncialis*, wild population
at Santeramo in Colle
(Italy); (**c**) *Ae. geniculata*
(left) and *Ae. ventricosa*
(right) growing together in
Garda (Italy); (**d**) *Ae.
tauschii* at CIMMYT
screenhouse (Texcoco,
Mexico); (**e**) x-ray scan of
a spikelet of *Ae. biuncialis*,
a dimorphic pair of seeds
can be noticed in the basal
fertile spikelet; (**f**) x-ray
scan of a spike of *Ae.
cylindrica*, in some of the
spikelets composing the
spike a pair of dimorphic
seeds can be noticed

abiotic stress tolerance [11]. The genetic diversity of species belonging to the GP-1
and GP-2 can be exploited to generate Synthetic Wheat Hexaploid (SWH) and chromosomal translocation introgressions. The most common SWH are produced by
hybridizing durum wheat with *Ae. tauschii*, as the latter is a huge source of diversity,
being adapted to a variety of environments in different subspecies and morphological varieties (see Chap. 18).

The GP-3 of wheat includes grass species of the genera *Agropyron*, *Elymus*,
Leymus and *Thinopyrum* (Fig. 17.1). Those species have been hybridized with cultivated wheat as genetic sources for disease resistance, salinity tolerance, and other
traits. Given the sexual barrier between cultivated wheat species and their tertiary
gene pool, to transfer traits from GP-3 species both physical and genetic methods
(causing random chromosome breaks and promoting recombination) have been
used, namely: spontaneous translocations, *in vitro* cultures, irradiation, and induced
homologous recombination [12] (see Chap. 18).

F. Guzzon et al.

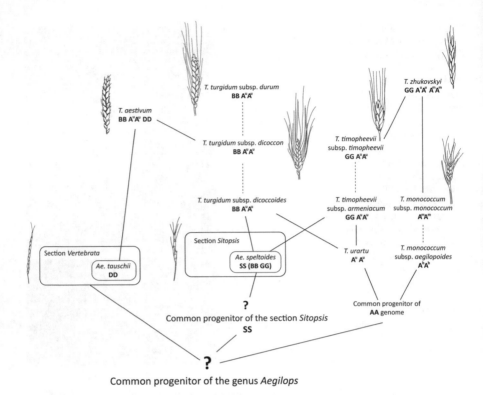

Fig. 17.3 Schematic representation of wheat evolution and domestication. Solid line represents spontaneous events of speciation and hybridization. Dashed line indicates human selection events. (Redrawn with permission from [10] by Marco Canella, Padua, Italy)

17.4 Wheat Genetic Resources Conservation

17.4.1 In situ Conservation

Some wheat wild relatives are considered endangered by the International Union for Conservation of Nature (IUCN) at global level and therefore their conservation is considered priority: i.e. *Amblyopyrum muticum* (EN-endangered), *Aegilops sharonensis* (VU-vulnerable), *Agropyron dasyanthum* (EN) and *Agropyron cimmericum* (EN). Other species, even if labeled as of "least concern" are showing populations declines in their natural habitats (e.g. *Aegilops longissima*). At continental level some species are recognized as endangered, e.g. in Europe *Ae. tauschii* is considered EN and *Ae. bicornis* is VU [13]. Considering the importance of wheat wild relatives for wheat breeding, it is also important to guarantee the conservation of species and populations that are not threatened but that have a great impact on wheat improvement as carriers of useful traits.

In this context, the implementation of *in situ* conservation strategies for wheat wild relatives is necessary. Indeed, even if *ex situ* conservation of genetic resources is easy and cost effective, *in situ* conservation has the advantage of allowing species to evolve in their original place and to retain a higher genetic diversity compared to seed bank accessions.

Maxted et al. [14] and Phillips et al. [15] identified regional diversity hot spots of *Aegilops* in which conservation reserves should be established: Syria and north Lebanon, central Israel, north-west Turkey, the Hatay region of Turkey, Turkmenistan and south France.

In Table 17.4 are listed the existing in situ reserves that conserve wild wheats.

In situ conservatories for crop wild relatives are also called genetic reserves and are generally located where protected areas have been established to conserve also other aspects of biodiversity, and so the additional resource requirements to conserve wild wheats may be minimal. Nevertheless, some specific actions are suggested to enhance the conservation of those species, for example: (I) reduce over-grazing, (II) decrease fire frequency and intensity, (III) reduce use of herbicides and pesticide (e.g. on field margins and roadsides), (IV) perform systematic monitoring of threatened populations, (V) carry out population reinforcement measures of the threatened populations, using seeds of the same populations conserved in genebanks [16]. National parks, military reserves, mountainous and controlled pastoral areas are often ideal locations for *in situ* reserves. Climate change will probably decrease, in the next few decades, the range of many wild wheats in core areas of WCWR diversity such as: North Africa, Middle East and southern Europe [17]. This underlines the importance of protecting populations of WCWR and of complementing *in situ* reserves with *ex situ* conservation to prevent the loss of many of these populations.

The *in situ* conservation of landraces and old cultivars is known as on-farm conservation, defined as: "the management of genetic diversity of locally developed crop varieties by farmers within their own agricultural systems" [18]. While in the abovementioned genetic reserves wild populations of WCWR are conserved in their natural habitats, on-farm conservation consists in the cultivation by farmers of

Table 17.4 *In situ* reserves for wheat and other cereals genetic resources conservation

Reserve name	Country	Taxa
Erebuni	Armenia	Wild wheats (*T. urartu, Triticum monococcum* subsp. *aegilopoides* and *T. timopheevii* subsp. *armeniacum*), goatgrasses (*Aegilops* spp.); also conserving: Vavilov's rye (*Secale vavilovii*), wild barley (*Hordeum* spp.)
Ammiad Project	Israel	*Triticum* spp. (also conserving *Hordeum* spp.)
Ham	Lebanon	*Triticum* spp. (also conserving *Hordeum* spp.)
Wadi Sweid	Lebanon	*Ae. biuncialis, Ae. geniculata, Ae. triuncialis, T. urartu*
Sale-Rsheida	Syria	*T. dicoccoides* (also conserving *Hordeum* spp.)
Ceylanpinar State Farm	Turkey	*Triticum* spp., *Aegilops* spp., (also conserving *Avena* spp. and *Hordeum* spp.)

locally developed, domesticated wheat varieties (landraces and/or old cultivars) to prevent their genetic erosion and eventual extinction. Strengthen value chains and therefore market opportunities for these varieties is likely the best incentive to promote their on-farm conservation by farmers.

On-farm conservation of wheat landraces and old cultivars is being put in place to enhance conservation as well as revival of those entities in several areas of the world. In particular, in some regions (e.g. East Shewa, Ethiopia; Emilia-Romagna, Italy; New England, USA; Czechia) wheat landraces are being rediscovered and re-introduced in cultivation often starting from *ex situ* collections.

17.4.2   *Ex situ Conservation*

Seed banking is currently considered as the most suitable *ex situ* conservation strategy for plants, like wheat, with orthodox seeds, i.e. seeds that can tolerate drying to low moisture content and subsequent freezing. The Commission on Genetic Resources for Food and Agriculture of the FAO proposed a series of standards for *ex situ* conservation of PGRFA that are currently followed by many international genebanks [19].

Ex situ seed conservation in genebanks can be divided into seven main activities: acquisition, seed drying, seed storage, viability monitoring, regeneration, characterization and distribution.

17.4.2.1   Acquisition

Materials can be acquired either from genebanks or from research or breeding programs. Wild relatives or landraces can be collected in the wild or obtained from farmers, respectively. When collecting populations of wild relatives in their natural habitat, it is important not to exceed the 20% of total seeds available in the sampled population not to affect the natural recruitment of natural populations.

Materials must be acquired legally, in accordance with local, national and international regulations. Materials must be described with Multi-crop Passport Descriptor data [20] and characterization data. A seed sample and its related passport data is defined as a seed accession.

17.4.2.2   Drying

Seed drying is one of the most crucial steps in seed conservation. High seed moisture content detrimentally affects seed storage viability. Seeds are dried to equilibrium in controlled environments ('drying rooms') with a temperature of 5–20 °C and 10–25% of relative humidity. Seed moisture content is regularly monitored until the seeds reach equilibrium, i.e. the moisture content of the seeds is in equilibrium with the relative humidity of the surrounding air. Wheat seeds are conserved in genebanks when they reach a moisture content between 5% and 8%. It is fundamental that, after the drying phase, seeds are stored in airtight containers to maintain the low moisture content. In some national and regional seed banks, equilibrium drying in drying rooms is not possible due to lack of infrastructure or capacity. In those cases, desiccants such as silica gel or zeolite beads can be used for seed drying [21].

17.4.2.3 Seed Storage

High temperatures also detrimentally affect seed longevity in storage. For long-term conservation, it is recommended to store dried seed accessions at a temperature of -18 ± 3 °C. In addition to the long-term ('base' collection), some banks have duplicate samples in an active short-medium term collection stored at a temperature range between -5 and 10 °C. Seed conserved in this 'active' collection are generally employed for regeneration, distribution and characterization, not to decrease the stocks conserved in the base collection.

It is important that seed accessions conserved in a germplasm bank are safety duplicated, e.g. the same accession is stored at other locations to provide an insurance against loss of material. Many genebanks duplicate their accessions at the Svalbard Global Seed Vault, located in the Artic Island of Spitsbergen, a seedbank that currently holds more than one million (with a capacity of 4.5 millions) of store duplicates (backups) of seed samples from the world's crop collections [22].

17.4.2.4 Viability Monitoring

Initial and regular seed viability testing is required to evaluate the quality of a seed lot. Seed germination is generally tested using standard protocols [23] with light and temperature-controlled incubators, using agar or filter paper as the germination medium. International standards recommend that initial germination percentage should exceed 85% for crop seed accessions stored for conservation purposes. As some specific wild relatives' accessions do not reach this threshold a lower viability can be accepted. The International Seed Testing Association (ISTA) suggests that the most suitable temperature to test wheat seed germination is 20 °C [23], while some *Aegilops* species were demonstrated to reach a higher germination when incubated at alternating temperature (e.g. 20/10 ° C) [24]. The germination of some wheat wild relatives can also be elicited by after-ripening, a period of dry storage during which seeds lose dormancy (i.e. the inability of viable seeds to germinate under optimal environmental conditions).

Many wheat wild relatives species show seed heteromorphism, defined as the production, within a spike, of two or more seed types that differ in morphological and/or eco-physiological traits. Indeed, within the genera *Aegilops* and *Triticum*, a dimorphic pair of seeds is often present in each of the spikelets composing the spike, with one seed being larger and brighter-colored than the other (Fig. 17.2). In the field, larger seeds germinate few weeks after dispersal, while the smaller ones remain dormant for several months due to the presence of a germination inhibitor in the glume. Due to this complex germination strategy, seeds of wild wheats need to be extracted from the spikelets and manually dehulled prior to the germination testing. Seed heteromorphism has implications also in longevity and conservation: it has been observed that smaller seeds of several *Aegilops* and wild *Triticum* species are longer-lived than their larger paired seeds when subjected to artificial ageing, having a greater endowment of antioxidant compounds, these being possibly involved in protection against ageing-related oxidative stress. Preliminary results revealed that smaller seeds of wild wheats are longer-lived also in *ex situ* conservation within genebanks [25].

Seed germination of stored accessions must be tested at regular intervals (e.g. every 10-15 years) to understand the loss of viability in storage and to plan re-collection or schedule regeneration activities. Walters et al. [26] found that the p_{50} (i.e. the time for seed viability to fall by 50%) for wheat seed accessions conserved in genebank conditions was 54 years. When the viability of an accessions falls below the 85% of the initial, regeneration or recollection activities need to be carried out in order to maintain available an accession with a high viability.

17.4.2.5 Regeneration

Seed multiplication is required when seed germination drops below 85% of the initial value, or when the quantity of seeds has been depleted due to frequent use of the accession. A sufficient number of seeds needs to be used for regeneration activities in order to maintain the genetic variability within the accessions. Commonly used approach is to employ between 7 and 10 g of seeds (approximately 140 to 250 seeds) for regeneration of wheat varieties. 100–130 plants should be regenerated for each accessions of wheat wild relatives. As wild wheats are considered as possible noxious weeds outside their native range, accessions belonging to those taxa are regenerated in controlled environments (i.e. screenhouses).

17.4.2.6 Characterization

A detailed description of different important traits is fundamental to ensure the maximum usability of the accessions by plant breeders. The characterization stage is often carried out during regeneration when several morphological, phenological and agronomical descriptors are assessed, also in order to confirm accessions' trueness to type. Regarding wheat genetic resources, these descriptors can be grouped as follows:

1. Seed traits, comprising morphological traits (e.g. germination, color, size, weight, vitreousness, number of shriveled seeds) but also grain quality (e.g. protein content and suitability for food processing) and agronomical traits (e.g. preharvest sprouting).

2. Spike morphology, with a characterization of the awns, glumes and spikelets.
3. Plant morphology, considering traits such as: plant height, young plant habit (e.g. upright or prostrate), straw color, leaf pubescence and tillering capacity.
4. Phenological traits, such as growth classes, i.e. classifying if an accession is a spring, winter or intermediate wheat. Inflorescence traits are also considered, e.g. days to flowering and daylength sensitivity (i.e. extent to which long days hasten flowering).
5. Stress susceptibility, considering the effects on plant growth of abiotic stresses (e.g. cold/high temperatures, drought, salinity) as well as biotic ones in terms of fungi (e.g. rust, powdery mildew, glume blotch, eye spot), pests (e.g. nematodes, hessian fly) and viruses (e.g. barley yellow dwarf virus).

Beside the morphological and agronomical traits, physiological and molecular descriptors are often employed to achieve the most reliable and complete characterization of wheat germplasm collections: this allows to evaluate trueness-to-type, to understand and organize the diversity of large germplasm collections and to mine collections for useful traits for breeding.

Some of the most used molecular techniques in wheat genotyping are:

- Studies based on restriction fragment length polymorphisms (RFLP), randomly amplified polymorphic DNA (RAPD), amplified fragment length polymorphisms (AFLP).
- Use of wheat microsatellites (WMS), simple sequence repeats (SSR), commonly known as microsatellites, have been shown to be very useful markers for trueness-to-type evaluation in wheat germplasm, being highly polymorphic both in cultivated and wild species. SSR can be genomic or 'expressed sequence tag' (EST-SSR), the latter having the advantage of possessing good generality between species.
- DArTseq genotyping, in-depth and robust technique to estimate genetic diversity among germplasm accessions. Single nucleotide polymorphisms (SNPs) detected through DArTseq can be investigated by assessing their allelic effects (i.e., genome wide association study, GWAS) and subsequently exploited for breeding.

17.4.2.7   Distribution

Germplasm distribution consists in the shipment of a sample of a seed accession conserved in a genebank in response to a request from a germplasm user. The accessibility of PGR accessions is strictly linked with the existence and updating of information databases, where the users can search the different conserved accession and linked passport data and order seed samples of the accessions they are interested in. The major database of PGR accessions conserved worldwide is Genesys PGR (https://www.genesys-pgr.org/). It brings together four million accessions located in over 450 genebank around the globe and allows the users to quickly search for and request germplasm accessions. Distribution is a fundamental activity for genebanks, involving a great number of accessions, for example the genebank of the International Maize and Wheat Improvement Center (CIMMYT, Mexico) sends worldwide, on average, more than nine thousand seed samples of WGR in more than 100 shipments annually, those seed samples are employed by the users mainly for research activities, breeding and direct cultivation.

Acquisition and distribution of germplasm across borders must follow international rules on phytosanitary certification and adhere to international treaties and conventions. Two main international treaties regulate the access and share of PGR: the Convention on Biological Diversity (CBD) and the International Treaty on Plant Genetic Resources for Food and Agriculture (ITPGRFA). The CBD of 1992 has three main aims: (1) the conservation of biological diversity; (2) the sustainable use of the components of biological diversity; (3) the fair and equitable sharing of the benefits arising out of the utilization of genetic resources. The Nagoya Protocol on Access to Genetic Resources and the Fair and Equitable Sharing of Benefits Arising from their Utilization to the Convention on Biological Diversity, also known as the Nagoya Protocol on Access and Benefit Sharing is a 2010 supplementary agreement to the CBD, it is an international agreement which aims at sharing the benefits arising from the utilization of genetic resources in a fair and equitable way. The ITPGRFA, adopted in 2001, aims at promoting the conservation of plant genetic resources and protecting farmers' rights to access and have fair and equitable sharing of benefits arising from the use of PGR. ITPGFRA established a multilateral system to exchange plant germplasm of a pool of 64 species of crops (Annex I species), through a Standard Material Transfer Agreement (SMTA). The SMTA is a private contract with standard terms and conditions that ensures that the relevant provisions of the ITPGRFA are followed by providers and recipients of material of plant genetic resources.

17.4.3   Wheat Genetic Resources Collections Worldwide

Since the end of nineteenth century, researchers highlighted the importance for breeding of the conservation and availability of landraces and crop wild relatives, especially witnessing the risk of genetic erosion of landraces due to their substitution with high-yielding improved varieties. The present concept of a genebank, as a facility for the long-term conservation of PGR, was first concretized, at the beginning of twentieth century, at the N. I. Vavilov Institute of Plant Industry in Saint Petersburg by its director R. Regel and especially its successor N.I. Vavilov, who personally focused a significant part of his research activity in collecting, conserving and studying wheat genetic resources. After the World War II, many genebanks were established in several country of the world to conserve and keep available wheat genetic resources and prevent the loss of landraces [27].

Currently, according to FAO (2010), there are more than eight hundred and fifty thousand accessions of wheat and wheat wild relative stored worldwide in genebanks. Accordingly, in our dataset there are 784,753 accessions of the genera

Table 17.5 The ten largest wheat genebanks (by number of accessions) worldwide

Institution code	Institution name	Country	Number of accessions
AUS 165	AGG	Australia	48,065
CHN001	ICGR-CAAS	China	43,039
IND001	NBPGR	India	32,154
ITA436	IBBR-CNR	Italy	32,751
LBN002	ICARDA	Lebanon	47,152
JPN183	NARO	Japan	37,907
MAR088	CRRA	Morocco	42,191
MEX002	CIMMYT	Mexico	141,759
RUS001	N.I. Vavilov Research Institute of Plant Industry	Russia	41,679
USA029	NSGC: USDA-ARS	USA	62,119

Data extracted from Genesys PGR [7], WIEWS and USDA databases and FAO [28]

Triticum and *Aegilops* recorded in the databases: Genesys PGR, FAO-WIEWS and USDA-GRIN (when the same accession is recorded in more than one of these databases, it is counted only once). Considering individual genebanks, CIMMYT holds the greatest number of accessions worldwide (with more than 140 thousand accessions) followed by the National Small Grains Germplasm Research Facility, USDA-ARS (USA) and the Australian Grain Genebank (Table 17.5).

However, it is difficult to estimate the number of unique accessions conserved *ex situ* as in many cases information about duplication is not recorded in passport data, although it is possible to do it. A study genotyping a sample of accessions of *Ae. tauschii* from 3 genebanks found that over 50% of the accessions in the sample were redundant [29].

To assess the representativeness of the diversity of the germplasm conserved *ex situ*, as opposed to the one existing (or that existed) in cultivation or in the wild, different approaches have been used, considering: the total size of collections, taxonomic coverage (number of genera and species), and ecogeographic coverage. A recent gap analysis conducted by the CGIAR Genebank Platform divided the diversity within the wheat genepool in hierarchical clusters (https://www.genesys-pgr.org/c/wheat) based on literature and experts' opinion, and estimated the number of accessions conserved *ex situ* for each group. This methodology was originally suggested by Van Treuren et al. 2009 to assess the composition of a germplasm collection. The results of this analysis suggested that in *ex situ* there are gaps of Durum wheat landraces from arid areas of Mali, Chad, Niger, Sudan, Libya, and Mauritania as well as *T. aestivum* subsp. *tibeticum* and *T. aestivum* subsp. *yunnanense* from China. Several gaps were also found in the coverage of the geographical distribution of wild and domesticated emmer.

When dealing with very large seed collections, in order to increase the accessibility of the conserved material, it is useful to cluster the accessions in core collections, grouping accessions with similar characteristics in terms of e.g. taxonomy, distribution, breeding history, characterization data.

Given the importance of wheat for agriculture worldwide, the seed conservation of wheat genetic resources is important not only for international and national genebanks but also for much smaller institutions, like community seed banks (CSB): i.e. small-scale local organizations that conserve seeds of landraces and wild useful plants on a medium-term basis and serve the needs of local communities [30]. For example, wheat accessions are conserved in CSB in Guatemala, Palestine and India.

17.5 Key Concepts

- Genetic resources of wheat are represented by: (1) WCWR, (2) landraces, (3) old cultivars, (4) modern cultivars and (5) special stocks.
- *In situ* conservation is the conservation of the diversity at the location where it is found, it consists in genetic reserves for WCWR and on farm programs for landraces and old cultivars. This conservation strategy allows genetic resources to evolve in their original area of distribution under selection by farmers and environmental factors and to retain a higher genetic diversity compared to seed bank accessions.
- *Ex situ* conservation of WGR consists in the storage of dry seeds at cold temperatures in germplasm banks. It is currently the most employed conservation strategy for WGR because it allows the long-term storage of many samples in relatively small spaces.

17.6 Conclusions

- To enhance the conservation of WGR it will be increasingly important to complement *ex situ* long-term conservation of seed accessions within genebanks with *in situ* conservation strategies both as genetic reserves for wheat wild relatives and on farm programs for landraces.
- To increase the usability of WGR collections, genebanks need to provide users with the most complete possible passport data, integrating information about collecting sites and phenotypic characterization with novel molecular data.
- Due to this increasing amount of passport information, genebanks need to invest in database systems that can efficiently store and keep available these data.
- Due to the increasing age of historical genebanks and therefore the storage time of many wheat seed accessions, the number of accessions that needs regeneration is going to increase. For this reason, is fundamental to characterize seed longevity of wheat genetic resources to prioritize accessions for viability monitoring and regeneration and avoid losses of germplasm.
- Safety duplication of seed accessions of WGR in external sites is a top-priority for genebanks in order to reduce the risk of losing the collections.

References

1. Food and Agriculture Organization of the United Nations (2009) International treaty on plant genetic resources for food and agriculture. Food and Agriculture Organization of the United Nations, Rome
2. Gepts P (2006) Plant genetic resources conservation and utilization: the accomplishments and future of a societal insurance policy. Crop Sci 46:2278–2292. https://doi.org/10.2135/cropsci2006.03.0169gas
3. Camacho Villa T, Maxted N, Scholten M, Ford-Lloyd B (2005) Defining and identifying crop landraces. Plant Genet Resour 3:373–384. https://doi.org/10.1079/PGR200591
4. Hammer K, Heller J, Engels J (2001) Monographs on underutilized and neglected crops. Genet Resour Crop Evol 48:3 5. https://doi.org/10.1023/A:1011253924058
5. Zaharieva M, Monneveux P (2014) Cultivated einkorn wheat (Triticum monococcum L. subsp. monococcum): the long life of a founder crop of agriculture. Genet Resour Crop Evol 61:677–706. https://doi.org/10.1007/s10722-014-0084-7
6. Li D, Pritchard H (2009) The science and economics of ex situ plant conservation. Trends Plant Sci 14:614–621. https://doi.org/10.1016/j.tplants.2009.09.005
7. Genesys PGR (2020) Gateway to genetic resources. https://www.genesys-pgr.org/
8. Harlan J, De Wet J (1971) Toward a rational classification of cultivated plants. Taxon 40:509–517. https://doi.org/10.2307/1218252
9. van Slageren M (1994) Wild wheats: a monograph of Aegilops L. and Amblyopyrum (Jaub. and Spach) Eig (Poaceae). Wageningen Agric Univ Pap 94
10. Kilian B, Mammen K, Millet E, Sharma R, Graner A, Salamini F, Hammer K, Özkan H (2011) Aegilops. In: Kole C (ed) Wild crop relatives: genomic and breeding resources. Cereals. Springer, Berlin/Heidelberg, pp 1–76
11. Kishii M (2019) An update of recent use of Aegilops species in wheat breeding. Front Plant Sci 10:Article 585. https://doi.org/10.3389/fpls.2019.00585
12. Rasheed A, Mujeeb-Kazi A, Ogbonnaya F, He Z, Rajaram S (2018) Wheat genetic resources in the post-genomics era: promise and challenges. Ann Bot 121:603–616. https://doi.org/10.1093/aob/mcx148
13. International Union for Conservation of Nature (2020) National Red List. https://www.nationalredlist.org/
14. Maxted N, White K, Valkoun J, Konokpa J, Hargreaves S (2008) Towards a conservation strategy for Aegilops species. Plant Genet Resour 6:126–141. https://doi.org/10.1017/S147926210899314X
15. Phillips J, Whitehouse K, Maxted N (2019) An in situ approach to the conservation of temperate cereal crop wild relatives in the Mediterranean Basin and Asian centre of diversity. Plant Genet Resour 17:185–195. https://doi.org/10.1017/S1479262118000588
16. Perrino E, Wagensommer R, Medagli P (2014) Aegilops (Poaceae) in Italy: taxonomy, geographical distribution, ecology, vulnerability and conservation. Syst Biodivers 12:331–349. https://doi.org/10.1080/14772000.2014.909543
17. Ostrowski M, Prosperi J, David J (2016) Potential implications of climate change on Aegilops species distribution: sympatry of these crop wild relatives with the major European crop Triticum aestivum and conservation issues. PLoS One 11:e0153974. https://doi.org/10.1371/journal.pone.0153974
18. Veteläinen M, Negri V, Maxted N (2009) European landraces: on-farm conservation management and use. Bioversity Technical Bulletin No. 15
19. Food and Agriculture Organization of the United Nations (2014) Genebank standards for plant genetic resources for food and agriculture
20. Alercia A, Diulgheroff S, Mackay M (2015) FAO/Bioversity multi-crop passport descriptors V.2.1 [MCPD V.2.1]. Food and Agriculture Organization of the United Nations (FAO), Rome

21. Bradford K, Dahal P, Van AJ, Kunusoth K, Bello P, Thompson J, Wu F (2018) The dry chain: reducing postharvest losses and improving food safety in humid climates. Trends Food Sci Technol 71:84–93. https://doi.org/10.1016/j.tifs.2017.11.002
22. Crop Trust (2020) Svalbard Global Seed Vault. https://www.croptrust.org/our-work/svalbard-global-seed-vault/
23. International Seed Testing Association (2021) ISTA 2021 Rules. ISTA, Bassersdorf
24. Guzzon F, Müller J, Abeli T, Cauzzi P, Ardenghi N, Balestrazzi A, Rossi G, Orsenigo S (2015) Germination requirements of nine European Aegilops species in relation to constant and alternating temperatures. Acta Bot Gall 162:349–354. https://doi.org/10.1080/1253807 8.2015.1088793
25. Gianella M, Balestrazzi A, Pagano A, Müller J, Kyratzis A, Kikodze D, Canella M, Rossi G, Mondoni A, Guzzon F (2020) Heteromorphic seeds of wheat wild relatives show germination niche differentiation. Plant Biol 22:191–202. https://doi.org/10.1111/plb.13060
26. Walters C, Wheeler L, Grothenius J (2005) Longevity of seeds stored in a genebank: species characteristics. Seed Sci Res 15:20. https://doi.org/10.1079/SSR2004195
27. Lehmann C (1981) Collecting European land-races and development of European gene banks-historical remarks. Kulturpflanze 29:29–40
28. Food and Agriculture Organization of the United Nations (2010) The Second Report on the state of the world's plant genetic resources for food and agriculture, Rome
29. Singh N, Wu S, Raupp WJ, et al. (2019) Efficient curation of genebanks using next generation sequencing reveals substantial duplication of germplasm accessions. Sci Rep 9:650. https://doi.org/10.1038/s41598-018-37269-0
30. Vernooy R, Sthapit B, Bessette G (2017) Community seed banks concept and practice – facilitator handbook. Bioversity International, Rome

Chapter 18
Exploring Untapped Wheat Genetic Resources to Boost Food Security

Julie King, Surbhi Grewal, John P. Fellers, and Ian P. King

Abstract Increasing the genetic diversity of wheat is key to its future production in terms of increasing yields, resistance to diseases and adaptability to fluctuations in global climate. The use of the progenitor species of wheat and also its wild relatives uniquely provides a route to vastly increase the genetic variation available to wheat breeders for the development of new, superior wheat varieties. The introduction of genetic variation from the wild relatives of wheat in the form of introduced chromosome segments or introgressions, has taken place for hundreds of years, albeit largely unintentionally in farmers' fields. However, the use of the wild relatives became more systematic from the 1950s onwards. The work has previously been hampered due to a lack of technology for the identification and characterisation of the introgressions and consequently the strategic use of the wild relatives. The advances in molecular biology over recent years now make it possible to generate wheat/wild relative introgressions on a scale not previously possible. In fact, the greatest threat to this area of work is now the lack of scientists/breeders with the understanding of chromosomes and their manipulation.

Keywords Wheat wild relatives · Recombination · Introgressions · Genomic *in situ* hybridisation · Single nucleotide polymorphisms (SNPs)

J. King (✉) · S. Grewal · I. P. King
School of Biosciences, University of Nottingham, Loughborough, UK
e-mail: julie.king@nottingham.ac.uk; surbhi.grewal@nottingham.ac.uk; ian.king@nottingham.ac.uk

J. P. Fellers
USDA-ARS-Hard Winter Wheat Genetics Research Unit, Manhattan, KS, USA
e-mail: john.fellers@usda.gov

© The Author(s) 2022
M. P. Reynolds, H.-J. Braun (eds.), *Wheat Improvement*,
https://doi.org/10.1007/978-3-030-90673-3_18

18.1   Learning Objectives

- What is an introgression?
- Chemical and radiation versus recombination.
- How to generate introgressions via homologous recombination and homoeologous recombination.
- How to generate introgressions from addition and substitution lines.
- How to use molecular tools for the detection of wheat/wild relative introgressions.
- Why is the phenotyping of introgression lines important?
- Understanding the case study.

18.2   Introduction

The rapidly increasing global population, set to pass the nine billion mark by 2050, presents one of the greatest challenges that humanity has faced – how to feed all the extra people? Major crops such as bread wheat, which provides 20% of the world's total calories and protein [1], will have to play a major role in feeding the population of the future. However, instead of increasing, wheat yields have recently been starting to plateau.

The plateauing of yields presently observed is most likely due to two compounding factors. Intensive breeding in the past, although very successful, has led to the exploitation and erosion of a proportion of the genetic variation available. This gradual erosion of genetic variation means that in time it will become increasingly more difficult for breeders to generate and identify new gene combinations to develop higher yielding varieties. In addition, the slowing of production increases is being further exacerbated by adverse environmental conditions resulting from climate change, e.g. heat, drought etc.

A major game changer for the production of wheat varieties adapted to climate change that would meet the needs of the increasing global population, is to dramatically increase the available gene pool. In order to achieve this, a new source of donor genetic variation that can be transferred into wheat, needs to be identified.

Wheat is related to a large number of wild species that grow in a wide range of very varied environments, e.g. in fields of cereals, deserts, salt inundated sand dunes, at high and low altitudes etc. These wild relatives, many of which evolved millions of years ago [2], provide a vast reservoir of genetic variation for potentially most, if not all, traits of agronomic and scientific importance, e.g., they carry completely novel forms and levels of genetic variation above and beyond that observed in cultivated wheat (See Chap. 17).

The transfer of genetic variation from the wild relatives in the past, while limited, has had a major impact on wheat production. While conventional breeding produces slow, but gradual, increases in yield production, the successful introgression of genetic material/genes from the wild relatives frequently results in substantial jumps in production and improvement. As a result, many commercial breeders believe that the transfer of genetic variation from the wild relatives may be the only way by which the increases in yield production required by 2050 can be achieved.

There are a number of examples of previously successful introgressions from wild relatives into wheat [3]. These include (1) An introgression from *Aegilops umbellulata* that carried a resistance gene to the disease leaf rust [4]. In 1960, this introgression saved US wheat production from catastrophic failure. (2) A spontaneous introgression from rye to wheat resulted in a substantial increase in grain yield and disease resistance [5]. The advantages that this introgression conferred over normal wheat were such that it was present in most wheat varieties in the 1990s (it is still present in many modern-day varieties). (3) A high proportion of present-day varieties carry an introgression from *Aegilops ventricosa* [6] that confers a yield advantage of circa 4% and also carries the only effective sources of resistance to the diseases eye spot and wheat blast [7]. As a result, the *Ae. ventricosa* introgression is present in nearly 90% of all new CIMMYT varieties. (4) Recent work has revealed that many past and present-day varieties carry *Triticum dicoccoides* introgressions that have been unconsciously selected for over time due to the advantage they confer over lines that lack them, e.g. the variety Robigus. (5) 30% of all wheat lines bred at CIMMYT are derived from crosses between normal wheat and "synthetic" wheat [8]. The latter is derived from crosses between *Aegilops tauschii* (DD genome) and tetraploid durum wheat (AABB genome) followed by chromosome doubling using colchicine. Since synthetic and bread wheat have the same genomic constitution, they can be readily hybridized to transfer novel alleles and genes from different accessions of *Ae. tauschii*, the D-genome progenitor.

From a physiological perspective, there have also been some clear benefits associated with introgressions. The erectophile leaf trait originated from *Triticum sphaerococcum* can be seen in many modern wheats, especially high yielding spring durums [9]. Evidence for genetic variation in source:sink balance and its importance in boosting yield and radiation use efficiency (RUE) has come from various sources, including studies with cytogenetic stocks [10]. Substitution of the long arm of chromosome 7D in hexaploid wheat with the homologous chromosome from *Agropyron elongatum* resulted in a significant increase in yield and biomass in six elite lines associated with increased spike fertility and post-anthesis RUE [10]. Synthetic wheats are also present in the pedigrees of lines with high yield potential and have contributed to outstanding expression of stress adaptive traits under heat and drought stress including more vigorous root systems and accumulation of stem carbohydrate reserves [11, 12].

Even though genetic variation from the wild relatives has delivered dramatic increases in wheat improvement, to date only a tiny fraction of the genetic variation available has been exploited. The reasons for this are the direct result of the difficulty in transferring genetic variation from the wild relatives to wheat (specific crossing schemes are required) and the difficulty in identifying plants which carry introgressions. In addition, introgressions carrying genes of interest also frequently carry undesirable genes. The removal of these genes using past technology proved extremely difficult. As a result of these difficulties, the use of wild relatives for wheat improvement went into decline. Thus, where there were many 100s of researchers in the field in the 1970s and 1980s, today only a handful of scientists globally, with the requisite expertise required to transfer genetic variation from the wild relatives to wheat, now remain. However, the advent of new molecular genetic technologies is reinvigorating the exploitation of wild relatives, i.e. it is now possible to transfer and characterise large numbers of introgressions, from a wide range of species, into wheat. These technologies coupled with specific crossing strategies are facilitating, for the first time, the large scale and systematic transfer of genetic variation of genetic variation from the gene pools of the wild relatives into wheat [13–19].

18.2.1 Different Classes of Wheat/Wild Relative

Before discussing introgression further it is first essential to establish the relationship between the different types of wild relative and wheat. Hexaploid wheat is an allohexaploid with 42 chromosomes (2n = 6x = AA BB DD). It has seven pairs of A genome chromosomes derived from *Triticum urartu* [20], seven pairs of B chromosomes from a species thought to be related to *Aegilops speltoides* [21–23] and seven pairs of D chromosomes derived from *Ae. tauschii* [24].

The wild relatives of wheat effectively fall into three classes or gene pools. The primary gene pool contains species which in several cases could more correctly be called ancestral species, have the same or very similar genomes to wheat. These species include *T. urartu* and *Triticum monococcum* (AA genome), *Triticum turgidum* (AABB genomes) and *Ae. tauschii* (DD genome). Species in the secondary gene pool also carry at least one genome very closely related to wheat although they show modifications, e.g. they might carry translocations or inversions relative to wheat. Species in the secondary gene pool include *Ae. speltoides* (SS genome) and *Triticum timopheevii* (AAGG genomes).

The genomes of the species in the primary and secondary gene pools thus have the equivalent gene content to that of a wheat genome although there may be some allelic differences as well as the structural changes. Thus, the genomes in the primary and secondary gene pools are said to be homologous to the genomes of wheat.

The genomes/chromosomes of the tertiary gene pool of wild relatives, although related, have diverged significantly from those of wheat, often with regard to both DNA content and chromosome structure and morphology. Thus, the chromosomes of these species are said to be homoeologous to those of wheat, i.e. related but not identical. There are a large number of these species from several different genera, e.g. *Aegilops caudata* (CC), *Ae. umbellulata* (UU), *Aegilops uniaristata* (NN), *Amblyopyrum muticum* (TT), *Secale cereale* (RR), *Thinopyrum bessarabicum* (E^bE^b), *Thinopyrum elongatum* (E^eE^e), *Thinopyrum intermedium* ($StStJ^rJ^rJ^{vs}J^{vs}$) and *Thinopyrum ponticum* ($E^eE^eE^bE^bE^xE^xStStStSt$).

18.2.2   Transferring Genetic Variation from Wild Relatives into Wheat

How is genetic variation from wild relatives transferred to wheat? The first step in the process requires that wheat is hybridised with a wild relative to produce an F_1 interspecific hybrid, e.g. pollen from a wild relative is used to pollinate wheat. These F_1 hybrids carry the haploid genomes of wheat and the haploid genome(s) of the wild relative and provide the starting point in the transfer of genetic variation from wild relatives.

Transfer of genetic variation from wild relatives has to date been achieved using two different methods. The first involves the use of chemicals or radiation to induce random breakage of chromosomes in the F_1 interspecific hybrids or their derivatives (e.g. addition or substitution lines – see later) (e.g. [25, 26]). These broken chromosome segments are said to have sticky ends and they can re-join with other broken chromosome segments. Wheat/wild relative translocations occur when a wheat chromosome segment fuses with a chromosome segment from a wild relative and thus results in the production of interspecific translocations. This process was used very successfully in the past by Ernie Sears to transfer leaf rust from A*e. umbellulata* into wheat [4]. However, chromosome breakage induced by chemicals or radiation occurs at random, i.e. translocations frequently occur between completely unrelated chromosomes. A direct result of this is that the progeny derived from translocations are frequently genetically unbalanced, e.g. they carry gene deletions (from the lost wheat chromosome(s)) and duplications (from the added wild relative chromosome(s)) which consequently have deleterious effects on plant vigour.

The second method is via recombination, i.e. the chromosomes of wheat and those of a wild relative recombine in the gametes of the F_1 interspecific hybrids or their derivatives at meiosis to produce interspecific wheat/wild relative chromosomes commonly known as introgressions*. These introgressions are then transmitted to the next generation through the gametes. Unlike translocations, because recombination occurs between related chromosomes, they are less likely to give rise to gene deletions and duplications (although deletions and duplications have been observed if the genomes of the wild relatives are translocated relative to wheat or unequal crossing over occurs).

Much of the work discussed in this chapter is applicable to the generation of introgressions either via translocation or recombination. However, because of the problems associated with the production of unbalanced gametes derived from translocations, the remainder of this paper is directed at the induction of introgressions via recombination.

*It should be noted that sometimes introgressions that have been generated via recombination are referred to as translocations which is not strictly correct. The term translocation refers to the phenomenon of chromosome breakage and reunion that is not associated with recombination at meiosis. Thus, introgressions generated via recombination should be referred to as "interspecific recombinant chromosomes" or "recombinant chromosomes."

18.3   Generation of Introgressions

The transfer of genetic variation to wheat from wild relatives whose genomes are homologous to one or more of those of wheat is relatively straight forward. In F_1 hybrids and their derivatives, the chromosomes of wheat and those of the wild relative are able to pair and recombine during meiosis, leading to the generation of interspecific recombinant chromosomes/introgressions which are recovered in the progeny.

For example: if hexaploid wheat (AABBDD) is crossed with diploid *T. urartu* (A^uA^u) the resulting interspecific hybrid's genomic constitution will be A^uABD (Fig. 18.1). At meiosis, recombination between the A and A^u chromosomes would be expected to be nearly normal. Thus, a large proportion of the gametes would be expected to carry a balanced number of 7 A/A^u recombinant chromosomes (although the A and A^u genomes are homologous a level of chromosome failure would still be expected in such crosses). In contrast to the A genome, the B and D genome chromosomes of wheat will not have homologous partners to pair with to form bivalents at meiosis, i.e. they will form univalents and not segregate normally to the spindle poles at anaphase I. Thus, their inclusion in the nuclei at telophase and the resulting gametes will occur at random. As a result, there will be large variations in the number of B and D genome chromosomes carried by the individual gametes, so that many will be genetically unbalanced and inviable. However, a small percentage of sufficiently balanced gametes will be produced. In order to address the high level of infertility, very large numbers of crosses are made to the F_1 interspecific hybrid, using the F_1 as the female parent, with normal wheat. This increases the likelihood that any viable gametes are fertilised, and the progeny produced will normally carry large numbers of introgressions. These plants are then recurrently crossed to wheat until lines carrying only a single A/A^u introgression are isolated.

Single introgressions, generated via backcrossing, will be in a heterozygous state and hence if they are self-fertilised, they will segregate for plants that carry the introgression and those that do not. Thus, plants homozygous for the introgressions need to be generated in order to ensure that they are stably inherited to the next generation. This can be achieved by taking heterozygous lines and either using the doubled haploid (DH) procedure [27] or simply by self-fertilizing and screening subsequent progenies with genetic markers for the presence of introgression homozygotes (see Sect. 18.3).

In contrast to species that have genomes homologous to wheat, the generation of introgressions from species with genomes that are homoeologous to those of wheat is more complicated. This is because recombination between homoeologous chromosomes is inhibited at meiosis by the *Ph1* locus located on the long arm of chromosome 5B [28] (See Chap. 16). One strategy to overcome this problem is to use lines that lack chromosome 5B or more commonly to use a mutant line in which the wild type *Ph1* locus has been deleted, i.e. the *ph1* mutant. In F_1 interspecific hybrids derived from crosses between a wild relative and wheat homozygous for the *ph1* mutation, recombination can occur between the chromosomes of wheat and

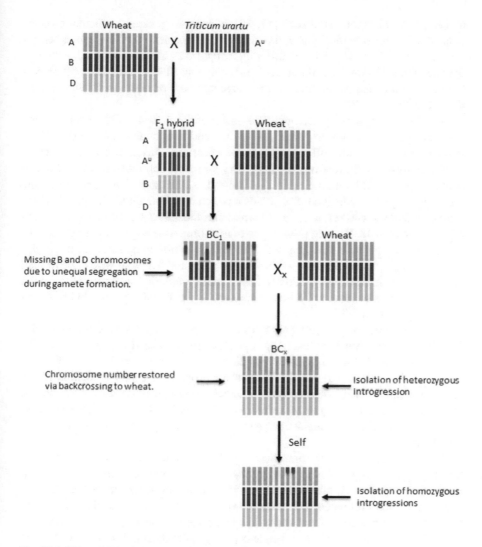

Fig. 18.1 Wheat/wild relative crossing strategy where the genome of the wild relative is homologous to one of the wheat genomes. The example shown is *T. urartu*, genome A^u, which is homologous to the A genome of wheat

chromosomes from the wild relative resulting in the generation of introgressions. However, the level of recombination observed is generally very much lower than that seen in interspecific hybrids between wild relatives with homologous genomes, with the result that the frequency of genetically unbalanced and hence inviable gametes is very much higher. The fertility of F₁ interspecific hybrids between wheat and wild species with homoeologous genomes is can be extremely low depending on the species, e.g. 16% of crosses to F₁ hybrids produced between wheat and *Am. muticum* generated seed while 29% of crosses to F₁ hybrids produced between wheat and *Ae. speltoides* generated seed [17, 18]. In order to generate sufficient progeny from the F_1 interspecific hybrids involving these species, very large numbers of crosses need to be made with wheat, using the hybrid as the female parent. Although the use of these F_1 interspecific hybrids is very labour intensive, their recent exploitation has led to the generation of very large numbers of new wheat/wild relative introgressions [17, 29].

An alternative strategy to generate introgressions from wild relatives with homoeologous genomes to wheat, is to use addition and/or substitution lines. Addition lines carry the full complement of wheat chromosomes + a pair of chromosomes from a wild relative, i.e. they carry 42 wheat chromosomes + 2 chromosomes from a wild relative = 44. In contrast, substitution lines have a single homologous pair of wheat chromosomes replaced by a homoeologous pair of chromosomes from a wild relative, i.e. 40 wheat chromosomes + 2 chromosomes from a wild relative = 42. Both of these types of lines are initially generated by chromosome doubling (using colchicine) a F_1 interspecific hybrid to generate an amphidiploid, e.g. an amphidiploid between hexaploid wheat and a diploid wild relative such as rye has 56 chromosomes AABBDDRR. These amphidiploids normally show a significantly higher level of fertility compared to the F_1 interspecific hybrids they were derived from.

Addition lines are generated (Fig. 18.2) by repeatedly backcrossing an amphidiploid to wheat until lines carrying a single chromosome from the wild relative have been isolated. These monosomic additions (42 wheat + 1 wild relative chromosome) are then allowed to self-fertilise and the progeny screened to identify plants carrying a pair of wild relative chromosomes. Although these disomic additions carry a pair of homoeologous wild relative chromosomes they are relatively unstable and thus require checking at each generation for their presence in order to maintain them.

The generation of substitution lines (Fig. 18.3) first requires a line of wheat that has lost a single copy of one pair of chromosomes (a monosomic line), e.g. a wheat line monosomic for chromosome 1A would have 40 chromosomes + 1 x 1A = 41. This line is then pollinated with a wheat/wild relative disomic addition line where the pair of chromosomes from the wild relative are homoeologous to the chromosome of wheat present only as a single copy, e.g. (40 wheat + 1 x 1A) x (42 wheat + 2 x 1R). The progeny of this cross will all carry a copy of the chromosome from the wild relative (1R) but will segregate for the presence or absence of the wheat chromosome present as only a single copy in the monosomic line (e.g. +1A or −1A). The progeny are then screened to select plants that have lost the single wheat chromosome, e.g. 40 wheat + 1R = 41 chromosomes while plants still carrying the single wheat chromosome (40 wheat + 1A + 1R = 42 chromosomes) are discarded. The selected plants are called monosomic substitution lines and once identified are self-fertilised to produce disomic substitution lines, e.g. 40 wheat + 2x1R = 42 chromosomes. Even though disomic substitution lines have lost a complete pair of homologous wheat chromosomes, the homoeologous wild relative chromosomes are frequently able to compensate for their absence (providing that the homoeologous chromosomes carry a related gene compliment etc.). Substitutions have a big

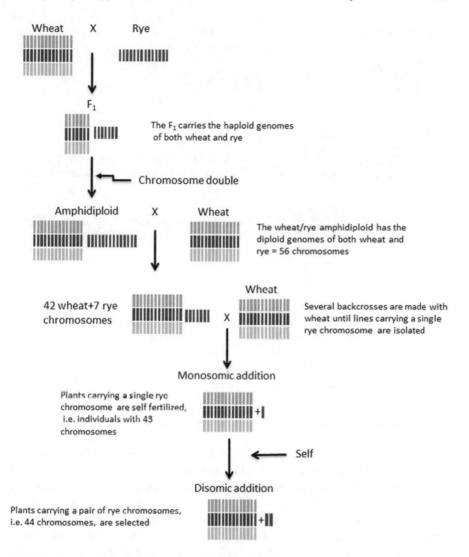

Fig. 18.2 Strategy for the production of a disomic addition line

advantage over addition lines in that they are normally stably inherited from one generation to the next.

The generation of introgression lines from substitution and addition lines can be achieved by crossing them twice to the wheat *ph1* mutant line followed by selection for lines that have lost the *Ph1* wild type locus but still retain the wild relative chromosome (Fig. 18.4). In the absence of the *Ph1* locus, homoeologous recombination can occur between the chromosomes of wheat and the wild relative leading to the generation of introgressions which can then be recovered in the progeny of crosses to normal wheat.

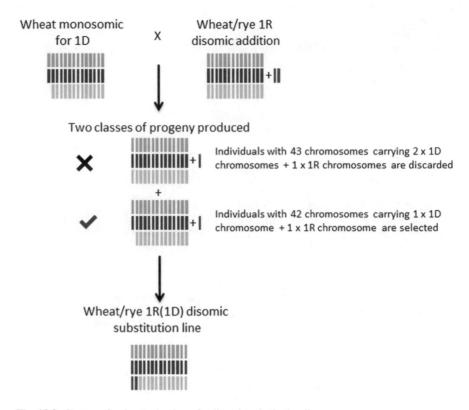

Fig. 18.3 Strategy for the production of a disomic substitution line

The decision to generate introgressions directly from interspecific F_1 hybrids or from addition/substitution lines is dependent on what is trying to be achieved. Both strategies have advantages and disadvantages. One advantage of using interspecific F_1 hybrids is that large numbers of introgressions can be quickly generated, potentially from the entire genome of a wild relative, without the need to generate a complete set of addition and/or substitution lines. Addition and substitution lines, however, are very useful if you are attempting to introduce genetic variation from a known area of the wild relative genome as they are considerably more fertile than F_1 interspecific hybrids and efforts can be focussed on the required chromosome.

While the removal of the *Ph1* locus is required for the induction of homoeologous recombination between wheat and the chromosomes of the majority of wild relatives there are exceptions, e.g. *Am. muticum*, *Ae. speltoides* and *Aegilops geniculata* [17, 18, 30]. These species carry a gene or genes that induce homoeologous recombination even in the presence of the wild *Ph1* locus. The efficacy of the genes responsible in *Am. muticum* and *Ae. speltoides* has been demonstrated through the generation of very large numbers of wheat/*Am. muticum* and wheat/*Ae. speltoides* introgressions from interspecific F_1 hybrids [17, 18] while chromosome $5M^g$ of *Ae.*

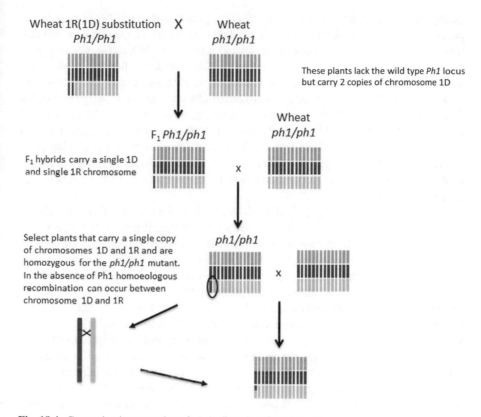

Fig. 18.4 Generating introgressions from a disomic substitution line

geniculata recombined with both chromosome 5D of wheat and also with the group 5 chromosomes of other wheat wild relatives [30].

18.4   Tools for Detection of Wheat/Wild Relative Introgressions

The recent advent of next generation sequencing and the concomitant development of new genetic marker technologies has resulted in a revolution in the field of wheat/ wild relative introgression. Previously, the lack of genetic markers was a major limiting factor in the detection and characterisation of introgressions. As a result, many introgressions could only be detected via phenotypic analysis and without characterisation, many were frequently very large and carried deleterious genes affecting plant vigour as well as the genes of agronomic importance.

Today the development of sequencing technologies is resulting in the generation of 1000s of molecular markers that can be exploited in wheat/wild relative introgression programmes. Single nucleotide polymorphism (SNPs) markers, in particular, have been valuable, i.e. SNP markers are based on a single base pair difference between wheat and a wild relative at a specific DNA sequence. The presence of an introgression is determined by screening individual plants to ascertain which bases are present. There are a number of platforms that can be used to screen plants with SNPs. The Axiom Wheat-Relative Genotyping array (Affymetrix), for example, has allowed large numbers of plants to be screened for the presence of circa 35,000 SNPs [14]. Alternatively, plants can be screened for introgressions via SNPs with Kompetitive Allele Specific PCR (KASP) markers, a genotyping technology based on allele-specific oligo extension and fluorescence resonance energy transfer for signal generation and found to be more cost-effective for large-scale projects (Fig. 18.5).

Recently, KASP markers polymorphic between wheat and ten of its wild relative species were developed [31]. These markers were designed not just to be polymorphic between a wild relative and wheat but to be polymorphic between a wild relative and a specific chromosome of wheat (Fig. 18.5). Thus, in addition to detecting the wild relative segment in the introgression line, these markers have the additional functionality in that they can indicate whether the segment is heterozygous or homozygous through the loss of wheat alleles for the KASP markers. These markers therefore firstly reduce the need for more labour-intensive laboratory techniques such as GISH (see end of this section) but secondly, and more importantly in a breeding programme, remove the need for logistically demanding progeny testing necessary to distinguish between plants with heterozygous or homozygous introgressions (In progeny testing, 10 to 20 progeny seed are germinated and tested with markers for the presence of the segment. Where all the progenies are found to contain the introgressed segment, it is assumed that the original plant was homozygous. Progeny showing segregation for the presence/absent will have been derived from a heterozygous parent. Testing of a second generation will validate the result). Moreover, in homozygous introgression lines these chromosome-specific KASP markers can indicate which genome of wheat (A, B or D) the recombination with the wild relative species has occurred. These markers have recently been used to characterise *Ae. caudata* introgressions in bread wheat [32] and D-genome introgressions from bread into durum wheat [33].

The continued developments in high-throughput sequencing and the reduction in costs are now allowing the generation of sequence data from wild relative species, e.g. *Ae. tauschii* [34], *T. urartu* [35], *S. cereale* [36]. Thus, chromosome-specific KASP markers are now being developed based on SNPs between wheat and the wild relative in single-copy regions of the wheat genome (unpublished results) taking away the cumbersome need to anchor the KASP primers to chromosome-specific alleles during their design. The sequencing of individual introgression lines is also enabling the detection of the site of recombination between wheat and the wild relative (unpublished data).

In addition to sequencing and marker technology, molecular cytological techniques and microscopy systems have also developed significantly. In the past, the detection of introgressions depended on cytological techniques which were very

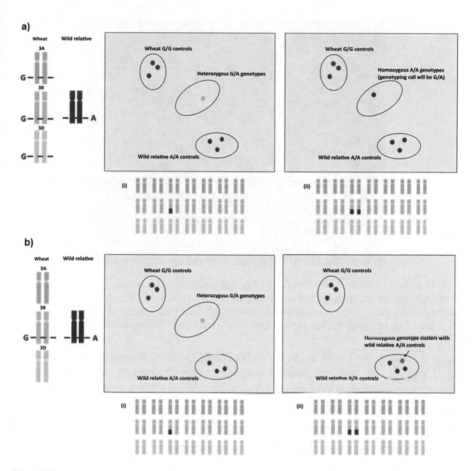

Fig. 18.5 (**a**) KASP marker designed to be polymorphic for a SNP found on all three genomes of wheat and a wild relative. The signals for both **a** (**i**) heterozygous introgression and **a** (**ii**) homozygous introgression cluster between the signals for the wheat controls and the wild relative controls. (**b**) KASP marker designed to be polymorphic between a wheat chromosome specific SNP (in this instance the SNP occurs on 3B) and a wild relative. The signal for heterozygous introgressions (**i**) will cluster between the signals for the wheat controls and the wild relative controls. The signal for homozygous introgressions (**ii**) will cluster with the wild relative controls

labour intensive and frequently provided limited information. However, techniques such as genomic *in situ* hybridisation, can now be used routinely to detect introgressed chromosome segments in wheat, while at the same time distinguishing the chromosomes of the three genomes of wheat from each other (Fig. 18.6). Furthermore, many systems have high levels of automation, e.g. large numbers of slides can be screened remotely for chromosome spreads and multiple fluorescent images taken.

In combination, cytological analysis, markers and sequence data are increasingly providing new information within this field of research. For example, until recently it was thought that the majority of wild relative introgressions were very large and

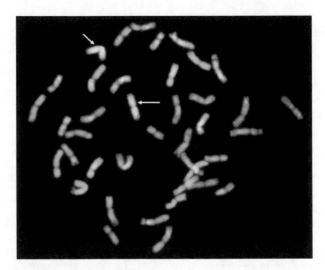

Fig. 18.6 Multi-colour GISH analysis of a homozygous introgression line. The wheat A genome chromosomes are shown in green, wheat B-genome chromosomes in blue and wheat D-genome chromosomes in red. The homozygous introgression from *Am. muticum* (white arrows) is shown in yellow. This introgression has recombined at both ends with the D-genome

recombinant events restricted to the ends of chromosomes. However, recent research has clearly shown that recombination appears to occur throughout the length of the chromosomes and that very small introgressions are not uncommon [29, 30].

18.5   Reducing the Size of Introgressions

When introgressions are very large, they carry deleterious genes affecting plant vigour as well as the genes of agronomic importance. Thus, further work needs to be carried out to reduce the size of the introgression to as small as possible carrying the wild relative gene of interest. Ernie Sears, the father of wheat cytogenetics and wheat wild relative introgression, developed a strategy to reduce the size of large introgressions [37]. He used this strategy to remove deleterious genes from two wheat/*Ae. umbellulata* introgressions while retaining a gene for resistance to leaf rust. Essentially this strategy involved the inter-crossing of plants containing over-lapping introgressions where the target gene was located in the overlap (Fig. 18.7). In the presence of *Ph1*, recombination freely occurred between the overlapping introgressions, resulting in some individuals among the progeny that had a significantly smaller introgression but still retained the gene for resistance to leaf rust. This strategy, although ground-breaking, was ahead of its time because the marker technology required to identify large numbers of overlapping introgressions and to identify smaller modified introgressions was not available at the time.

Fig. 18.7 Strategy to reduce the size of a large introgressed segment

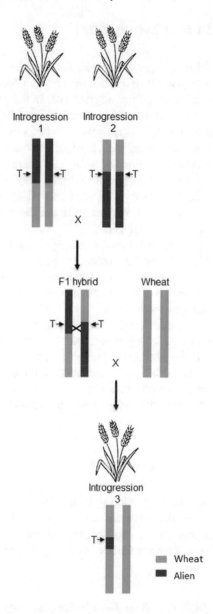

Alternative strategies to reduce the size of introgressions such as the induction of further homoeologous recombination between wheat and a wild relative chromosome segment in the absence of *Ph1* [38] have also previously proved difficult to undertake, again largely due to the lack of markers available to detect rare small recombinants.

18.6 Phenotyping

While this paper concentrates on the process of wheat/wild relative introgression, it is important to discuss the role of phenotyping in the exploitation of introgression lines in breeding programmes. In the past, wild relatives and in particular addition and substitution lines, have been screened to identify genetic variation for a specific trait. Introgressions have then been made to transfer genetic variation for the target trait. However, the resulting introgressions have been found to carry important genetic variation for additional characters. For example, an introgression from *Ae. ventricosa* was introduced into wheat that carried resistance to the disease eyespot [39]. However, at the time of writing, it has been shown that it also possesses the only source of resistance to the disease wheat blast and also confers a significant yield advantage [7].

Others have used a strategy where they have transferred the entire genome of a wild relative into wheat by generating large numbers of introgressions. Although the wild relatives used were selected as they were known to carry genetic variation for several target traits, the emphasis of the work was on screening any resulting introgressions for as many traits as possible. Irrespective of how they have been generated, in order for the potential of the 1000s of new wheat/wild relative introgressions being generated to be realised, it is essential that each is screened for a broad spectrum of traits in a wide range of environments. This will make it possible to determine what agronomic characters are affected by the genes present in each introgression. This large-scale screening is critical because without it each introgression will just be seeds in a packet of unknown agronomic value.

18.7 Case Study

In order to provide an insight into the workflow, disciplines and logistics required to undertake a present day introgression programme we here describe a case study of a wheat/wild relative programme carried out at the Nottingham BBSRC Wheat Wild Relative Centre involving *Am. muticum* (Fig. 18.8).

18.7.1 Step 1 – Generation of Introgression Lines

Am. muticum was used to pollinate wheat (variety Paragon), which carried the *ph1* mutation, to produce F_1 interspecific hybrids. It is important to note two key facts with regard to these hybrids. Firstly, because each F_1 lacked the wild type *Ph1* locus, homoeologous recombination could occur during meiosis and in addition, *Am. muticum* carries a gene(s) which promotes homoeologous recombination. Secondly,

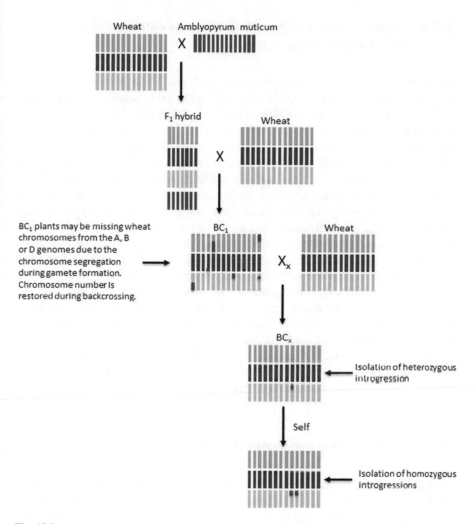

Fig. 18.8 Case study: generating introgressions from *Am. muticum*

since the interspecific hybrids only possessed the haploid genomes of wheat and *Am. muticum*, i.e. ABDT, and thus homologous pairs of chromosomes were not present, only homoeologous recombination could occur. As a result of this strategy large numbers of introgressions were generated. However, the drawback to this strategy was that the F_1 hybrids showed very low fertility. In order to obtain progeny and hence isolate introgressions, each of the interspecific hybrids was extensively crossed to wheat.

18.7.2 Step 2 – Molecular Identification of Introgressions and Their Characterisation

The resulting progeny were screened using the Axiom® Wheat-Relative Genotyping Array (the array carries circa 35,000 SNP markers that show polymorphism between 10 wild relatives and wheat varieties such as Paragon) to detect potential introgressions. Plants carrying introgressions were recurrently crossed to wheat and the presence or absence of introgressions was confirmed using genomic *in situ* hybridisation.

The ultimate aim of the programme is to generate lines that are homozygous for a single introgression so the lines can be multiplied and distributed to collaborators for phenotypic analysis. As described earlier this can be achieved either by the DH procedure or simply by self-fertilisation. In order to test the potential of the DH procedure, several plants heterozygous for an introgression were selected from the BC$_3$ generation and crossed with maize. In the resulting hybrids, the maize chromosomes were eliminated, and the resulting haploid plants were chromosome doubled to give rise to DH lines that were potentially homozygous for each introgression. The presence of homozygous introgressions was confirmed via GISH and by genome specific KASP markers, i.e. this also allowed the determination of which wheat chromosomes were involved in each introgression. This work led to the isolation of 66 wheat/*Am. muticum* introgressions which were multiplied and have been distributed to both the public and private sectors free of IP [27]. This programme is presently generating many additional new wheat/*Am. muticum* homozygous introgressions, via self-fertilisation, prior to their distribution.

18.7.3 Step 3 – Making Use of the Introgression Lines

An initial series of phenotypic analyses have been undertaken on 20 wheat/A*m. muticum* introgression lines for resistance to leaf, stem and yellow rusts by collaborators from the USDA at Kansas State University [40]. In each case introgressions were identified that conferred resistance. Furthermore, resistance to *Wheat streak mosaic virus* and powdery mildew was also observed. These introgressions are presently being introduced into US adapted germplasm for further testing. At the time of writing large numbers of further introgressions have been sent to Kansas, and distributed within the UK for large scale phenotypic analyses on a wide range of traits (all of the available homozygous lines developed at Nottingham are listed at the following web sites: Nottingham https://www.nottingham.ac.uk/wrc/home.aspx and Norwich https://www.jic.ac.uk/research-impact/germplasm-resource-unit/).

18.8   Key Concepts

This chapter has discussed the different strategies for the generation of introgressions from the wild relatives of wheat and their detection and characterisation.

1. Definition of an introgression and understanding the difference between an introgression and a translocation.
2. The benefits of generating an introgression via recombination rather than via random breakage and joining of chromosomes.
3. Advantages and disadvantages of generating introgressions via homologous recombination compared to homoeologous recombination.
4. The generation and use of addition and substitution lines.
5. How the developments in molecular biology over the last decade have enabled the detection and characterisation of introgressions.
6. The importance of phenotyping

18.9   Conclusions

- Wheat evolved only once or twice about 8 to 10,000 years ago while many of its wild relatives, of which there are hundreds of different accessions, evolved several millions of years ago.
- Wild relatives therefore provide a vast reservoir of genetic variation above and beyond anything seen in wheat for potentially all traits of agronomic importance.
- Due to a lack of adequate technologies, it has been very difficult to exploit genetic variation in the wild relatives for wheat improvement. Recently, the development of new technologies, has enabled the large-scale transfer of genetic variation from the wild relatives into wheat. These technologies, combined with large-scale phenotypic analyses, will enable the genetic variation from wild relatives to have a major global impact on wheat production.
- Future technological advances, such as a greater understanding of the genomes of the wild relatives, will further enhance our ability to transfer genetic variation into wheat.
- A major concern for the exploitation of the wild relatives is the lack of scientists with the prerequisite expertise in wheat chromosome manipulation.

References

1. Braun H, Atlin G, Payne T (2010) Multi-location testing as a tool to identify plant response to global climate change. In: Reynolds M (ed) Climate change and crop production. CIMMYT, Mexico City, pp 115–138

2. Bernhardt N, Brassac J, Kilian B, Blattner FR (2017) Dated tribe-wide whole chloroplast genome phylogeny indicates recurrent hybridizations within Triticeae. BMC Evol Biol 17:141. https://doi.org/10.1186/s12862-017-0989-9

3. Ortiz R, Sayre KD, Govaerts B, Gupta R, Subbarao GV, Ban T, Hodson D, Dixon JM, Ortiz-Monasterio JI, Reynolds MP (2008) Climate change: can wheat beat the heat? Agric Ecosyst Environ 126:46–58. https://doi.org/10.1016/j.agee.2008.01.019

4. Sears E (1956) The transfer of leaf-rust resistance from Aegilops umbellulata to wheat. In: Genetics in plant breeding. Brook-haven symposia in biology 1956, pp 1–22

5. Villareal RL, Rajaram S, Mujeeb-Kazi A, Del Toro E (1991) The effect of chromosome 1B/1R translocation on the yield potential of certain spring wheats (Triticum aestivum L.). Plant Breed 106:77–81. https://doi.org/10.1111/j.1439-0523.1991.tb00482.x

6. Maia N (1967) Obtention de blés tendres résistants au piétin-verse (Cercosporella herpotrichoides) par crôisements interspécifiques blés × Aegilops. Comptes Rendus l'Académie d'Agriculture Fr 53:149–155

7. Cruz CD, Peterson GL, Bockus WW, Kankanala P, Dubcovsky J, Jordan KW, Akhunov E, Chumley F, Baldelomar FD, Valent B (2016) The 2NS translocation from Aegilops ventricosa confers resistance to the Triticum pathotype of Magnaporthe oryzae. Crop Sci 56:990–1000. https://doi.org/10.2135/cropsci2015.07.0410

8. Dreisigacker S, Kishii M, Lage J, Warburton M (2008) Use of synthetic hexaploid wheat to increase diversity for CIMMYT bread wheat improvement. Aust J Agric Res 59:413–420

9. Fischer R (2007) Understanding the physiological basis of yield potential in wheat. J Agric Sci 145:99–113. https://doi.org/10.1017/S0021859607006843

10. Reynolds MP, Calderini DF, Condon AG, Rajaram S (2001) Physiological basis of yield gains in wheat associated with the LR19 translocation from Agropyron elongatum. Euphytica 119:139–144. https://doi.org/10.1023/a:1017521800795

11. Lopes MS, Reynolds MP (2011) Drought adaptive traits and wide adaptation in elite lines derived from resynthesized hexaploid wheat. Crop Sci 51:1617–1626. https://doi.org/10.2135/cropsci2010.07.0445

12. Cossani CM, Reynolds MP (2015) Heat stress adaptation in elite lines derived from synthetic hexaploid wheat. Crop Sci 55:2719–2735. https://doi.org/10.2135/cropsci2015.02.0092

13. Xu J, Wang L, Deal KR, Zhu T, Ramasamy RK, Luo M-C, Malvick J, You FM, McGuire PE, Dvorak J (2020) Genome-wide introgression from a bread wheat × Lophopyrum elongatum amphiploid into wheat. Theor Appl Genet 133:1227–1241. https://doi.org/10.1007/s00122-020-03544-w

14. Winfield MO, Allen AM, Burridge AJ, Barker GLA, Benbow HR, Wilkinson PA, Coghill J, Waterfall C, Davassi A, Scopes G, Pirani A, Webster T, Brew F, Bloor C, King J, West C, Griffiths S, King I, Bentley AR, Edwards KJ (2016) High-density SNP genotyping array for hexaploid wheat and its secondary and tertiary gene pool. Plant Biotechnol J 14:1195–1206. https://doi.org/10.1111/pbi.12485

15. Grewal S, Hubbart-Edwards S, Yang C, Scholefield D, Ashling S, Burridge A, Wilkinson PA, King IP, King J (2018) Detection of T. urartu introgressions in wheat and development of a panel of interspecific introgression lines. Front Plant Sci 9:1565. https://doi.org/10.3389/fpls.2018.01565

16. Grewal S, Yang C, Edwards SH, Scholefield D, Ashling S, Burridge AJ, King IP, King J (2018) Characterisation of Thinopyrum bessarabicum chromosomes through genome-wide introgressions into wheat. Theor Appl Genet 131:389–406. https://doi.org/10.1007/s00122-017-3009-y

17. King J, Grewal S, Yang C-Y, Hubbart S, Scholefield D, Ashling S, Edwards KJ, Allen AM, Burridge A, Bloor C, Davassi A, da Silva GJ, Chalmers K, King IP (2017) A step change in the transfer of interspecific variation into wheat from Amblyopyrum muticum. Plant Biotechnol J 15:217–226. https://doi.org/10.1111/pbi.12606

18. King J, Grewal S, Yang C, Hubbart Edwards S, Scholefield D, Ashling S, Harper JA, Allen AM, Edwards KJ, Burridge AJ, King IP (2018) Introgression of Aegilops speltoides segments

in Triticum aestivum and the effect of the gametocidal genes. Ann Bot 121:229–240. https://doi.org/10.1093/aob/mcx149

19. Devi U, Grewal S, Yang C, Hubbart-Edwards S, Scholefield D, Ashling S, Burridge A, King IP, King J (2019) Development and characterisation of interspecific hybrid lines with genome-wide introgressions from Triticum timopheevii in a hexaploid wheat background. BMC Plant Biol 19:183. https://doi.org/10.1186/s12870-019-1785-z

20. Dvořák J, di Terlizzi P, Zhang H-B, Resta P (1993) The evolution of polyploid wheats: identification of the A genome donor species. Genome 36:21–31. https://doi.org/10.1139/g93-004

21. Sarkar P, Stebbins GL (1956) Morphological evidence concerning the origin of the B genome in wheat. Am J Bot 43:297–304. https://doi.org/10.1002/j.1537-2197.1956.tb10494.x

22. Riley R, Unrau J, Chapman V (1958) Evidence on the origin of the B genome of wheat. J Hered 49:91–98. https://doi.org/10.1093/oxfordjournals.jhered.a106784

23. Marcussen T, Sandve SR, Heier L, Spannagl M, Pfeifer M, Jakobsen KS, Wulff BBH, Steuernagel B, Mayer KFX, Olsen O-A (2014) Ancient hybridizations among the ancestral genomes of bread wheat. Science 345:1250092. https://doi.org/10.1126/science.1250092

24. McFadden ES, Sears ER (1946) The origin of Triticum spelta and its free-threshing hexaploid relatives. J Hered 37(81):107. https://doi.org/10.1093/oxfordjournals.jhered.a105590

25. Mukai Y, Friebe B, Hatchett JH, Yamamoto M, Gill BS (1993) Molecular cytogenetic analysis of radiation-induced wheat-rye terminal and intercalary chromosomal translocations and the detection of rye chromatin specifying resistance to Hessian fly. Chromosoma 102:88–95. https://doi.org/10.1007/BF00356025

26. Brown-Guedira GL, Singh S, Fritz AK (2003) Performance and mapping of leaf rust resistance transferred to wheat from Triticum timopheevii subsp. armeniacum. Phytopathology 93:784–789. https://doi.org/10.1094/PHYTO.2003.93.7.784

27. King J, Newell C, Grewal S, Hubbart-Edwards S, Yang C, Scholefield D, Ashling S, Stride A, King IP (2019) Development of stable homozygous wheat/Amblyopyrum muticum (Aegilops mutica) introgression lines and their cytogenetic and molecular characterization. Front Plant Sci 10:34. https://doi.org/10.3389/fpls.2019.00034

28. Riley R, Chapman V (1958) Genetic control of the cytologically diploid behaviour of hexaploid wheat. Nature 182:713–715. https://doi.org/10.1038/182713a0

29. Baker L, Grewal S, Yang C, Hubbart-Edwards S, Scholefield D, Ashling S, Burridge AJ, Przewieslik-Allen AM, Wilkinson PA, King IP, King J (2020) Exploiting the genome of Thinopyrum elongatum to expand the gene pool of hexaploid wheat. Theor Appl Genet 133:2213–2226. https://doi.org/10.1007/s00122-020-03591-3

30. Koo D-H, Liu W, Friebe B, Gill BS (2017) Homoeologous recombination in the presence of Ph1 gene in wheat. Chromosoma 126:531–540. https://doi.org/10.1007/s00412-016-0622-5

31. Grewal S, Hubbart-Edwards S, Yang C, Devi U, Baker L, Heath J, Ashling S, Scholefield D, Howells C, Yarde J, Isaac P, King IP, King J (2020) Rapid identification of homozygosity and site of wild relative introgressions in wheat through chromosome-specific KASP genotyping assays. Plant Biotechnol J 18:743–755. https://doi.org/10.1111/pbi.13241

32. Grewal S, Othmeni M, Walker J, Hubbart-Edwards S, Yang C, Scholefield D, Ashling S, Isaac P, King IP, King J (2020) Development of wheat-Aegilops caudata Introgression lines and their characterization using genome-specific KASP markers. Front Plant Sci 11:606. https://doi.org/10.3389/fpls.2020.00606

33. Othmeni M, Grewal S, Hubbart-Edwards S, Yang C, Scholefield D, Ashling S, Yahyaoui A, Gustafson P, Singh PK, King IP, King J (2019) The use of pentaploid crosses for the introgression of Amblyopyrum muticum and D-genome chromosome segments into durum wheat. Front Plant Sci 10:1110. https://doi.org/10.3389/fpls.2019.01110

34. Luo M-C, Gu YQ, Puiu D, Wang H, Twardziok SO, Deal KR, Huo N, Zhu T, Wang L, Wang Y, McGuire PE, Liu S, Long H, Ramasamy RK, Rodriguez JC, Van SL, Yuan L, Wang Z, Xia Z, Xiao L, Anderson OD, Ouyang S, Liang Y, Zimin AV, Pertea G, Qi P, Bennetzen JL, Dai X, Dawson MW, Müller H-G, Kugler K, Rivarola-Duarte L, Spannagl M, Mayer KFX, Lu F-H, Bevan MW, Leroy P, Li P, You FM, Sun Q, Liu Z, Lyons E, Wicker T, Salzberg SL, Devos KM,

Dvořák J (2017) Genome sequence of the progenitor of the wheat D genome Aegilops tauschii. Nature 551:498–502. https://doi.org/10.1038/nature24486

35. Ling H-Q, Ma B, Shi X, Liu H, Dong L, Sun H, Cao Y, Gao Q, Zheng S, Li Y, Yu Y, Du H, Qi M, Li Y, Lu H, Yu H, Cui Y, Wang N, Chen C, Wu H, Zhao Y, Zhang J, Li Y, Zhou W, Zhang B, Hu W, van Eijk MJT, Tang J, Witsenboer HMA, Zhao S, Li Z, Zhang A, Wang D, Liang C (2018) Genome sequence of the progenitor of wheat A subgenome Triticum urartu. Nature 557:424–428. https://doi.org/10.1038/s41586-018-0108-0

36. Bauer E, Schmutzer T, Barilar I, Mascher M, Gundlach H, Martis MM, Twardziok SO, Hackauf B, Gordillo A, Wilde P, Schmidt M, Korzun V, Mayer KFX, Schmid K, Schön C-C, Scholz U (2017) Towards a whole-genome sequence for rye (Secale cereale L.). Plant J 89:853–869. https://doi.org/10.1111/tpj.13436

37. Sears E (1981) Transfer of alien genetic material to wheat. In: Evans L, Peacock W (eds) Wheat science - today and tomorrow. Cambridge University Press, Cambridge, pp 75–89

38. Wan W, Xiao J, Li M, Tang X, Wen M, Cheruiyot AK, Li Y, Wang H, Wang X (2020) Fine mapping of wheat powdery mildew resistance gene *Pm6* using 2B/2G homoeologous recombinants induced by the *ph1b* mutant. Theor Appl Genet 133:1265–1275. https://doi.org/10.1007/s00122-020-03546-8

39. Doussinault G, Delibes A, Sanchez-Monge R, Garcia-Olmedo F (1983) Transfer of a dominant gene for resistance to eyespot disease from a wild grass to hexaploid wheat. Nature 303:698–700. https://doi.org/10.1038/303698a0

40. Fellers JP, Matthews A, Fritz AK, Rouse MN, Grewal S, Hubbart-Edwards S, King IP, King J (2020) Resistance to wheat rusts identified in wheat/Amblyopyrum muticum chromosome introgressions. Crop Sci 60:1957–1964. https://doi.org/10.1002/csc2.20120

Further Readings

1. Griffiths S, Sharp R, Foote TN, Bertin I, Wanous M, Reader S, Colas I and Moore G (2006). Molecular characterization of *Ph1* as a major chromosome pairing locus in polyploid wheat. *Nature* 439: 749–752

2. Kishii M (2019). An update of recent use of Aegilops species in wheat breeding. *Front Plant Sci* 10.585. https://doi.org/10.3389/fpls.2019.00585

Chapter 19
Disease Resistance

Michael Ayliffe, Ming Luo, Justin Faris, and Evans Lagudah

Abstract Wheat plants are infected by diverse pathogens of economic significance. They include biotrophic pathogens like mildews and rusts that require living plant cells to proliferate. By contrast necrotrophic pathogens that cause diseases such as tan spot, *Septoria nodurum* blotch and spot blotch require dead or dying cells to acquire nutrients. Pioneering studies in the flax plant-flax rust pathosystem led to the 'gene-for-gene' hypothesis which posits that a resistance gene product in the host plant recognizes a corresponding pathogen gene product, resulting in disease resistance. In contrast, necrotrophic wheat pathosystems have an 'inverse gene-for-gene' system whereby recognition of a necrotrophic fungal product by a dominant host gene product causes disease susceptibility, and the lack of recognition of this pathogen molecule leads to resistance. More than 300 resistance/susceptibility genes have been identified genetically in wheat and of those cloned the majority encode nucleotide binding, leucine rich repeat immune receptors. Other resistance gene types are also present in wheat, in particular adult plant resistance genes. Advances in mutational genomics and the wheat pan-genome are accelerating causative disease resistance/susceptibility gene discovery. This has enabled multiple disease resistance genes to be engineered as a transgenic gene stack for developing more durable disease resistance in wheat.

Keywords Wheat diseases · Resistance · Gene · Effector · Biotroph · Necrotroph

M. Ayliffe (✉) · M. Luo · E. Lagudah
CSIRO, Canberra, ACT, Australia
e-mail: michael.ayliffe@csiro.au; ming.luo@csiro.au; evans.lagudah@csiro.au

J. Faris
United States Department of Agriculture (USDA), Fargo, ND, USA
e-mail: justin.faris@usda.gov

© The Author(s) 2022
M. P. Reynolds, H.-J. Braun (eds.), *Wheat Improvement*,
https://doi.org/10.1007/978-3-030-90673-3_19

19.1   Learning Objectives

- An overview of the contrasting genetic and molecular interactions that occur between wheat and its pathogens.

19.2   Pioneering Studies in Model Biotrophic Pathosystems

Plant diseases reduce crop yield potential. Resistant cultivars provide the most cost effective and environmentally friendly means for disease control. A major ongoing problem with disease resistance has been that once widely deployed, its effectiveness is lost due to genetic changes in the respective pathogens. Consequently, an ongoing search for ever-more host resistance genes is required. The problem of resistance gene failure has become so significant that in some countries, growers routinely use fungicides and pesticides, which can lose effectiveness and also cause potential health and environmental issues.

The challenge in using genetically conferred disease resistance is therefore to find ways of prolonging the useful period ('durability') of a particular resistance source. This requires a thorough understanding of the genetic interactions between the host plant and pest/pathogen system. A major advance was made in the biotrophic flax plant/flax rust pathosystem by Flor in the 1950s. Biotrophic pathogens like rust diseases can only grow on living plant tissue in contrast to nectotrophic pathogens, which live on dead or dying host tissue (described below). Flor developed the gene-for-gene hypothesis, which states that for every plant resistance gene there is a corresponding gene, encoding an avirulence gene product, in the pathogen that is recognised [1]. Modern molecular biology has confirmed Flor's insightful research and shown that when pathogens infect their hosts they secrete an array of effector molecules into the plant. Plant hosts have evolved specific receptor molecules, encoded by resistance genes, that each directly or indirectly recognise a particular pathogen effector molecule (i.e. the molecular basis for the gene-for-gene hypothesis). Upon recognition of a pathogen molecule the resistance protein activates a resistance response called effector triggered immunity (ETI). Loss or change in recognised pathogen effectors, which are also called avirulence proteins, leads to no pathogen recognition by the plant and hence a loss of resistance to biotrophic pathogens. Changing or losing recognised avirulence effector molecules is how new races of pathogens evolve to overcome plant resistance.

Plants also possess a second type of resistance called PAMP (pathogen associated molecular pattern)-triggered immunity (PTI). PTI differs from ETI in that all microbes possess some conserved molecules (e.g. chitin in fungal cell walls, bacterial flagellin protein) that the plant can recognise with pattern recognition receptors (PRRs) leading to the activation of PTI. This resistance protects plants against most potential biotrophic pathogens. For biotrophic pathogens, such as wheat rusts and mildews, to be able to grow on a particular plant species it must suppress the PTI response of the host, which it does by introducing effector molecules. The suite of effectors that each pathogen possesses and the ability of these molecules to effectively target specific plant proteins plays a large role in determining what plant species the pathogen can colonise. As described above, plants in turn have evolved R proteins that can recognise specific effectors and activate ETI, thereby making the host resistant. In turn the pathogen loses or alters recognised effectors in an ongoing arms race between the host and the pathogen [2].

Flor worked on flax rust, a disease caused by an autoecious, dikaryotic fungus that infects flax, a self-pollinating diploid host plant. A detailed knowledge of the life cycle and breeding behaviour of each organism and an ability to perform genetic crosses on both were essential for his discoveries. Flor's discoveries were not only applicable to the flax/flax rust system but also held true in many other pathosystems as well. He established two basic principles, firstly, the genetic interaction that led to incompatibility (i.e. resistance in the host and avirulence in the pathogen) involved dominant genes in both the plant and the pathogen. Dominance is a strong indicator of a functional gene in contrast to recessive, often loss of function mutations. Secondly, genetic knowledge of one organism enabled the genotype of the other to be determined. This is the genetic basis of pathogen race surveys where isolates are screened against a panel of plants with known resistance genes, enabling what avirulence and virulence genes are present in the pathogen to be determined. This information in turn informs about the resistance or susceptibility of elite wheat cultivars, that contain known resistance genes, to each pathogen isolate. It also provides information on the genetic relationships and evolutionary pathways existing between different pathogen races.

Advances in molecular genetics over the last 25 years (50 years post-Flor) have enabled gene cloning in plant and pathogen species, which has confirmed Flor's work. In the case of rust disease resistance, a maize transposable element (Ac) was used to generate insertional mutants of flax rust resistance genes thereby enabling their cloning (see Fig. 19.1). The molecular structure of several rust resistance genes and their products was then determined. Transposon tagging was also used to identify the tobacco mosaic virus N resistance gene in tobacco and maize Rp1 rust resistance gene, while map-based cloning, which uses linked DNA markers as entry points to scan overlapping large DNA fragments to identify gene candidates (Fig. 19.1), enabled the isolation of the Arabidopsis RPS2 bacterial resistance. From these studies which used fungal, viral and bacterial species, respectively, the flax rust L6, tobacco N and Arabidopsis RPS2 resistance genes were shown to encode proteins with a similar modular structure of an N-terminal nucleotide binding site and C-terminal leucine rich repeats (NLR) [3].

NLR genes were subsequently identified in all plant species and shown to be the largest class of disease resistance genes present in plants, including wheat. The current pan genome of bread wheat, which is derived from 16 cultivars, contains around 2500 NLR genes, 31–34% of which are shared across all genomes. The number of unique NLR's ranges from 22 to 192 per cultivar [4]. The NLR gene family is highly diverse, although some genes appear orthologous across species as well as homoeologous within some of the Triticeae species. NLR proteins function in two

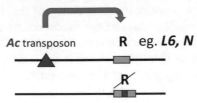

•R gene inactivation (by transposon tagging)

Ac transposon   R   eg. *L6, N*

•Map-based cloning (by "chromosome walking or landing')
                                          eg. *Pto, RPS2, RPM1, mlo*

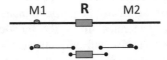

M1   R   M2

Fig. 19.1 Classic methods of resistance gene isolation. Transposon tagging using heterologous transposons was used to isolate a number of resistance genes such as the flax *L6* gene, tobacco *N* gene and maize *Rp1* gene. Susceptible mutants arising from transposon insertions in the causative R gene were sought from active transposon lines. The transposon insertion then acted as a molecular tag to enable isolation of the R gene. An alternative approach was map-based cloning where markers closely linked to an R gene were sought. These markers then enabled the isolation of large DNA fragments from the locus by screening large insert BAC, PAC and YAC libraries. Overlapping DNA inserts that spanned the locus were then sought and analysed for R gene candidates. Technology advances have created new methods of R gene isolation based on exome capture or chromosome isolation as detailed in Table 19.1

ways, by either directly recognising a single specific pathogen effector molecule introduced into the plant cell upon which ETI is activated, or alternatively, by recognising the modification of a host protein targeted by a pathogen effector (guard model) again leading to ETI [5]. After effector recognition by an NLR protein a complex defense cascade is activated that can be accompanied by host cell hypersensitive cell death in some instances.

19.3   Genetics of Resistance to Wheat Biotrophic Pathogens- Rusts and Mildew

Increasing sophistication and advances in both genomic and marker technologies (Table 19.1) have led to a rapid increase in the cloning of powdery mildew and rust (see Chap. 7) resistance genes from the complex wheat genome (Table 19.2). The majority of these genes are race-specific all plant developmental stage resistance (ASR) genes that usually encode NLR proteins (Table 19.2). In general, each resistance specificity is conferred by a single NLR gene at the locus although the genomic organisation of NLR loci is variable. The wheat *Pm3* locus consists of a single gene that encodes at least 56 allelic sequences of which 17 have been shown to be

Table 19.1 Marker and cloning technologies used in wheat for mapping and isolation of resistance genes

Marker/technology name	Abbreviation	Technology Platform
Morphological		Visual
Restriction fragment length polymorphisms	RFLP	DNA hybridisation
Random amplified polymorphic sequences	RAPD	PCR/gel electrophoresis
Simple sequence repeat markers	SSR	PCR/gel electrophoresis
Cleaved amplified polymorphism	CAPS	PCR/gel electrophoresis
Kompetitive allele specific PCR	KASP	PCR/fluorescence
90,000 single nucleotide polymorphism array	90K SNP chip	Fluorescent hybridisation
Resistance gene enrichment sequencing	RenSeq	Exome capture/deep sequencing
Mutation resistance gene enrichment sequencing	MutRenSeq	Mutagenesis/RenSeq
AgRenSeq	AgRenSeq	GWAS RenSeq
Chromosome isolation	ChromSeq	Chromosome isolation and sequencing
Mutation chromosome sequencing	MutChromSeq	Mutation, chromosome isolation and sequencing

functional [6] (examples shown in Table 19.2). The proteins encoded by these alleles share at least 97% identity. This gene is therefore analogous to the barley *Mla* powdery mildew locus, the Arabidopsis *RPP13* downy mildew resistance locus and flax *L* resistance locus which all encode large allelic series. However, most cloned wheat rust and mildew resistance genes are members of small, tandem gene families with a single member conferring the resistance phenotype. In some instances, orthologous sequences in different species have been shown to encode functional resistance. For example, orthologues of the barley *Mla* gene present in *T. monococcum*, *Secale secalis* and *Ae. tauschii* encode the *TaMla1*, *Sr50* and *Sr33* resistance genes, respectively [7].

Both single gene, multi-allele loci and tandem gene families evolve new resistance specificities by intra or intergenic recombination to encode new protein sequences. Central to this recombination process is sequence diversification arising from mutation and selection favouring protein diversification, rather than conservation, at specific regions in the NLR protein. These mechanisms enable new plant resistance specifies to evolve which are required to combat pathogen populations that can rapidly evolve new virulences by loss or alteration of avirulence genes.

An exception in wheat to the generalisation that only a single NLR protein is required for resistance activation is the *Lr10* locus where two sequence unrelated NLR genes are needed to give resistance [9]. Resistance loci that encode dual NLR genes required for resistance have been identified in other plant species [5]. In some of these dual NLR systems one protein contains an "integrated decoy" domain that interacts directly with a pathogen effector. The decoy domain shows similarity to

Table 19.2 Examples of cloned wheat resistance genes and their corresponding pathogen effectors molecules (modified with permission from [8])

Gene	Genpept	[a]Pathogen(s)	Product	Resistance	Corresponding effector	Genpept
Pm1	ERZ1467246	Bgt	NBS-LRR	ASR	AvrPm1	MT773601 MT773602
Pm2	CZT14023	Bgt	NBS-LRR	ASR	AvrPm2|BgsE-5845|BgtriticaleE-5845	APO13050
Pm3a	AAY21626	Bgt	NBS-LRR	ASR	AvrPm3$^{a2/f2}$	ALJ53448
Pm3b	AAQ96158	Bgt	NBS-LRR	ASR	AvrPm3$^{b2/c2}$	VDB86352
Pm3c	ABB78077	Bgt	NBS-LRR	ASR	AvrPm3$^{b2/c2}$	VDB86352
Pm3CS	ABB78080	Bgt	NBS-LRR	ASR	–	–
Pm3d	AAY21627	Bgt	NBS-LRR	ASR	AvrPm3^{d3}	VDB92447
Pm3e	ABB78078	Bgt	NBS-LRR	ASR	–	–
Pm3f	AAZ23113	Bgt	NBS-LRR	ASR	AvrPm3$^{a2/f2}$	ALJ53448
Pm3g	ABB78079	Bgt	NBS-LRR	ASR	–	–
Pm3k	ABY58665	Bgt	NBS-LRR	ASR	–	–
Pm8	AGY30894	Bgt	NBS-LRR	ASR	–	–
Pm17	AYD60116	Bgt	NBS-LRR	ASR	–	–
Pm21	MF370199	Bgt	NBS-LRR	ASR		
Pm60	AUO29720	Bgt	NBS-LRR	ASR	–	–
Pm60a	AUO29719	Bgt	NBS-LRR	ASR	–	–
Pm60b	AUO29721	Bgt	NBS-LRR	ASR	–	–
PmR1	AUO29718	Bgt	NBS-LRR	ASR	–	–
Fhb1	ARJ54140	Fg	Pore forming toxin?			
	QCX35088		Histidine rich Ca^{2+} binding protein			
Stb6	AUF72930	St	WAK	ASR	AvrStb6	MG018996
Stb16q	QOP58831	St	Cysteine rich receptor kinase			
Snn1	AKO62681	Pn	WAK	ASS[b]	SnTox1	AEX93351

Gene	Genpept	[a]Pathogen(s)	Product	Resistance	Corresponding effector	Genpept
Snn3-D1	MW054700	*Pn*	MSP kinase	ASS	*SnTox3*	AGE15690
Sr13	ATE88995	*Pgt*	NBS-LRR	ASR	–	–
Sr21	AVK42833	*Pgt*	NBS-LRR	ASR	–	–
Sr22	CUM44200	*Pgt*	NBS-LRR	ASR	–	–
Sr26		*Pgt*	NBS-LRR	ASR	–	–
Sr33	AGQ17382	*Pgt*	NBS-LRR	ASR	–	–
Sr35	AGP75918	*Pgt*	NBS-LRR	ASR	*AvrSr35*	AUI41041
Sr45	CUM44213	*Pgt*	NBS-LRR	ASR	–	–
Sr46	AYV61514	*Pgt*	NBS-LRR	ASR	–	–
Sr50	ALO61074	*Pgt*	NBS-LRR	ASR	*AvrSr50*	KAA1079665
Sr60	QEM39042	*Pgt*	Protein kinase	ASR	–	–
Sr61		*Pgt*	NBS-LRR	ASR	–	–
TmMLA1	ADX06722	*Bgt*	NBS-LRR	ASR		
TaRCR1	AOV81588	*Rc*	NBS-LRR	ASR		
Tsn1	ADH59425	*Pn, Ptr, Bs*	Protein kinase-NBS-LRR	ASS	*ToxA*	ABD85141
Yr5	QEQ12705	*Pst*	NBS-LRR	ASR	–	–
Yr7	QEQ12704	*Pst*	NBS-LRR	ASR	–	–
Yr10	AAG42168	*Pst*	NBS-LRR	ASR	–	–
Yr15	AXC33067	*Pst*	Protein kinase	ASR	–	–
Yr36	ACF33187	*Pst*	START kinase	APR	–	–
YrSP	QEQ12706	*Pst*	NBS-LRR	ASR	–	–
YrU1	QIM55694	*Pt*	NBS-LRR	ASR	–	–
Lr1	ABU54404	*Pt*	NBS-LRR	ASR	–	–
Lr10/ RGA2	AAQ01784/ ACG63521	*Pt*	NBS-LRR pair	ASR	–	–

(continued)

Table 19.2 (continued)

Gene	Genpept	[a]Pathogen(s)	Product	Resistance	Corresponding effector	Genpept
Lr21	ACO53397	Pt	NBS-LRR	ASR	–	–
Lr22a	ARO38245	Pt	NBS-LRR			
Lr34	ACN41354	Pt, Pst, Pgt, Pm	ABC transporter	APR	–	–
Lr67	QEA08561	Pt, Pst, Pgt, Pm	Hexose transporter	APR	–	–

[a]Bgt – Blumeria graminis f. sp. tritici, Bs – Bipolaris sorokiniana, Fg – Fusarium graminearum, Pn – Parastagonospora nodorum, Pgt – Puccinia graminis f. sp. tritici, Pst – Puccinia striiformis f. sp. tritici, Pt – Puccinia triticina, Ptr – Pyrenophora tritici-repentis, Rc – Rhizoctonia cerealis, St – Septoria tritici
[b]ASS – All Stage Susceptibility

other plant proteins that are likely the true host targets of the effector [5]. The second NLR member contributes to signalling a defense response upon effector recognition by the first member [5]. Approximately 10% of plant NLR genes encode integrated decoy domains [10] and 28 different integrated domains have been identified in wheat NLR genes. However, an integrated decoy domain has not been identified in either NLR gene required for *Lr10* resistance.

As described above, during biotrophic pathogen colonisation large numbers of effector proteins are introduced into host cells to suppress plant defence responses and alter cell homeostasis to benefit the pathogen [11]. Each effector targets particular plant proteins or processes within the plant cell. The genomes of the wheat powdery mildew pathogen, *Blumeria graminis* f. sp. *tritici* (*Bgt*) and wheat rust pathogens *Puccinia triticina* (*Pt*) and *P. graminis* f. sp. *tritici* (*Pgt*) encode in excess of 800, 2000 and 1700 predicted, secreted effector protein genes, respectively. The function of most effectors and their host targets is unknown, however, some wheat pathogen effectors have been shown to target transcription factors, mRNA processing apparatus, chloroplasts and suppress plant defense responses. Similar to plant resistance genes, diversifying selection also occurs in effector genes.

For both wheat stem rust and powdery mildew several pathogen avirulence genes and their cognate resistance genes have been cloned (Table 19.2). Of the two wheat stem rust effector genes cloned, *AvrSr35* and *AvrSr50*, both encode proteins that directly bind to their corresponding resistance proteins, Sr35 and Sr50, respectively [12, 13]. Similarly, some mildew effector proteins have also been shown to directly bind to their corresponding NLR protein encoded by the barley *Mla* locus. However, in many other pathosystems this is not the case and these effectors are assumed to target and modify host proteins, with the R protein then recognising the modified host protein that it guards. While *Sr35* and *Sr50* recognise single effectors (*AvrSr35* and *AvrSr50*, respectively) some members of the *Pm3* allelic series can recognise multiple effectors (Table 19.2).

However, not all wheat ASR genes encode NLR proteins with *Yr15* and *Sr60* being exceptions. Both genes encode tandem protein kinases [14, 15] making them structurally distinct and their mode of action has yet to be determined. The Yr15 and Sr60 proteins are structurally similar to the barley stem rust resistance protein, Rpg1, which is a dual kinase protein, albeit with one kinase domain no longer functional.

Several APR genes have also been cloned from wheat. Both the *Lr34/Yr18/Sr57 /Pm38/Ltn1* APR gene (hereafter *Lr34*) and the *Lr67/Yr46/Sr55/Pm46/Ltn3* APR gene (hereafter *Lr67*) provide broad spectrum resistance to *Pt*, *Pst*, *Pgt* and *Bgt* pathogens [16, 17]. Interestingly both genes also induce premature senescence of mature leaf tips (leaf tip necrosis (*ltn*)) suggesting a mechanistic commonality. Consist with this hypothesis is the lack of additivity of these two partial adult plant resistance genes. However, the products of these two genes are different. *Lr34* encodes an ABC transporter protein that is suggested to transport abscisic acid while *Lr67* encodes a hexose transporter protein no longer capable of sugar transport.

How these two genes provide resistance has not yet been fully established; however, a remarkable feature of *Lr34* is its durability having been used in agriculture for many decades without being overcome. A third wheat gene, *Yr36*, is unlike *Lr34* and *Lr67* in that it confers partial APR to *Pst* only. *Yr36* encodes a START domain containing kinase protein that is believed to interact with a chloroplast peroxidase protein resulting in elevated levels of hydrogen peroxide production [18]. Interestingly *Yr36* shows additive resistance with both *Lr34* and *Lr67* suggesting a different mode of action to these other two genes.

However, not all APR genes are broad-spectrum. A number of race-specific genes (e.g. *Lr12*, *Lr22a*, *Lr22b*, *Yr49*) have been identified that do not provide resistance until later in plant development [19]. One of these genes, *Lr22a*, has been cloned and shown to encode an NLR protein [20]. Given the tendency of NLR genes to be overcome by pathogen mutation to virulence it seems unlikely that this latter type of APR will remain durable.

19.4 Genetics of Resistance to Wheat Necrotrophic Pathogens

Necrotrophic pathogens (see Chap. 8) differ from biotrophic pathogens, like rusts (see Chap. 7) and mildews, in that they require dead or dying cells to acquire nutrients, whereas biotrophs require living plant cells to proliferate. Initially it was thought that necrotrophs exuded a barrage of cell wall-degrading enzymes to kill their host and allow them to acquire nutrients. However, research over the past two decades on interactions between wheat and necrotrophic fungal pathogens *Pyrenophora tritici-repentis* and *Parastagonospora nodorum* has shown that more complex mechanisms are involved. In these pathosystems, pathogen-produced effectors, which are called necrotrophic effectors (NEs) (formerly called host-selective toxins) are recognized by the products of specific plant genes in an inverse gene-for-gene manner (Fig. 19.2). In other words, recognition of a fungal NE by the product of a dominant host gene leads to a compatible interaction (disease susceptibility), and the lack of recognition of the pathogen leads to resistance. Therefore, in plant-necrotroph interactions, plant genes that actively recognize the pathogen are considered susceptibility genes as opposed to plant-biotroph interactions where they act as resistance genes (Fig. 19.2) [21].

The first NE to be identified from a wheat pathogen was Ptr ToxA, which was first discovered in the foliar disease tan spot pathogen *Pyrenophora tritici-repentis* and later in pathogens that cause Septoria nodorum blotch (SNB) (*Parastagonospora nodorum*) and spot blotch (*Bipolaris sorokianiana*). ToxA is a 13.2 kDa protein encoded by a single gene and was the first proteinaceous NE identified in a plant pathogen [22].

Early work in the wheat-tan spot system demonstrated that ToxA is a significant virulence factor and that host resistance is governed by a single recessive gene on chromosome 5BL designated *Tsn1* [23]. Subsequent work evaluating Chinese Spring nullisomic-tetrasomic stocks and chromosome deletion lines demonstrated

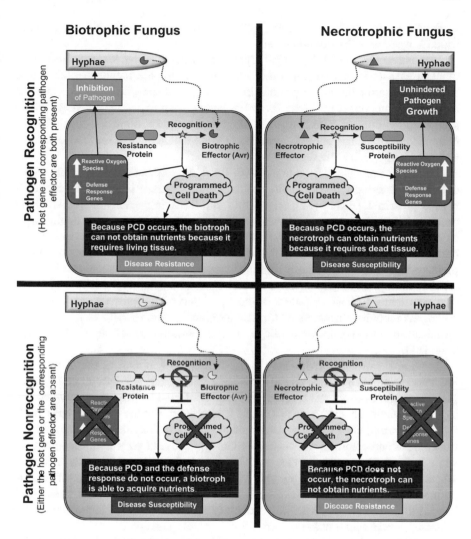

Fig. 19.2 Gene-for-gene interactions occurring between plant hosts and biotrophic pathogens and inverse gene-for-gene interactions with necrotrophic pathogens. (Top left panel) When a biotrophic pathogen introduces a recognised effector (avirulence effector) into a plant cell containing the corresponding resistance gene a defense response is activated that inhibits further pathogen development and can lead to programmed cell death of the infected cell. (Bottom left panel) Conversely if the host plant lacks the appropriate resistance gene or the biotrophic pathogen lacks the recognised effector plant disease susceptibility occurs. (Top right panel) In contrast, in necrotrophic interactions a recognised pathogen effector leads to plant cell death that is required for pathogen development. In the absence of the appropriate pathogen effector or plant resistance protein, which in this case is a susceptibility factor, resistance to the necrotrophic pathogen occurs

that resistance was not governed by an actively expressed recessive gene, but rather the lack of a dominant susceptibility gene, as plants null for chromosome 5B were insensitive to ToxA. This finding showed that pathogen recognition by the host plant was necessary to confer susceptibility, and all wheat gene-NE interactions in the wheat-*Pa. nodorum* and wheat-*Py. tritici-repentis* pathosystems have since been found to operate in the same manner.

Map-based cloning of *Tsn1* [23] showed it encodes an NLR similar to typical plant disease resistance proteins that recognize biotrophic pathogens, but it also contained a serine/threonine protein kinase (PK) domain (Table 19.2). Only the barley stem rust resistance gene *Rpg5* has been shown to encode an NLR with an integrated PK domain. The two genes differ in that the PK domain is located at the N terminus in *Tsn1* and at the C terminus in *Rpg5* indicating that they arose through independent gene fusion events.

Tsn1 does not physically interact with ToxA directly but rather interacts with a chloroplast-localized plastocyanin protein called ToxABP1 and a pathogenesis-related protein 1 (PR-1) [22], suggesting that Tsn1 is a guard of ToxABP1, as described above in the guard hypothesis. The Tsn1-ToxA interactions requires light to manifest necrosis because *Tsn1* transcription is strongly downregulated under darkness. There is also evidence that a compatible Tsn1-ToxA interaction leads to photosystem alterations. These findings suggest that ToxA recognition by Tsn1 leads to disruption of photosynthesis. However, a compatible Tsn1-ToxA interaction also leads to host responses typically observed in resistance to biotrophic pathogens including production of phenylpropanoids and reactive oxygen species, lignification and electrolyte leakage, which indicates hijacking of ETI pathways [21].

ToxA is functionally redundant in the three pathogen species that it has been identified in and acquisition of the *ToxA* gene likely occurred through interspecific gene transfer. However, the level of virulence conferred by ToxA can vary depending on the pathogen and the host genetic background. For the disease SNB caused by *Pa. nodorum*, the Tsn1-ToxA interaction plays a major role in susceptibility of both durum and common wheat, but for tan spot caused by *Py. tritici-repentis*, the same interaction has never been associated with disease susceptibility in durum wheat and its contribution to tan spot virulence in common wheat can range from nothing to major depending on the host genetic background. More studies are needed to determine the potential variability of the Tsn1-ToxA interaction in the wheat-*B. sorokiniana* pathosystem.

In addition to the Tsn1-ToxA inverse gene-for-gene relationship in the wheat-*Pa. nodorum* pathosystem, eight additional sensitivity genes that recognize proteinaceous pathogen-produced NEs to confer SNB have been identified [22]. These include the wheat genes *Snn1*, *Snn2*, homoeologous genes *Snn3-B1* and *Snn3-D1*, *Snn4*, *Snn5*, *Snn6*, and *Snn7*. *Snn1*, *Snn4*, and *Snn5* reside on chromosome arms 1BS, 1AS, and 4BL, and recognize the NEs SnTox1, SnTox4, and SnTox5, respectively. *Snn3-B1* and *Snn3-D1* are homoeologous genes located on 5BS and 5DS, respectively, and both recognize the NE SnTox3. Genes *Snn2*, *Snn6*, and *Snn7* are located on 2DS, 6AL, and 2DL, respectively, and were originally thought to recognize different NEs (SnTox2, SnTox6, SnTox7). However, it was recently shown that these three NEs were actually the same molecule and therefore designated as SnTox267. Among these wheat genes, *Snn1* and *Snn3-D1* have been cloned as have their corresponding NE-producing genes *SnTox1* and *SnTox3*.

The *Snn1* gene was the first wheat gene identified to confer sensitivity to a *Pa. nodorum* NE, and the *Snn1*-SnTox1 interaction plays a significant role in the development of SNB in wheat. SnTox1 is a 10.3 kDa protein that facilitates infection of mesophyll cells, and also protects the fungal mycelium from plant-produced chitinases by binding chitin [22]. Therefore, SnTox1 serves a dual function, which provides an explanation as to why the *SnTox1* gene is highly prevalent among isolates.

Snn1 was isolated from Chinese Spring by positional cloning [24] and encodes a wall-associated kinase (WAK), which is a receptor kinase class that harbor intracellular PK domains and extracellular galacturonan binding (GUB_WAK) and epidermal growth factor-calcium binding (EGF_CA) domains. WAKs usually serve as PRRs that recognize pathogen or damage-associated molecular patterns (PAMPs or DAMPs) leading to PTI. Whereas the WAK gene *Snn1* confers SNB susceptibility, recent research has revealed that other biotroph or hemibiotroph resistance genes are also members of the WAK class of receptor kinases. Therefore, whereas ToxA hijacks a *Tsn1*-associated ETI pathway, SnTox1 hijacks a PTI pathway thus revealing the capability of necrotrophic fungal pathogens to subvert diverse host resistance pathways for pathogenesis.

The SnTox1 protein interacts directly with the extracellular portion of the Snn1 protein [24]. Upon recognition, the mitogen-activated protein kinase (MAPK) gene *TaMAPK3* is rapidly upregulated, which is followed by an oxidative burst, DNA laddering, *PR* gene expression, and cell death, all of which are hallmarks of a defense response to biotrophic pathogens. Like *Tsn1*, *Snn1* transcription is tightly light regulated and a compatible Snn1-SnTox1 interaction is light-dependent suggesting again that photosynthetic pathways are likely involved.

The *Snn3* gene recognizes the *Pa. nodorum* NE SnTox3, which leads to SNB susceptibility. SnTox3 is a 25.8 kDa protein that does not interact directly with the *Snn3* protein but, like ToxA, does interact with PR-1 protein family members. Like SnTox1, SnTox3 appears to also have a dual function by hijacking a host necrosis-inducing pathway through the activation of *Snn3* and suppressing PR-1-mediated defenses [25].

The identification of *Aegilops tauschii* accessions that were sensitive to SnTox3 led to the mapping of the SnTox3-sensitivity gene on chromosome arm 5DS. Comparative mapping experiments revealed that the 5BS and 5DS SnTox3 sensitivity genes were homoeologous, and henceforth referred to as *Snn3-B1* and *Snn3-D1*, respectively. The *Ae. tauschii Snn3-D1* gene was subsequently isolated by positional cloning and found to encode integrated PK and major sperm protein (MSP) domains [26]. Therefore, the cloning of a third *Pa. nodorum* NE sensitivity gene revealed a third class of gene targeted by the pathogen to induce necrosis. Although genes with MSP domains exist in dicots, genes with integrated PK and MSP domains are specific to monocots [26]. An MSP domain was recently shown to be present in the orange blossom wheat midge resistance gene *Sm1* where it occurred along with PK and NLR domains. However, the function of MSP domains in plants is yet to be determined.

Three necrotroph susceptibility genes have now been cloned from wheat, and although they all represent a different class of gene, they all harbor a PK domain. It is likely that the PK is necessary for signaling to induce the development of necrosis. The other domains (i.e. NLR, GUB_WAK, EGF_CA, MSP) are likely necessary for NE recognition either directly as is the case with *Snn1*-SnTox1, or indirectly as observed with *Tsn1*-ToxA and *Snn3*-SnTox3. The cloning of additional NE sensitivity genes will shed light on whether additional gene classes are targeted by necrotrophic pathogens.

The wheat-*Pa. nodorum* pathosystem includes at least nine host gene-NE interactions mentioned above, and the wheat-*Py. tritici-repentis* pathosystem has three known interactions, i.e. *Tsn1*-Ptr ToxA, *Tsc1*-Ptr ToxC, and *Tsc2*-Ptr ToxB [22]. In addition to these inverse gene-for-gene interactions, numerous resistance QTL have been identified in both systems, and it is unknown whether some QTL may represent additional NE sensitivity genes. In the wheat-*Pa. nodorum* pathosystem, no qualitative dominant resistance has been identified. However, a single dominant tan spot resistance gene, designated *Tsr7*, which gives broad-spectrum, race-nonspecific resistance has been identified [27]. *Tsr7* was discovered in wild emmer wheat but confirmed to be present in several modern wheat varieties and provide high levels of tan spot resistance. The mechanisms underlying *Tsr7*-mediated resistance are unknown, but it's cloning will provide more insights into the molecular interactions between wheat and necrotrophic pathogens.

Tsr7 is a good target for breeding tan spot resistant wheat varieties, but, in addition breeders should also focus on removing NE sensitivity genes for both *P. nodorum* and *Py. tritici-repentis* NEs to obtain resistance. However, removal of NE sensitivity genes may come with a couple of caveats. First, because NE sensitivity genes are 'resistance gene-like' they may also confer resistance to a biotrophic pathogen as well as susceptibility to a necrotrophic pathogen. So far, there is no indication that any of the identified wheat NE sensitivity genes also provide resistance to any modern races of biotrophic pathogens, but it is not without precedence. The oat *Vb* gene, which conditions susceptibility to the necrotrophic pathogen that causes Victoria blight, and the *Pc-2* gene, which gives resistance to oat crown rust, have never been separated despite much effort suggesting they may be the same gene [28]. Second, breeders must be careful not to inadvertently introgress necrotroph susceptibility genes when breeding for biotrophic resistance. Germplasm breeding material should be characterized to know what susceptibility genes are present, and which molecular markers (see Chap. 28) are used to eliminate, or select against, NE sensitivity genes.

19.5   Genetics of Resistance to Hemi-Biotrophic Pathogens

Resistance genes to the foliar disease Septoria tritici blotch (STB) caused by the apoplastic fungus *Zymoseptoria tritici* (see Chap. 9) have been categorized into two groups. Qualitative, race-specific *Z. tritici* resistance genes with strong phenotypic effects have been reported for over 20 genes. Similar to rust ASR genes, many quali-tative STB resistance genes are independent of plant growth stage. By contrast, the quantitative class of resistance genes generally show small to moderate effect and over 80 QTL have been described. An STB gene originating from *Aegilops tauschii*, *Stb16q*, was assigned to the quantitative class, however it is one of the few genes that are effective at the seedling stage with major effect against all *Z tritici* isolates tested. Its classification as a quantitative gene has been called into question and there are suggestions that it be designated as a qualitative gene.

Of the 22 catalogued *Stb* genes, two have been cloned by positional cloning, namely *Stb6* and *Stb16q*. *Stb6* exhibits a 'gene-for-gene' relationship and encodes a WAK protein that detects the presence of a matching apoplastic effector, AvrStb6, from *Z. tritici*. *Stb16q* encodes a 684 amino acid cysteine rich receptor kinase (CRRK) protein with two extracellular copies of a domain with unknown function (DUF26) characterised by conserved C-X8-C-X2-C motifs, a predicted transmem-brane domain and an intracellular serine/threonine (Ser/Thr) protein kinase domain [29]. The DUF26 domain of the *Gingko biloba* Gnk2 protein shows mannose bind-ing activity suggesting STB16q may therefore recognize apoplastic plant or fungi-derived mannose or derivatives to activate broad-spectrum defense (Saintenac et al. 2020).

Another hemi-biotrophic fungal pathosystem of significance in wheat is *Fusarium graminearum* which causes head blight (Fusarium head blight-FHB; bleaching of the spike, also referred to as scab) (see Chap. 9). In addition to FHB's effects on yield loss its ability to contaminate grains with trichothecene mycotoxins pose significant human health problems. Genetics of FHB resistance has largely been reported as quantitative due to the complicated process of infection, confound-ing plant morphological attributes, plant phenology and growth conditions. Over 200 FHB QTL have been reported in wheat with *Fhb1*, the first documented and most studied major quantitative trait. From three independent high resolution genetic studies a cluster of genes were identified at the *Fhb1* locus. Among the *Fhb1* candidate genes, Rawat et al. [30] concluded a pore-forming toxin-like gene as the causative gene while Su et al. [31] and Li et al. [32] reported a histidine-rich calcium-binding protein as the *Fhb1* gene product but disagreed about the mode of action. Notwithstanding these contradictory findings, which require clarification, molecular markers generated from the *Fhb1* locus have been excellent tools for selecting *Fhb1* resistance.

A much clearer definition of the gene underlying FHB resistance in plants carry-ing *Fhb7* derived from the wheat grass species, *Thinopyrum elongatum* (E genome), has been described [33]. *Fhb7* was introgressed into wheat as 7E/7B and 7E/7D chromosomal translocations and subsequently shown to encode a glutathione S-transferase (GST) encoding gene that also conferred crown rot resistance. *Fhb7* detoxifies deoxynivalenol (DON- trichothecene toxin) by enzymatic conjugation of a glutathione (GSH) residue onto the toxic epoxide moiety of DON. Unexpectedly *Fhb7* appears to originate from the fungal species *Epichloë aotearoae* and has been transferred to *Th elongatum* by horizontal gene transfer, i.e. a natural fungus-to-plant gene acquisition. What remains unknown is why *Epichloë aotearoae* evolved a DON detoxification gene? It has been suggested that it may detoxify the fungus's own toxins or help it to compete against *Fusarium* species for grass colonization.

19.6   Resistance Gene Stacks-Progress Towards Durable Resistance

As already indicated, obtaining durable disease resistance in wheat to rust and mildew pathogens is difficult because ASR genes are often overcome when deployed singularly due to rapid pathogen effector evolution resulting in virulence. In addition, while some APR genes are both durable and broad-spectrum, individual APR genes don't provide agronomically acceptable levels of disease protection. An obvious solution is to combine multiple ASR genes and/or multiple additive APR genes. However, maintaining polygenic disease resistance combinations throughout breeding programs is an expensive and laborious task, despite modern molecular marker technologies (see Chap. 28).

Genetic engineering offers a potential solution to this problem by enabling multiple cloned disease resistance genes to be introduced into a single locus in the wheat genome. We have recently developed a transgene cassette that encodes four ASR to wheat stem rust and one broad spectrum APR gene i.e. (*Sr22/Sr35/Sr45/Sr 50/Lr67*) [34]. This large gene cassette (approximately 40 kb in size) has been successfully introduced into wheat using Agrobacterium-mediated transformation and transgenics selected that contain all 5 genes at a single locus. In field trials these wheat lines have shown very high levels of disease resistance and additional experiments have confirmed that at least 4 of the 5 genes are functional.

Transgene cassettes therefore offer possibilities to improve resistance durability in wheat because for virulence evolution many pathogen isolates will require multiple mutations to overcome the multiple ASR genes assembled at the locus. A similar approach has also been undertaken in potato where three ASR genes were combined that each provide isolate specific resistance to the potato late blight pathogen *Phytophora infestans* [35]. Future gene stacks also offer the possibility of combining additive APR genes at a single locus to develop high levels of polygenic APR that can be bred with single locus inheritance.

Another advantage of gene stacking technology is it can help prevent single gene deployment of ASR genes which can lead to their rapid breakdown. Conversely, however, if single gene deployment of members of a gene stack occurs it has the potential to erode the polygenic resistance at this locus. Ideally ASR genes in a gene stack would not have previously been deployed in wheat and be broadly effective against most pathogen isolates. Finally, an obvious impediment to the use of resistance gene stacks in the immediate future is that they are a GM solution. However, a recent amendment in the US ruling that cisgenic transgenics can be considered nonGM may create future opportunities (US Dept Agriculture, 2020).

19.7 Key Concepts

An ongoing challenge with disease resistance when widely deployed is the general loss of its effectiveness. Accordingly, an ongoing search for new host resistance genes is required. Both single gene, multi-allele loci and tandem gene families evolve new resistance specificities by a variety of mechanisms outlined in the chapter to produce new resistance proteins. These new host resistance specificities play an important role in combating pathogen populations that can rapidly evolve new virulences via loss or alteration of avirulence genes. In plant-necrotroph interactions, plant genes that actively recognize the pathogen are considered susceptibility genes in contrast to plant-biotroph interactions where they serve as resistance genes; a key concept that underlies 'gene-for-gene' and 'inverse gene-for-gene' relationships.

19.8 Conclusion

Advances in wheat genome and pan-genome sequencing, mutational genomics and gene capture are facilitating the discovery of resistance genes that are effective against biotrophic and hemibiotrophic pathogens, but act as susceptibility factors in necotrophic pathogen infections. More efficient gene capture technologies are likely to be developed combined with comprehensive pan genome sequences which will further facilitate mutational genomics strategies such as MutRenSeq in isolating resistance genes encoded by NLR immune receptors. Such technologies are likely to broaden beyond NLR's to include other resistance gene classes such as receptor kinases. These isolated genes will continue to provide perfect markers for breeders and enable further gene stacks to be developed. While gene stacks offer a potential solution for biotrophs and hemi-biotrophs, their applicability to necrotrophic pathogens is far less given the inverse-gene-for-gene relationship; unless more dominant necrotrophic genes like *Tsr7* can be found. Other options are likely to include gene editing (see Chap. 29) for rapid knockout of necrotrophic susceptibility genes in elite cultivars. Ultimately these translational research outputs will provide precision breeding using resistance gene toolkits which will augment durable disease resistance strategies.

References

1. Flor HH (1971) Current status of the gene-for-gene concept. Annu Rev Phytopathol 9:275–296. https://doi.org/10.1146/annurev.py.09.090171.001423
2. Jones JDG, Dangl JL (2006) The plant immune system. Nature 444:323–329. https://doi.org/10.1038/nature05286

3. Jones JDG (1996) Plant disease resistance genes: structure, function and evolution. Curr Opin Biotechnol 7:155–160. https://doi.org/10.1016/S0958-1669(96)80006-1

4. Walkowiak S, Gao L, Monat C, Haberer G, Kassa MT, Brinton J, Ramirez-Gonzalez RH, Kolodziej MC, Delorean E, Thambugala D, Klymiuk V, Byrns B, Gundlach H, Bandi V, Siri JN, Nilsen K, Aquino C, Himmelbach A, Copetti D, Ban T, Venturini L, Bevan M, Clavijo B, Koo D-H, Ens J, Wiebe K, N'Diaye A, Fritz AK, Gutwin C, Fiebig A, Fosker C, Fu BX, Accinelli GG, Gardner KA, Fradgley N, Gutierrez-Gonzalez J, Halstead-Nussloch G, Hatakeyama M, Koh CS, Deek J, Costamagna AC, Fobert P, Heavens D, Kanamori H, Kawaura K, Kobayashi F, Krasileva K, Kuo T, McKenzie N, Murata K, Nabeka Y, Paape T, Padmarasu S, Percival-Alwyn L, Kagale S, Scholz U, Sese J, Juliana P, Singh R, Shimizu-Inatsugi R, Swarbreck D, Cockram J, Budak H, Tameshige T, Tanaka T, Tsuji H, Wright J, Wu J, Steuernagel B, Small I, Cloutier S, Keeble-Gagnère G, Muehlbauer G, Tibbets J, Nasuda S, Melonek J, Hucl PJ, Sharpe AG, Clark M, Legg E, Bharti A, Langridge P, Hall A, Uauy C, Mascher M, Krattinger SG, Handa H, Shimizu KK, Distelfeld A, Chalmers K, Keller B, Mayer KFX, Poland J, Stein N, McCartney CA, Spannagl M, Wicker T, Pozniak CJ (2020) Multiple wheat genomes reveal global variation in modern breeding. Nature 2020:1–7. https://doi.org/10.1038/s41586-020-2961-x

5. Cesari S (2018) Multiple strategies for pathogen perception by plant immune receptors. New Phytol 219:17–24. https://doi.org/10.1111/nph.14877

6. Krattinger SG, Keller B (2016) Molecular genetics and evolution of disease resistance in cereals. New Phytol 212:320–332. https://doi.org/10.1111/nph.14097

7. Mago R, Zhang P, Vautrin S, Šimková H, Bansal U, Luo M-C, Rouse M, Karaoglu H, Periyannan S, Kolmer J, Jin Y, Ayliffe MA, Bariana H, Park RF, McIntosh R, Doležel J, Bergès H, Spielmeyer W, Lagudah ES, Ellis JG, Dodds PN (2015) The wheat Sr50 gene reveals rich diversity at a cereal disease resistance locus. Nat Plants 1:15186. https://doi.org/10.1038/nplants.2015.186

8. Kourelis J, Kamoun S (2020) RefPlantNLR: a comprehensive collection of experimentally validated plant NLRs. bioRxiv:2020.07.08.193961. https://doi.org/10.1101/2020.07.08.193961

9. Loutre C, Wicker T, Travella S, Galli P, Scofield S, Fahima T, Feuillet C, Keller B (2009) Two different CC-NBS-LRR genes are required for Lr10-mediated leaf rust resistance in tetraploid and hexaploid wheat. Plant J 60:1043–1054. https://doi.org/10.1111/j.1365-313X.2009.04024.x

10. Sarris PF, Cevik V, Dagdas G, Jones JDG, Krasileva KV (2016) Comparative analysis of plant immune receptor architectures uncovers host proteins likely targeted by pathogens. BMC Biol 14:8. https://doi.org/10.1186/s12915-016-0228-7

11. Franceschetti M, Maqbool A, Jiménez-Dalmaroni MJ, Pennington HG, Kamoun S, Banfield MJ (2017) Effectors of filamentous plant pathogens: commonalities amid diversity. Microbiol Mol Biol Rev 81. https://doi.org/10.1128/mmbr.00066-16

12. Salcedo A, Rutter W, Wang S, Akhunova A, Bolus S, Chao S, Anderson N, De Soto MF, Rouse M, Szabo L, Bowden RL, Dubcovsky J, Akhunov E (2017) Variation in the AvrSr35 gene determines Sr35 resistance against wheat stem rust race Ug99. Science 358:1604–1606. https://doi.org/10.1126/science.aao7294

13. Chen J, Upadhyaya NM, Ortiz D, Sperschneider J, Li F, Bouton C, Breen S, Dong C, Xu B, Zhang X, Mago R, Newell K, Xia X, Bernoux M, Taylor JM, Steffenson B, Jin Y, Zhang P, Kanyuka K, Figueroa M, Ellis JG, Park RF, Dodds PN (2017) Loss of AvrSr50 by somatic exchange in stem rust leads to virulence for Sr50 resistance in wheat. Science 358:1607–1610. https://doi.org/10.1126/science.aao4810

14. Klymiuk V, Yaniv E, Huang W, Raat D, Distelfeld A, Korol A, Dubcovsky J, Schulman AH, Fahima T (2018) Cloning of the wheat Yr15 resistance gene sheds light on the plant tandem kinase-pseudokinase family. Nat Commun. Online early. https://doi.org/10.1038/s41467-018-06138-9

15. Chen S, Rouse MN, Zhang W, Zhang X, Guo Y, Briggs J, Dubcovsky J (2020) Wheat gene Sr60 encodes a protein with two putative kinase domains that confers resistance to stem rust. New Phytol 225:948–959. https://doi.org/10.1111/nph.16169

16. Krattinger SG, Lagudah ES, Spielmeyer W, Singh RP, Huerta-Espino J, McFadden H, Bossolini E, Selter LL, Keller B (2009) A putative ABC transporter confers durable resistance to multiple fungal pathogens in wheat. Science 323:1360–1363. https://doi.org/10.1126/science.1166453

17. Moore JW, Herrera-Foessel S, Lan C, Schnippenkoetter W, Ayliffe M, Huerta-Espino J, Lillemo M, Viccars L, Milne R, Periyannan S, Kong X, Spielmeyer W, Talbot M, Bariana H, Patrick JW, Dodds P, Singh RP, Lagudah E (2015) A recently evolved hexose transporter variant confers resistance to multiple pathogens in wheat. Nat Genet 47:1494–1498. https://doi.org/10.1038/ng.3439

18. Gou J-Y, Li K, Wu K, Wang X, Lin H, Cantu D, Uauy C, Dobon-Alonso A, Midorikawa T, Inoue K, Sánchez J, Fu D, Blechl A, Wallington E, Fahima T, Meeta M, Epstein L, Dubcovsky J (2015) Wheat stripe rust resistance protein WKS1 reduces the ability of the thylakoid-associated ascorbate peroxidase to detoxify reactive oxygen species. Plant Cell 27:1755–1770. https://doi.org/10.1105/tpc.114.134296

19. Ellis JG, Lagudah ES, Spielmeyer W, Dodds PN (2014) The past, present and future of breeding rust resistant wheat. Front Plant Sci 5:1–13. https://doi.org/10.3389/fpls.2014.00641

20. Thind AK, Wicker T, Šimková H, Fossati D, Moullet O, Brabant C, Vrána J, Doležel J, Krattinger SG (2017) Rapid cloning of genes in hexaploid wheat using cultivar-specific long-range chromosome assembly. Nat Biotechnol 35:793–796. https://doi.org/10.1038/nbt.3877

21. Faris JD, Friesen TL (2020) Plant genes hijacked by necrotrophic fungal pathogens. Curr Opin Plant Biol 56:74–80. https://doi.org/10.1016/j.pbi.2020.04.003

22. Friesen TL, Faris JD (2021) Characterization of effector-target interactions in necrotrophic pathosystems reveals trends and variation in host manipulation. Annu Rev Phytopathol. https://doi.org/10.1146/annurev-phyto-120320-012807

23. Faris JD, Zhang Z, Lu H, Lu S, Reddy L, Cloutier S, Fellers JP, Meinhardt SW, Rasmussen JB, Xu SS, Oliver RP, Simons KJ, Friesen TL (2010) A unique wheat disease resistance-like gene governs effector-triggered susceptibility to necrotrophic pathogens. Proc Natl Acad Sci U S A 107:13544–13549

24. Shi G, Zhang Z, Friesen TL, Raats D, Fahima T, Brueggeman RS, Lu S, Trick HN, Liu Z, Chao W (2016) The hijacking of a receptor kinase-driven pathway by a wheat fungal pathogen leads to disease. Sci Adv 2:e1600822

25. Sung Y-C, Outram MA, Breen S, Wang C, Dagvadorj B, Winterberg B, Kobe B, Williams SJ, Solomon PS (2021) PR1-mediated defence via C-terminal peptide release is targeted by a fungal pathogen effector. New Phytol 229:3467–3480. https://doi.org/10.1111/nph.17128

26. Zhang Z, Running KLD, Seneviratne S, Peters Haugrud AR, Szabo-Hever A, Shi G, Brueggeman R, Xu SS, Friesen TL, Faris JD (2021) A protein kinase–major sperm protein gene hijacked by a necrotrophic fungal pathogen triggers disease susceptibility in wheat. Plant J 106:720–732. https://doi.org/10.1111/tpj.15194

27. Faris JD, Overlander ME, Kariyawasam GK, Carter A, Xu SS, Liu Z (2020) Identification of a major dominant gene for race-nonspecific tan spot resistance in wild emmer wheat. Theor Appl Genet 133:829–841. https://doi.org/10.1007/s00122-019-03509-8

28. Wolpert TJ, Dunkle LD, Ciuffetti LM (2002) Host-selective toxins and avirulence determinants: what's in a name? Annu Rev Phytopathol 40:251–285. https://doi.org/10.1146/annurev.phyto.40.011402.114210

29. Saintenac C, Cambon F, Aouini L, Verstappen E, Smt G, Poucet T, Marande W, Berges H, Xu S, Jaouannet M, Favery B, Alassimone J, Sanchez-Vallet A, Faris J, Kema G, Robert O, Langin T (2021) A wheat cysteine-rich receptor-like kinase confers broad-spectrum resistance against Septoria tritici blotch. Nat Commun 12

30. Rawat N, Pumphrey MO, Liu S, Zhang X, Tiwari VK, Ando K, Trick HN, Bockus WW, Akhunov E, Anderson JA, Gill BS (2016) Wheat Fhb1 encodes a chimeric lectin with agglutinin domains and a pore-forming toxin-like domain conferring resistance to Fusarium head blight. Nat Genet 48:1576–1580. https://doi.org/10.1038/ng.3706

31. Su Z, Bernardo A, Tian B, Chen H, Wang S, Ma H, Cai S, Liu D, Zhang D, Li T, Trick H, St. Amand P, Yu J, Zhang Z, Bai G (2019) A deletion mutation in TaHRC confers Fhb1 resistance to Fusarium head blight in wheat. Nat Genet 51:1099–1105. https://doi.org/10.1038/s41588-019-0425-8

32. Li G, Zhou J, Jia H, Gao Z, Fan M, Luo Y, Zhao P, Xue S, Li N, Yuan Y, Ma S, Kong Z, Jia L, An X, Jiang G, Liu W, Cao W, Zhang R, Fan J, Xu X, Liu Y, Kong Q, Zheng S, Wang Y, Qin B, Cao S, Ding Y, Shi J, Yan H, Wang X, Ran C, Ma Z (2019) Mutation of a histidine-rich calcium-binding-protein gene in wheat confers resistance to Fusarium head blight. Nat Genet 51:1106–1112. https://doi.org/10.1038/s41588-019-0426-7

33. Wang H, Sun S, Ge W, Zhao L, Hou B, Wang K, Lyu Z, Chen L, Xu S, Guo J, Li M, Su P, Li X, Wang G, Bo C, Fang X, Zhuang W, Cheng X, Wu J, Dong L, Chen W, Li W, Xiao G, Zhao J, Hao Y, Xu Y, Gao Y, Liu W, Liu Y, Yin H, Li J, Li X, Zhao Y, Wang X, Ni F, Ma X, Li A, Xu SS, Bai G, Nevo E, Gao C, Ohm H, Kong L (2020) Horizontal gene transfer of Fhb7 from fungus underlies Fusarium head blight resistance in wheat. Science 368:eaba5435. https://doi.org/10.1126/science.aba5435

34. Luo M, Xie L, Chakraborty S, Wang A, Matny O, Jugovich M, Kolmer JA, Richardson T, Bhatt D, Hoque M, Patpour M, Sørensen C, Ortiz D, Dodds P, Steuernagel B, Wulff BBH, Upadhyaya NM, Mago R, Periyannan S, Lagudah E, Freedman R, Lynne Reuber T, Steffenson BJ, Ayliffe M (2021) A five-transgene cassette confers broad-spectrum resistance to a fungal rust pathogen in wheat. Nat Biotechnol 39:561–566. https://doi.org/10.1038/s41587-020-00770-x

35. Ghislain M, Byarugaba AA, Magembe E, Njoroge A, Rivera C, Román ML, Tovar JC, Gamboa S, Forbes GA, Kreuze JF, Barekye A, Kiggundu A (2019) Stacking three late blight resistance genes from wild species directly into African highland potato varieties confers complete field resistance to local blight races. Plant Biotechnol J 17:1119–1129. https://doi.org/10.1111/pbi.13042

Chapter 20
Insect Resistance

**Wuletaw Tadesse, Marion Harris, Leonardo A. Crespo-Herrera,
Boyd A. Mori, Zakaria Kehel, and Mustapha El Bouhssini**

Abstract Studies to-date have shown the availability of enough genetic diversity in the wheat genetic resources (land races, wild relatives, cultivars, etc.) for resistance to the most economically important insect pests such as Hessian fly, Russian wheat aphid, greenbug, and Sun pest. Many R genes – including 37 genes for Hessian fly, 11 genes for Russian wheat aphid and 15 genes for greenbug – have been identified from these genetic resources. Some of these genes have been deployed singly or in combination with other genes in the breeding programs to develop high yielding varieties with resistance to insects. Deployment of resistant varieties with other integrated management measures plays key role for the control of wheat insect pests.

Keywords Breeding · Gene introgression · Insect resistance

The original version of this chapter was revised. The correction to this chapter is available at https://doi.org/10.1007/978-3-030-90673-3_33

W. Tadesse (✉) · Z. Kehel
International Center for Agricultural Research in the Dry Areas (ICARDA), Rabat, Morocco
e-mail: w.tadesse@cgiar.org; z.kehel@cgiar.org

M. Harris
North Dakota State University, Fargo, ND, USA
e-mail: marion.harris@ndsu.edu

L. A. Crespo-Herrera
International Maize and Wheat Improvement Center (CIMMYT), Texcoco, Mexico
e-mail: l.crespo@cgiar.org

B. A. Mori
University of Alberta, Edmonton, AB, Canada
e-mail: bmori@ualberta.ca

M. El Bouhssini
Mohamed VI Polytechnic University, Benguerir, Morocco
e-mail: Mustapha.ElBouhssini@um6p.ma

© The Author(s) 2022, Corrected Publication 2022
M. P. Reynolds, H.-J. Braun (eds.), *Wheat Improvement*,
https://doi.org/10.1007/978-3-030-90673-3_20

20.1 Learning Objectives

- To understand the most important wheat insect pests, their geography, the mechanisms of insect resistance and breeding for insect resistance.

20.2   Introduction

The demand for wheat is increasing along with the increasing human population, and is expected to surge to one billion tons by the year 2050 [1]. Fulfilling this demand with the increasing impact of climate change, likely environmental and resources degradation, reduced supply and increasing cost of inputs, and emergence of new virulent pests will be challenging unless we deploy efficient strategies, methods and policies to develop climate smart wheat technologies with high genetic gain and minimum yield losses. Globally, wheat yield losses of 5.1 and 9.3% have been reported due to insect pests during the pre-and post- green revolution era, respectively [2]. This chapter summarizes the most important wheat insect pests, their geographic distribution and economic importance, sources and mechanisms of resistance, gene introgression, breeding methods and approaches for insect resistance.

20.3   Major Wheat Insect Pests, Geographic Distribution and Economic Importance

There are many insects affecting wheat production at global and regional levels. The major ones are Hessian fly, sunn pest, Cereal leaf beetle; Wheat stem sawfly, Russian wheat aphid, Greenbug, Bird Cherry-Oat Aphid, English grain aphid and Orange wheat blossom midge (Fig. 20.1). The biology, geographic distribution and their economic importance are indicated in Sections 20.3.1 to 20.3.9.

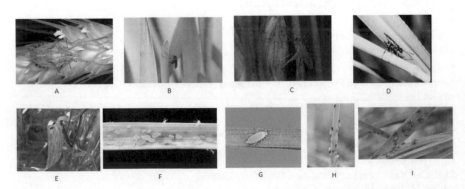

Fig. 20.1 Major wheat insect pest: (**a**) Sunn pest; (**b**) Hessian fly; (**c**) Cereal leaf beetle; (**d**) Wheat stem sawfly; (**e**) Russian wheat aphid; (**f**) Greenbug; (**g**) Bird Cherry-Oat Aphid; (**h**) English grain aphid; (**i**) Orange wheat blossom midge

20.3.1 Hessian Fly (Diptera: Cecidomyiidae)

Mayetiola destructor (Say) is an important pest of wheat in North Africa, North America, Southern Europe, Northern Kazakhstan, Northwestern China, and New Zealand. Yield losses of 30% are common but there can be complete crop failure if infestation coincides with young stage of the wheat crop [3]. Hessian fly adults are small (less than 1/8 inch long) and do not feed and do not live long. Females lay from 100 to 300 eggs. The pest has three larval instars. The first induces a gall nutritive tissue at its feeding site. The second grows rapidly. The third completes its development in a puparium that looks like a flax seed, which is where the pupa also lives. Hessian fly has a facultative diapause during the third instar and overwinters in wheat stubble or volunteer wheat. Depending on environmental field conditions, Hessian fly can complete 2–3 generations/year.

20.3.2 Sunn Pest

Sunn pest (Fig. 20.2) refers to several species in two genera *Eurygaster* and *Aelia*. The most widespread and damaging species to wheat is *Eurygaster integriceps* Puton (Hemiptera: Scutelleridae) which is about 12 mm long in size. Sunn pest is widespread throughout South and East Europe, North Africa, Near East, West and South-Central Asia. Yield losses attributable to direct feeding typically range between 50% and 90% [4, 5]. Prolyl endoproteases injected into the grain during feeding severely compromise the quality of the resulting flour by degrading the vital gluten proteins [6]. Sunn pest has one generation per year. Adults overwinter mainly in mountains and hills surrounding wheat fields. In early spring, mature

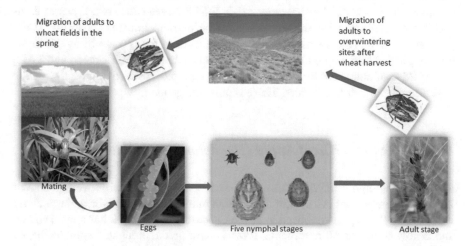

Migration of adults to wheat fields in the spring

Migration of adults to overwintering sites after wheat harvest

Mating

Eggs

Five nymphal stages

Adult stage

Fig. 20.2 Life cycle of Sunn pest

adults migrate from the overwintering locations to cereal fields. The pest has five nymphal stages, only four of which feed (2nd to 5th). Sunn pest incidence occurs in multi-year cycles, and when it happens, it is devastating. In a 2 ton wheat crop, the presence of 2 insects per m^2 can destroy baking quality of the total harvest [Dr. Hans-Joachim Braun, personal communication, May 24, 2021].

20.3.3   Cereal Leaf Beetle (Coleoptera: Chrysomelidae)

Oulema melanopus (L.) with average size of five millimeters (3|16 inch) length, is a significant pest of wheat in Europe and Central Asia but has been reported causing damage in several other parts of the world including Algeria, Tunisia, Morocco, Syria, India, Pakistan, Iran, United States and Canada. Yield losses in a single field can be as high as 55% in spring wheat and 23% in winter wheat [7], though damage is mostly locally confined. Larvae are more damaging than adults and have been reported to consume plant biomass 1 to 10 times their body weight. Cereal leaf beetle has one generation per year. Adult beetles overwinter in protected areas such as wind rows, crop stubble and tree bark crevices. Larvae go through four instars.

20.3.4   Wheat Stem Sawfly (Hymenoptera: Cephidae)

Wheat stem sawfly (WSSF) is the common name of a number of sawfly wheat pest species in North America, Europe, North Africa and Asia with average adult insect size of 3/4 inch length. In North America, the most important species is *Cephus cinctus* Norton. Yield losses inflicted by this pest in the Northern Great Plains exceed $350 million a year [8]. In Europe, North Africa and West Asia, the most common species is *Cephus pygmaeus* (L.). WSSF larvae cause two types of damage; larval feeding inside the stem reduces the nutrient transfer capability of the plant and weakens the stems. The most severe form of loss is caused by the stems that are girdled and topple to the ground just before harvest. In Morocco and Syria, 40 and 28% of stems cut by WSSF have been reported, respectively [8, 9]. Larvae pass through four or five instars. There is only one generation per year.

20.3.5   Russian Wheat Aphid (Hemiptera: Aphididae)

Russian wheat aphid (RWA), *Diuraphis noxia* (Kurdjumov), with average size of 2 mm, is an important insect pest of wheat in many parts of the world, particularly in dry areas. Its origin is believed to be the Caucasus region, but it has spread widely and is now found on all continents. RWA is light green, elongated spindle-shaped aphid, with distinguishing double tail. Feeding on young wheat leaves causes a number of symptoms including longitudinal chlorotic streaking with a convoluted rolling of the leaf. Rolling of the leaves reduces photosynthetic area and protects aphids from contact insecticides and natural enemies. Yield losses of 20–90% have been reported in different parts of the world [10–12]. Depending on environmental conditions, RWA is reported to have sexual as well as asexual reproduction.

20.3.6 Greenbug (Hemiptera: Aphididae)

The species *Schizaphis graminum* (Rondani), commonly known as greenbug with average adult size of 1.6 mm length, has an uncertain origin, however it is considered of paleartic origin, probably from the Middle East or Central Asia. Its distribution encompasses Asia, Southern Europe, Africa, North and South America. The apterous individuals are light green with dark-tipped siphunculi, and typically with a green longitudinal stripe on their abdomen. *S. graminum* feeds on various genera of Poaceae, e.g. *Agropyron, Avena, Bromus, Dactylis, Eleusine, Festuca, Hordeum, Lolium, Oryza, Panicum, Poa, Sorghum, Triticum* and *Zea*. It is capable of transmitting Barley Yellow Dwarf Virus (BYDV) especially the Schizaphis graminum virus (SGV) strain. Feeding of *S. graminum* on susceptible plants causes chlorosis and necrotic spots at the feeding site.

20.3.7 Bird Cherry-Oat Aphid (Hemiptera: Aphididae)

Commonly known as the bird cherry-oat aphid, *Rhopalosiphum padi (L.)*, with a size range of 1–2 mm length, has an origin difficult to trace, since it is currently distributed worldwide. Its sexual phase takes part on various *Prunus* species, e.g. in Europe it overwinters on *Prunus padus* L., and in North America on *P. virginiana* L. Among all aphid species mentioned in this chapter, *R. padi* is the only one able to overwinter on a species other than from the *Poaceae* family. Based on phylogenetic studies using SCAR markers on nuclear DNA, mitochondrial DNA (cyt.b) markers, and tracking life history of aphids, it has been shown that there are two lineages differing in their life cycle: (1) holocyclic, with the sexual phase on the primary host (*P. padus*) and a parthenogenetic phase during summer in Poaceae species; (2) anholocyclic, with only the parthenogenetic phase on grasses; this occurs in places where the winter is mild. The damage caused by this aphid in wheat is not evident to the human eye until plants are seriously damaged, by then plants become yellowish, stunted and most often dead. Even though the economic losses caused by this aphid in the absence of virus are not reported, it can significantly reduce yield by 31% and up to 62% when damage is combined with BYDV infection [13].

20.3.8   English Grain Aphid (Hemiptera: Aphididae)

Sitobion avenae (F.), commonly known as the English grain aphid, probably origi-
nates from Europe, and it is currently present in Europe, Northern and Southern
Africa, Eastern India and Nepal and North and South America. This species is a
yellow-green or reddish-brown aphid, small to medium sized and broadly elongated
(1.9–3.5 mm). It has a pale cauda and typically black knees and cornicles, the latter
twice as long as the cauda. This aphid species overwinters on Poaceae species where
also the sexual cycle occurs, even though aphids can continue reproducing parthe-
nogenetically the whole year. It is a vector of BYDV, particularly the strains
Macrosiphum avenae virus (MAV) and *Padi avenae* virus (PAV). Similar to *R. padi*,
this aphid species does not cause visible symptoms on the wheat plants when feeding, but it can reduce spring wheat yields by 20% at only 300 aphid-days.

20.3.9   Orange Wheat Blossom Midge (Diptera: Cecidomyiidae)

The wheat midge *Sitodiplosis mosellana* (Géhin) is a small (approximately 3 mm
long), delicate, mosquito-like orange color fly distributed throughout many wheat-
growing regions of the Northern Hemisphere, especially between 42°N and 62°N
latitude. From Eurasia – where it is a pest today – it spread to also become a serious
pest in North America and China in the 1800s and 1900s, respectively [14–16].
There is a single generation each year. Adult emergence coincides with anthesis.
The first two larval instars feed on the developing seed, thereby harming both wheat
yield and quality. The seed can be entirely consumed. Infested seeds that are large
enough to be harvested exhibit undesirable changes in germination, protein and
dough strength. The third instar stays inside the floret until high moisture conditions
trigger its departure. Larvae burrow to a depth of a few centimetres and overwinter
inside a cocoon. During outbreaks, wheat losses are large. In 1983, an estimated 30
million in Canadian dollars was lost in Saskatchewan. In 2004, an estimated one
million tonnes were lost in the United Kingdom [17].

20.4   Mechanisms of Plant Resistance to Wheat Pests

In the middle of the twentieth century, Painter in his classic book Insect Resistance
in Crop Plants [18] proposed two types of plant resistance. The first is Antixenosis
(also called Non-preference). Here the resistant plant trait interferes with arthropod
behavior. Many aspects of behavior contribute to colonization of the plant and existence on the plant, thereafter, including egg-laying by adult females and feeding by
larvae. The second is Antibiosis wherein plant traits interfere with the arthropod's
physiological processes after it arrives on the plant, including digestion or

maturation of eggs. In screening tests, manifestations of plant antixenosis and anti-biosis are failure of arthropod survival, growth and/or reproduction, the most extreme form of which is immediate death soon after attack begins.

Plants have physical traits conferring antixenosis and antibiosis – such as tri-chomes, slippery surfaces, and fortified tissues. However, the most notable plant resistance traits are chemicals. 'Primary plant chemistry' supports basic physiologi-cal processes such as growth and reproduction. 'Secondary plant chemistry' sup-ports more specialized functions, including defense against predators and parasites. Some defense chemicals are produced by the plant all the time as 'constitutive defences' whereas others are produced only when they are needed as 'induced defences'. In addition to deployment of induced chemicals in 'direct defence' against arthropods, induced chemicals also are deployed in 'indirect defence' against arthropods by attracting natural enemies of the attacking herbivorous arthro-pod in order to assist the plant in harming its enemies.

Clearly, plants have evolved traits that allow them to actively resist predators and parasites. It is generally assumed that active resistance traits have a cost for the plant. Resistance traits evolve when the benefit is greater than the cost. Especially prized by agriculturalists are resistance traits that entirely exclude the arthropod from colonizing a particular wheat cultivar. In such cases, we expect strong selec-tion pressure and the possibility that the pest will evolve to overcome the resistance. Now, the resistance trait must be replaced by a different resistance trait. This ongo-ing cycling of resistance deployment followed by pest or pathogen adaptation is an example of the "arms races" occurring in agriculture.

Painter [18] described an option that reduces selection pressure for the "arms race". Plants have traits that allow them to 'tolerate' the pest. The pest is given a place to live, but the resources it is given are more restricted compared with a geno-type lacking the tolerance trait. Traits conferring 'tolerance' have the advantage of less selection pressure but also have the disadvantage of allowing pest populations to persist, albeit at a lower level. Tolerance traits are identified in a screening test in which all plant genotypes are subjected to the same level of attack (usually a low rather than high level). Subsequently, the relative degree of damage exhibited by the various genotypes is scored. Tolerant genotypes are better at growing and reproduc-ing in the presence of the arthropod. The arthropod population grows more slowly on a tolerant versus non-tolerant genotype.

20.5 Genetic Diversity and Gene Mining
         for Insect Resistance

Wheat genetic diversity, defined as the total number of genetic characteristics pres-ent in the *Triticum* species, is the most important factor for wheat improvement in terms of adaptation, yield potential, end-use quality, drought and heat tolerance, resistance to diseases and insect pests. Large number of wheat genetic resources including land races, old cultivars, wild relatives and elite breeding lines are available in the gene banks at CIMMYT and ICARDA and other international and national institutions [19]. However, only a limited amount (about 10%) of the available genetic resources have been utilized for improvement purposes by breeders globally due to (a) gene bank accessions are too obsolete, clumsy and wild with difficulty to breed and even if successful, it may lead into linkage drags, (b) the germplasm is poorly characterized and the available data might not be accessible and match the interest of breeders, (c) enough genetic diversity might be available in the elite breeding lines and varieties. Deployment of effective strategies and tools to undertake gene mining and introgression is highly important to increase the utilization of genetic resources in the wheat breeding programs. Some of these strategies and techniques are indicated in Sections 20.5.1 to 20.5.3.

20.5.1   Focused Identification of Germplasm Strategy (FIGS)

Distribution of genetic resources is a key and core gene bank activity aiming at responding to requests from various users including breeders, researchers, farmers, etc. [20]. When the request does not specify the germplasm and traits sought, a random sample is selected and sent to requesters. Core collections, proposed originally by Brown in 1989 [21], were developed for major crops which include 10% of holdings representing the geographic- characterization- or genetic-based diversity.

To effectively respond to inquiries that directly meet the needs of the users, the focused identification of the germplasm strategy FIGS has been developed at the International Center for Agriculture Research in the Dry Areas (ICARDA) in the last decade. FIGS has become a better alternative to random sampling and the use of core collections since it is specific to each trait and is selecting manageable size subsets with higher probability of finding the desired traits. It is based on finding the relationship between the environmental conditions of collection sites and the traits requested by users.

FIGS uses two approaches: filtering and modeling; both of which select best-bet environments that are likely to have imposed selection pressure for specific traits on *in-situ* populations over time. Developing a FIGS filtering strategy requires deep understanding of the ecology and the optimal conditions of the expression of the trait under study, how these conditions affect the crop, and how this will relate to a selection pressure on an *in-situ* population. The FIGS modeling pathway explores the mathematical relationship between the adaptive trait of interest and the long-term climatic and/or soil characteristics of collection sites. The mathematical conceptual framework of FIGS is based on the paradigm that the trait as a response variable depends on the environment attributes considered as the covariates. The quantification process leads to the generation of *a priori* information, which is used in the prediction of accessions that would carry the desired trait.

Previous success in using FIGS has been reported for example in the identification for sources of resistance to Sunn pest in wheat in Syria and for Russian wheat aphid in bread wheat [22].

Here we represent an example of how a filtering approach was used to select best bet subset for selecting FIGS subset for Sunn pest: 1. Start with all georeferenced landraces for which a suite of monthly agro-climatic data was available from WorldClim (8376 Accessions); 2. Collection sites from a geographic region between latitudes 30° to 45° and longitudes 35°-80° where progressed to the next step to represent areas where Sunn pest has been reported as an historic pest; 3. Sites in China, Pakistan and India were also excluded because there have been only recent reports of Sunn pest in these countries; 4. Accessions collected from sites whose long term average annual rainfall was less than 280 mm per year were excluded as Sunn pest populations are not particularly dense in very arid environments; 5. Accessions from sites that experience long term average minimum monthly temperatures of less than −10 ° C were also excluded as it was hypothesized that areas experiencing particularly harsh winters would not favor high population densities of Sunn pest; 6. Maximizing agroecological diversity which resulted into 534 accessions of which half were from Afghanistan.

The evaluation of this Sunn pest FIGS subset yielded 9 accessions that were resistant to the juvenile stage of the pest (1 from Tajikistan and 8 from Afghanistan), which was an excellent result considering that 1000s had been screened previously without success [4]. This example demonstrates that (1) even a very simple filter, using just monthly temperature and annual rainfall, can be effective at capturing invaluable genotypes, and (2) it is essential to understand something about the biology of the organism in question when designing a filter.

20.5.2   Screening Techniques for Resistance to Wheat Pests

When screening plant materials for resistance to a particular arthropod – whether in the field or the greenhouse – two observations signal the possibility that a particular genotype is resistant. The first is the complete absence of the insect, whereas it is clearly present on other genotypes. The second is reduced presence relative to its greater presence on other genotypes. Conclusions based on such observations are more reliable if a variety of plant genotypes are tested simultaneously. Highly susceptible genotypes must always be included. They act as 'controls', providing proof that the absence of the pest from a particular genotype resulted from its ability to resist attack rather than because it escaped attack due to a failure of testing conditions. The screening techniques described below for resistance to Hessian fly, Sunn pest, Cereal leaf beetle, Wheat stem sawfly and Russian wheat aphid are in use at ICARDA [22], whereas those presented for the greenbug, bird cherry-oat aphid and English grain aphid are commonly used at CIMMYT.

20.5.2.1 Hessian Fly

Screening for Hessian fly can be carried out in hotspots in the field under natural infestation but also in the greenhouse. In the field, planting date needs to be adjusted so that 1–2 leaf stage of the crop coincides with the emergence of the flies. For example, in North Africa, a delayed planting date creates strong pest pressure on the tested plants. Evaluation of plant genotypes for resistance is usually made 3–4 weeks after infestation in the greenhouse or when symptoms are clearly seen on the susceptible check. Selection is straight forward, since susceptible plants show stunted growth and a dark green color and contain live larvae, whereas the resistant plants exhibit normal growth and a normal light green color and contain either mostly or only dead first-instar larvae.

20.5.2.2 Sunn Pest

Screening is conducted only in the field under artificial infestation. Test entries are planted under mesh screen cages. Plants are infested at the time of Sunn pest's once yearly migration to wheat fields using insects collected by sweep nets. The evaluation is based on vegetative stage damage either 4 weeks after infestation or when symptoms are clearly visible on the susceptible check. The following rating scale of 1–6 is used to assess shoot and leaf damage (and plant stunting): 1 = no damage and no stunting; 2 = 1–5% damage, with very little stunting; 3 = 6–25% damage with low level of stunting; 4 = 26–50% damage, with moderate level of stunting; 5 = 51–75% damage with high level of stunting, and 6 = >75% damage, with severe stunting.

20.5.2.3 Cereal Leaf Beetle

Because cereal leaf beetle has one generation/year, screening of germplasm is carried out in hotspots in the field under natural infestation. When severe damage is seen on the flag leaf of the susceptible check, the evaluation is conducted using the following rating scale: 1 = no damage, 2 = 10% or less of leaves damaged, 3 = 25% or less of leaves damaged, 4 = 50% or less of leaves damaged, 5 = 75% or less of leaves damaged, 6 = more than 75% of leaves damaged, including the flag leaf.

20.5.2.4 Wheat Stem Sawfly

Wheat stem saw fly produces one generation/year. Screening of germplasm for resistance to this pest is mostly carried out in hotspots in the field under natural infestation. At the end of the season, just prior to harvest, evaluation for resistance is based on the % stems cut by larvae: >30% = susceptible, 20–30% = moderately susceptible, <10% = moderately resistant, <5% = resistant.

20.5.2.5   Russian Wheat Aphid

Screening for Russian wheat aphid is carried out in hotspots in the field under natural and/or artificial infestation but also in the greenhouse. Evaluation is made when symptoms of leaf rolling and leaf chlorosis are clearly visible on susceptible checks using a 1–3 scale for leaf rolling (LR), where: 1 = no rolling, 2 = trapping or curling in one or more leaves, and 3 = rolling in one or more leaves. For leaf chlorosis (LC) a 1–6 scale is used where: 1 = no LC, 2 = <33% of leaf area with LC, 3 = 33–66% area with LC, 4 = >66% area with LC, 5 = necrosis in at least one leaf, and 6 = plant death [23].

20.5.2.6   Greenbug

Because of the symptoms caused by *S. graminum* it is possible to perform massive screenings, allowing the identification of resistant germplasm in short spans (10–14 days). Protocols consist of sowing row or hill plots of eight to ten seeds in flats; 3 days after emergence plants are infested by placing infested leaves on the plots with an average density of four to five aphids per plant; scores of symptoms in percent of chlorosis are taken 10–14 days after infestation, or using a 0–9 damage scale where: 0 = No damage and 9 = dead. However, more quantitative and eye-independent measurements, is the evaluation of chlorophyll content, which has been successfully used in wheat to identify resistance sources and map chromosomic regions associated with the resistance [24].

20.5.2.7   Bird Cherry-Oat Aphid & English Grain Aphid

Evaluating resistance to these two aphid species is more challenging, since none of these cause visible symptoms on the plants. One option is to conduct the typical life table assessments, where the intrinsic rate of increase is calculated, however, this is time consuming and the number of plant materials that can be evaluated is limited. Another option is to determine the aphid growth, this allows a somewhat larger number of genotypes to be evaluated. One more option is to assess the biomass loss of the seedlings in an infested vs. non-infested setup. There is one additional complication, in the case of the EGA it is fundamental to asses the germplasm at the adequate phenological stage, since evaluations at other stages can result in false positive results.

20.5.2.8   Orange Wheat Blossom Midge

Field screening methods have enabled resistance scoring of hundreds or even thousands of genotypes in a single season. Two to three weeks after egg-laying occurs, an evaluator threshes a wheat spike (5 per plot), noting the presence of the bright orange mature larva. Genotypes that exclude larvae or support significantly fewer larvae compared to susceptible controls are classified as resistant. Genotypes can be misclassified as resistant if planted too early or too late, making the spike either no longer attractive to the egg-laying female or not suitable for larval colonization of the seed embryo. Bad weather can also prevent infestation. Multiple planting dates help reduce these problems but restrict the number of genotypes that can be screened each season. Screening in the greenhouse does not have these problems but requires establishment of a laboratory colony, each generation of which requires a 5-month long period of obligatory diapause in the cold.

20.5.3   Identification and Introgression of Insect Resistant Genes

Using the FIGS approach and the different screening protocols of screening for insect resistance both in the field under hot spot locations and in the greenhouse using artificial inoculation, important *Resistance (R)* genes have been identified and mapped for each of the important wheat insect pests including the Hessian fly, Wheat midge, Greenbug, Russian wheat aphid and Wheat curl mite. According to Harris et al. [25], out of the total 479 R genes reported in wheat, only 69 R genes are targeted for insects and mites, mainly for Hessian fly (37 genes), Russian wheat aphid (11 genes) and Greenbug (15 genes). Most of the resistance genes for Hessian fly were identified from *Triticum aestivum* accessions such as Grant, Patterson, 86981RC1-10-3, 8268G1-19-49, KS89WGRC3 (C3), and KS89WGRC6 (C6). Similarly, the majority of the Russian wheat aphid genes were identified from *Triticum aestivum* genotypes (PI137739, PI262660, PI 294994, PI 372129, PI 243781). *Aegilops tauschii* accessions have been identified as excellent sources of resistance for Greenbug, Russian wheat aphid and Hessian fly [26–28]. Rye (*Secale cereale*) has been reported as the source of *H25* for Hessian fly, *Gb2* and *Gb6* for Greenbug while *Aegilops triuncialis* has been reported as the source of H30 gene of Hessian resistance.

Because of the co-evolution between wheat and insects, stacking of major R genes is very important for the development of durable resistance. This is mainly feasible for Hessian fly, Greenbug, Russian wheat aphid and the Wheat curl mite since there are R genes clustered around the same chromosome intervals. For example, for Hessian fly, there are 15 genes reported on the short arm of chromosome 1A (*H3, H5, H6, H9, H10, H11, H12, H14, H15, H16, H17, H19, H28, H29* and *Hdic*); three genes on the long arm of chromosome 3D (*H24, H26, H32*) and three genes on the short arm of chromosome 6D (*H13, H23, H_{WGRC4}*). Similarly, for Russian Wheat Aphid six *Dn* genes are clustered on the short arm of chromosome 7D (*Dn1, Dn2, Dn5, Dn6, Dn8* and *Dnx*) (Dweikat et al. 1997) and for Greenbug resistance 8 *Gb* genes are clustered on the long arm of chromosome 7D (*Gbx1, Gba, Gbb, Gbc, Gbd, Gbz, Gb3* and *Gbx2*) (Zhu et al. 2005). Some of these genes such as H9 and

H10; H26 and H32; *Dn1*, *Dn2*, *Dn5*, *Dn6* and *Dnx*; *Gbz* and *Gb3* are tightly linked and hence they can be easily introgressed simultaneously during the gene pyramiding process.

The resistance sources from wheat relatives and land races do not have all the desired traits to be a variety by themselves. They can only serve as gene sources for traits of interest such as insect resistance, drought and heat tolerance, disease resistance, etc. Introgression of such genes from wild relatives into common wheat is very difficult and requires efficient introgression techniques and approaches (for more details see Chap. 18). Though there are successful natural gene introgressions as exemplified by wheat–rye translocations of 1BL.1RS and 1AL.1RS, which arose spontaneously from centromeric breakage and reunion, gene introgression/transfer in pre-breeding programs can be carried out using gene transfer through hybridization and chromosome- mediated gene transfer approaches or through direct gene transfer using molecular approaches. The most successful and highly used gene introgression techniques is the development of primary synthetic wheats (2n =6x = 42, AABBDD) which is an amphiploidy developed by crossing the *T. turgidum* spp. *durum* (2n = 4x = 28, BBAA) with *Ae. tauschii* (2n = 2x = 14, DD) and chromosome doubling of the F1 through colchicine treatment [29]. The primary synthetic wheats have served as a bridge to transfer important genes such as resistance to Hessian fly, aphids, Sunn pest and many other important genes for resistance to abiotic and biotic stresses [22–24, 30, 31]. Recently, screening of synthetic wheats for resistance to HF and Sunn pest has resulted in the identification three synthetic hexaploid wheat lines possessing resistance to both Moroccan Hessian fly biotype and Syrian Sunn pest [32].

20.6 Breeding for Insect Resistance

The main objective of any breeding programme is to develop high yielding, better quality and adapted varieties with resistance to the major abiotic and biotic stresses prevailing in the target region. Breeding for insect resistance should be carried out in combination with other important traits targeting the regions where the insect pest is economically important. The wheat programs at CIMMYT and ICARDA undertake intensive characterization of parents for different traits such as yield potential, disease (root and foliar) resistance, heat and drought tolerance, insect resistance (Hessian fly, Sunn pest and aphids) and better nutritional quality. Once the progenitors are characterized the breeding programs assemble crossing blocks targeting wheat growing regions in developing economies. High yielding and adapted hall mark wheat cultivars representing the major-agro-ecologies, synthetic derived hexaploid wheats, and elite lines are included in the different crossing blocks. Simple, three-way and back crosses are carried out commonly with the application of diagnostic markers for gene pyramiding in the F2, F1top, and BC1 F1 populations [19, 33]. Selection of the segregating generation for different traits from F2 to F4 is carried out using the selected bulk or modified pedigree selection schemes as indicated (Fig. 20.3).

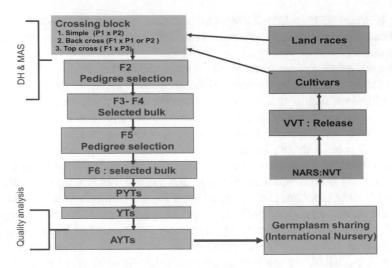

Fig. 20.3 Germplasm development and distribution scheme for Hessian Fly and Sunn Pest resistance at ICARDA; *P* parent, *F* Filial generation, *DH* Doubled Haploids, *MAS* Marker Assisted Selection, *PYTs* Preliminary Yield Trials, *YTs* Yield Trials, *AYTs* Advanced Yield Trials, *NVT* National Variety Trial, *VVT* Variety Verification Trial

In addition to the scheme indicated in Fig. 20.3, ICARDA has developed a modified speed breeding for elite x elite crosses whereby we manage crosses and F1s in the greenhouse and segregating generations and head-rows in the field at Merchouch in Morocco using summer x winter shuttle approach. Elite genotypes at F7 stage are evaluated in hotspot locations at Jemmaa Shaim in Morocco for Hessian fly and at Terbol station in Lebanon for Sunn pest resistance following the screening techniques indicated earlier in this chapter. The elite genotypes are also evaluated across key locations for yellow and stem rusts resistance at Kulumsa (Ethiopia) and Izmir (Turkey), for heat tolerance at Wadmedani (Sudan) and for root diseases, drought tolerance at Merchouch and Sid Al Aydi stations in Morocco. Elite genotypes with high yield potential, yellow rust resistance, drought and heat tolerance with 100% resistance to the Moroccan Hessian fly biotype have been identified and distributed to national programs in the CWANA region through ICARDA's international nursery distribution system for direct release and parentage purposes [19].

Similarly, breeding programs in the USA have developed resistant varieties for the major insects such as Hessian fly, Russian wheat aphid, Greenbug, and Wheat stem saw fly. More than 60 Hessian fly resistant wheat varieties have been released in the USA between 1950 and 1983 and less than 1% yield loss have been reported in areas where resistant cultivars have been deployed [34]. Resistance conferred by the *Sm1* gene has revolutionized management of Wheat midge [35]. Discovered by Canadian researchers in 1996, *Sm1* is now deployed in many parts of the world. Larvae die without causing damage to developing seeds. To ensure long-term durability of *Sm1*, the Canadian Wheat Board took the unusual step of requiring *Sm1* be deployed in a 90:10% mixture of resistant to susceptible seeds.

20.7   Summary

Genetic diversity for resistance to biotic and abiotic stresses is the backbone for the success of any breeding program. Studies to-date have shown the availability of enough genetic diversity in the wheat genetic resources (land races, wild relatives, cultivars, etc.) for resistance to the most economically important insect pests such as Hessian fly, Russian wheat aphid, Greenbug and Sunn pest. Many R genes – including 37 genes for Hessian fly, 11 genes for Russian wheat aphid and 15 genes for Greenbug – have been identified from these genetic resources. Some of these genes have been deployed singly or in combination in the breeding programs to develop high yielding varieties with resistance to insects. Gene pyramiding using marker assisted selection is important to stack two or more R genes in an adapted cultivar in order to increase the durability of insect resistance. Breeding for tolerance traits would exert less selection pressure on insect pests to evolve the ability to overcome the deployed trait. It is also important to develop and deploy resistant varieties in a given agro-ecology instead of using a given variety across a large mega-environment along with integrated pest management options in order to slow down the development and spread of virulent biotypes of the insect pests.

20.8   Review Questions

1. Describe the most important insect pests of wheat.
2. What are the mechanisms of insect resistance?
3. Explain the most common sources for insect resistance and the most efficient and widely used strategies for gene introgression from wild relatives of wheat.
4. Explain the breeding methods to develop and deploy high yielding varieties with Hessian fly resistance.

20.9   Key Concepts

Identification, development and deployment of insect resistant wheat varieties in integrated pest management scheme (IPM) is the most economical, socially feasible and environment friendly approach.

20.10   Conclusions

Wheat genetic resources are reservoirs for different genes including for resistance to insects. Identification and introgression of these insect resistant genes into adapted cultivars using both classical and molecular approaches is key for successful development of high yielding and widely adapted wheat varieties with resistance to major insect pests.

References

1. Hunter MC, Smith RG, Schipanski ME, Atwood LW, Mortensen DA (2017) Agriculture in 2050: recalibrating targets for sustainable intensification. Bioscience 67:386–391. https://doi.org/10.1093/biosci/bix010
2. Dhaliwal GS, Jindal V, Dhawan AK (2010) Insect pest problems and crop losses: changing trends. Indian J Ecol 37:1–7
3. Insects H of small grain (2007) Hessian fly. In: Buntin G, Pike K, Weiss M, Webster J (eds) Ratcliffe, RH. Entomological Society of America, p 120
4. Bouhssini ME, Street K, Joubi A, Ibrahim Z, Rihawi F (2009) Sources of wheat resistance to Sunn pest, Eurygaster integriceps Puton, in Syria. Genet Resour Crop Evol 56:1065. https://doi.org/10.1007/s10722-009-9427-1
5. Miller R, El-Bouhssini M, Lahloui M (2007) Insect pest of small grains outside of North America. In: Buntin G, Pike K, Weiss M, Webster J (eds) Handbook of small grain insects. Entomological Society of America, p 120
6. Darkoh C, El-Bouhssini M, Baum M, Clack B (2010) Characterization of a prolyl endoprotease from Eurygaster integriceps Puton (Sunn pest) infested wheat. Arch Insect Biochem Physiol 74:163–178. https://doi.org/10.1002/arch.20370
7. Kher SV, Dosdall LM, Cárcamo HA (2011) The cereal leaf beetle: biology, distribution and prospects for control. Prairie Soils Crop 4:32–41
8. Morrill W, Weiss M (2007) Wheat stem sawfly. In: Buntin G, Pike K, Weiss M, Webster J (eds) Handbook of small grain insects. Entomological Society of America, p 120
9. Parker B, El-Bouhssini M, Skinner M (2001) Field guide: insect pests of wheat and barley in North Africa, West and Central Asia. International Center for Agricultural Research in the Dry Areas, Aleppo
10. Peairs F (2007) Russian wheat aphid. In: Buntin G, Pike K, Weiss M, Webster J (eds) Handbook of small grain insects. Entomological Society of America, p 120
11. Stoetzel MB (1987) Information on and identification of Diuraphis noxia (Homoptera: Aphididae) and other aphid species colonizing leaves of wheat and barley in the United States. J Econ Entomol 80:696–704
12. Voss TS, Kieckhefer RW, Fuller BW, McLeod MJ, Beck DA (1997) Yield losses in maturing spring wheat caused by cereal aphids (Homoptera: Aphididae) under laboratory conditions. J Econ Entomol 90:1346–1350. https://doi.org/10.1093/jee/90.5.1346
13. Simon JC, Blackman RL, Le Gallic JF (1991) Local variability in the life cycle of the bird cherry-oat aphid, Rhopalosiphum padi (Homoptera: Aphididae) in western France. Bull Entomol Res 81:315–322. https://doi.org/10.1017/S0007485300033599
14. Barnes HF (1956) Gall midges of economic importance, vol VII: Cereal crops. Crosby Lockwood & Son, London
15. Lamb RJ, Wise IL, Olfert OO, Gavloski J, Barker PS (1999) Distribution and seasonal abundance of Sitodiplosis mosellana ITODIPLOSIS (Diptera: Cecidomyiidae) in spring wheat. Can Entomol 131:387–397. https://doi.org/10.4039/Ent131387-3
16. Olfert OO, Mukerji MK, Doane JF (1985) Relationship between infestation levels and yield caused by wheat midge, Sitodiplosis Mosellana (Géhin) (Diptera: Cecidomyiidae), in spring wheat in Saskatchewan. Can Entomol 117:593–598. https://doi.org/10.4039/Ent117593-5
17. Oakley JN, Talbot G, Dyer C, Self MM, Freer JBS, Angus WJ, Barett JM, Feuerhelm G, Snape J, Sayers L, Bruce TJA, Smart LE, Wadhams LJ (2005) Integrated control of wheat blossom midge: variety choice, use of pheromone traps and treatment thresholds. Project Report No. 363
18. Painter H (1951) Insect resistance in crop plants. Macmillan Education, New York
19. Tadesse W, Sanchez-Garcia M, Tawkaz S, El-Hanafi S, Skaf P, El-Baouchi A, Eddakir K, El-Shamaa K, Thabet S, Gizaw Assefa S, Baum M (2019) Wheat breeding handbook at ICARDA. ICARDA

20. Anglin NL, Amri A, Kehel Z, Ellis D (2018) A case of need: linking traits to Genebank accessions. Biopreserv Biobank 16:337–349. https://doi.org/10.1089/bio.2018.0033

21. Brown AHD (1989) The case for core collections. In: Brown AHD, Frankel OHD, Marshall DR, Williams JT (eds) The use of plant genetic resources. Cambridge University Press, pp 136–156

22. El Bouhssini M, Street K, Amri A, Mackay M, Ogbonnaya FC, Omran A, Abdalla O, Baum M, Dabbous A, Rihawi F (2011) Sources of resistance in bread wheat to Russian wheat aphid (Diuraphis noxia) in Syria identified using the Focused Identification of Germplasm Strategy (FIGS). Plant Breed 130:96–97. https://doi.org/10.1111/j.1439-0523.2010.01814.x

23. Crespo-Herrera LA, Smith CM, Singh RP, Åhman I (2013) Resistance to multiple cereal aphids in wheat-alien substitution and translocation lines. Arthropod Plant Interact 7. https://doi.org/10.1007/s11829-013-9267-y

24. Crespo-Herrera L, Singh RP, Reynolds M, Huerta-Espino J (2019) Genetics of greenbug resistance in synthetic hexaploid wheat derived germplasm. Front Plant Sci 10:782. https://doi.org/10.3389/fpls.2019.00782

25. Harris MO, Friesen TL, Xu SS, Chen MS, Giron D, Stuart JJ (2015) Pivoting from Arabidopsis to wheat to understand how agricultural plants integrate responses to biotic stress. J Exp Bot 66:513–531. https://doi.org/10.1093/jxb/eru465

26. Dweikat I, Ohm H, Patterson F, Cambron S (1997) Identification of RAPD markers for 11 Hessian fly resistance genes in wheat. Theor Appl Genet 94:419–423

27. Joukhadar R, El-Bouhssini M, Jighly A, Ogbonnaya FC (2013) Genome-wide association mapping for five major pest resistances in wheat. Mol Breed 32:943–960. https://doi.org/10.1007/s11032-013-9924-y

28. Zhu LC, Smith CM, Fritz A, Boyko E, Voothuluru P, Gill BS (2005) Inheritance and molecular mapping of new greenbug resistance genes in wheat germplasms derived from Acgilops tauschii. Theor Appl Genet 111:831–837. https://doi.org/10.1007/s00122-005-0003-6

29. Mujeeb-Kazi A, Hettel GP (1995) Utilizing wild grass biodiversity in wheat improvement: 15 years of wide cross research at CIMMYT. CIMMYT Res. CIMMYT, Mexico

30. Ogbonnaya FC, Abdalla O, Mujeeb-Kazi A, Kazi AG, Xu SS, Gosman N, Lagudah ES, Bonnett D, Sorrells ME, Tsujimoto H (2013) Synthetic hexaploids: harnessing species of the primary gene pool for wheat improvement. In: Plant breeding reviews. Wiley, pp 35–122

31. El Bouhssini M, Ogbonnaya FC, Chen M, Lhaloui S, Rihawi F, Dabbous A (2013) Sources of resistance in primary synthetic hexaploid wheat (Triticum aestivum L.) to insect pests: Hessian fly, Russian wheat aphid and Sunn pest in the fertile crescent. Genet Resour Crop Evol 60:621–627. https://doi.org/10.1007/s10722-012-9861-3

32. Sabraoui A, Emebiri L, Tadesse W, Ogbonnaya FC, El Fakhouri K, El Bouhssini M First combined resistance to Hessian fly and Sunn pest identified in synthetic hexaploid wheat. J Appl Entomol n/a. https://doi.org/10.1111/jen.12903

33. Tadesse W, Nachit M, Abdalla O, Rajaram S (2016) Wheat breeding at ICARDA: achievements and prospects in the CWANA region. In: Bonjean A, Angus B, van Ginkel M (eds) The world wheat book, vol 3: A history of wheat breeding. Lavoiseier, Paris

34. Berzonsky WA, Ding H, Haley SD, Harris MO, Lamb RJ, McKenzie RIH, Ohm HW, Patterson FL, Peairs FB, Porter DR, Ratcliffe RH, Shanower TG (2003) Breeding wheat for resistance to insects. Plant Breed Rev 22:221–296

35. Harris MO, Anderson K, El Bouhssini M, Peairs F, Hein G, Xu S (2017) Wheat pests: insects, mites, and prospects for the future. In: Achieving sustainable cultivation of wheat, vol 1: Breeding, quality traits, pests and diseases. Burleigh Dodds, Cambridge, pp 467–544

Chapter 21
Yield Potential

M. John Foulkes, Gemma Molero, Simon Griffiths, Gustavo A. Slafer, and Matthew P. Reynolds

Abstract This chapter provides an analysis of the processes determining the yield potential of wheat crops. The structure and function of the wheat crop will be presented and the influence of the environment and genetics on crop growth and development will be examined. Plant breeding strategies for raising yield potential will be described, with particular emphasis on factors controlling photosynthetic capacity and grain sink strength.

Keywords Yield potential · Grain sink strength · Radiation-use efficiency · Trait-based breeding

21.1 Learning Objectives

- Identify the developmental stages and underlying processes that limit yield potential in modern wheats

M. J. Foulkes (✉)
School of Biosciences, University of Nottingham, Leicestershire, UK
e-mail: john.foulkes@nottingham.ac.uk

G. Molero
KWS Momont SAS, Mons-en-Pévèle, France
e-mail: gemma.molero@kws.com

S. Griffiths
The John Innes Centre, Norwich, UK
e-mail: simon.griffiths@jic.ac.uk

G. A. Slafer
ICREA (Catalonian Institution for Research and Advanced Studies), AGROTECNIO (Center for Research in Agrotechnology), University of Lleida, Lleida, Spain
e-mail: gustavo.slafer@udl.cat

M. P. Reynolds
International Maize and Wheat Improvement Center (CIMMYT), Texcoco, Mexico
e-mail: m.reynolds@cgiar.org

M. P. Reynolds, H.-J. Braun (eds.), *Wheat Improvement*,
https://doi.org/10.1007/978-3-030-90673-3_21

- Understand the reasons for yield variation between modern wheat genotypes according to the expression of traits determining source and sink strength
- Suggest pre-breeding crossing strategies to optimise the source-sink dynamic and increase yield potential

21.2 Rationale for Raising Yield Potential

Wheat (*Triticum aestivum* L.) is globally grown on more than 220 million hectares of land with a global average yield of 3.43 t ha^{-1} determining a current global annual production of c. 750 Mt. ([1]; Fig. 21.1). Wheat is the most widely grown crop and contributes c. 20% of calories and proteins to human beings [2]. The current level of production was achieved over a period with a stable global area over the last 25 years, and therefore the critical increase in production was due to the yield per unit area (Fig. 21.1). At least 30–50% of the critical increase in yield observed was due to the improved yield potential through breeding; and, due to environmental and economic reasons, future growth in production will depend more on improving yield potential through breeding than in the past [2].

Crop yield potential (YP) is defined as the maximum attainable yield per unit land area that can be achieved by a particular crop cultivar in an environment to which it is adapted when pests and diseases are effectively controlled and nutrients and water are non-limiting. Attainable yield (AY) may be defined as the yield a skilful farmer should reach when taking judicious account of economics and risk, i.e. it would be close to YP under irrigated conditions and to water-limited YP in rainfed conditions. The exploitable yield gap (i.e. gap between farm yield and attainable yield) has been estimated at 30% for winter wheat in the UK and 50% for spring wheat in Mexico [2]. Given these yield gaps, at first sight it may not appear cost effective to invest in increasing genetic yield potential. However, the

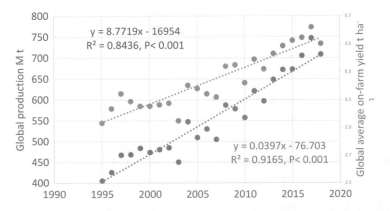

Fig. 21.1 Global production and average yield for bread wheat from 1995 to 2018. (Prepared with data from [1])

implementation of improved agronomic practices is much less straightforward – both practically and economically – for farmers than changing cultivars. Furthermore, increasing yield potential is the only avenue to improve productivity where growers have fully closed the exploitable gap. Strong precedents for yield improvement through wheat breeding started with and have extend well beyond the Green Revolution (e.g. [3]). The fast-growing fields of both genetics with the availability of the wheat genome sequence and high-throughput field phenotyping platforms (Chap. 27) offer considerable promise for more efficient screening of genetic resources, parental characterization and progeny selection to accelerate breeding progress. The existence of well-established national and international crop improvement networks, such as those coordinated by CIMMYT, will enable new genotypes to be rapidly and extensively tested in and delivered to representative target regions.

In addition, an important outcome of breeding for yield potential is higher attainable yields under moderate abiotic stresses. Selection for greater yield potential has frequently resulted in higher production in environments subject to abiotic stress (usually water and heat) in wheat.

In the following sections of the present chapter the physiological traits associated with current rates of yield gains are examined and then the major breeding challenges for raising future yield potential are considered. For concision in these following sections we will show and discuss the most common descriptions of physiological traits summarizing where necessary conflicting results that naturally can be always found in the literature.

21.3 Current Rates of Progress in Yield Potential and Associated Traits

The current annual rate of genetic gain in wheat yield potential from datasets reported globally averages 0.6% (0.3%–1.1%) [2]. Annual genetic gains for grain yield of wheat in CIMMYT international Elite Spring Wheat Yield Trials were 0.5% in optimally irrigated environments [4]. However, in different regions with relevance for global wheat production there seems to have been no genetic gains in yield over the last few decades (e.g. [5]). Moreover, the rate of yield gains required to meet predicted global demand for wheat in 2050 at *ca.* 1.3% per annum is higher than the present rates of genetic gains [6], even those in the regions where gains are still apparent. The levelling off of yield in some countries and regions may occur because: (i) farmers cannot achieve the crop and soil management required to reach attainable yield and/or (ii) crop response to additional inputs exhibits a diminishing marginal yield benefit as yield approaches the ceiling; and/or (iii) genetic progress has been counteracted by climate change (particularly by heat stress).

During the Green Revolution in the 1960s and 1970s, yield progress was associated with gains in harvest index (grain dry matter as a proportion of the above-ground dry matter; HI) due to the introduction of semi-dwarf (*Rht:* Reduced Height) genes. Field studies on sets of historic cultivars show grain yield progress in recent decades has been associated with greater above-ground biomass in the UK [7], Australia [8], China [9] and NW Mexico [3]. Yield progress was also associated with continued progress in HI in China [9] and Argentina [10]. Overall, this evidence indicates that a simultaneous increase of photosynthetic capacity and grain partitioning in modern wheat cultivars is a crucial task for wheat breeders for future gains in yield potential.

21.4 Opportunities for Future Gains in Yield Potential

Wheat crops harvest light – they convert solar energy, carbon dioxide and water into biomass. Water is required in proportion to the energy captured. Under light-limited conditions wheat yield potential depends on the following (Eq. 21.1):

$$Yield\left(g\,m^{-2}\right) = Incident\ radiation\left(MJ\,m^{-2}\right) \times Radiation\ capture\left(\%\,/\,100\right)$$
$$\times Radiation-use\,efficiency\left(g\,MJ^{-1}\right) \times Harvest\ Index \tag{21.1}$$

The physiological processes determining radiation capture and conversion and grain dry matter partitioning, as well as water and nutrient capture, are summarized in Fig. 21.2. These processes are discussed further below.

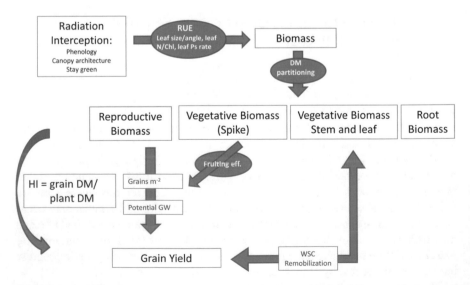

Fig. 21.2 Schematic diagram of physiological processes determining yield potential in wheat (under light limited conditions)

21.4.1 Optimize Root Traits

Breeding for enhanced biomass may be partly dependent on breeding for deeper or denser root systems to increase soil water and nutrient uptake in the absence of improvements in above-ground water and nutrient-utilization efficiencies, particularly in rainfed environments. Nevertheless, breeding for root characteristics has been seldom implemented to date, principally because of the difficulties of screening root phenotypes directly. Lower canopy temperatures might be taken as an indirect indication of a greater root water uptake capacity. Genetic variation in root system size has been widely reported in wheat. There is some evidence that root size of wheat landraces is larger compared with that of modern cultivars [11]. There is also evidence that derivatives of primary synthetic spring wheats (resynthesized hexaploid wheat lines by crossing modern durum wheat (*Triticum turgidum* L.), donor of the AB genome, with the wild progenitor goat grass (*Aegilops triuncialis* L.), donor of the D genome), have greater root biomass compared to bread wheat recurrent parents [12]. Future genetic progress could potentially be accelerated by the development of markers for marker-assisted selection. To develop such markers there is a need for a high precision root phenotyping because the genetic differences may be small, and detailed root physiological measurements are difficult when large numbers of genotypes are involved. A detailed summary of how today's non-invasive phenotyping technologies that measure roots can be strategically combined to speed up germplasm enhancement of roots is beyond the scope of this chapter; however, fortunately comprehensive recent reviews are available, e.g. Watt et al. [13].

21.4.2 Optimize Phenology

Crop phenology must be conducive firstly to avoid catastrophic climatic effects on productivity (frost immediately before anthesis, severe heat during grain filling). This is known as "adaptation" through modifying the duration to anthesis to avoid such extreme events. Secondly, improvements in crop phenology could also contribute to spike fertility as well as being tailored to different photoperiod and temperature regimes. Physiologically, the following stages are usually distinguished: plant emergence, tillering, terminal spikelet (mostly coinciding with onset of stem elongation in field conditions), initiation of booting, spike emergence, anthesis and maturity. These stages may be grouped into: emergence to onset of stem extension (1); onset of stem extension to initiation of booting (2); initiation of booting to anthesis (3); and anthesis to maturity (4). The time-span of each development phase essentially depends on temperature, day-length and genotype (as affected by sowing date) and genetic sensitivity to these two environmental factors.

Once the terminal spikelet is formed, stem elongation starts and slightly later the spike begins to grow. Floret initiation occurs during this phase from the onset of stem elongation to booting and determines maximum number of floret primordia.

This process is not responsive to spike growth (as the metabolic cost of initiating floret primordia is extremely low, the process may be largely independent of availability of resources); and the maximum number of floret primordia does not correlate with the final numbers of fertile florets and grains. Spike growth, slow in its early stages, increases greatly about the time of booting. Floret abortion starts in the booting stage due to competition for carbohydrates during this phase and finishes at anthesis. It has been shown that lengthening the duration of the stem-elongation phase improves grain number through allowing a larger biomass accumulation during this critical phase and consequently increasing assimilate supply to the juvenile spike determining the proportion of floret primordia as competent florets at anthesis [14].

The dynamics of tillering and tiller mortality in wheat are also strongly linked to the timing of developmental stages. The timing of tiller emergence is linked to leaf appearance. When plants experience an increase in shading of lower tiller buds in the canopy changing the red – far red ratio of light coinciding with onset of stem extension, tillering ceases. Under field conditions tiller mortality starts coinciding with the onset of stem elongation; as stems start to be dominant sinks reducing the availability of assimilates to late-formed tillers. Mortality of tillers stops at anthesis, stabilising the number of tillers that will reach maturity. Large genetic variation has been identified in the potential amount of dry matter wasted by non-surviving shoots that could potentially be exploited to minimise their detrimental effects on spike DM partitioning and increase grain number [15].

21.4.3   Increase Radiation-Use Efficiency

Radiation-use efficiency (RUE), defined as the solar energy conversion into above-ground biomass, is a major bottleneck to improve grain yield potential in breeding. It is expected that future genetic gains in wheat yield will rely on improved biomass production [2] whilst achieving a stable expression of HI at values of 0.50 and above; and modest increases in biomass have been reported in recent years [3, 7, 9]. Photosynthesis is the primary determinant of plant biomass with more than 90% of biomass derived directly from photosynthetic products. Compelling evidence that increasing photosynthesis does increase yield, considering that other constraints do not become limiting, comes from the 30 years of free-air carbon dioxide enrichment (FACE) experiments.

RUE together with light interception, both components that determine biomass, are the most integrative estimates of photosynthesis and can be used directly to boost yield through their combination with positive expression of sink traits such as harvest index. Molero et al. [16] proposed the use of exotic material (landrace and synthetic derivative lines) as a valuable resource to increase RUE among other traits.

21.4.3.1 Case-Study 1: Genetic Variation in RUE Was Characterized in a Modern Panel of Spring Wheat. Results Indicated Significant Underutilized Photosynthetic Capacity in Existing Wheat Germplasm

Unpublished data on RUE evaluated at different growth stages in CIMMYT spring wheat cultivars released from 1966 until 2014 shows genetic gains in RUE during grain filling while a negative trend was observed for RUE evaluated pre-grain filling during the critical phase when grain number is determined (Fig. 21.3). These findings, together with the genetic variation observed for RUE expressed at different growth stages [16], strongly support the case for significantly underutilized photosynthetic capacity in existing wheat germplasm and that gains in grain yield may come from increasing RUE particularly in the pre-anthesis period to increase grain number.

However, as part of a translational research approach, stacking of different traits that significantly boost genetic gains needs to be combined in a common platform. For example, as alternative strategies to increase RUE, recent studies propose to exploit natural existing variation in elite material for spike [17], leaf lamina [18] and leaf sheath [19] photosynthesis, pigment composition [20] and carboxylation capacity of Rubisco [21], among others. Prins et al. [22] recently demonstrated the potential benefit of replacing Rubisco of *T. aestivum* with Rubisco from *Hordeum vulgare* or the wild *Aegilops cylindrica*, in terms of achieving higher assimilation rates. McAusland et al. [18] identified a wide variation for flag-leaf photosynthesis rate that was accession and not species dependent.

In parallel with these "steady-state" approaches, recent interest in evaluating dynamic responses of photosynthesis in a fluctuating light environment identified photosynthesis induction as a critical trait for improving productivity in rice [23]. Taylor and Long [24] proposed that slow photosynthesis induction rates in wheat

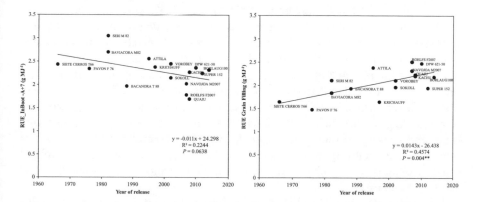

Fig. 21.3 Radiation-use Efficiency (RUE) measured from initiation of booting until 7 days after anthesis (RUE_InBoot-A+7) and from 7 days after anthesis until physiological maturity (RUE_GF) versus year of release from 16 varieties evaluated during 2015–2016 and 2016–2017 under yield potential conditions in NW Mexico. (Unpublished data from G. Molero)

could reduce daily carbon accumulation by up to 20% for a single leaf. In addition, genetic variation for photosynthetic induction has been recently identified in wheat [25]. However, the link between rapid induction and increased RUE or yield is yet to be demonstrated. Field phenotyping methods for RUE and related traits in wheat have been recently reviewed by Reynolds et al. [26]. The development of new methodologies based on remote sensing techniques will be crucial in selecting lines with high RUE together with other photosynthetic-related traits to accelerate genetic gains (Chap. 27).

21.4.4 Increase Spike Partitioning and Fruiting Efficiency

Grain yield improvement is highly associated with grain number per unit area in wheat [7, 27]. Current evidence suggests grain sink strength remains a critical yield-limiting factor and that improving the balance between source and sink is critical for further raising yield potential [2]. Grain growth of modern wheat cultivars is in general little limited by the source during grain filling [27], although co-limitation by source may occur in some cases [3, 7].

The period of stem elongation is critical for yield determination when grains per unit area is determined. Grain number is far more responsive to crop growth during this phase than the preceding phase from emergence to onset of stem extension [2]. During the stem-elongation phase, stem and spike growth overlaps affecting assimilate supply to the spike hence floret survival and grain number (e.g. Rivera-Amado et al. [28]). Since stem and spike growth mainly overlaps during the rapid spike growth phase from booting to anthesis, the extent of competition between the spike and stem differs between stem internodes. A recent investigation on CIMMYT spring wheat elite lines showed decreased DM partitioning to stem internodes 2 (top down, peduncle −1) and 3 was most effective in enhancing spike dry matter partitioning, spike growth and grain number per unit area [28].

The fruiting efficiency (FE; number of grains set per unit of spike dry weight at anthesis) is a key trait which reflects the efficiency with which resources allocated to the growing juvenile spike are used to set grains. The fruiting efficiency subsumes the dynamics of floret production, floret survival and grain abortion which determines the grain number. There is clear genetic variability in FE among modern wheat cultivars which is well correlated with grains per unit area with genetic loci identified (e.g. Gerard et al. [29]). Improvements in FE could be associated with better intra-spike partitioning, for example, by reduced partitioning to the rachis or awns [28]. Alternatively genetic variation in FE may be influenced by levels of spike cytokinins, which play a key role in the stimulation of cell division, from booting to anthesis (see Sect. 21.5).

It is important that higher FE should not be achieved at the expense of having smaller florets with smaller potential grain weight (see Sect. 21.4.5). However, evidence suggests FE can be improved independently of effects on ovary size [30] likely through an improved partitioning of DM within the spike. In addition, it has been suggested that anatomical structure of the vascular system within the spikelet could be modified to increase FE by favouring translocation of assimilate to the distal floral primordia within a spikelet [31]. The florets closer to the rachis node are directly supplied by the principal vascular bundles of the rachilla, while the distal florets lack a direct connection to the vascular bundle and therefore might not have an equal chance of accessing assimilates from the source [31]. Fruiting efficiency should be amenable for breeding as it is heritable and responds to selection. Another avenue for increasing FE may be improving the loading of sucrose in the phloem in the vascular system for more efficient moving of photo-assimilates from source to sink tissues to enhance grain number [27].

21.4.4.1 Case-Study 2: Genetic Variation in Spike Partitioning Index (SPI) and FE and Related Traits in a Modern Spring Wheat Panel Was Characterized by Rivera-Amado et al. [28]. Variation Was Highly Correlated with HI

The genetic variation in novel grain partitioning traits was characterized in a panel of 26 CIMMYT spring wheat cultivars: stem internode 2 and 3 dry matter partitioning at anthesis was correlated with spike dry matter partitioning index (SPI) and rachis specific weight was correlated with FE [28] (Table 21.1). These results indicated that there is sufficient variation within modern CIMMT spring wheat cultivars for these traits alone to achieve a step-change in HI in CIMMYT spring wheat to 0.60 by combining within a novel plant ideotype the largest expression of target traits for grain partitioning.

Table 21.1 Grain partitioning traits (mean of 26 genotypes of CIMMYT CIMCOG spring wheat panel and value for best genotype) and relevant correlations with spike partitioning index, FE and HI. Values represent means 2011–2012 and 2012–2013

Trait	Spike partitioning index	Fruiting efficiency	Stem internode 2 + 3 partitioning index	Rachis specific weight	Florets per spikelet
Units	Unitless	grains g^{-1}	Unitless	g cm^{-1}	Florets spklt^{-1}
Mean expression	0.236	85.51	0.165	13.6	2.40
Best expression	0.266	123.81	0.133	11.0	2.75
Relevant corr. (r)	0.37 *** (with HI)	0.36 *** (with HI)	0.61 *** (with SPI)	0.46 *** (with FE)	0.81 *** (with HI)

***$P < 0.001$

21.4.5   Increase Potential Grain Weight

Although average grain weight is frequently negatively related to grains per m^2, evidence indicates that in the vast majority of conditions wheat grains do not experience a shortage of assimilates to be filled. These assimilates include not only: (i) actual crop photosynthesis, which over the first half of grain filling is predominantly in excess of demands (as grains start growing slowly and the canopy photosynthetic capacity is at its maximum and exposed to increasing radiation levels), but also (ii) water soluble carbohydrates accumulated in stems and leaf sheaths before the onset of grain filling that can be remobilised to complement current photosynthesis. This lack of source limitation for grain growth is supported by evidence that: (i) grain weight does not respond (or responds only slightly) to severe manipulations of source strength (e.g. to defoliations) during the effective period of grain filling and (ii) sizeable amounts of water soluble carbohydrates often remain in the stem when measured at physiological maturity [19]. Thus, in most circumstances, grain filling is sink-limited; i.e. the capacity of the grains to grow largely determines their final weight. This explains why grain weight is much less plastic (and has higher heritability) than grain number.

Therefore, yield potential can be genetically increased by increasing post-anthesis sink-strength given by the number of grains set by the crop and their potential weight. Thus, genetic gains in yield potential would be also achieved through improving potential grain weight (i.e. the capacity of the grains to accumulate resources). As grain growth is largely sink-limited, the potential size of the grains would have been established before the actual growth: the storage capacity is firstly set and then that capacity is filled with dry matter. Indeed, the timing of determination of potential grain weight seems to comprise pre- and post-anthesis processes. As elegantly described recently by Calderini et al. [32], the capacity of the grains to grow is chiefly defined by the size of the carpels of the florets and by the number of endosperm cells.

The floret carpel will become the pericarp after grain set, thus likely setting an upper limit for grain weight realisation during the effective grain filling. Carpels grow for a short period (c. 7–15 days, depending on temperature) immediately before anthesis [32]. The relationship between the size of the carpels at anthesis and the final weight of the grains developed in them after pollination has been shown for a wide range of different genotypes and background environmental conditions (e.g. Reale et al. [33]). This is commensurate with the fact that grain weight has been related to the amount of pericarp dry matter [34].

Endosperm cells are the actual units where starch will be stored, thus their number may also limit the capacity of the grain to store dry matter. The association of grain weight with the number of endosperm cells, developed over the first c. 7–15 days (depending on temperature) immediately after anthesis, is well established. Indeed, reductions in grain weight potential due to the effect of heat were related to reductions in endosperm cell number (e.g. Kaur et al. [35]).

Thus, breeding for improved potential grain weight is a real alternative to grain number that can be exploited, if the increased potential grain weight is not linked to a reduced number of grains [14]. Indeed, genetic factors controlling potential grain weight, without representing a compensation due to reductions in grain number, have been identified; and transgenic lines over-expressing expansins (proteins relaxing cell walls) produced significant increases in yield of field-grown wheat through increasing potential grain size [32].

21.5 Plant Signalling Approaches to Increase Yield Potential

There is increasing evidence that variation in grain number is regulated by plant growth regulators during the rapid spike growth phase from booting to anthesis in wheat. Cytokinins play a key role in the stimulation of cell division and nucleic acid metabolism. Altering spike cytokinin concentration through expression level of two cytokinin oxidase genes has been shown to increase grain number in wheat [36]. Cytokinin levels are regulated by a balance between biosynthesis enzymes (e.g. isopentenyl pyrophosphate transferase) and degradation enzymes (e.g. cytokinin oxidase/dehydrogenase). The grain sink strength of the spike meristem could therefore be enhanced by altering cytokinin homeostasis through the upregulation or the downregulation of these enzymes, respectively, to coordinate growth and floret fertility.

In addition, it has been observed that excessive ethylene production results in wheat grain abortion under high temperature stress, suggesting that reduced grain accumulation of ethylene in wheat may be a desirable trait. A negative association was observed between spike dry weight at anthesis and ethylene production in a GWAS population at high temperatures in the field and genetic bases were indicated [37]. Stress ethylene production, for example under soil compaction or drought, can also induce grain abortion. High ethylene levels also inhibit grain-filling rates by restricting assimilate partitioning to developing grains resulting in low starch biosynthesis and high accumulation of soluble carbohydrates, ultimately decreasing grain yield. In addition, there is evidence that the ABA/ethylene ratio is positively related to grain filling rate by regulating starch synthesis [27]. Pinpointing the plant hormone signals underlying grain set/abortion and their genetic basis in wheat should therefore permit the development of genotypes with less conservative strategies for determination of grain number.

An alternative plant signalling avenue to increase grain sink strength may be to increase the concentration of trehalose-6-phosphate (T6P), a sugar signal that regulates growth and development, and increases starch synthesis in spikes. Genetic modification of trehalose-6-phosphate phosphatase and chemical intervention approaches have been used to modify the T6P pathway and improve crop performance under favourable conditions in the wheat [27].

21.6   Trait-Based Breeding for Yield Potential

Increased genetic yield potential is a key driver of both productivity and variety replacement. Some of the key traits have already been discussed in this chapter and it is important that crossing strategies achieve an effective balance among them. For example, increasing RUE alone does not guarantee increased yield unless additional assimilates result in more and/or larger grains. The fact that increased photosynthetic potential does not necessarily optimize yield is supported by the negative association observed between harvest index and biomass [3]. Therefore, to achieve full expression of yield potential, it is necessary to optimize the source:sink dynamic by ensuring that expression of grain set matches the photosynthetic potential of current and future genotypes (Fig. 21.3).

Evidence for genetic variation in source:sink balance (SSB) and its importance in boosting yield and radiation-use efficiency in field-grown plots has come from various sources. Experiments in wheat have shown that a high demand for assimilates —determined by sink strength of the grains – can stimulate the supply of photo-assimilates based on light treatments, as well as studies with cytogenetic stocks [38]. More recently, a cross designed to combine high sink strength in high RUE backgrounds resulted in doubled-haploid lines expressing exceptional yield and biomass in a high yielding environment in Southern Chile [30].

However, for novel approaches to be adopted, proofs-of-concept must be demonstrated in a breeding context. This necessarily involves translational research via pre-breeding that demonstrates genetic gains from new innovations across an appropriate range of target environments, and in lines that also contain the component agronomic traits essential to make new cultivars marketable. The pre-breeding steps include: (i) designing crosses to combine promising yield-boosting traits; (ii) identifying the best sources of those traits among diverse genetic resources using phenotypic and where available genomic data; (iii) validating new trait combinations through crossing and trialing the best new progeny; and (iv) sharing the new germplasm and breeding technologies with breeding programs for validation globally. Results of the CIMMYT Wheat Yield Collaboration Yield Trial (WYCYT) have shown significant increases in yield potential across international wheat targets in the selected progeny of crosses designed to combine favourable sources of source and sink traits. In summary, stacking "source" and "sink" related traits (Fig. 21.4) via strategic crossing seems to be a viable way to boost genetic yield gains while at the same time involving intuitively valuable traits for increasing for potential yield.

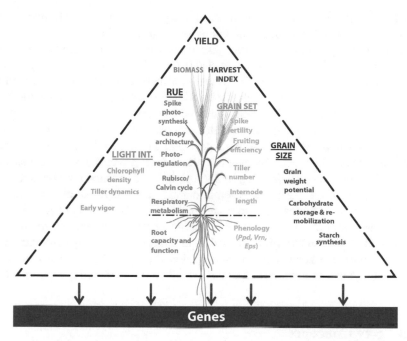

Fig. 21.4 Trait hierarchy in relation to approximate degree of integration, depicting some of the established drivers of biomass (source) on the left of the plant, and harvest index (sink) on the right side. Abbreviations: *Int* interception. (Reprinted with permission from [26])

21.7 Genetic Regulation of Grain Number and Yield Potential

The genetics of phenology in wheat are relatively well understood. The genes controlling winter/spring growth habit (*Vrn-1*) and photoperiod response (*Ppd-1*), which are responsible for coarse-tuning time to anthesis, are often completely fixed in breeders' gene pools targeting a specific environment. QTL with smaller effects on phenology are collectively recognized as earliness *per se* genes and also are critical for fine-tuning time to anthesis as well as for the duration of particular subphases composing time to anthesis. There is an increasing body of evidence for the role of these phenology genes in increasing grain number and yield. With regard to plant height, beyond GA-insensitive *Rht-B1b* and *Rht-D1b* genes that have been extensively deployed to increase yield potential, other dwarfing genes such as *Rht13* or *Rht18* have also been shown to increase grain yield, and others such as *Rht8* may also increase yield but just under particular environmental conditions. As plant height has been already optimized in most growing regions, further increases in the availability of assimilates for spike growth may require reductions limited to small specific stem internodes to favour spike growth as proposed by Rivera-Amado et al. [28]. Several studies have identified QTL which control height by disproportionate reduction in the length of specific internodes, e.g., Cui et al. [39]. However, these studies did not include the measurement of SPI or grain yield.

Grain yield improvement is highly associated with grain number per unit area in wheat as described in Sects. 21.4.4 and 21.4.5. Outside the major adaptive genes for phenology and plant height, the QTLs and trait marker associations described for grain number are generally of small effect and subject to strong environmental interactions. This results in the low heritability of this trait. Nevertheless, a few genes have been found to be robust and validated. For example, GNI-1A on chromosome 2AL encodes a homeodomain leucine zipper class I (HD-Zip I) transcription factor, the expression of which was highest in the distal floret primordia of the spikelet and in parts of the rachilla [40]. In tetraploid wheat, reduced function mutations resulted in increased grain set per spikelet, grain number, and yield. Another example is TaAPO-A1 which is the wheat orthologue of Aberrant Panicle Organization in rice [41] on chromosome 7A in wheat. A mutation in the F-box domain defines two common alleles in modern global bread wheat which are strongly associated with spikelet number. Further study and manipulation of these pathways provides targets for the deployment of induced and natural variation for increased grain number.

21.8 Key Concepts

Under light limited conditions wheat yield potential depends on the following:

$$Yield\left(g\,m^{-2}\right) = Incident\ radiation\left(MJ\,m^{-2}\right) \times Radiation\ capture\ (\%/100)$$
$$\times Radiation-use\ efficiency\left(g\,MJ^{-1}\right) \times Harvest\ Index \qquad (21.2)$$

Current evidence suggests grain sink strength remains the critical yield-limiting fact and that improving the balance between source and sink is critical for further raising yield potential. Thus, in most circumstances, grain filling is sink-limited; i.e. the capacity of the grains to grow largely determines their final weight. Therefore, yield potential can be genetically increased by increasing post-anthesis sink-strength given by the number of grains set by the crop and their potential weight. There is significantly underutilized photosynthetic capacity in existing wheat germplasm and gains in grain number could come from increasing pre-anthesis RUE. Alternatively, grain number can be increased through enhancing partitioning to spikes at anthesis through optimized phenology and/or favouring partitioning of assimilates to spikes at the expense of specific stem internodes. In addition, grain sink strength may be raised by increasing potential grain weight via increasing carpel weight at anthesis or endosperm cell number and/or size. Simultaneous increases in these source and sink traits are required to accelerate rates of genetic gain. Stacking "source" and "sink" related traits via strategic crossing in trait-based breeding is a crucial task to boost genetic yield gains while at the same time involving intuitively valuable traits for increasing for potential yield.

21.9 Summary

Crop yield potential is defined as the maximum attainable yield per unit land area that can be achieved by a particular crop cultivar in an environment to which it is adapted when pests and diseases are effectively controlled and nutrients and water are non-limiting. Under light limited conditions wheat yield potential depends on: Incident radiation (MJ m^{-2}) × Radiation Capture (%/100) × Radiation-use efficiency (g MJ^{-1}) × Harvest Index. Yield potential can be genetically increased by increasing post-anthesis sink-strength given by the number of grains set and their potential weight, and grain sink strength remains a critical yield-limiting factor. The period of stem elongation is critical for yield determination when grains per unit area is determined. There is scope to exploit natural existing variation in elite material for spike, leaf lamina and leaf sheath photosynthesis, pigment composition and carboxylation capacity of Rubisco to increase RUE during stem elongation and hence grain number. Furthermore, grain number may be increased by fine-tuning of the phenological phases using phenology genes to favour spike growth during stem elongation, or optimizing the trade-off between partitioning of assimilates to spikes versus stem internode growth. Complementary to these avenues for increasing grain number, fruiting efficiency can be increased through modifying spike hormone regulation or intra-spike partitioning to maximize grains set per unit spike weight. Finally, potential grain weight is an alternative trait to increase grain sink strength that can be exploited through increasing the carpel weight at anthesis or endosperm cell number and/or size. Achieving a simultaneous increase of photosynthetic capacity and grain partitioning in modern wheat cultivars is a crucial task for breeders. Stacking these "source" and "sink" related traits via strategic crossing in trait-based breeding is a viable way to boost genetic yield gains while at the same time involving intuitively valuable traits for increasing for potential yield.

References

1. FAOSTAT (2018) Crop production statistics
2. Fischer RA, Byerlee D, Edmeades GO (2014) Crop yields and global food security: will yield increase continue to feed the world? ACIAR Monograph No. 158. Australian Centre for International Agricultural Research, Canberra
3. Aisawi KAB, Reynolds MP, Singh RP, Foulkes MJ (2015) The physiological basis of the genetic progress in yield potential of CIMMYT spring wheat cultivars from 1966 to 2009. Crop Sci 55:1749–1764. https://doi.org/10.2135/cropsci2014.09.0601
4. Lala C-H, Crossa J, Huerta-Espino J, others (2017) Genetic yield gains in CIMMYT's international Elite Spring Wheat Yield Trials by modeling the genotype - environment interaction. Crop Sci 57:789–801
5. Maeoka RE, Sadras VO, Ciampitti IA, Diaz DR, Fritz AK, Lollato RP (2020) Changes in the phenotype of winter wheat varieties released between 1920 and 2016 in response to in-furrow fertilizer: biomass allocation, yield, and grain protein concentration. Front Plant Sci 10:1786. https://doi.org/10.3389/fpls.2019.01786

6. Hall AJ, Richards RA (2013) Prognosis for genetic improvement of yield potential and water-limited yield of major grain crops. Field Crop Res 143:18–33. https://doi.org/10.1016/j.fcr.2012.05.014

7. Shearman VJ, Sylvester-Bradley R, Scott RK, Foulkes MJ (2005) Physiological processes associated with wheat yield progress in the UK. Crop Sci 45:175–185. https://doi.org/10.2135/cropsci2005.0175

8. Sadras VO, Lawson C (2011) Genetic gain in yield and associated changes in phenotype, trait plasticity and competitive ability of South Australian wheat varieties released between 1958 and 2007. Crop Pasture Sci 62:533–549. https://doi.org/10.1071/CP11060

9. Xiao YG, Qian ZG, Wu K, Liu JJ, Xia XC, Ji WQ, He ZH (2012) Genetic gains in grain yield and physiological traits of winter wheat in Shandong Province, China, from 1969 to 2006. Crop Sci 52:44–56. https://doi.org/10.2135/cropsci2011.05.0246

10. Lo Valvo PJ, Miralles DJ, Serrago RA (2018) Genetic progress in argentine bread wheat varieties released between 1918 and 2011: changes in physiological and numerical yield components. Field Crop Res 221:314–321. https://doi.org/10.1016/j.fcr.2017.08.014

11. Waines JG, Ehdaie B (2007) Domestication and crop physiology: roots of green-revolution wheat. Ann Bot 100:991–998. https://doi.org/10.1093/aob/mcm180

12. Reynolds MP, Dreccer F, Trethowan R (2007) Drought-adaptive traits derived from wheat wild relatives and landraces. J Exp Bot 58:177–186. https://doi.org/10.1093/jxb/erl250

13. Watt M, Fiorani F, Usadel B, Rascher U, Muller O, Schurr U (2020) Phenotyping: new windows into the plant for breeders. Annu Rev Plant Biol 71:689–712. https://doi.org/10.1146/annurev-arplant-042916-041124

14. Slafer GA, Savin R, Sadras VO (2014) Coarse and fine regulation of wheat yield components in response to genotype and environment. Field Crop Res 157:71–83. https://doi.org/10.1016/j.fcr.2013.12.004

15. Berry PM, Spink JH, Foulkes MJ, Wade A (2003) Quantifying the contributions and losses of dry matter from non-surviving shoots in four cultivars of winter wheat. Field Crop Res 80:111–121. https://doi.org/10.1016/S0378-4290(02)00174-0

16. Molero G, Joynson R, Pinera-Chavez FJ, Gardiner L, Rivera-Amado C, Hall A, Reynolds MP (2019) Elucidating the genetic basis of biomass accumulation and radiation use efficiency in spring wheat and its role in yield potential. Plant Biotechnol J 17:1276–1288. https://doi.org/10.1111/pbi.13052

17. Molero G, Reynolds MP (2020) Spike photosynthesis measured at high throughput indicates genetic variation independent of flag leaf photosynthesis. Field Crop Res 255:107866. https://doi.org/10.1016/j.fcr.2020.107866

18. McAusland L, Vialet-Chabrand S, Jauregui I, Burridge A, Hubbart-Edwards S, Fryer MJ, King IP, King J, Pyke K, Edwards KJ, Carmo-Silva E, Lawson T, Murchie EH (2020) Variation in key leaf photosynthetic traits across wheat wild relatives is accession dependent not species dependent. New Phytol 228:1767–1780. https://doi.org/10.1111/nph.16832

19. Rivera-Amado C, Molero G, Trujillo-Negrellos E, Reynolds M, Foulkes J (2020) Estimating organ contribution to grain filling and potential for source upregulation in wheat cultivars with a contrasting source-sink balance. Agronomy 10:1–21. https://doi.org/10.3390/agronomy10101527

20. Joynson R, Molero G, Coombes B, Gardiner L-J, Rivera-Amado C, Piñera-Chávez FJ, Evans JR, Furbank RT, Reynolds MP, Hall A Uncovering candidate genes involved in photosynthetic capacity using unexplored genetic variation in Spring Wheat. Plant Biotechnol J n/a. https://doi.org/10.1111/pbi.13568

21. Silva-Pérez V, De Faveri J, Molero G, Deery DM, Condon AG, Reynolds MP, Evans JR, Furbank RT (2020) Genetic variation for photosynthetic capacity and efficiency in spring wheat. J Exp Bot 71:2299–2311. https://doi.org/10.1093/jxb/erz439

22. Prins A, Orr DJ, Andralojc PJ, Reynolds MP, Carmo-Silva E, Parry MAJ (2016) Rubisco catalytic properties of wild and domesticated relatives provide scope for improving wheat photosynthesis. J Exp Bot 67:1827–1838. https://doi.org/10.1093/jxb/erv574

23. Acevedo-Siaca LG, Coe R, Wang Y, Kromdijk J, Quick WP, Long SP (2020) Variation in photosynthetic induction between rice accessions and its potential for improving productivity. New Phytol 227:1097–1108. https://doi.org/10.1111/nph.16454
24. Taylor SH, Long SP (2017) Slow induction of photosynthesis on shade to sun transitions in wheat may cost at least 21% of productivity. Philos Trans R Soc B Biol Sci 372:20160543. https://doi.org/10.1098/rstb.2016.0543
25. Salter WT, Merchant AM, Richards RA, Trethowan R, Buckley TN (2019) Rate of photosynthetic induction in fluctuating light varies widely among genotypes of wheat. J Exp Bot 70:2787–2796. https://doi.org/10.1093/jxb/erz100
26. Reynolds M, Chapman S, Crespo-Herrera L, Molero G, Mondal S, Pequeno DNL, Pinto F, Pinera-Chavez FJ, Poland J, Rivera-Amado C, Saint-Pierre C, Sukumaran S (2020) Breeder friendly phenotyping. Plant Sci 295:110396. https://doi.org/10.1016/j.plantsci.2019.110396
27. Reynolds M, Atkin OK, Bennett M, Cooper M, Dodd IC, Foulkes MJ, Frohberg C, Hammer G, Henderson IR, Huang B, Korzun V, McCouch SR, Messina CD, Pogson BJ, Slafer GA, Taylor NL, Wittich PE (2021) Addressing research bottlenecks to crop productivity. Trends Plant Sci 26:607–630. https://doi.org/10.1016/j.tplants.2021.03.011
28. Rivera-Amado C, Trujillo-Negrellos E, Molero G, Reynolds MP, Sylvester-Bradley R, Foulkes MJ (2019) Optimizing dry-matter partitioning for increased spike growth, grain number and harvest index in spring wheat. Field Crop Res 240:154–167. https://doi.org/10.1016/j.fcr.2019.04.016
29. Gerard GS, Alqudah A, Lohwasser U, Börner A, Simón MR (2019) Uncovering the genetic architecture of fruiting efficiency in bread wheat: a viable alternative to increase yield potential. Crop Sci 59:1853–1869. https://doi.org/10.2135/cropsci2018.10.0639
30. Bustos DV, Hasan AK, Reynolds MP, Calderini DF (2013) Combining high grain number and weight through a DH-population to improve grain yield potential of wheat in high-yielding environments. Field Crop Res 145:106–115. https://doi.org/10.1016/j.fcr.2013.01.015
31. Wolde GM, Mascher M, Schnurbusch T (2019) Genetic modification of spikelet arrangement in wheat increases grain number without significantly affecting grain weight. Mol Gen Genomics 294:457–468. https://doi.org/10.1007/s00438-018-1523-5
32. Calderini DF, Castillo FM, Arenas-M A, Molero G, Reynolds MP, Craze M, Bowden S, Milner MJ, Wallington EJ, Dowle A, Gomez LD, McQueen-Mason SJ (2021) Overcoming the trade-off between grain weight and number in wheat by the ectopic expression of expansin in developing seeds leads to increased yield potential. New Phytol 230:629–640. https://doi.org/10.1111/nph.17048
33. Reale L, Rosati A, Tedeschini E, Ferri V, Cerri M, Ghitarrini S, Timorato V, Ayano B, Porfiri O, Frenguelli G, Ferranti F, Benincasa P (2017) Ovary size in wheat (Triticum aestivum L.) is related to cell number. Crop Sci 57:914–925
34. Herrera J, Calderini DF (2020) Pericarp growth dynamics associate with final grain weight in wheat under contrasting plant densities and increased night temperature. Ann Bot 126:1063–1076. https://doi.org/10.1093/aob/mcaa131
35. Kaur V, Behl RK, Singh S, Madaan S (2011) Endosperm and pericarp size in wheat (Triticum aestivum L.) grains developed under high temperature and drought stress conditions. Cereal Res Commun 39:515–524. https://doi.org/10.1556/CRC.39.2011.4.6
36. Zhang J, Liu W, Yang X, Gao A, Li X, Wu X, Li L (2011) Isolation and characterization of two putative cytokinin oxidase genes related to grain number per spike phenotype in wheat. Mol Biol Rep 38:2337–2347. https://doi.org/10.1007/s11033-010-0367-9
37. Valluru R, Reynolds MP, Davies WJ, Sukumaran S (2017) Phenotypic and genome-wide association analysis of spike ethylene in diverse wheat genotypes under heat stress. New Phytol 214:271–283. https://doi.org/10.1111/nph.14367
38. Reynolds MP, Pellegrineschi A, Skovmand B (2005) Sink-limitation to yield and biomass: a summary of some investigations in spring wheat. Ann Appl Biol 146:39–49. https://doi.org/10.1111/j.1744-7348.2005.03100.x

39. Cui F, Li J, Ding A, Zhao C, Wang L, Wang X, Li S, Bao Y, Li X, Feng D, Kong L, Wang H (2011) Conditional QTL mapping for plant height with respect to the length of the spike and internode in two mapping populations of wheat. Theor Appl Genet 122:1517–1536. https://doi.org/10.1007/s00122-011-1551-6
40. Sakuma S, Golan G, Guo Z, Ogawa T, Tagiri A, Sugimoto K, Bernhardt N, Brassac J, Mascher M, Hensel G, Ohnishi S, Jinno H, Yamashita Y, Ayalon I, Peleg Z, Schnurbusch T, Komatsuda T (2019) Unleashing floret fertility in wheat through the mutation of a homeobox gene. Proc Natl Acad Sci 116:5182–5187. https://doi.org/10.1073/pnas.1815465116
41. Muqaddasi QH, Brassac J, Koppolu R, Plieske J, Ganal MW, Röder MS (2019) TaAPO-A1, an ortholog of rice ABERRANT PANICLE ORGANIZATION 1, is associated with total spikelet number per spike in elite European hexaploid winter wheat (Triticum aestivum L.) varieties. Sci Rep 9:13853. https://doi.org/10.1038/s41598-019-50331-9

Chapter 22
Heat and Climate Change Mitigation

Dirk B. Hays, Ilse Barrios-Perez, and Fatima Camarillo-Castillo

Abstract High temperature stress is a primary constraint to maximal yield in wheat, as in nearly all cultivated crops. High temperature stress occurs in varied ecoregions where wheat is cultivated, as either a daily chronic metabolic stress or as an acute episodic high heat shock during critical periods of reproductive development. This chapter focuses on defining the key biochemical processes regulating a plant's response to heat stress while highlighting and defining strategies to mitigate stress and stabilize maximal yield during high temperature conditions. It will weigh the advantages and disadvantages of heat stress adaptive trait breeding strategies versus simpler integrated phenotypic selection strategies. Novel remote sensing and marker-assisted selection strategies that can be employed to combine multiple heat stress tolerant adaptive traits will be discussed in terms of their efficacy. In addition, this chapter will explore how wheat can be re-envisioned, not only as a staple food, but also as a critical opportunity to reverse climate change through unique subsurface roots and rhizomes that greatly increase wheat's carbon sequestration.

Keywords Climate change mitigation · Respiration · Heat shock · Ethylene · Leaf epicuticular wax · Source and sink relationships

22.1 Learning Objectives

- Identify factors responsible for yield loss during acute high temperature stress heat shock.
- Define alternate hormonal yield pathways that maximize and/or limit wheat yield during acute high temperature stress.

D. B. Hays (✉) · I. Barrios-Perez
Soil and Crop Sciences, Texas A&M University, College Station, TX, USA
e-mail: dbhays@tamu.edu; ibarrios@tamu.edu

F. Camarillo-Castillo
International Maize and Wheat Improvement Center (CIMMYT), Texcoco, Mexico
e-mail: f.camarillo@cgiar.org

M. P. Reynolds, H.-J. Braun (eds.), *Wheat Improvement*,
https://doi.org/10.1007/978-3-030-90673-3_22

- Define unique adaptive traits that suppress the induction of yield-limiting signal transduction pathways during yield formation.
- Define the impact of nighttime high temperature stress on respiration-derived yield limitations in wheat, and explore current strategies to minimize it.
- Identify key traits that increase wheat's climate change mitigation capacity.
- Contrast the breeding efficiency of employing physiological and idiotypic based trait introgression versus integrated yield based selection strategies.

22.2   Introduction

Global warming – The steadily growing concentration of atmospheric carbon dioxide (CO_2) has currently reached an excess of 409 ppm or 720 gigatons of carbon (C) (GtC), its highest point in more than 800,000 years (NOAA). As is widely known, increasing atmospheric CO_2, along with methane (CH_4), and nitrous oxide (N_2O) are the primary factors raising global temperatures. Warming temperatures are already constraining local agricultural production, thus inflating food prices nationally and globally. If rapid mitigation strategies are not implemented on a global scale, we can expect constraints on our food systems to increase in frequency and severity, leading to regional conflicts and migration events. The gravity of these negative consequences partly depends on our ability to adapt current cultivars to increasing temperatures, while also modifying our agro-ecological practices to mitigate these challenges. In the past decade alone (2008–2017) anthropogenic activity has resulted in the release of 9.4 GtC y^{-1} via fossil fuel combustion. Land use changes, including modern agricultural cultivation, emit an additional 1.3 GtC y^{-1}. Oceans and terrestrial lands sequester more than half of these anthropogenic emissions (2.4 and 3.0 GtC y^{-1}, respectively) as primary carbon sinks. The remaining net emissions are lingering in the atmosphere, further contributing to global warming at an annual increase of 4.7 GtC y^{-1} [1]. Under current global practices, agriculture thus contributes a sizeable (12%) fraction to the net atmospheric CO_2 emissions [2]. Given high temperature stress damage to wheat is a common issue globally, agriculture scientists must adapt practices and varieties to mitigate its impact on yield, while also transforming agriculture from a source of C emissions to compensated sinks.

Heat stress can reduce wheat yields throughout the crop's life cycle, either as an acute, chronic and nighttime stress. During early seedling establishment, heat stress reduces coleoptile elongation, impairing emergence. During vegetative development it accelerates the transition to flowering, increasing frost damage risks. While during inflorescence heat stress can accelerate grain development and ablate tillering meristems reducing grain number and yield [3]. During microgametogenesis and embryo development, acute heat stress can sterilize pollen and result in embryo abortion resulting in reduced grain number [4]. While heat stress during grain maturation can hasten the transition to the dry seed stage negatively impacting end-use quality.

The aforementioned effects are regulated in many instances by a heat stress induced increase in the plant hormone ethylene [5]. As such, breeding for heat tolerance in wheat should focus on the introgression of traits that moderate heat stress in wheat's internal tissues, thus allowing the maximal expression of growth and yield conducive regulatory pathways, while minimizing the induction of yield limiting pathways regulated by ethylene. This approach is in line with Passiourra's focus on optimizing traits that confer growth conducive conditions versus yield limiting plant survival traits [6].

High night temperatures (HNT) are important concerns of global warming. Data suggest that nighttime temperatures are rising at 1.4 times the rate of daytime temperatures [7]. HNT are often longer in duration during crop development and occur over broader geographic regions compared to chronic and episodic daytime heat stress. The physiological basis for HNT yield decline and current strategies to select for increased yield in response to HNT will be explored. In addition, this chapter will also explore how the idiotype of wheat could be re-envisioned into an asset in the world's arsenal of climate change mitigation through wheat root-derived CO_2 sequestration.

22.3 Factors Responsible for Yield Loss During Acute High Temperature Stress

Optimal temperatures for wheat growth and development have been defined by numerous studies to range between 17 to 23 °C [8]. Above 37 °C, growth stops, while temperatures above 48 °C are lethal to most wheat genotypes. When occurring prior to or shortly after anthesis, temperatures above 30 °C cause pollen and floret sterility through early embryo abortion, both of which reduce grain number. Additionally, temperatures above 30 °C during maturation can reduce overall grain fill, and end-use quality by reducing starch deposition and altering high and low molecular weight glutenin and gliadin ratios [9]. The expression of genes involved in carbohydrate metabolism during grain maturation such as sucrose synthase, soluble starch synthase, phosphoglycerate kinase, starch branching enzyme have been shown to be highly sensitive to acute heat stress. Genes for several α-amylase inhibitors also are down-regulated, as are a number of genes for gluten proteins, including α-gliadins, LMW-GSs and a few HMW-GSs and γ-gliadins [9].

Also detrimental to wheat yield is chronic high temperature stress, which is confounded with high night temperatures. Chronic heat stress is differentiated from acute heat shock injury which occurs as a rapid increase in temperatures over a few days. The physiological basis for lower yields due to chronic heat stress is poorly understood, yet in other crops enhanced development is recognized as a function of growing degree days. In addition, beyond screening for increased yield, few trait that confer tolerance to chronic heat have been defined.

In temperate wheat growing climates such as North America, where fluctuations in warm versus cold fronts are common, chronic heat stress or early acute heat stress can hasten transitions to reproductive development leaving wheat vulnerable to frost damage and reproductive sterility. In these environments selection of earliness to avoid acute periods of heat stress may be worthwhile to combat continuing trends in global warming, however in the near-term earliness may not be a viable option.

22.4 The Role of Ethylene in Regulating High Temperature Stress Responses in Wheat

Ethylene is a gaseous plant hormone, that has been shown to exhibit time and dose-dependent effects on plants during heat stress. Under low concentrations, the developmental induction of ethylene positively regulates leaf and cotyledon expansion, lateral root growth and dormancy release in seeds and buds. However, high concentrations of ethylene can also impose an inhibitory, or senescence effect which is deleterious to growth and yield expression. Heat stress induced ethylene has been shown to regulate early leaf senescence, and result in reduced spikelet fertility through both pollen sterility and embryo abortion. Its production in embryos has been shown to accelerate senescence [5] and reduce total seed progeny, likely as an important conserved mechanism to optimize resource allocation during stress [10]. Similarly, heat stress ethylene has been reported in developing pollen and flowers where it induces sterility [11]. In other systems, ethylene has been shown to regulate leaf abscission in response to reduced auxin flux and to increase in response to reduced glucose levels, which promote the synthesis of abscisic acid (ABA) through glucose signalling [12]. In wheat responses to acute heat stress, it is unclear whether the yield limiting impacts of increased ethylene are induced pathways in embryos and grains themselves, an ethylene elevation resulting from the reduction in photosynthetic glucose, or a reduction in glucose and fructose from diminished sucrose hydrolysis via invertases or sucrose synthases in sink tissues (Fig. 22.1). In maize, this latter scenario is referred to as the Shannon hypothesis for sink strength. In this hypothesis, sucrose is hydrolysed by cell wall and vacuolar invertases in the phloem unloading zone at the pedicel and placenta-chalazal connection to the seed nucellus [13]. These enzymes have been shown to be sensitive to drought stress in early developing maize kernels, which results in plant cell death (PCD) at this connective abscission zone [14]. An analogous event regulates the tapetum connections to developing pollen microspores. In this case, drought and heat stress reduce cell wall invertase gene expression in rice and wheat tapetum resulting in premature PCD in the tapetum's connection to microspores and loss of sugar (glucose and fructose) translocation [15, 16]. Both early embryo, seed and microspore development are dependent on a sink-signalled supply of sucrose to facilitate early developmental growth, cell expansion, and starch deposition to maintain viability. Loss of invertase gene expression in pedicel and tapetum may be a critical feature of heat and drought

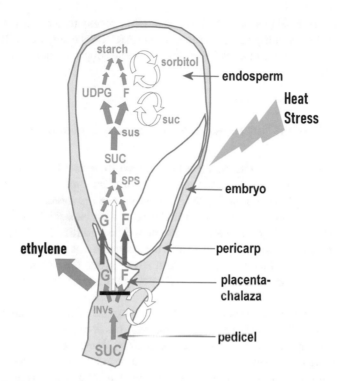

Fig. 22.1 The Shannon hypothesis (in green) with the addition of heat stress induced ethylene (in black) proposes that sucrose (SUC) first cleaved by cell wall invertases (INV) in the pedicel, the placenta chalazal, and the basal endosperm transfer layer. Hexoses glucose (G) and fructose (F) enter the endosperm across the endosperm transfer layer. In the endosperm and embryo, SUC is resynthesized by sucrose phosphate synthase (SPS), and transferred within the endosperm and embryo where it is re-hydrolyzed by sucrose synthase (SUS) for metabolism into starch. Reduction in INV or SUS by heat stress in either in the placenta-chalazal, endosperm, or embryo could reduce glucose derived suppression of ethylene synthesis regulated by ABA resulting in abscission at the placenta chalazal and an early developmental senescence of the endosperm and embryo. (Modified with permission from Ref. [13])

sensitive wheat genotypes [4]. The role of invertases in the basal pedicel regions of developing kernels is also consistent with phloem unloading being an important central feature in promoting turgor gradients and pressure driven sucrose movement and establishing sink strength in developing seed and microspore sinks [17]. At present, the sequence of regulatory responses to acute heat stress leading to increased ethylene, decreased invertase activity, reduced starch deposition and early microspore and kernel senescence requires clarification.

The negative role ethylene can play in response to acute heat stress in wheat and other cereals, can however be inhibited using competitive inhibitors of ethylene response such as 1-methyl cyclopropane (1-MCP) or ethylene biosynthesis inhibitor. Wheat genotypes have also been identified that either don't show increased ethylene in response to heat stress or are insensitive to ethylene during reproductive development as it relates to yield [6]. In addition, recent studies have shown that negative regulation of the ethylene response via ARGOS8 using CRISPR-CAS9 or other genetic engineering technologies can be used to improve yield in maize under drought [18].

22.5   Traits that Suppress Stress Pathway Induction

Because responses to heat stress have been shown to be regulated by independent and convergent signal transduction pathways, a prudent approach should also seek to combine multiple heat adaptive traits that moderate the internalization of ethylene inducing high temperature. Traits that reduce both excess photosynthetic and high temperature conferring solar radiation should be prioritized. This approach has been used inadvertently during humanity's domestication of wheat, by regional and multi-locational breeding programs in heat stress prone environments, and more recently through targeted introgression of specific traits conferring adaptive advantages [4].

Targeted screening for lower canopy temperatures (CT) or high canopy temperature depression (CTD) during high temperature stress is one such approach that has been demonstrated to confer improved yields [19]. However, measuring CT is a proxy for both water use traits such as stomatal conductance and root depth and solar radiation avoidance traits such as incident leaf angle, leaf rolling, and leaf epicuticular wax. Each of the traits contributing to lower CT are developmentally regulated and highly responsive to time of day and microclimate fluctuations such as cloud cover changes and wind gusts, which can render measurements unstable [4].

An alternate approach is to focus on the selection and pyramiding of traits that reduce CT, by increasing the crops *albedo,* which is the ratio of incidence to reflected radiation, in a manner that reduces water use for transpirational cooling during heat stress. Specific traits with these qualities include increased epicuticular wax (EW), erect leaves and in some cases, enhanced leaf rolling. Improved root morphology such as deeper root angles can extend transpirational cooling and growth during heat and drought stress, yet should be selected in combination with traits that reduce solar radiation and preserve limited soil water for the duration of growth and grain development.

Selection for increased EW has been shown to double the proportion of solar radiation that is reflected, reducing photo-inhibition and leaf burning, while significantly reducing leaf and spike temperatures. Increased leaf and glume EW has also been associated with an increased harvest index (HI), residual leaf water content while lowering transpiration and stomatal conductance in a manner that improves water and radiation use efficiencies (WUE and RUE) [20].

Visual selection for glaucousness, which is the light blue grey bloom on leaves, stems and glumes, is often used as a proxy for high EW or as a heat tolerance trait itself. The genetic control of glaucousness, is however independent of total EW content or reflectance. EW is a complex mixture of long chain hydrocarbons of varying length with unique functional groups consisting of alcohols, aldehydes, ketones, and methyl groups. Glaucousness is a function of a higher ratios of alcohols, aldehydes and ketone functional groups arranged as erect glaucous rods and plates [21] versus reduced methyl functional groups arranged as non-glaucous flat plates. As such, glaucousness is not a suitable selection proxy for high EW content, and the associated physiological benefits. Wheat genotypes with higher EW, cooler CT and higher reflectance can be identified with high or low degrees of glaucousness (Fig. 22.2). Accurate selection of high EW wheat genotypes should utilize new hyperspectral indices for EW [22] or the chemical extraction of EW content assay developed by Ebercron [23]. Recent studies have used the latter method to identify novel genetic loci regulating increased EW in wheat. Increased EW QTL are also co-associated with measures of cooler leaf and spike temperatures (CT), and improved yield component stability during heat and drought stress (referred to as heat or drought susceptibility index (HSI or DSI) which is the ratio of the mean

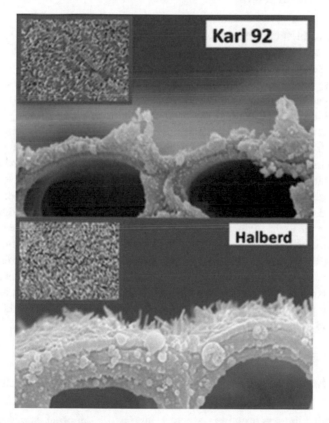

Fig. 22.2 Scanning electron microscopy cross section of leaves of two glaucous wheat lines with varying epicuticular content. The glaucous wheat cultivar 'Karl 92' is heat susceptible with a thin epicuticular wax layer, while the glaucous wheat cultivar 'Halbred' is heat tolerant with a thick epicuticular wax

yield under stress versus control conditions). The QTL identified are located on chromosomes 1A, 2B, 3B, 6B, 7A and 7B [24]. While a gene cluster regulating glaucousness has been identified on wheat chromosomes 2BS and 2DS where S refers to the short arm [25]. It should be noted that the inheritance of high EW and glaucousness on flag leaves and the glumes appears independent and care is needed to select for high EW on both leaves and glumes.

Genotypes with erect leaves that reduce high incidence solar radiation or that express leaf rolling in contrast to prostrate or droopy flag leaves have been suggested to be protected from heat stress injury in wheat, barley and other cereals [3]. However few studies have found strong associations between the trait and improved yield under low water or high temperature stress.

22.6 Nighttime High Temperature Stress Impacts on Wheat Yield

The impact of high day temperature on wheat yield has been well documented in numerous studies. However, fewer studies beyond selection for yield, have identified critical selection targets for tolerance to HNT. The biochemical basis for HNT yield decline has been predominately attributed to increasing nighttime respiration which is presumed to decrease the amount of stored photoassimilates available for plant growth and grain filling. In this case, metabolite profiling showing increases in tricarboxylic acid (TCA) intermediates in leaves exposed to HNT is highlighted as evidence [7]. It should be noted that high night respiration may however represent an increase in nighttime ATP dependent photoassimilate deposition and an accelerated seed maturation rate.

Yield declines of 4% per increase in °C above 14 °C night temperatures have been defined. These studies attributed the decline in yield and yield components to alterations in the grain maturation process and overall shortening of the grain development period rather than photosynthetic or photoassimilate source limitations that might be attributed to elevated night respiration [26]. Discriminating this difference is an important consideration for developing the most effective HNT tolerance screening protocols. The current available literature suggests focusing on grain filling rate as a probable target for selection for HNT in addition to yield and kernel specific yield components.

22.7 Climate Change Mitigation via High Root Biomass

Internationally, there is growing consensus among the scientific, governmental, and political establishment that managed agricultural soils represent a viable opportunity to reverse climate change by sequestering atmospheric CO_2 into soil C via underground root biomass [2]. The known synergistic benefits of soil C include improved soil health, increased crop resilience to abiotic stressors (such as those caused by climate change), and reduced soil erosion and runoff. The C input from plants is cycled into the surrounding soils through two main mechanisms: the continual growth and dieback of small-diameter roots (root turnover), and the final deposition of the shoot and root systems following senescence. The root turnover mechanism is temperature dependent, with higher and lower rates of turnover being related to tropical and low-temperature environments, respectively. This turnover represents a continual deposition of C in soils. Modeling results, as well as the high presence of suberin (protective layer of plant roots) in the soil organic matter (SOM), indicate that a majority of long-lasting material is derived from roots rather than shoots [27]. Emission of CO_2 from these stocks into the atmosphere via soil organisms is accelerated by increasing temperatures [28], with the C sources that are inherently resistant to degradation generally being labeled as 'recalcitrant' [29]. The sequestered material requires maintenance, as decomposition will release the stored C back into the atmosphere. Given this understanding, a reasonable course of action for successful sequestration is to focus efforts on directing root growth into deeper subsoils and larger structures such as rhizomes, as the residence time for C below 30 cm has been shown to be on the order of millennia owing to the spatial separation between microorganisms and SOM [30].

In order to gain a true understanding of the efficacy of this strategy, a quick means of measuring belowground plant traits is needed. Most current methods are slow, cumbersome (e.g., full destructive excavation) or sacrifice accuracy of measurements for speed. As a compromise between speed and accuracy, researchers have pivoted towards ground penetrating radar (GPR) or new technologies such as magnetic resonance imaging (MRI) or X ray computed tomography (X ray CT) as a proxy measure for both high root, and rhizome biomass and high soil organic C.

22.8 Climate Change Mitigation and Potential of High-Root Biomass Grain Crops

The morphology of wheat is highly variable with heights ranging from 0.5 to 1.5 m. The species is generally noted for its extensive fibrous root system. Unfortunately, wheat along with other fibrous root annual grain crops (such as annual sorghum, annual maize, wheat, rice and millet) can sequester only ~0.48 GtC y^{-1} on a global basis, primarily through crop root biomass. This sequestration potential is not sufficient to mitigate the current increase in C emissions as it represents roughly 10% of the current 4.7 GtC y^{-1} net emissions [2].

Rhizomes are an alternative subsurface anatomical feature common to wild ancestors of our staple crops such as rice, wheat, and sorghum that could provide the solution to increased agricultural C sequestration. Rhizomes are axillary shoots that develop from axillary buds. Axillary shoots that develop from buds in aerial nodes are known as branches, while those that develop from buds close to the soil surface in the grasses are known as tillers or rhizomes. Transcriptome analysis have shown that expressed genes in the rhizome tip were more similar to above ground stems in *Sorghum propinquum* and *S. halepense* [31]. Thus, rhizomes appear to function as stems. This has important benefits for subsurface carbon storage. As subsurface stems, their apical meristem likely supply auxin and as such are not targeted for senescence following above ground biomass harvest. Rhizomes thus persist from season to season, resulting in maintenance of sequestered C both in the rhizomes and in the rhizome node derived roots. Similarly, because source-sink relationships are diverged within a rhizomatous grain system, and increase in biomass following grain harvest, they may have less photoassimilate competition to grain yield than vascular and source-sink-connected large root biomass.

While not known to occur in wheat per se, close relatives of wheat, which have been used as donors for traits like disease resistance or perenniality, are also known to possess significant rhizome biomass. These include perennial rye's *Leymus triticoides* and *Leymus cinereus* [32]. In addition, perennial wheat grass *Thinopyrum* has been used to produce rhizomatous hybrids of wheat [33] (Fig. 22.3).

As an example, hybrids of cultivated Sorghum (*Sorghum bicolor*) with *S. halepense* (PS) or *S. propinquum* (RAS) are being used to derive grain- and forage-type hybrids that produce rhizomes with a diameter of 1 cm. Root and rhizome biomass in these hybrids has been recorded to reach 24 MT ha^{-1}, with 90% of it derived from the rhizome. Millet hybridized to Napier grass produces rhizome root biomass in excess of 70 MT ha^{-1} [34]. The rhizomes increase subsurface C sequestration 10 to 30-fold compared to their non-rhizomatous annual crop counterparts. If grown globally as replacements for current annuals, high root: rhizome biomass grain crops have the potential to sequester more than 4.7 GtC y^{-1} (unpublished extrapolations), and could negate and begin to reverse current net C emissions. Given the global urgency, the challenge is to devise the most efficient method to transfer the high soil C-sequestration potential in rhizomes from perennial wheat relatives into a yield-competitive perennial or annual grain-types.

Genetic analysis have identified two complementary dominant genes, Rhz2 and Rhz3, that control tillering and rhizome development – including rhizome number, branching, internode number and length in *O. longistaminata* [35]. Loss of function of either Rhz2 or Rhz3 inhibits rhizome development. However, to our knowledge, Rhz2 and Rhz3 are not yet cloned. QTLs that control rhizome development in *S. propinquum* correspond to most of the *O. longistaminata* QTLs. Modern remote sensing-based selection, gene editing and gene-based, marker-assisted selection tools and strategies can be employed to breed yield competitive rhizomatous wheat and other grain crops as high C sequestration replacements for current low root biomass grain crop cultivars.

Fig. 22.3 (**a**) Plots of wildryes *Leymus triticoides* and *Leymus cinereus*; (**b**) Rhizomes in a hybrid of *L. triticoides* and *L. cinereus*; (**c**) Rhizome in *L. triticoides*; (**d**) Roots of *Thinopyrum intermedium* (intermediate wheatgrass) with thicker rhizomes. (Figure reprinted with permission from Ref. [36])

22.9 High Throughput Phenotyping Selection Strategies to Introgress Multiple Heat Stress Adaptive Traits

Independent conceptual models for grain yield (GY) under heat and drought have been proposed based on the following main drivers: light interception (LI), RUE, partitioning of total assimilates, WUE and harvest index [37]. Each of these main drivers contains genetically determined traits that can potentially lead to an additive genetic effect for resilience to heat when combined through strategic crossing [38]. Physiological traits such as canopy temperature (CT) are already utilized as a selection criteria. Other key traits such as EW remain underutilized except for selection of donors for favorable alleles because of the expensive and laborious methods for phenotyping.

Limitations on field phenotyping restrict our capacity to unravel complex morphological and physiological traits. One important consideration that requires understanding is that numerous physiological traits that have been defined as key targets for selection are developmentally programmed and exhibit temporal variation. As such, high throughput phenotyping selection trials should employ strategies that compensate for temporal and phenological variation (in this case defined as variations in flowering time). This requires partitioning trials into genotypes with narrow ranges of flowering times and conducting high throughput phenotyping scans in narrow temporal windows. This can help avoid confounding phenotypic measures associated with phenology or variations in time of day measured rather than the intended physiological selection.

In breeding programs, high-precision phenotyping can enable the screening of segregating material, advanced lines and germplasm [39]. Increasing the accuracy of phenotyping can provide more reliable estimates of heritability and variance components, facilitate gene discovery and enable prediction of complex traits using genomic selection. The strong association of spectral secondary traits with GY and in season biomass highlights the potential of hyperspectral canopy reflectance to increase productivity in wheat.

New UAV systems are available that combine both hyperspectral, thermal and LIDAR sensors, which should open the door to ideal integrative selection regimes. These are ideal systems for physiological breeding, however at present they remain cost prohibitive even for well-funded breeding programs (Fig. 22.4).

Spectral vegetation indices (SVI) are a quick, easy and inexpensive method of transforming light reflectance into simple indicators of photosynthetic and canopy variations. The simple ratio index (SR) and the normalized difference vegetation index (NDVI) are two of the first SVIs developed for detecting green vegetation. Both indices combine the percentage of reflectance at the wavelengths where plants absorb (~750 to 800 nm) and reflect (800 to 2500 nm) light. Several other SVIs have been derived for sensing the water content of plants, photosynthetic radiation, carotenoid pigments, plant height, leaf area, and diseases [22].

Recent studies have developed new spectral indices for selection of EW with accuracies reaching 65% [22]. EW indices combine narrow wavelength at 625 nm, with narrow wavelength at 736 nm and 832 nm reflectance for direct measures of leaves. Other indices were derived for canopy level reflectance indices derived from 2 narrow bands at 617 and 718 nm with EW prediction accuracies of 71%. It is worth emphasizing that selection for ideal leaf and glume EW should target genotypes which express high levels of EW in response to developmentally programmed cues that are irrespective and prior to heat or drought stress cues and not in response to stress cues. The rationale is that genotypes that express high EW in response to developmental cues will be more heat tolerant, while genotypes which increase EW content in response to stress have already incurred the yield penalty through induction of the heat and drought stress pathways regulated by ethylene. Given this understanding, selection for ideal EW requires critical attention to regularly monitor EW from flag leaf development to late maturation using multi-location selections.

Fig. 22.4 VNIR/SWIR Hyperspectral spectrometer with integrated LIDAR for quantifying leaf wax and other unique heat stress adaptive traits

22.10   High Throughput Phenotyping Selection to Introgress Roots and Rhizomes

Unlike above ground foliar traits, roots are obscured by their growth medium. As such, measurement of root system traits has historically been an invasive process. Invasive techniques include the excavation of root systems, or 'shovelomics', soil coring, and rhizotrons. Noninvasive observations are attractive as it does not interfere with plant growth. Several techniques have been proposed, tested, and validated to different extents. These methods include MRI, X ray CT, and GPR. As high throughput selection tools the development of a field-based MRI or X ray CT are still in their infancy in terms of commercially available solutions. While coarser in terms of root imaging, both high throughput field compatible GPR instruments and the data processing methods needed to phenotype both total root biomass and root architecture traits are well developed and commercially available (see Crop Phenomics at cropphenomics.com).

MRI is an imaging technique most widely known for its application in medicine and chemical analysis. Signals are largely derived from ^{1}H protons which are abundant in water molecules and thus living tissues. Results of root trait (root length, root mass, root diameter, growth angles) quantification using MRI have been shown to be comparable to conventional methods.

X ray CT scans operate by projecting EM radiation in the X ray region of the spectrum through a sample from multiple angles. Elements within the sample attenuate the signal to different degrees depending on the density of their electrons. The resulting data is then used to construct a 3D image of belowground root architecture. Both X ray CT scans and MRI scans have mainly been used to evaluate root traits within pot-grown plants. A direct comparison of the two methods showed that CT scans tended to have a higher spatial resolution than MRI.

GPR also uses electromagnetic (EM) radiation. Returned information is similar in nature to seismic data, and data has traditionally been examined using seismic analysis methods. GPR has several characteristics that make it ideal for examining belowground biological materials: It is non-invasive, non-destructive, and data collection is rapid. The emitted waves reside in the radio/microwave portion of the EM spectrum and can record data at a range of depths depending on the frequency of the output signal, the soil matrix environment, and the antenna design. As a rule of thumb, higher frequencies (greater than 1.0 GHz) result in lower penetration depths. This is due to greater energy absorption of free water and scattering. Penetration depth and frequency are inversely related, and range from 100.0 m at 50 MHz to 0.1 m at 50 GHz. Penetration depths of agricultural soil subsurfaces are complicated by higher average water content signal attenuation. One benefit of higher soil water content is an increase in resolution in the time domain. As the signal velocity varies with dielectric permittivity, smaller distances are traversed with the same number of collected samples. High-intensity reflections are created when the EM energy encounters a media interface with a high difference in relative dielectric. Water, with a relative dielectric of 81, creates a high contrast with the surrounding soil which typically has a dielectric of ~4. This fact can be leveraged for small root detection.

22.11 Ground Penetrating Radar Application in Life Sciences

Due to their larger size, the coarse roots of trees are more easily detectable by GPR. As such, the majority of early root studies using GPR have been used to mensurate tree root biomass or the biomass of large tuber-like roots of cassava (*Manihot esculenta*) [40]. However, Liu et al. [41] recently performed a study which used GPR to detect the roots of wheat (*Triticum aestivum* L.) in field conditions, while Wolfe 2021 [34] developed both a novel GPR instrument, and new data processing methods to quantitatively discriminated fibrous root versus rhizome biomass.

22.12 Trait Introgression Versus Integrated Yield Selection Strategies for Heat Stress Tolerance

As discussed in detail in Chap. 23 on drought, a careful analysis of the efficacy of breeding specific adaptive traits versus direct selection for yield components is a worthwhile exercise. In traditional breeding programs, selection for disease resistance and flowering time attributes are prioritized while selection for heat stress adaptive physiological and morphological traits are rare. Direct selection for yield under multi-locational trials is considered the most efficient method for combining the best adaptive traits for heat stress and water-stress. However, identifying novel physiological and morphological traits in adapted genotypes for introgression into elite breeding lines is a valid method to improve overall heat stress tolerance in breeding programs. Following introgression, high throughput remote sensing based phenotypic selection of novel heat adaptive traits can be used as a compliment to direct yield selection to discriminate the potential yield benefit of the given traits. Markers for key adaptive traits such as high EW and glaucousness have been identified. Markers linked to developmentally regulated EW deposition should be prioritized. As well, introgression of high EW can utilize new hyperspectral EW spectral indices [22]. When selecting for yield, attention should be given to selecting for both high yield and ideal high single kernel weight and kernel dimension stability across multi-location high temperature trials. Lines which show yield instability and high variation in single kernel weight and dimensions should be discarded as heat susceptible and responsive to heat stress ethylene regulated early transitions to the dry seed stage. Novel high throughput digital kernel weight and dimension instruments can be used to improve the accuracy of direct yield selection for heat stress tolerance.

22.13 Key Concepts

- High confidence predictions state that increasing use of fossil fuels and inorganic fertilizers will continue to increase global temperatures, further challenging wheat productivity and its capacity to meet a growing population's food needs.
- Both acute and chronic high temperature stress in addition to high night temperatures are recognized as important limiting factors affecting wheat productivity.
- Heat stress impairs emergence by reducing coleoptile elongation, speeds the transition to flowering thus exposing wheat to spring frost injury. During reproductive development acute heat stress can suppress sucrose hydrolyzing invertases reducing assimilate translocation to developing pollen and embryos resulting in pollen sterility, kernel abortion, or a transition to early dry seed stage thus negatively effecting yield and end-use quality.
- Many of the negative impacts of heat stress are regulated by the plant hormone ethylene.

- Development or selection of heat stress ethylene insensitive genotypes for improving heat tolerance has demonstrated efficacy in improving heat tolerance in other cereals.
- Novel traits which reflect excess solar radiation, such as increased EW and glaucousness, or avoid direct solar radiation, such as erect leaves, have been shown to be useful in reducing wheat leaf and glume temperatures and are important breeding targets for improving heat tolerance.
- Both molecular markers for marker-assisted selection and hyperspectral crop indices which select for leaf EW have been developed. Both can be used to discriminate between heat tolerant, developmentally regulated high EW genotypes versus genotypes exhibiting stress induced EW.
- Novel ground penetrating radar tools and data processing software have been developed to aid in the selection of deep root wheat lines, with high root biomass for enhanced CO_2 sequestration into soils.
- Traits such as rhizomes should be transferred from wheat relatives to significantly increase wheat's soil carbon sequestration potential and contribution to climate change mitigation.

22.14 Summary

Novel strategies can be employed to define unique traits that confer improved adaptation to heat stress in combination with improved drought stress. Traits such as deeper root architectures, erect leaves and increase epicuticular leaf and glume wax help reflect or avoid heat stress by reflecting excess photosynthetic solar and thermal infrared radiation. These traits common to heat adapted species when optimized in wheat, help moderate internal plant temperatures, avoid induction of yield limiting hormone stress pathways regulated by ethylene while conserving excess water loss through transpirational cooling. In this manner, heat stress adaptive traits help conserve water for optimal growth and yield. When combined with traits that increase wheat's soil carbon sequestration potential, they may improve wheat's role as an essential food staple for the earth's growing population.

References

1. Le Quéré C, Andrew RM, Canadell JG, Sitch S, Korsbakken JI, Peters GP, Manning AC, Boden TA, Tans PP, Houghton RA, Keeling RF, Alin S, Andrews OD, Anthoni P, Barbero L, Bopp L, Chevallier F, Chini LP, Ciais P, Currie K, Delire C, Doney SC, Friedlingstein P, Gkritzalis T, Harris I, Hauck J, Haverd V, Hoppema M, Klein Goldewijk K, Jain AK, Kato E, Körtzinger A, Landschützer P, Lefèvre N, Lenton A, Lienert S, Lombardozzi D, Melton JR, Metzl N, Millero F, Monteiro PMS, Munro DR, Nabel JEMS, Nakaoka S, O'Brien K, Olsen A, Omar AM, Ono T, Pierrot D, Poulter B, Rödenbeck C, Salisbury J, Schuster U, Schwinger J, Séférian R, Skjelvan I, Stocker BD, Sutton AJ, Takahashi T, Tian H, Tilbrook B, van der

Laan-Luijkx IT, van der Werf GR, Viovy N, Walker AP, Wiltshire AJ, Zaehle S (2016) Global carbon budget 2016. Earth Syst Sci Data 8:605–649. https://doi.org/10.5194/essd-8-605-2016

2. Bossio DA, Cook-Patton SC, Ellis PW, Fargione J, Sanderman J, Smith P, Wood S, Zomer RJ, von Unger M, Emmer IM, Griscom BW (2020) The role of soil carbon in natural climate solutions. Nat Sustain 3:391–398. https://doi.org/10.1038/s41893-020-0491-z

3. Hunt JR, Hayman PT, Richards RA, Passioura JB (2018) Opportunities to reduce heat damage in rain-fed wheat crops based on plant breeding and agronomic management. Field Crop Res 224:126–138. https://doi.org/10.1016/j.fcr.2018.05.012

4. Powell N, Ji X, Ravash R, Edlington J, Dolferus R (2012) Yield stability for cereals in a changing climate. Funct Plant Biol 39:539–552

5. Hays DB, Do JH, Mason RE, Morgan G, Finlayson SA (2007) Heat stress induced ethylene production in developing wheat grains induces kernel abortion and increased maturation in a susceptible cultivar. Plant Sci 172:1113–1123. https://doi.org/10.1016/j.plantsci.2007.03.004

6. Passioura JB (1977) Grain yield, harvest index, and water use of wheat. J Aust Inst Agric Sci 43:117–120

7. Sadok W, Jagadish SVK (2020) The hidden costs of nighttime warming on yields. Trends Plant Sci 25:644–651. https://doi.org/10.1016/j.tplants.2020.02.003

8. Burke JJ, Mahan JR, Hatfield JL (1988) Crop-specific thermal kinetic windows in relation to wheat and cotton biomass production. Agron J 80:553–556. https://doi.org/10.2134/agronj1988.00021962008000040001x

9. Altenbach SB (2012) New insights into the effects of high temperature, drought and post-anthesis fertilizer on wheat grain development. J Cereal Sci 56:39–50. https://doi.org/10.1016/j.jcs.2011.12.012

10. Kaur H, Ozga JA, Reinecke DM (2021) Balancing of hormonal biosynthesis and catabolism pathways, a strategy to ameliorate the negative effects of heat stress on reproductive growth. Plant Cell Environ 44:1486–1503. https://doi.org/10.1111/pce.13820

11. Jegadeesan S, Beery A, Altahan L, Meir S, Pressman E, Firon N (2018) Ethylene production and signaling in tomato (Solanum lycopersicum) pollen grains is responsive to heat stress conditions. Plant Reprod 31:367–383. https://doi.org/10.1007/s00497-018-0339-0

12. Cheng W-H, Endo A, Zhou L, Penney J, Chen H-C, Arroyo A, Leon P, Nambara E, Asami T, Seo M, Koshiba T, Sheen J (2002) A unique short-chain dehydrogenase/reductase in Arabidopsis glucose signaling and abscisic acid biosynthesis and functions. Plant Cell 14:2723–2743. https://doi.org/10.1105/tpc.006494

13. Bihmidine S, Hunter C, Johns C, Koch K, Braun D (2013) Regulation of assimilate import into sink organs: update on molecular drivers of sink strength. Front Plant Sci 4:177. https://doi.org/10.3389/fpls.2013.00177

14. McLaughlin JE, Boyer JS (2004) Glucose localization in maize ovaries when kernel number decreases at low water potential and sucrose is fed to the stems. Ann Bot 94:75–86. https://doi.org/10.1093/aob/mch123

15. Koonjul PK, Minhas JS, Nunes C, Sheoran IS, Saini HS (2004) Selective transcriptional down-regulation of anther invertases precedes the failure of pollen development in water-stressed wheat. J Exp Bot 56:179–190. https://doi.org/10.1093/jxb/eri018

16. Ji X, Shiran B, Wan J, Lewis DC, Jenkins CLD, Condon AG, Richards RA, Dolferus R (2010) Importance of pre-anthesis anther sink strength for maintenance of grain number during reproductive stage water stress in wheat. Plant Cell Environ 33:926–942. https://doi.org/10.1111/j.1365-3040.2010.02130.x

17. Turgeon R (2006) Phloem loading: how leaves gain their independence. Bioscience 56:15–24. https://doi.org/10.1641/0006-3568(2006)056[0015:PLHLGT]2.0.CO;2

18. Shi J, Gao H, Wang H, Lafitte HR, Archibald RL, Yang M, Hakimi SM, Mo H, Habben JE (2017) ARGOS8 variants generated by CRISPR-Cas9 improve maize grain yield under field drought stress conditions. Plant Biotechnol J 15:207–216. https://doi.org/10.1111/pbi.12603

19. Pinto RS, Reynolds MP (2015) Common genetic basis for canopy temperature depression under heat and drought stress associated with optimized root distribution in bread wheat. Theor Appl Genet 128:575–585. https://doi.org/10.1007/s00122-015-2453-9

20. Mohammed S, Huggins TD, Beecher F, Chick C, Sengodon P, Mondal S, Paudel A, Ibrahim AMH, Tilley M, Hays DB (2018) The role of leaf epicuticular wax in the adaptation of wheat (Triticum aestivum L.) to high temperatures and moisture deficit conditions. Crop Sci 58:679–689. https://doi.org/10.2135/cropsci2017.07.0454

21. Samuels L, Kunst L, Jetter R (2008) Sealing plant surfaces: cuticular wax formation by epidermal cells. Annu Rev Plant Biol 59:683–707. https://doi.org/10.1146/annurev.arplant.59.103006.093219

22. Camarillo-Castillo F, Huggins TD, Mondal S, Reynolds MP, Tilley M, Hays DB (2021) High-resolution spectral information enables phenotyping of leaf epicuticular wax in wheat. Plant Methods 17:58. https://doi.org/10.1186/s13007-021-00759-w

23. Ebercon A, Blum A, Jordan WR (1977) A rapid colorimetric method for epicuticular wax contest of sorghum leaves1. Crop Sci:17, cropsci1977.0011183X001700010047x. https://doi.org/10.2135/cropsci1977.0011183X001700010047x

24. Mohammed S, Huggins T, Mason E, Beecher F, Chick C, Sengodon P, Paudel A, Ibrahim A, Tilley M, Hays D (2021) Mapping the genetic loci regulating leaf epicuticular wax, canopy temperature, and drought susceptibility index in Triticum aestivum. Crop Sci 61:2294–2305. https://doi.org/10.1002/csc2.20458

25. Hen-Avivi S, Savin O, Racovita RC, Lee W-S, Adamski NM, Malitsky S, Almekias-Siegl E, Levy M, Vautrin S, Bergès H, Friedlander G, Kartvelishvily E, Ben-Zvi G, Alkan N, Uauy C, Kanyuka K, Jetter R, Distelfeld A, Aharoni A (2016) A metabolic gene cluster in the wheat W1 and the barley Cer-cqu loci determines β-diketone biosynthesis and glaucousness. Plant Cell 28:1440–1460. https://doi.org/10.1105/tpc.16.00197

26. García GA, Serrago RA, Dreccer MF, Miralles DJ (2016) Post-anthesis warm nights reduce grain weight in field-grown wheat and barley. Field Crop Res 195:50–59. https://doi.org/10.1016/j.fcr.2016.06.002

27. Rasse DP, Rumpel C, Dignac M-F (2005) Is soil carbon mostly root carbon? Mechanisms for a specific stabilisation. Plant Soil 269:341–356. https://doi.org/10.1007/s11104-004-0907-y

28. Crowther TW, Todd-Brown KEO, Rowe CW, Wieder WR, Carey JC, Machmuller MB, Snoek BL, Fang S, Zhou G, Allison SD, Blair JM, Bridgham SD, Burton AJ, Carrillo Y, Reich PB, Clark JS, Classen AT, Dijkstra FA, Elberling B, Emmett BA, Estiarte M, Frey SD, Guo J, Harte J, Jiang L, Johnson BR, Kröel-Dulay G, Larsen KS, Laudon H, Lavallee JM, Luo Y, Lupascu M, Ma LN, Marhan S, Michelsen A, Mohan J, Niu S, Pendall E, Peñuelas J, Pfeifer-Meister L, Poll C, Reinsch S, Reynolds LL, Schmidt IK, Sistla S, Sokol NW, Templer PH, Treseder KK, Welker JM, Bradford MA (2016) Quantifying global soil carbon losses in response to warming. Nature 540:104–108. https://doi.org/10.1038/nature20150

29. Kleber M (2010) What is recalcitrant soil organic matter. Environ Chem 7:320–332

30. Rumpel C, Kögel-Knabner I (2011) Deep soil organic matter—a key but poorly understood component of terrestrial C cycle. Plant Soil 338:143–158. https://doi.org/10.1007/s11104-010-0391-5

31. Jang CS, Kamps TL, Skinner DN, Schulze SR, Vencill WK, Paterson AH (2006) Functional classification, genomic organization, putatively cis-acting regulatory elements, and relationship to quantitative trait loci, of sorghum genes with rhizome-enriched expression. Plant Physiol 142:1148–1159. https://doi.org/10.1104/pp.106.082891

32. Yun L, Larson SR, Mott IW, Jensen KB, Staub JE (2014) Genetic control of rhizomes and genomic localization of a major-effect growth habit QTL in perennial wildrye. Mol Gen Genomics 289:383–397. https://doi.org/10.1007/s00438-014-0817-5

33. Cox TS, Bender M, Picone C, Van TDL, Holland JB, Brummer EC, Zoeller BE, Paterson AH, Jackson W (2002) Breeding perennial grain crops. CRC Crit Rev Plant Sci 21:59–91. https://doi.org/10.1080/0735-260291044188

34. Wolfe M (2021) Application of a prototype ground penetrating radar array for the detection of bioenergy grass root systems. Texas A&M University
35. Hu FY, Tao DY, Sacks E, Fu BY, Xu P, Li J, Yang Y, McNally K, Khush GS, Paterson AH, Li Z-K (2003) Convergent evolution of perenniality in rice and sorghum. Proc Natl Acad Sci 100:4050–4054. https://doi.org/10.1073/pnas.0630531100
36. The Plant and Soil Sciences eLibrary (2021) Lessons. https://passel2.unl.edu/list/lesson
37. Cossani CM, Reynolds MP (2012) Physiological traits for improving heat tolerance in wheat. Plant Physiol 160:1710–1718. https://doi.org/10.1104/pp.112.207753
38. Reynolds M, Langridge P (2016) Physiological breeding. Curr Opin Plant Biol 31:162–171. https://doi.org/10.1016/j.pbi.2016.04.005
39. Haghighattalab A, González Pérez L, Mondal S, Singh D, Schinstock D, Rutkoski J, Ortiz-Monasterio I, Singh RP, Goodin D, Poland J (2016) Application of unmanned aerial systems for high throughput phenotyping of large wheat breeding nurseries. Plant Methods 12:35. https://doi.org/10.1186/s13007-016-0134-6
40. Delgado A, Hays DB, Bruton RK, Ceballos H, Novo A, Boi E, Selvaraj MG (2017) Ground penetrating radar : a case study for estimating root bulking rate in cassava (Manihot esculenta Crantz). Plant Methods 1–11. https://doi.org/10.1186/s13007-017-0216-0
41. Liu H, Xing B, Zhu J, Zhou B, Wang F, Xie X, Liu QH (2018) Quantitative stability analysis of ground penetrating radar systems. IEEE Geosci Remote Sens Lett 15:522–526. https://doi.org/10.1109/LGRS.2018.2801827

Chapter 23
Drought

Richard A. Richards

Abstract Established breeding methods for wheat in dry environments continue to make gains. It will remain the cornerstone for wheat improvement. This Chapter discusses proven methods to make additional gains. It discusses a way to benchmark yield potential in dry environments and how this can be used to determine whether unexpected agronomic or genetic factors are limiting yields. It examines opportunities, advantages and disadvantages of trait-based selection methods for dry environments, and it presents a framework by which important traits can be selected. Both high throughput and marker-based methods of selection are examined for their success and feasibility of use in breeding. It also highlights the importance of agronomic approaches in combination with breeding to continue to improve yield potential in water limited environments. Finally, the elements of success of translation from research to the delivery of new varieties is examined.

Keywords Water use · Water use efficiency (WUE) · Harvest index · Water-limited yield potential · Trait-based selection

23.1 Learning Objectives

- Identify factors responsible for yield gap before improving yield potential under drought.
- Establishing a water-limited framework to improve yield.
- Identification of physiological traits that can improve performance under drought.
- Combining trait-based selection with management practices to improve grain yield.
- Breeding and selection of physiological traits.

R. A. Richards (✉)
CSIRO Agriculture and Food, Canberra, ACT, Australia
e-mail: Richard.Richards@csiro.au

• Translation from pre-breeding to new cultivars – the elements of success.

23.2 Introduction

Drought is a recurring feature in most parts of the world where wheat is grown. Around 75% of the area sown to wheat is rainfed and of this 46% has low to moderate rainfall and 29% high rainfall. The remaining 24% of the land is irrigated. However, the high rainfall and irrigated regions will have either sub-optimal rainfall in some years or insufficient irrigation water to meet the crops water requirement for maximum yield [1]. Accordingly, water limitations are a regular occurrence in almost all wheat growing regions. This will be exacerbated as pressure mounts on water for irrigation to be used for higher value crops than wheat, as well as for cities, industrial use and for the environment. With increasing population growth and increasing demand for food this places greater importance on increased productivity with less water.

Wheat improvement in water-limited environments has always been a challenge. Wheat breeders have struggled to make genetic gain and although they have been successful progress has been slow. This is because every drought is different in terms of intensity, duration and timing and so genotype x year interactions are large, and this slows genetic gain. Agronomists have also struggled to understand the complex underlying limitations of rainfed cropping environments and there is the complex and unpredictable seasonal variability to contend with. This seasonal variability can make management decisions difficult.

Maximising grain yield in dry environments depends on the ability of the crop to use as much of the available soil water as possible in a time frame where other constraints such as heat and more severe drought is avoided as much as possible. Thus, breeders who selected for earlier flowering in an environment where terminal drought was a common feature provided the first successful varieties in dry environments. This was because crops avoided flowering during the more severe dry and hot periods. It also resulted in a higher harvest index.

Important yield improvements have relied on a better understanding of the cropping environment. A startling example of the complexity of dryland cropping environments comes from studies in Australia that examined the on-farm relationship between seasonal rainfall and grain yield [2]. The expectation is that grain yield will be closely related to rainfall. But in semi-arid environments this was often not observed. Instead, to our surprise there was almost no relationship (Fig. 23.1). Although with enough data points an upper boundary line between rainfall and yield emerged. The slope of this boundary line in fact defines the upper limit to water use efficiency (WUE). In the French and Schultz study [2] it was around 20 kg grain per mm of rainfall. It was also found that the intersect on the rainfall axis was about 100 mm. In other words, about 100 mm of rainfall is required before grain is formed, which demonstrates that precious rainfall is squandered through often unavoidable evaporation from the soil surface.

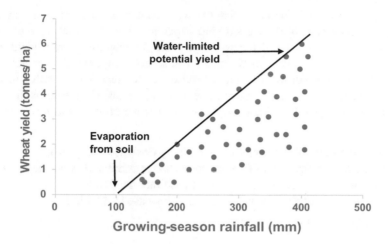

Fig. 23.1 Schematic representation of the relationship between wheat yield and growing season rainfall. Circles represent examples of individual farm paddock yields. (Modified with permission from [3])

There are many reasons for the surprising finding that rainfall had little bearing on crop yield in water-limited environments. The most important ones are as follows:

(i) There are many soil constraints other than rainfall such as soils may be too acid, too salty, or too hard, which limit the growth of the roots and hence water and nutrient uptake.
(ii) Soil-borne diseases are common which also limits the growth of an effective and healthy root system for the uptake of water.
(iii) And as a result of the above farmers may be reluctant to use adequate fertilisers because they are not cost effective. This will further limit yields if other constraints are minor.

Accordingly, improving performance in water-limited environments may not necessarily come from introducing particular physiological traits associated with water uptake and water-use efficiency because water may not be the main limiting factor for yield. Rather, improved yields may come from changing management practices that reduce soil-borne diseases of wheat or lessen soil constraints. It may also come from breeding for tolerance to soil-borne diseases or soil chemical constraints as these limit water uptake from the soil. These may have a greater impact than improving traits more directly involved with water uptake or the efficiency of water use.

This work by French and Shultz has provided a much more objective assessment of how and what changes to cropping systems and breeding are required for increased yield in water-limited conditions. It has been adopted widely by farmers and agronomists as a benchmark for measuring yield potential in rainfed regions in Australia, Argentina, USA and China [4].

Around the time French and Schultz were collecting data for their surprising findings Passioura published a seminal paper in 1977 [5] which simplified our understanding of the critical factors responsible for increasing grain yield in dry environments. He identified three factors that determines crop performance in dry environments. These provided a more precise way to identify factors that form the basis of genetic or agronomic improvement in crop yields when water is limited.

The three factors Passioura proposed to improve crop performance were as follows:

(i) Transpire more of the limited water supply (increase water use),
(ii) Increase the efficiency of this transpired water during the exchange of CO_2 for water to produce biomass (increase transpiration efficiency).
(iii) Convert more of the biomass into grain (increase harvest index)

This is simply stated as:

$$\text{Grain Yield} = \text{Water Use} \times \text{Water Use Efficiency} \times \text{Harvest Index}$$

Passioura suggested that an increase in any one of these three determinants should increase grain yield in water-limited environments. Furthermore, he suggested that unlike the yield components (spike number, grain number, grain size, etc.), each component is likely to be largely independent of the other enabling breeders to focus on selection for one or all determinants.

This framework was a radical departure from earlier thinking on ways to improve the growth and yield of water-limited crops. One of the most important aspects of this identity is that it focuses on crop productivity and not drought tolerance or drought resistance and it also removes the focus away from survival, which for crop plants, has little relevance. This latter point has been important as most candidate transgenes for drought have been identified from selecting for survival in managed conditions yet this bears no relationship to crop performance [6]. The focus on crop productivity also turned our attention to longer term processes associated with crop production and to resource limitations. It drew attention to the importance of practices pre-crop (stubble retention, fallow weed control, crop sequence, sowing time) and in-crop (weed control, fertiliser application) to improve available water use and water use efficiency so as to increase grain yield [7].

This identity provided a very important framework for improving wheat productivity in water-limited environments through genetic improvement.

23.3   Breeding and Selection for Yield in Water-Limited Environments

Wheat breeding is generally empirical – that is guided by experience. It is similar world-wide with selection during early generations for highly heritable traits such as flowering time, plant height, some disease resistances and some grain attributes.

After selection and selfing more homozygous germplasm is available for evaluation in larger field plots. Identification of more elite material is then made with a focus on grain yield, disease resistance and grain quality, if the grain is used for making end products. This elite material is then tested at multi-locations in the target region. Eventually, after consideration of yield performance, disease resistance and grain quality new cultivars are released. Molecular markers are likely to be included in the selection process for traits controlled by few loci. So far there are none that specifically target dry conditions. However, there are markers that can help optimise the time of flowering to avoid drought and markers for acid soil tolerance and nematode resistance that are important to improve the growth of root systems where soil acidity and nematodes are problems.

Gains in breeding under water-limited conditions are likely to be slower than under favourable conditions as year-to-year variation is highly unpredictable and can vary substantially. Accordingly, genotype x environment interactions for yield are high making yield progress slow. This raises the question as to whether further gains may be possible by selecting for specific physiological traits which influence water use, water-use efficiency or harvest index as well as grain yield.

23.4 Direct Selection for Grain Yield or Trait-Based Selection to Improve Performance Under Drought?

A discussion which is important is whether trait-based selection for drought is worthwhile or whether direct selection for grain yield is always going to be more effective. It is common to select for obvious defects in early generations such as grain sterility and susceptibility to disease; it is also common to select for appropriate flowering time or plant height and certain grain quality attributes. But it is rare for breeders to select for physiological traits that may be related to yield under drought. This is largely because easily selectable traits have already been selected and fixed in breeding germplasm; it is also because breeders believe they make more gain using direct selection for yield. It is generally assumed that direct selection for the highest yielding lines in water-limited environments will automatically combine the most favourable traits. Furthermore, the efficiency of direct selection for grain yield has improved in recent decades. Machinery for sowing and harvesting has vastly improved, robotics for seed packaging large trials speeds up the process and reduces errors, and improved herbicides has led to large trials where thousands of lines are evaluated in multi-locations. In addition, statistical tools to manage spatial variability and trial analysis have become outstanding. Improved understanding of limiting factors associated with soils or nutrition have also resulted in better agronomy of breeding trials. Overall, the efficiency of breeding and the direct selection for yield, which integrates all physiological processes, has resulted in very efficient breeding programs (see Chap. 2). Thus, one may ask what is the value of trait-based selection?

Trait-based selection does have highly appealing features for breeding. It is designed to complement existing breeding programs and is not dissimilar to approaches taken to improve specific resistances/tolerances to diseases, soil chemical constraint or for components of grain quality. Possible advantages of this trait-based approach to breeding have previously been enunciated [8]. They are briefly listed here with examples or specific comments given in italics.

1. The desirable expression or appropriate genetic variability for important physiological traits may not be present in breeding programs. Thus, genotypes with greater expression of important traits must be identified for use in breeding; this can lead to faster and greater genetic gain for important traits.
   Long coleoptiles for better emergence in dry soils in a semi-dwarf background are generally not found in breeding programs [8] – as are other proven traits such as early vigour, xylem vessel diameter.
2. The physiological trait may have a higher heritability than grain yield and so selection for it may lead to faster genetic gain in yield
   E.g. coleoptile length, early vigour, transpiration efficiency
3. Selection for the trait may be more cost-effective than selection for yield
   This must be the case for all traits if they are to be successful. It is worth pointing out that the cost per field plot for yield is not cheap.
4. Out-of-season selection or selection in controlled environments may be possible resulting in multiple cycles of selection per year and faster genetic gain.
   This is the case for most of the traits given in Table 23.1.
5. The trait may be amenable to marker-assisted selection, whereas grain yield is not.
   See also Table 23.1.
6. Multiple yield enhancing traits may be pyramided.
   A good example of this is coleoptile length and early seedling vigour [9].

23.5 Which Physiological Traits?

Flowering time is the most important trait in almost all dry environments. Fortunately, it is also one of the most heritable traits in wheat and it is easy to select visually. Ideally flowering must occur whilst conditions are still favourable and before it gets too dry or too hot. It is all to do with getting timing right. Time of flowering has been the single most important trait in most dry environments as it marks the transition between further growth of leaves, stems and tillers and the growth of grains. In many regions drought commonly occurs during grain-filling at the end of the season (i.e. terminal drought) when temperatures are higher and so evapotranspiration is also higher. In these circumstances the earlier flowering occurs then the more favourable conditions will be for grain filling. It is worth noting that since the beginning of wheat improvement in dryland Australia in the late 1800s breeders were selecting for greater yields but they were achieving this by inadvertently selecting

Table 23.1 Summary of the most important traits, selection environment and selection method for improving yield of temperate cereals in water-limited environments

Trait	Selection environment – favourable or droughted	Markers or genomic regions identified	Most efficient selection method
Time of flowering	Either	Yes	Phenotype and marker
Seedling establishment	Favourable	Yes	Phenotype and marker
Shoot vigour	Favourable	Yes	Phenotype
Root vigour	Favourable	Yes	Phenotype
Root architecture	Favourable	No	Phenotype
Transpiration efficiency (CID)	Favourable	Yes	Phenotype
Stomatal conductance (transpiration)	Favourable	Yes	Phenotype
Stem carbohydrate remobilization (WSC)	Favourable	Yes	Phenotype
Tillering	Favourable	Yes	Phenotype or marker
Glaucousness	Favourable	Yes	Phenotype
Leaf rolling	Favourable	Yes	Phenotype
Floret sterility	Non-droughted	Yes	Phenotype
Canopy architecture	Favourable	Yes	Phenotype

Modified with permission from [13]

for earlier maturity; the importance of phenology was probably not evident at the time.

Selection for physiological traits to indirectly improve yields started to receive attention around the time of the Green Revolution and the time that the dwarfing genes *Rht-B1b* and *Rht-D1b* were being widely recognised in breeding as a way of increasing grain yield and this drew attention to other possible physiological traits that may be important. For example, what role do awns play in wheat [10]? Are there root system traits, that should be important under drought, available to incorporate into wheats in dry regions and is there genetic variation available [11]? Also, much information was available in the ecological literature on how indigenous plants coped with chronic dry conditions and there was substantial interest in understanding the mechanisms involved as it was proposed that they may also be applied to crops. However, the reality is there are few similarities between plants growing in dry conditions in the wild and crops on farms. Indigenous plants in dry conditions must survive dry conditions whereas crops on farms must be managed so that they produce income for farmers. Survival tactics generally means very slow growth or the cessation of it and this limits the ability of the crop to respond to rainfall.

One of the important features of the Passioura identity was the focus away from survival and towards productivity. Each of the components of the identity are focused on crop growth that results in grain production when water is limited. It has become an important guide to identify traits in breeding as any increase in grain yield must come from an improvement in one of the three components. A corollary of this is that if breeders observe genetic variation for a trait in their populations then it will only be important for yield if it alters one of the three components. Thus, the identity can be used effectively to do a reality check on whether an observed trait will influence yield or not.

Table 23.1 shows a list of the most important traits that have been recommended to improve the grain yield of wheat where water is limited (e.g. [8, 12]). These traits may not be universally important in all rainfed environments as some may have greater impact in specific environments. Indeed, some traits listed may negatively impact on yield in some dry environments. A good example of this is fast early vigour which is considered highly desirable to increase the proportion of transpiration relative to evapotranspiration when the soil surface is exposed and mostly moist during the early vegetative phase as this increases crop water use and increases biomass. However, if the crop is growing on stored soil moisture the extra leaf area growth associated with early vigour is likely to deplete soil water such that little would be available for grain filling and yield would be lower. Further discussion on each of these traits is given in Richards et al. [13].

Several important features are apparent from Table 23.1. Firstly, the most effective environment to select for traits associated with performance under drought is under favourable moisture conditions. Favourable conditions maximise the phenotypic variance and heritability of each trait whereas dry conditions reduce them to slow genetic advance. Secondly, molecular markers or genomic regions (quantitative trait loci – QTL) have been identified for most of the key traits linked to improved performance under drought (Table 23.1). A third notable feature is that, currently, the accurate measurement of the actual phenotype rather than a molecular marker or QTL is the most efficient and fastest method of selection for almost all traits. This is because most traits are controlled by many genes.

There are several drawbacks to using QTL. Firstly, they vary with genetic background and so the identification of QTL is often specific only to the population being studied. QTL x environment interactions are extremely widespread. Finally, all QTL may only account for 30–70% of the total phenotypic variation whereas accurate measurement of the phenotype, even for polygenic traits, may be close to 100% of the phenotypic variation.

For the reasons above a considerable research investment into discovering ways to maximise repeatable phenotypic variation and ways to hasten the time taken for the measurement of the phenotype remains of utmost importance to make effective genetic gain.

It is worth noting that many of these traits will also be important for other abiotic stresses – in particular, adaptation to heat. The best examples here are: (i) time of flowering to adjust phenology, (ii) seedling establishment, (iii) glaucousness, (iv) leaf rolling, (v) canopy erectness. See Hunt et al. [14] for more detailed information on these traits in relation to heat.

Many of the traits shown in Table 23.1 are unlikely to be universally important as was mentioned earlier with the example of early vigour. Thus, some will be critical for some rainfall patterns and not for others. Some physiological traits may also require a particular crop management to obtain maximum benefit. Understanding these interactions will be important to capture the value in new varieties.

The same traits are given in Table 23.2 together with an assessment as to whether they are likely to be region specific and the management that may be important to increase their impact or expression. It is evident from Table 23.2 that if any of these traits are incorporated into released varieties then management practices could also be modified to further enhance their value on-farm. This point is particularly important as the greatest successes in breeding have often been associated with a particular management. The best example of this is the Green Revolution where wheats with the dwarfing genes were able to respond to better management and higher inputs because they did not lodge.

Table 23.2 Traits currently being studied or in breeding programs [13] that have been identified to improve yield in dry environments and an assessment of which management practices may influence their impact

Genetically altered trait	Region specific or universal	Agronomic condition or management practice that could influence trait impact
Time of flowering	Universal	Sowing time, prevalence of frost around flowering.
Seedling establishment (long coleoptile)	Universal	Timely sowing, stored soil water, pre-emergent herbicides
Early shoot vigour	Region specific	Late sowing, herbicide resistant weeds, reduced tillage, plant density and row spacing, nitrogen, sowing depth
Root vigour	Universal	Hard soil, nutrient deficient, hostile soil, cultivation, herbicides
Root depth	Region specific	Sowing density, row spacing, cultivation, seed dressings
Reduced tillering	Region specific	Sowing density, early sowing, nitrogen management, sowing depth
Transpiration efficiency/ stomatal conductance	Universal	Stored soil water at sowing, crop duration, sowing date, nitrogen management
Crop duration	Universal	Sowing date, nitrogen management, sowing density, row spacing, availability of grazing animals
Floret sterility	Universal	Sowing date, nitrogen management
Glaucousness	Universal	None identified
Stem carbohydrate storage and remobilisation	Universal	Sowing density, nitrogen management, fungicides
Stay green	Region specific (?)	Nitrogen management, fungicides
Canopy architecture at flowering	Universal	Sowing density, row spacing, nitrogen management

Modified with permission from [15]

23.6   Trait Validation and Translation to Breeding Programs

Once traits have been identified the next step is the most important. It is to translate the trait discovery to a product for farmers. It involves the incorporation of the trait into a breeding program and to validate the impact on yield. This can be done at the same time. There are several ways this can be accomplished, and it depends on the trait. If the trait is already in the breeding program and its expression is satisfactory then active selection for the trait is possible as lines progress through the breeding pipeline. If the expression of the trait is known (measured) for each line in a yield trial, then the relationship between trait expression and yield can be assessed. In these trials it is essential to also score height and flowering time on each line to ensure that these factors are equivalent for each trait and that they are not responsible for trait or yield variation.

When the expression of the physiological trait in a breeding population is inadequate and needs to be enhanced then new parental material is required to inject into the breeding program. Under these circumstances a more directed breeding program is required and the nature of it will depend on the inheritance and heritability of the physiological trait. Ideally, a backcrossing program is used to introduce the trait into a desirable background which will be suitable for release to farmers. This will also provide yield information on the high or low expression of the trait in the same genetic background. Conducting a backcrossing program for a complex trait is feasible providing the phenotype can be screened quickly and effectively. More detail on trait validation and incorporation of different traits into breeding programs is also described by Richards et al. [13]. An example of breeding for a complex physiological trait, which is also complex genetically, is given in the case study below.

In general, success in breeding depends upon being able to screen large numbers effectively, it also makes a substantial difference if the selectable trait has a high heritability and that breeders have substantial genetic variation in their breeding population so that selection can occur. But this can still result in slow progress because of large genotype x season interactions.

23.7   A Case Study of Translational Research: Breeding Wheat Varieties with High Transpiration Efficiency Using Carbon Isotope Discrimination

An improvement in transpiration efficiency (TE), i.e. the ratio of the rates of photosynthesis to transpiration, will be important in all water-limited environments provided it is not negatively associated with factors that increase water use or harvest index. During photosynthesis plants discriminate against the rarer $^{13}CO_2$ and prefer the more abundant $^{12}CO_2$. Farquhar and Richards [16] demonstrated that the degree of discrimination against ^{13}C was indeed related to TE in wheat and that there were genetic differences. They proposed that a measure of discrimination denoted as $\Delta^{13}C$ of plant material was a robust measure of TE as it was an integrated measure of photosynthesis and transpiration during the growth of that plant material. Thus, it is not a spot measure like leaf photosynthesis but a time integrated measure over the life of the plant sample measured. It was proposed that selecting for a low $\Delta^{13}C$ could increase TE of crops.

After investigating rainfall patterns throughout the wheat growing regions in Australia we targeted the northern wheat growing region as the region that low $\Delta^{13}C$ should be most effective. This region has less in-season rainfall as a proportion of total rainfall than other parts of Australia and hence relies more on water stored in the soil than other regions. Low $\Delta^{13}C$ can be associated with a lower stomatal conductance and so there may be an extra benefit for low $\Delta^{13}C$ in water-limited environments where there is a terminal drought, such as in Australia's northern region, as a lower conductance may conserve soil moisture for use during grain filling which is likely to increase harvest index [17].

For regions of Australia with a larger proportion of in-season rainfall, particularly during the winter, we believe greater progress in yield could be made by selecting for greater early vigour [8, 18, 19]. Lines with low $\Delta^{13}C$ may be at a disadvantage due to a possible negative association between early growth and low $\Delta^{13}C$ [19]. We undertook a detailed study on how carbon isotope discrimination ($\Delta^{13}C$) varies with season, genotype, growth conditions and the tissue to measure. This is described in Condon et al. [20]. This information was essential to establish the most effective way to screen germplasm for $\Delta^{13}C$. This aspect of the work took several years of research. It established that the $\Delta^{13}C$ was not expressed satisfactorily under controlled conditions and that it had to be measured in the field and that single plants or single rows could be used as they had the same value of $\Delta^{13}C$ as plots. It was also established that the measurement of $\Delta^{13}C$ is ideally done at the early mid-tillering stage of growth and that the soil moisture conditions should be favourable. If conditions are unfavourable, then this can alter stomatal conductance and hence alter the $\Delta^{13}C$ value. These factors established that optimal conditions were important to maximise the genetic component of $\Delta^{13}C$ variation and hence the heritability.

There was substantial risk involved in selecting for $\Delta^{13}C$ in a breeding program as it is a complex trait and, while QTL for $\Delta^{13}C$ have also been identified in several wheat populations, each of these QTL have a small effect and therefore unlikely to be useful in breeding [21]. On the other hand, earlier work established that the measurement of $\Delta^{13}C$ was highly repeatable and heritable and genotype x year interactions were small and it is an integrative measure over time [18].

The research described above was conducted at the same time as an extensive search was made for the most suitable donor of high TE (low $\Delta^{13}C$) to use in the breeding program. An older commercial variety from the southern part of Australia called Quarrion was chosen. It was a winter wheat, but a spring wheat was required for the target region. Despite some limitations Quarrion already had a reasonable 'package' of adaptation, disease resistance and grain quality to the target region and so this variety was unlikely to introduce too many undesirable features into the breeding program. A backcross program was embarked upon and the reason for this is that the recurrent parent from the target region that already possessed highly desirable attributes could be chosen. In this case Hartog was chosen as the recurrent parent. It was already very well adapted to the target region in terms of yield. It was accepted by growers because of its yield and it also had robust disease resistance and very good grain quality and most important it had a relatively low TE (high $\Delta^{13}C$). A breeding program was commenced to backcross low $\Delta^{13}C$ (high TE) from the donor parent Quarrion into the variety Hartog. Another commercial wheat was also chosen to be a recurrent parent that had very high yield but poor grain quality and a low TE. Over time the importance of grain quality in this region increased and so the focus on the Hartog background increased.

Time was clearly important as during the backcrossing disease resistances can break down and further breeding progress in yield can mean the recurrent parent is superceded. The initial generations were speeded up in the glasshouse and we conducted our first screen in the field on F_3 lines. Multiple low $\Delta^{13}C$ lines were selected and immediately crossed several times to Hartog and BC_2F_4 lines were developed in the glasshouse. Large numbers of these lines were grown in the field to select for low $\Delta^{13}C$. A substantial number of $BC_2F_{4.6}$ lines were then yield tested over several years at multiple locations as well as extensive grain quality and disease resistance testing. Limited backcrossing was done to retain as much variation as possible in agronomic and grain quality traits so that selection for these traits could also be carried out.

Studies demonstrated that in south-eastern Australia lines selected for low $\Delta^{13}C$ resulted in a 2 to 15% yield advantage at yield levels between 5 t ha^{-1} and 1 t ha^{-1} when compared with high-$\Delta^{13}C$ sister lines [22]. Subsequently the varieties Drysdale and Rees were released commercially. These varieties combined high TE with broad spectrum disease resistance and with high grain quality suitable for international markets. Unfortunately, soon after their release, a new exotic strain of stripe rust entered Australia that was virulent on Drysdale and Rees and this has limited the adoption of these varieties. Backing up the breeding program a more-recent spring wheat variety, LPB Scout, derived from parents with low $\Delta^{13}C$ was also released in Australia.

Clearly, $\Delta^{13}C$ is a complex trait and, while QTL for $\Delta^{13}C$ have also been identified in several wheat populations, each of these QTL have had a small effect and therefore unlikely to be useful in breeding [21].

23.8   The Elements of Success

Retrospectively it is evident that the approach enunciated by Passioura [5] to increase the yield of water-limited crops has been enlightening and has provided clear guidelines to both breeders and agronomists (see also [3]). It has been successful because it proposed a resource-driven approach linked to crop productivity instead of associating yield with drought resistance. A further extension to these ideas, developed by French and Schultz [2], identified a practical upper limit to the yield of field grown crops in water-limited environments. This upper limit, linearly related to the water supply, was adopted as a benchmark by agronomists and farmers, and has been particularly important to improving the management of water-limited crops worldwide.

The main elements of success have been to identify physiological traits to improve performance under drought and the following points are suggested as essential for success:

1. A clear physiological framework complimented by a rigorous understanding of the target environment.
2. A strong focus on wheat improvement for a target set of environments.
3. An integrated stable team with skills in agronomy, physiology, molecular biology, genetics and breeding that are mainly located together and who have daily dialogue.
4. A focus on precise phenotyping.
5. A commitment to field research and field validation using appropriate populations fixed for height and maturity but varying for the target trait(s).
6. Stability in funding and a long-term commitment to maintaining a broad skills base.
7. A commitment to the application of results and germplasm to commercial plant breeders, combined with a regular dialogue with breeders.
8. An interaction with farmers and knowledge of the broader cereal industry.

However, success in delivering to breeders and then breeders delivering new varieties to farmers is rare. Failure is where the trait is not adopted in breeding programs. There can be many reasons for failure and some are:

1. The hands-on commercial breeder does not have the time or commitment to the trait as does the pre-breeder. The breeder is more committed to his/her own material where they designed the cross and have nurtured the material through the breeding process.
2. There may be more immediate priorities for the breeder such as more robust disease resistance or better grain quality that will be more readily adopted by farmers.
3. The breeder may receive unadapted parental material from the pre-breeder which means that the breeder has to make the initial crosses and make selections in subsequent generations in unadapted material.
4. Where the breeder does receive adapted material such as in a BC_2F_3 material the genetic background may not be suitable to the breeder's target environment.
5. If the breeder has to make selection for the trait then she/he may not have the resources nor the intimate knowledge of the physiological trait to make effective selection.
6. There could be IP issues which may discourage commitment by the breeder.

It is proposed that for delivery of new varieties to farmers the best solution is for the pre-breeder to work side-by-side with the breeder throughout every part of the breeding process. This starts with the breeder having input into the most suitable genetic backgrounds to use in crossing and it may involve pre-release parental material from the breeding program. The breeder and pre-breeder may then guide the germplasm through early generation speed breeding to provide the pre-breeder with germplasm to conduct effective early generation selections for the desired trait. Later generations in the field then require input from both the breeder and pre-breeder.

23.9   Key Concepts

- Trait based selection can complement established breeding methods to improve yield in water-limited environments.
- The presence of limiting factors that impede the growth of an effective root system should first be explored and overcome if present e.g. root diseases and/or soil chemical constraints.
- Identification of important traits must be based on a crop productivity framework of water-use, water-use efficiency and harvest index. This must be in relation to the target environment.
- Management practices must be considered in relation to traits as they can be synergistic to yield.
- Most important traits are polygenic and unsuitable for marker-based selection. However, high throughput selection methods can generally be developed.
- A close working relationship with a commercial breeder is essential for success to develop an integrated varietal package for farmers and to validate traits in the field as quickly as possible.

23.10   Summary

A scientific understanding of factors underpinning adaptation to water-limited environments coupled with good genetics and breeding will deliver potential varieties and/or parents with potential for improved performance under drought in the target environments. Success in the delivery of new varieties with yield enhancing traits will finally depend on forming a strong relationship with a commercial breeder.

References

1. Fischer RA, Byerlee D, Edmeades GO (2014) Crop yield and global food security: will yield increase continue to feed the world (Monograph 158). Australian Centre for International Agricultural Research, Canberra

2. French RJ, Schultz JE (1984) Water-use efficiency of wheat in a Mediterranean-type environment. I. The relation between yield, water-use and climate. Aust J Agric Res 35:743–764
3. Passioura JB, Angus JF (2010) Improving productivity of crops in water-limited environments. Adv Agron 106:37–75
4. Sadras VO (2020) On water-use efficiency, boundary functions, and yield gaps: French and Schultz insight and legacy. Crop Sci 60:2187–2191
5. Passioura JB (1977) Grain-yield, harvest index, and water-use of wheat. J Aust Inst Agric Sci 43:117–120
6. Passioura J (2006) Increasing crop productivity when water is scarce - from breeding to field management. Agric Water Manag 80:176–196
7. Kirkegaard JA, Hunt JR (2010) Increasing productivity by matching farming system management and genotype in water-limited environments. J Exp Bot 61:4129–4143
8. Richards RA, Rebetzke GJ, Condon AG, van Herwaarden AF (2002) Breeding opportunities for increasing the efficiency of water use and crop yield in temperate cereals. Crop Sci 42:111–121
9. Rebetzke GJ, Richards RA (1999) Genetic improvement of early vigour in wheat. Aust J Agric Res 50:291–301. https://doi.org/10.1071/A98125
10. Evans LT, Bingham J, Jackson P, Sutherland J (1972) Effect of awns and drought on the supply of photosynthate and its distribution within wheat ears. Ann Appl Biol 70:67–76. https://doi.org/10.1111/j.1744-7348.1972.tb04689.x
11. Hurd EA (1974) Phenotype and drought tolerance in wheat. Agric Meteorol 14:39–55
12. Reynolds MP, Saint PC, Saad ASI, Vargas M, Condon AG (2007) Evaluating potential genetic gains in wheat associated with stress-adaptive trait expression in elite genetic resources under drought and heat stress. Crop Sci 47:S 172–S-189. https://doi.org/10.2135/cropsci2007.10.0022IPBS
13. Richards RA, Rebetzke GJ, Watt M, Condon AG, Spielmeyer W, Dolferus R (2010) Breeding for improved water productivity in temperate cereals: phenotyping, quantitative trait loci, markers and the selection environment. Funct Plant Biol 37:85 97
14. Hunt JR, Hayman PT, Richards RA, Passioura JB (2018) Opportunities to reduce heat damage in rain-fed wheat crops based on plant breeding and agronomic management. Field Crop Res 224:126–138. https://doi.org/10.1016/j.fcr.2018.05.012
15. Richards RA, Hunt JR, Kirkegaard JA, Passioura JB (2014) Yield improvement and adaptation of wheat to water-limited environments in Australia - a case study. Crop Pasture Sci 65:676–689. https://doi.org/10.1071/CP13426
16. Farquhar GD, Richards RA (1984) Isotopic composition of plant carbon correlates with water-use efficiency of wheat genotypes. Aust J Plant Physiol 11:539–552
17. Schoppach R, Fleury D, Sinclair TR, Sadok W (2017) Transpiration sensitivity to evaporative demand across 120 years of breeding of Australian wheat cultivars. J Agron Crop Sci 203:219–226
18. Richards RA, Lukacs Z (2002) Seedling vigour in wheat-sources of variation for genetic and agronomic improvement. Aust J Agric Res 53:41–50. https://doi.org/10.1071/AR00147
19. Condon A, Richards R, Rebetzke G, Farquhar G (2004) Breeding for high water-use efficiency. J Exp Bot 55:2447–2460. https://doi.org/10.1093/jxb/erh277
20. Condon AG, Richards RA, Farquhar GD (1992) The effect of variation in soil-water availability, vapor-pressure deficit and nitrogen nutrition on carbon isotope discrimination in wheat. Aust J Agric Res 43:935–947
21. Rebetzke GJ, Condon AG, Farquhar GD, Appels R, Richards RA (2008) Quantitative trait loci for carbon isotope discrimination are repeatable across environments and wheat mapping populations. Theor Appl Genet 118:123–137. https://doi.org/10.1007/s00122-008-0882-4
22. Rebetzke GJ, Condon AG, Richards RA, Farquhar GD (2002) Selection for reduced carbon isotope discrimination increases aerial biomass and grain yield of rainfed bread wheat. Crop Sci 42:739–745

Chapter 24
Micronutrient Toxicity and Deficiency

Peter Langridge

Abstract Micronutrients are essential for plant growth although required in only very small amounts. There are eight micronutrients needed for healthy growth of wheat: chlorine, iron, boron, manganese, zinc, copper, nickel and molybdenum. Several factors will influence the availability of micronutrients, including levels in the soil, and mobility or availability. Zinc deficiency is the most significant problem globally followed by boron, molybdenum, copper, manganese and iron. Deficiency is usually addressed through application of nutrients to seeds, or through foliar spays when symptoms develop. There is considerable genetic variation in the efficiency of micronutrient uptake in wheat, but this is not a major selection target for breeding programs given the agronomic solutions. However, for some micronutrients, the concentrations in the soil can be very high and result in toxicity. Of the micronutrients, the narrowest range between deficiency and toxicity is for boron and toxicity is a significant problem in some regions. Although not a micronutrient, aluminium toxicity is also a major factor limiting yield in many areas, usually associated with a low soil pH. Agronomic solutions for boron and aluminium toxicity are difficult and expensive. Consequently, genetic approaches have dominated the strategies for addressing toxicity and good sources of tolerance are available.

Keywords Micronutrients · Boron · Aluminium · Deficiency · Toxicity

P. Langridge (✉)
School of Agriculture Food and Wine, University of Adelaide, Adelaide, SA, Australia

Wheat Initiative, Julius-Kühn-Institute, Berlin, Germany
e-mail: peter.langridge@adelaide.edu.au

© The Author(s) 2022
M. P. Reynolds, H.-J. Braun (eds.), *Wheat Improvement*,
https://doi.org/10.1007/978-3-030-90673-3_24

24.1 Learning Objectives

- Recognizing the symptoms and possible causes of micronutrient deficiency or toxicity.
- An understanding of the agronomic or genetic strategies that can be used to correct the problems of micronutrient deficiency.
- Ability to decide when agronomic or genetic interventions may be needed.

24.2 Introduction

Seventeen elements have been identified as essential for healthy plant growth and development. These are usually grouped as major or macro-nutrients and micronutrients based on the amount required by the plants. The major elements and the concentrations (mmol/kg) needed for normal growth are: carbon (C, 40,000), oxygen (O, 30,000), hydrogen (H, 60,000), nitrogen (N, 1000), phosphorus (P, 60), potassium (K, 250), calcium (Ca, 125), magnesium (Mg, 80) and sulphur (S, 30). The demand for micronutrients is much lower reflecting their role in specific biological processes rather than as major building blocks for plant organs: chlorine (Cl, 3.0), iron (Fe, 2.0), boron (B, 2.0), manganese (Mn, 1.0), zinc (Zn, 0.3), copper (cu, 0.1), nickel (Ni, 0.05) and molybdenum (Mo, 0.001). Several studies indicate that silicon (Si) may be beneficial, but not essential, for wheat production. Other elements, particularly heavy metals such as cobalt (Co), required by legumes, and cadmium (Cd) can be taken up by wheat plants and deposited in the grain and, although they may have little effect on plant growth, they are highly undesirable for human consumption.

Although micronutrients are required in only very small amounts, their absence can have highly adverse effects on healthy growth and, consequently, on yield. In extreme case, the plants will not survive since these nutrients are essential. Low levels of micronutrients in grain will also reduce their nutritive value for humans.

The availability of nutrients for plants can be highly variable and dynamic and is influenced by a range of inputs including fertilizers, pollutants and the chemistry of the soil, in addition to losses through leaching, erosion and removal (harvesting) of plant material. Weathering and solubilisation of rock, soil and organic matter can all lead to the input of metal ions. A dynamic equilibrium will develop between pools of nutrients and the soil solution. This is influenced by the rate of replenishment of ions. The replenishment is also referred to as the capacity factor for a particular soil and the ion activity in the soil solution is called the intensity. The interactions between the capacity and intensity are strongly influenced by the soil pH and soil structure.

In addition to affecting the availability of micronutrients, extremes of soil pH can also lead to nutrient toxicities. Highly acidic soils can lead to Al and Mn toxicity and deficiency in Mo, while alkaline soils will often show B toxicity and Fe, Zn and Mn deficiency. For all micronutrients, there is a range of concentration in the soil that is ideal for growth; too little will limit growth, while too much can result in toxicity. The major toxicity problems for wheat production, apart from salinity, are due to aluminium, which is not a required micronutrient for wheat growth, and boron, which has the narrowest range of concentration for optimal growth of all micronutrients.

24.3   Deficiency

Micronutrient deficiencies can lead to a wide range of alterations in normal plant growth and development. Visual symptoms (Table 24.1) are usually only apparent under extreme deficiency, but mild deficiencies can result in substantial reductions in grain yield. Given the variable role of these elements, the symptoms of deficiency also vary greatly (Table 24.1). There are good images available on the internet for the symptoms of micronutrient deficiencies (see Exercise 24.9.1). The nutritional

Table 24.1 Micronutrients required for healthy plant growth, their role in plant metabolism and symptoms associated with deficiencies

Micronutrient	Pathway	Enzymes	Symptoms
Copper	Electron transport	Ascorbic acid oxidase, tyrosinase, monoamine oxidase, uricase, cytochrome oxidase, phenolase, laccase, and plastocyanin	Unlignified cell walls, permanent wilting and limp leaves
Chlorine	Photosynthetic reactions		Poor germination, chlorosis and nectrotic lesions
Manganese	Respiration	Some dehydrogenases, decarboxylases, kinases, oxidases and peroxidases	Reduced sugar and cellulose content, increased drought sensitivity, reduced fertility
Nickel	Unclear	Urease and hydrogenases	Impeded use of nitrogenous fertilisers
Molybdenum	Nitrogen use	Nitrogenase, nitrogen reductase	Nitrogen deficiency, chlorosis and necrosis on leaf margins. Leaves become pale ad malformed.
Boron	Cell division, growth and membrane function	Synthesis of uracil, cell wall structure	Problems related to cell wall formation including reduced shoot and root growth, infertility
Zinc	Electron transport and auxin biosynthesis	Alcohol dehydrogenase, glutamic dehydrogenase, and carbonic anhydrase	Interveinal chlorosis, and necrosis particularly in older leaves

Based on information from [4]

status of the plant will also affect its susceptibility to disease; in some cases, decreasing and in other, increasing disease susceptibility [1]. For example, Mn plays an important role in lignin and phenol biosynthesis and Mn application has been used to control a range of diseases including mildew, take-all and tan spot (for example, Simoglou and Dordas [2]). Zinc has also been found to reduce disease severity, but this may be due to a Zn effect on the pathogen rather than through changes to the plant metabolism [3]. Of the other micronutrients there is little clear evidence of an effect on disease response, although silicon may provide some protection to insect predation [3].

24.4   Areas of the World Most Susceptible to Nutrient Deficiencies or Toxicity

Several factors can lead to micronutrient deficiency in plants including low levels of the nutrients in the soil and low mobility or availability of the nutrients due to low solubility in the form required for uptake. Soil-microbe interactions can also influence the availability of the micronutrients. Where free $CaCO_3$ is abundant in the soil chemistry, this can fix micronutrient cations, at a high soil pH the solubility of many micronutrients is reduced, and replenishment can be low if there is little organic matter in the soil. The impact of pH on nutrient availability is represented in Fig. 24.1.

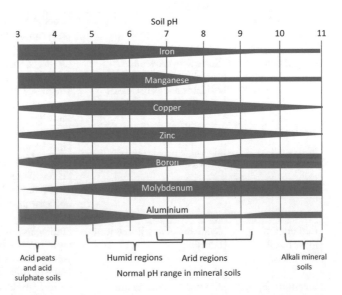

Fig. 24.1 Diagrammatic representation of the relationship between soil pH and micronutrient availability. (Modified with permission from Plants in Action [4] http://plantsinaction.science. uq.edu.au, published by the Australian Society of Plant Scientists)

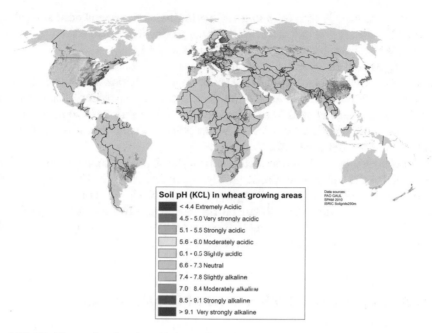

Fig. 24.2 World map showing the soil pH in wheat growing regions. (Dr. Kai Sonder, personal communication based on data from [5, 6])

The importance of soil pH in influencing both micronutrient deficiency and toxicity, is emphasized by the diversity of environments where wheat is grown. In Fig. 24.2 the soil pH in wheat growing regions is illustrated and shows that both high and low pH soil can be found. Acid soils are particularly prevalent in Europe, Eastern USA and southern Brazil while alkaline soils are found around the Mediterranean, the Middle East through to Western India, Northern China and Australia.

Estimating the full impact of micronutrient deficiencies is difficult. Although there has been extensive compositional analysis of soil in some regions, there are many areas where detailed information is lacking. A country level analysis found that "once the macronutrient deficiencies of soils are treated, Sillanpää [7] estimated that of the important agricultural soils of the world, 49% are deficient in zinc (Zn), 31% deficient in boron (B), 15% deficient in molybdenum (Mo), 14% deficient in copper (Cu), 10% deficient in manganese (Mn) and 3% deficient in iron (Fe)."

Globally, zinc deficiency is the most important for wheat production, and is particularly severe in Mediterranean-type and arid environments such as Turkey, Libya, and parts of India and Pakistan. Zn solubility in soils decreases with rising soil pH (Fig. 24.1) and high soil phosphorus can also induce Zn deficiency. Indeed, there is a link between Zn and P uptake, wheat plants under Zn deficiency will increase P uptake to a level that can be toxic [8]. This effect appears to be related to the observation that Zn deficiency up-regulates the expression of high affinity phosphate transporters [9].

Iron deficiency is seen in similar regions to Zn deficiency and occurs in calcareous soils, which cover extensive areas of crop production. As noted above, Fe availability is also strongly influenced by soil pH (Fig. 24.1).

Manganese deficiency can occur in coarse textured alkaline soils where it can be leached out of the soils. This can be a serious issue in regions where wheat is grown in rotation with rice and inundation of the soil can leach Mn into deep soil layers. Molybdenum deficiency is less widespread but can be serious in some regions, for example 44.67 million hectares in winter wheat production areas in China is regarded as Mo deficient [10].

In contrast, boron is a non-metal nutrient that is quite mobile in soils and can leach down the soil profile. Consequently, boron deficiency has been a problem in some humid climates, such as Bangladesh, Thailand and parts of China. Of all the micronutrients, boron has a particularly narrow range between deficiency and toxicity. Deficiency occurs where soluble B (boric acid) is below 0.5 mg/kg and toxicity occur at concentrations higher than 5.0 mg/kg.

24.5   Importance of Micronutrient Content of Grain for End Users

A wide range of factors influence the nutritional quality of the wheat grain and processed products (see Chap. 12). Not surprisingly, the ability of wheat to effectively take up nutrients from the soil will impact on the overall nutrient composition of the plant and the harvested grain. Ensuring a healthy and nutritionally balanced plant, is fundamental to producing nutritious grain for human and other animal consumption. Most effort in elevating micronutrients composition of wheat grain has focused on zinc and iron and, in addition to Chap. 12, there are several good reviews covering this topic (for example, Yu and Tian [11]).

24.6   Agronomic Approaches to Addressing Nutrient Deficiency

There are several options for managing potential micronutrient deficiencies. These include applying micronutrients directly to the soil, as a foliar pray or through seed treatments. Soil fertilization can suffer from problems of nutrient availability and may require high doses of fertilizer. Foliar sprays are generally regarded as the most effective in improving yield and the nutritional status of the grain. An advantage of spraying is that farmers can wait to see if symptoms of nutrient deficiency become visible before spraying but this also means that spraying will occur at late crop developmental stages and this may be too late for some deficiencies to be corrected. Spraying can be high cost and not easily applied for resource poor farmers. Overall, seed treatment is generally regarded as the best agronomic option for addressing micronutrient deficiency [12].

There are two basic approaches to treating seed to address micronutrient deficiency. A low technology and low-cost approach is known as seed priming, where wheat grains for sowing are soaked in a nutrient solution to partially rehydrate but avoiding allowing germination (reviewed in Farooq et al. [12]). Grain can then be redried to allow storage and transport. The simplicity of this method makes it suitable for on-farm application. Primed seed will usually germinate more rapidly and evenly than un-primed seed. This approach has been successfully used for zinc (use of 0.3% zinc sulphate), boron (0.008 M boric acid), manganese (0.1 M manganese sulphate), and copper (0.1 M copper sulphate) (reviewed in Farooq et al. [12]).

A more sophisticated approach to seed preparation is through seed coating (reviewed in Afzal et al. [13]). In recent years there has been significant improvement in seed coating technologies and, in addition to helping address micronutrient deficiencies, seed coating can also be used to apply fungicides, insecticides, nematicides and biostimulants. Wiatrak [14] evaluated polymer coating combined with a mixture of manganese, copper and zinc. The seed coating improved dry matter yield by 23%, N uptake by 25%, P uptake by 23% and grain yield was 2% higher than the control [14]. Seed coating does require some specialist equipment for the different methods of application: a dry powder applicator, rotary coater or drum coater. Seed dressing with a rotary coater is quite widely used on-farm and offers a simple method for applying micronutrients.

24.7   Genetic Approaches to Improving Nutrient Uptake

Nutrient use efficiency is defined as the ability of a cultivar to grow and yield well compared to a standard cultivar in soils deficient in the target nutrient. There does appear to be useful genetic variation in micronutrient efficiency for most micronutrients. Assessment and screening of germplasm has been primarily based on measuring yield of different cultivars in fields know to suffer from specific micronutrient efficiencies. In some case, controlled environment, greenhouse or growth rooms, or hydroponic systems have been used to evaluate uptake efficiency. Since micronutrients are required in such small amounts, screening can be complicated by the nature of the growth medium being used since very low levels of micronutrients present in water or on equipment can influence the results. Further, the level of micronutrient in the seed used for sowing, will have a significant impact. Careful characterization is needed to ensure that differences observed in the plant performance are indeed related to the target micronutrient or to variation in the nutrient content of the seed used for the experiments. In addition, to considering the chemical and structural properties of the soil, when using soil-based screening methods, it is also important to consider the possible influence of soil microorganisms on micronutrient availability [15]. Advances in genomics technologies has provided an opportunity to explore the diversity of the microbial populations associated with plant roots. The plant-microbe interactions we see in agricultural systems have resulted from co-evolution of plants and microbes in natural ecosystems and the combination of crop genomics with molecular microbiology offers options for modifying the interactions to improve the sustainability of crop production [16].

The most widely used approach has been to grow out diverse bread and durum wheat accessions in environments known to be deficient in specific micronutrients and assess their performance using plots fertilized with the deficient nutrient as controls. The nutrient content of the plants and the grain is usually also measured to provide an indication of the nutrient uptake efficiency. For example, a screening of 24 genotypes in India under manganese deficient or sufficient (based on foliar sprays) conditions was used to identify lines able to maintain yield under Mn deficient conditions [17]. In this case, grain yield was related to grain Mn content and uptake with Mn efficiency and Mn uptake accounting for 86% and 66% of the yield differences under low Mn [17]. In another screen of 61 cultivars, 18 were identified as inefficient in Mn uptake, 21 as slightly and 11 as moderately efficient [18]. Similar results are seen for molybdenum efficiency with Mo efficient lines yielding 90% while Mo inefficient lines yielded only 50% under Mo deficient conditions compared to the same lines under Mo fertilization [10]. Genotypic variation in performance under boron deficiency based on seed set also ranges from 97% for efficient lines compared to only 11% in inefficient germplasm [19].

Field-based approaches to screening for nutrient efficiency can be complicated by other environmental and edaphic factors. For some micronutrients, pot trials in greenhouses can be used. For example, variations in Mn efficiency can be detected in pot trials by measuring plant biomass accumulation. Hydroponics or a supported hydroponic system can be used in some cases although there can be issues related to differences in root architecture and structure compared to soil grown plants. Shen et al. [20] screened 26 wheat cultivars for variation in responses to iron deficiency using plants grown initially in quartz sand and then transferred to a hydroponic system. This system allowed measurements of a number of physiological and biochemical factors associated with iron uptake and use including siderophore release and resulted in the identification of lines particularly tolerant to iron deficiency [20].

While good variation has been found in wheat germplasm collections for the efficient uptake and utilization of most micronutrients, the level of efficiency offered may not be sufficient to deal with deficiency in some regions. For example, several studies have identified genetic variation in the severity of a number of symptoms associated with copper deficiency [21]. In such cases, there may be an opportunity to explore wild or close relatives of wheat as a source for high efficiency. Cereal rye (*Secale cereale*) has been identified a possible source of high efficiency since it is able to grow well in environments known to be highly deficient in micronutrients. In the case of copper efficiency, a gene on rye chromosome 5RL provided good Cu efficiency when transferred into a wheat background [22].

The genetic control of micronutrient efficiency has been studied primarily from the perspective of enhancing the grain micronutrient content and this is impacted by both the uptake of the micronutrients by the plant and the translocation to the grain (see Chap. 12). Relocation of nutrients to the grain does not appear to be related to specific nutrients since accessions showing good translocation of Zn to the grain also show high levels of other nutrients (Chap. 12). In contrast, the genetic control of micronutrient uptake appears to be specific for individual micronutrients since germplasm screening has not shown efficiency for multiple micronutrients although this may also be due to the lack of overlap between germplasm pools used in screening.

The broad spread in efficiency seen in germplasm screens, does suggest that efficiency is under complex genetic control, which could be due to multiple loci or high allelic diversity at a small number of loci. Results of genetic studies appear to be contradictory concerning the number of loci influencing micronutrient uptake efficiency. For example, a study of Zn accumulation using genome wide association study (GWAS) found seven loci associated with grain accumulation [23]. This complexity is reflected in the number of genes know to be associated with micronutrient uptake and transport with over 20 genes identified in wheat [24]. In contrast, single major genes have been identified as potential candidates for efficient uptake of copper, chloride and manganese where 42% of the total variation could be explained by a single locus in durum wheat [25].

Overall, our knowledge of the genetic control of micronutrient efficiency is largely based around work aimed at improving the micronutrient content of the grain (see Chap. 12) rather than uptake efficiency. Given the availability of alternative strategies for addressing micronutrient deficiency, largely through seed treatment or dressing, direct selection for micronutrient efficiency in breeding program is a generally a low priority.

24.8 Micronutrient Toxicity

Micronutrient toxicity occurs when the level of soluble nutrients in the soil exceeds a tolerance threshold. The most important micronutrient toxicities are aluminium, boron and manganese, with Al and B the most significant for wheat production areas. Salinity is also a major and increasing problem in many regions but is not regarded as a micronutrient toxicity. In contrast to nutrient deficiencies, there are few management or agronomic options for ameliorating toxicities. In the case of Al toxicity due to soil acidity, liming is an option but is largely used only in wealthy countries. Genetic solutions to micronutrient toxicity problems represent the primary option for control. This is reflected in the extensive work that has been undertaken into the elucidation of the genetic control of toxicity tolerance. For both B and Al tolerance, the genes controlling tolerance have been isolated and their mode of action extensively studied.

Mn toxicity does affect some wheat producing areas where soils are acid and waterlogged or poorly drained. The symptoms of Mn toxicity include reduced growth, interveinal chlorosis, leaf tip necrosis and brown spots on mature leaves [26]. There is genetic variation for Mn toxicity tolerance based on hydroponic screens and screening for tolerant germplasm in a breeding program is feasible [27]. However, Mn toxicity tends to be transient and is not considered a major breeding objective. In contrast, Al and B toxicity tolerance are significant breeding objectives is many wheat growing regions.

24.8.1 Boron Toxicity

Boron can accumulate to toxic levels in dry environments on alkaline soils of marine or volcanic origin and, in some cases, as a result of long-term irrigation [28]. The main form of boron is soil solution is as $B(OH)_3$ or boric acid. Globally, more areas are affected by boron deficiency than toxicity. However, toxicity occurs in many areas where wheat is grown, including, southern Australia, the Middle East from Turkey to Israel, areas in Peru and Chile, parts of Russia and central Asia, and on the ferralsols of India [28]. Boron toxicity symptoms are characterized by leaf necrosis moving from the leaf tips inwards due to the deposition of boron in tissues at the end of the plant transpiration stream (Fig. 24.3b). High soil boron also causes severe root stunting in susceptible lines (Fig. 24.3d). There are very few viable options for ameliorating boron toxicity apart from extensive leaching with low B water [28]. Fortunately, there is good genetic variation for boron tolerance in bread and durum wheat (Fig. 24.3a). In a study in Australia involving an extensive wheat germplasm

Fig. 24.3 Boron toxicity symptoms and screening. Genetic diversity in boron tolerance is illustrated through the images of leaves from plants grown in high boron soil (**a**). The lines shown, from left to rights, are India 126, G61450 (landraces from India and Greece respectively), Australian cultivars Halberd, Moray, Wyona, Warigul, Schomburgk, WI*MMC, Reeves and an African landrace, Kenya Farmer. The leaf symptoms of boron toxicity (**b**) are characterized by necrosis proceeding inward from the leaf tip. Screening for tolerance can be undertaken by growing seedlings in high boron soil boxes (**c**) or using a hydroponic screen. In boron sensitive lines, high boron severely inhibits root growth (**d**)

collection grown at 233 sites over 12 years, varieties tolerant to boron were found to yield around 16% more than intolerant genotypes in regions where boron toxicity was known to be a problem [29]. Since symptoms of susceptibility to high soil boron are visible in seedlings, with tolerant lines showing no or reduced symptoms, hydroponic screens (Fig. 24.3d and Exercise 24.9.1) or sowing seeds in seedling trays containing high boron soil (Fig. 24.3c) can be used as simple and rapid screens.

In bread wheat, tolerance is predominantly conferred by the *Bo1* gene which is thought to have originated in wheat varieties in Australia in the early twentieth century. This gene is located on chromosome 7BL in the bread wheat variety Halberd [30] and is also found on 7BL in durum wheat cultivar Lingzhi [31]. A further locus for tolerance was identified in a bread wheat landrace G61450 [32]. The underlying genes have been isolated and characterized [33]. The gene encodes a root-specific boron transporter that appears to function by pumping boron out of the root thereby preventing excess boron from entering the transpiration stream. Interesting, the tolerance locus found in cultivated wheat appears to have arisen via several genomic changes involving tetraploid introgression, dispersed gene duplication, and changes in gene structure resulting in variation in gene expression. The extensive allelic variation seen in the 7BL gene, has resulted in the range in tolerance responses represented in Fig. 24.3a.

A survey of allelic diversity in advanced breeding lines in Australian breeding germplasm, identified the deployment of four different alleles at the *Bo1* locus on 7BL. The allele *Bo1-B5b* was the most widely used in southern Australia where boron toxicity is an issue but was almost completely absent in advanced lines in the Northern regions where the *Bo1-B5g* allele dominated [33]. These results suggest that there is active selection against the boron tolerance allele in regions where soil boron is present at non-toxic levels and this likely reflects the narrow range between deficiency and toxicity for this element.

Through the isolation of the *Bo1* gene and characterization of allelic diversity at this locus, breeders can make use of diagnostic markers to ensure the appropriate level of tolerance or efficiency is present in their breeding lines [33].

24.8.2   Aluminium Toxicity

Aluminium is highly abundant in soils and under normal conditions it remains in an insoluble form as Al-oxyhydroxides or as clay minerals. However, at low pH (below 4.5) Al can become soluble as the highly toxic Al^{3+} cation. In this toxic form, Al can block root growth and severely hinder plant growth and development. Al toxicity is one of the most widespread limitations to crop production and ranks with salinity and water stress in the extent of its effect. Acid soils have been estimated to affect around 30% of the world's cropping area and in many regions, the area affected is increasing as a result of farming practices [34]. In Europe and North America, lime ($CaCO_3$) is widely used to reduce soil acidity. If the pH can be raised to 6 or 7, Al^{3+} will be insoluble and no longer a problem. However, in poorer regions, particularly in South America and Sub-Saharan Africa, liming is not an option and soil acidity is a major limitation to production.

Al toxicity primarily affects root growth with strong inhibition of root hair development and root branching (Fig. 24.4a). Seeds will often germinate and appear normal, but as the inhibition of root growth becomes more severe, plants will start to wilt. The strong impact of Al on root growth means that a simple hydroponic screen can be used to identify tolerant germplasm (Fig. 24.4b, c, d). The regions of the root affected by Al are areas where cells are dividing and expanding, around the root tip, and the elongation and root hair zones.

It is important to note that Al can also have a negative impact on the uptake and transport of a range of nutrients in wheat. There is also some evidence that the severity of Al toxicity can be influenced by the uptake efficiency of several nutrients, particularly iron [35].

There is considerable variation in tolerance to Al in both bread and durum wheat although the genetic control differs. The ability of some wheat cultivars to tolerate Al is related to the exclusion of Al from the root tip. A major locus for tolerance is

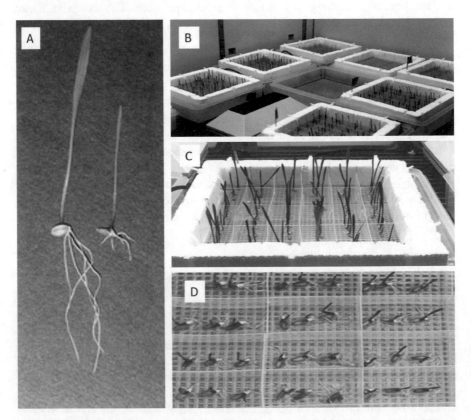

Fig. 24.4 Symptoms and screening for Al toxicity tolerance. The severe inhibitory effect of Al on root growth is shown (**a**). The reduced seedling growth is also apparent. Screening for Al tolerance can be readily undertaken using a hydroponic system shown in **b, c** and **d**

found on 4DL of bread wheat but most studies indicate that there are a number of other genes that can also influence the level of tolerance [36]. The gene at the 4DL locus, *TaALMT1*, encodes a transporter protein that serves as a ligand-activated anion channel [37]. This gene is constitutively expressed in root apices at a higher level in tolerant compared in intolerant genotypes. The mode of action is through the release of malate anions from the root apices which appears to chelate the Al^{3+} in the apoplast to render it harmless [38]. Several studies have explored the location and impact of other genes with possible loci on 5AS, 6AL, 7AS, 2DL and 3DL (reviewed in Ryan [26]). These additional loci may have potential in lifting the level of tolerance in wheat but currently, selection has focused on the *TaALMT1* locus on 4DL.

Ryan (2018) suggested a number of options for increasing the current level of tolerance found in wheat germplasm including the search for novel alleles, given the known diversity at this locus and evidence that rye (*Secale cereale*) has a far higher level of tolerance than its close relative wheat. Pyramiding Al-tolerance loci, and the possibility of using genetic engineering or gene editing to enhance expression of *TaALMT1*, are additional options.

24.9 Exercises

24.9.1 Support the Diagnosis of Micronutrient Deficiencies in Wheat

The internet provides a good resource for identifying the symptoms of micronutrient deficiency. Conduct an image search using the follow terms "wheat" plus "deficiency" plus "symptoms" plus "zinc" or "iron" or "boron", or "copper" or "nickel" or "chlorine" or "manganese" or "molybdenum". Assemble the images showing symptoms and prepare a description of the key phenotypes. Focus on the leaf symptoms and try and provide a description that allows differentiation of the symptom's characteristic for each deficiency.

24.9.2 Establish a Filter-Based System for Screening Wheat Accessions for Tolerance to Boron Toxicity

A simple procedure is described below for screening wheat accessions for boron tolerance. In selecting germplasm to screen, you will usually find that landraces from the eastern Mediterranean and North Africa and elite germplasm from Southern Australia have some level of tolerance, while European and North American cultivars are quite sensitive. This filter-paper method of screening can also be used to assess aluminium toxicity tolerance.

Use three treatment levels for the solution culture-root length assay at 100 mg B l^{-1} (B100), 50 mg B l^{-1} and 0 mg B l^{-1} (B0). Seedling root lengths of wheat varieties will respond consistently at the concentrations: 50, 100. A control treatment (B0) was included to account for genetic variation of root length in the absence of boron toxicity. Seeds of each line should be surface sterilized with 5.0% sodium hypochlorite and pre-germinated for 8 days at 4 °C in Petrie dishes on filter paper soaked in water. After the 8 days, take three evenly germinated seeds, for each accession, and place these embryo-downwards at a spacing of 2 cm across the middle of filter paper (Ekwip 32 x 46 cm grade R6) soaked in either the B0 or B100 solutions. The base solution used in both the control (B0) and high-concentration treatment (B100) must include 0.5 mM $Ca(NO_3)_2$, 0.0025 mM $ZnSO_4$ and 0.015 mM H_3BO_3, following the method of Chantachume et al. [39]. For the B50 and B100 treatments, add the appropriate additional H_3BO (50 or 100 mg per litre). The filter papers were rolled and covered with aluminium foil, then stored upright at 15 °C for 12 days. After the 12 day period, unroll the filter paper and measure the length of longest root of each seedling. Use the ration of the root length in the controls (B0) to the B50 and B100 treatments as the measure of boron toxicity tolerance.

24.10   Key Concepts

- Micronutrients are critical for plant growth and are not always easy to identify. Multiple strategies can be employed to address deficiency or toxicity problems.
- Deficiency is usually managed through seed priming or coating, or foliar sprays when symptoms first show.
- Many studies have identified extensive genetic variation in micronutrient uptake efficiency but use of this germplasm is not a high priority for most breeding program.
- The prime focus of micronutrient uptake and transport has been on enhancing the nutritional value of wheat grains for humans.
- Nutrient toxicity is most appropriately managed through genetic improvement of wheat since agronomic approaches are generally inefficient, short-term and expensive.
- The major genes controlling boron and aluminium toxicity tolerance have been cloned and their mode of action well characterized.

24.11   Conclusions

Micronutrients are essential for plant growth and development. There is also good evidence that several micronutrients play an important role in disease responses. Therefore, ensuring wheat plants have access to sufficient levels of all eight micronutrients is critical for production. Extensive genetic variation is known for both nutrient use efficiency, but agronomic approaches are often effective in dealing with deficiencies. Consequently, breeding for micronutrient efficiency generally takes a low priority relative to the many other traits assessed in a breeding program.

In contrast, breeding represents the main strategy for managing the impact of boron, or aluminium toxicity. Toxicity due to high levels of manganese can also be an issue in some regions but is not regarded as a major international problem for wheat production. Given the importance of boron and aluminium toxicity, there has been considerable effort in identifying sources of tolerance and defining the genetic and biochemical mechanisms of tolerance. The major genes controlling toxicity tolerance have been isolated and allelic diversity explored in large germplasm collections. Diagnostic markers are now available for the major tolerance loci and these are extensively deployed in breeding program that target regions susceptible to boron or aluminium toxicity.

References

1. Tripathi DK, Singh S, Singh S, Mishra S, Chauhan DK, Dubey NK (2015) Micronutrients and their diverse role in agricultural crops: advances and future prospective. Acta Physiol Plant 37:139. https://doi.org/10.1007/s11738-015-1870-3
2. Simoglou KB, Dordas C (2006) Effect of foliar applied boron, manganese and zinc on tan spot in winter durum wheat. Crop Prot 25:657–663. https://doi.org/10.1016/j.cropro.2005.09.007
3. Graham RD, Webb MJ (1991) Micronutrients and disease resistance and tolerance in plants. In: Micronutrients in agriculture. John Wiley & Sons, Ltd, pp 329–370
4. Attwell B, Kriedemann P, Turnbull C (1999) Plants in action Australian Society of Plant Scientists. Macmillan Education, Melbourne
5. IFPRI (2019) Global spatially-disaggregated crop production statistics data for 2010 version 2.0. In: Int. food policy res. Institute, Harvard Dataverse V4. https://doi.org/10.7910/DVN/PRFF8V
6. Hengl T, Leenaars JGB, Shepherd KD, Walsh MG, Heuvelink GBM, Mamo T, Tilahun H, Berkhout E, Cooper M, Fegraus E, Wheeler I, Kwabena NA (2017) Soil nutrient maps of sub-Saharan Africa: assessment of soil nutrient content at 250 m spatial resolution using machine learning. Nutr Cycl Agroecosyst 109:77–102. https://doi.org/10.1007/s10705-017-9870-x
7. Sillanpää M (1991) Micronutrients assessment at the country level: an international study. FAO Soils Bull 214
8. Webb MJ, Loneragan JF (1988) Effect of zinc deficiency on growth, phosphorus concentration, and phosphorus toxicity of wheat plants. Soil Sci Soc Am J 52:1676–1680. https://doi.org/10.2136/sssaj1988.03615995005200060032x
9. Huang C, Barker SJ, Langridge P, Smith FW, Graham RD (2000) Zinc deficiency up-regulates expression of high-affinity phosphate transporter genes in both phosphate-sufficient and -deficient barley roots. Plant Physiol 124:415–422. https://doi.org/10.1104/pp.124.1.415
10. Yu M, Hu C-X, Wang Y-H (2002) Molybdenum efficiency in winter wheat cultivars as related to molybdenum uptake and distribution. Plant Soil 245:287–293. https://doi.org/10.1023/A:1020497728331
11. Yu S, Tian L (2018) Breeding major cereal grains through the lens of nutrition sensitivity. Mol Plant 11:23–30. https://doi.org/10.1016/j.molp.2017.08.006
12. Farooq M, Wahid A, Siddique KHM (2012) Micronutrient application through seed treatments: a review. J Soil Sci Plant Nutr 12:125–142

13. Afzal I, Javed T, Amirkhani M, Taylor AG (2020) Modern seed technology: seed coating delivery systems for enhancing seed and crop performance. Agriculture 10. https://doi.org/10.3390/agriculture10110526

14. Wiatrak P (2013) Infuence of seed coating with micronutrients on growth and yield of winter wheat in Southeastern Coastal Plains. Am J Agric Biol Sci 8. https://doi.org/10.3844/ajabssp.2013.230.238

15. Rengel Z (2015) Availability of Mn, Zn and Fe in the rhizosphere. J Soil Sci Plant Nutr 15:397–409

16. Escudero-Martinez C, Bulgarelli D (2019) Tracing the evolutionary routes of plant–microbiota interactions. Curr Opin Microbiol 49:34–40. https://doi.org/10.1016/j.mib.2019.09.013

17. Jhanji S, Sadana US, Sekhon NK, Khurana MPS, Sharma A, Shukla AK (2013) Screening diverse wheat genotypes for manganese efficiency based on high yield and uptake efficiency. Field Crop Res 154:127–132. https://doi.org/10.1016/j.fcr.2013.07.015

18. Bansal RL, Nayyar VK, Takkar PN (1992) Field screening of wheat cultivars for manganese efficiency. Field Crop Res 29:107–112. https://doi.org/10.1016/0378-4290(92)90081-J

19. Rerkasem B, Netsangtip R, Lordkaew S, Cheng C (1993) Grain set failure in boron deficient wheat. Plant Soil 155:309–312. https://doi.org/10.1007/BF00025044

20. Shen J, Zhang F, Chen Q, Rengel Z, Tang C, Song C (2002) Genotypic difference in seed iron content and early responses to iron deficiency in wheat. J Plant Nutr 25:1631–1643. https://doi.org/10.1081/PLN-120006048

21. Owuoche JO, Briggs KG, Taylor GJ, Penney DC (1994) Response of eight Canadian spring wheat (Triticum aestivum L.) cultivars to copper: pollen viability, grain yield plant−1 and yield components. Can J Plant Sci 74:25–30. https://doi.org/10.4141/cjps94-006

22. Leach RC, Dundas IS (2006) Single nucleotide polymorphic marker enabling rapid and early screening for the homoeolocus of beta-amylase-R1: a gene linked to copper efficiency on 5RL. Theor Appl Genet 113:301–307. https://doi.org/10.1007/s00122-006-0296-0

23. Zhou Z, Shi X, Zhao G, Qin M, Ibba MI, Wang Y, Li W, Yang P, Wu Z, Lei Z, Wang J (2020) Identification of novel genomic regions and superior alleles associated with Zn accumulation in wheat using a genome-wide association analysis method. Int J Mol Sci 21. https://doi.org/10.3390/ijms21061928

24. Evens NP, Buchner P, Williams LE, Hawkesford MJ (2017) The role of ZIP transporters and group F bZIP transcription factors in the Zn-deficiency response of wheat (Triticum aestivum). Plant J 92:291–304. https://doi.org/10.1111/tpj.13655

25. Khabaz-Saberi H, Graham RD, Pallotta MA, Rathjen AJ, Williams KJ (2002) Genetic markers for manganese efficiency in durum wheat. Plant Breed 121:224–227. https://doi.org/10.1046/j.1439-0523.2002.00690.x

26. Ryan PR (2018) Assessing the role of genetics for improving the yield of Australia's major grain crops on acid soils. Crop Pasture Sci 69:242–264. https://doi.org/10.1071/CP17310

27. Moroni JS, Briggs KG, Taylor GJ (1991) Pedigree analysis of the origin of manganese tolerance in Canadian spring wheat (Triticum aestivum L.) cultivars. Euphytica 56:107–120. https://doi.org/10.1007/BF00042053

28. Nable RO, Bañuelos GS, Paull JG (1997) Boron toxicity. Plant Soil 193:181–198. https://doi.org/10.1023/A:1004272227886

29. McDonald GK, Taylor JD, Verbyla A, Kuchel H (2013) Assessing the importance of subsoil constraints to yield of wheat and its implications for yield improvement. Crop Pasture Sci 63:1043–1065. https://doi.org/10.1071/CP12244

30. Jefferies SP, Pallotta MA, Paull JG, Karakousis A, Kretschmer JM, Manning S, Islam AKMR, Langridge P, Chalmers KJ (2000) Mapping and validation of chromosome regions conferring boron toxicity tolerance in wheat (Triticum aestivum). Theor Appl Genet 101:767–777. https://doi.org/10.1007/s001220051542

31. Jamiod S (1996) Genetics of boron tolerance in durum wheat. University of Adelaide

32. Paull JG, Nable RO, Rathjen AJ (1992) Physiological and genetic control of the tolerance of wheat to high concentrations of boron and implications for plant breeding. Plant Soil 146:251–260. https://doi.org/10.1007/BF00012019
33. Pallotta M, Schnurbusch T, Hayes J, Hay A, Baumann U, Paull J, Langridge P, Sutton T (2014) Molecular basis of adaptation to high soil boron in wheat landraces and elite cultivars. Nature 514:88–91. https://doi.org/10.1038/nature13538
34. Jones D, Ryan P (2003) Aluminium toxicity. In: Thomas B, Murphy D, Murray B (eds) Encyclopedia of applied plant sciences, pp 656–664
35. Bityutskii N, Davydovskaya H, Yakkonen K (2017) Aluminum tolerance and micronutrient accumulation in cereal species contrasting in iron efficiency. J Plant Nutr 40:1152–1164. https://doi.org/10.1080/01904167.2016.1264591
36. Raman H, Stodart B, Ryan PR, Delhaize E, Emebiri L, Raman R, Coombes N, Milgate A (2010) Genome-wide association analyses of common wheat (Triticum aestivum L.) germplasm identifies multiple loci for aluminium resistance. Genome 53:957–966. https://doi.org/10.1139/G10-058
37. Ryan PR, Skerrett M, Findlay GP, Delhaize E, Tyerman SD (1997) Aluminum activates an anion channel in the apical cells of wheat roots. Proc Natl Acad Sci U S A 94:6547–6552. https://doi.org/10.1073/pnas.94.12.6547
38. Delhaize E, Ryan PR, Randall PJ (1993) Aluminum tolerance in wheat (Triticum aestivum L.) (II. Aluminum-stimulated excretion of malic acid from root apices). Plant Physiol 103:695–702. https://doi.org/10.1104/pp.103.3.695
39. Chantachume Y, Smith D, Hollamby GJ, Paull JG, Rathjen AJ (1995) Screening for boron tolerance in wheat (T. aestivum) by solution culture in filter paper. Plant Soil 177:249–254. https://doi.org/10.1007/BF00010131

Chapter 25
Pre-breeding Strategies

Sivakumar Sukumaran, Greg Rebetzke, Ian Mackay, Alison R. Bentley, and Matthew P. Reynolds

Abstract In general terms, pre-breeding links needed traits to new varieties and encompasses activities from discovery research, exploration of gene banks, phenomics, genomics and breeding. How does pre-breeding given its importance differ from varietal-based breeding? Why is pre-breeding important? Pre-breeding identifies trait or trait combinations to help boost yield, protect it from biotic or abiotic stress, and enhance nutritional or quality characteristics of grain. Sources of new traits/alleles are typically found in germplasm banks, and include the following categories of 'exotic' material: obsolete varieties, landraces, products of interspecific hybridization within the Triticeae such as chromosome translocation lines, primary synthetic genotypes and their derivatives, and related species mainly from the primary or secondary gene pools (Genus: *Triticum* and *Aegilops*). Genetic and/or phenotyping tools are used to incorporate novel alleles/traits into elite varieties. While pre-breeding is mainly associated with use of exotics, unconventional crosses or selection methodologies aimed to accumulate novel combinations of alleles or traits into good genetic backgrounds may also be considered pre-breeding. In the current chapter, we focus on pre-breeding involving research-based screening of genetic resources, strategic crossing to combine complementary traits/alleles and progeny selection using phenomic and genomic selection, aiming to bring new functional diversity into use for development of elite cultivars.

Keywords Simple and complex traits · Genetic diversity · Physiological breeding · Proof of concept · Genetic gains

S. Sukumaran (✉) · A. R. Bentley · M. P. Reynolds (✉)
International Maize and Wheat Improvement Center (CIMMYT), Texcoco, Mexico
e-mail: s.sukumaran@cgiar.org; a.bentley@cgiar.org; m.reynolds@cgiar.org

G. Rebetzke
CSIRO Agriculture and Food, Canberra, ACT, Australia
e-mail: greg.rebetzke@csiro.au

I. Mackay
SRUC, Edinburgh, UK
e-mail: ian.mackay@sruc.ac.uk

25.1 Learning Objectives

- Understand the rationale, objectives, approaches and tools used in wheat pre-breeding.

25.2 Introduction

Why is there a focus on pre-breeding? Plant breeders typically prefer to cross among elite lines [1] (see Chap. 7), except when specific and otherwise unavailable traits are needed such as disease resistance, this being the main route for introducing genetic diversity in conventional breeding. However, such repetitive use of elite breeding lines may limit the ability of new cultivars to adapt to emerging threats such as harsher climates and an ever-evolving spectrum of biotic threats. The use of well characterized primary synthetic hexaploids and landraces is a relatively straightforward way to widen genetic diversity and represents a key objective of pre-breeding. If tetraploid and hexaploid genomes lack genetic variation for biotic and abiotic stress tolerances, wild species can be used in interspecific hybridization (wide crossing) to add specific new diversity (see Chap. 18). Hence pre-breeding ensures continuity of supply for novel and diverse genetic variability in readily useful backgrounds that can enter breeding pipelines [2–4] and help broaden the wheat genepool generally. Physiological pre-breeding can be practiced by crossing with novel sources of traits as well as among elite material in order to deterministically stack complementary physiological traits to raise yield potential and adaptation to abiotic stress [5]. In general, the activities that precedes the development of a variety and initial reshuffling of genes by a breeder is termed 'pre-breeding' (Fig. 25.1).

25.3 Definitions

- Gene bank & Genetic resources: More details can be found in Chap. 17.
- Traits: Any physiological, morphological, biochemical, or genetic character of a plant including resistance/susceptibility to biotic stresses that can be used to differentiate two genotypes is called a trait.
- Simple traits and complex traits: Simple traits are often categorical, determined by few genes and are simple to phenotype and genotype. Complex traits are usually quantitative and determined by many genes with small effects. The heritability estimates of complex traits are commonly lower than the simple traits due to the many possibilities for interaction with genetic background, growth stage and environment [6].
- *In silico:* An experiment performed by computer or by computer simulations.

Fig. 25.1 A general sequence of events from gene bank to varieties

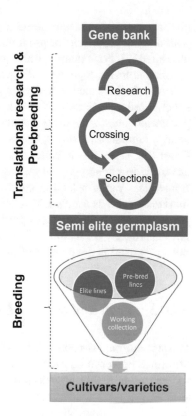

25.4 Aspects of Practical Pre-breeding

25.4.1 Access to Genetic Resources Through Gene Banks

The first step in pre-breeding is access to gene bank material (collection and conservation of germplasm; see Chap. 17) relevant to the breeding of the cultivars. Genetic resources broadly include modern cultivars in current use, obsolete cultivars, landraces, wild relatives, and genetic and cytogenetic stocks, breeding lines, and synthetic wheats etc. CIMMYT's wheat gene bank contains over 150,000 samples of wheat, the single largest collection of germplasm for any crop consisting of wild relatives, landraces, synthetics, cultivars, semi-elite lines, and mapping populations. More details can be found in Chap. 17.

25.4.2 Screening Genetic Resources

In practice, it is not feasible to screen 150,000 collections from a gene bank at a single time in the field or greenhouse for all traits of interest due to logistical limitations. The best approach when working with large numbers of accessions is to first phenotype for simple agronomic traits in the field while applying high throughput phenotyping via remote or proximal sensing to expand the range of traits that can be measured [7]. For example, traits such as plant height, phenology, lodging and other agronomic traits, together with grain yield, are the most important economically, and may be screened visually or at high throughput using proxies derived from spectral reflectance indices (SRI). SRIs are also used for a range of physiological parameters such as hydration status, photosynthetic pigments, in season biomass, canopy temperature, stay-green etc. For details of screening using SRIs please (see Chap. 27). Once this is done, smaller panels of lines (typically 150 to 300 entries) are made to phenotype and genotype in detail and determine marker trait associations/QTL. When selecting candidate parents for strategic crossing, major gene markers – such as for *Ppd, Vrn, Rht* and those for kernel weight – can provide key supplementary information to guide targeting and help avoid excessive segregation among progeny for height and maturity class. Genomic selection models have also been proposed in the context of parental selection [8] (Fig. 25.2).

25.4.3   Trait and Marker Discovery in Germplasm Panels

Germplasm panels need to be constructed in such a way that they have sufficient statistical power to be used to identify genetic markers associated with target traits as well as heritable phenotypes. Screening to characterize traits of interest may

Fig. 25.2 Process of utilizing the gene bank accessions (**a**) define a number of entries from the gene bank as initial set and reduce it to a number where detailed phenotyping can be done (**b**) examples of trait diversity present in the genebank (eg. spike length and size) (**c**) snapshot of phenotyping initiation to booting by growing 2000 accessions in the field at Sonora, Mexico, and (**d**) primary synthetic hexaploid panel formed by crossing durum wheat with *Aegilopsis*

include evaluation in environments to assess yield potential and response to biotic and abiotic stresses. The traits (Fig. 25.3), which may be genetically simple or complex in nature, need to be studied using various phenotypic screening approaches e.g., visual selection, high throughput phenotyping, and novel methods for screening. A detailed review of methods based on physiological parameters can be found elsewhere [9]. For details and strategies to breed specifically for diseases, drought stress, and nutrition, please refer to Chaps. 9, 10 and 11, respectively.

Gene discovery is based on use of populations constructed with the purpose of mapping traits into genomic regions of the wheat chromosomes. Basically, two main methods are followed: (1) genome-wide association mapping; and (2) QTL (Quantitative trait loci) mapping [10]. Genome-wide association mapping exploits linkage disequilibrium to provide high resolution and fast mapping. If the population is large enough, heritability high, and the trait architecture simple, this method can pinpoint the gene of interest. QTL mapping or linkage mapping is complementary to GWAS, where lines contrasting in a character of interest are used to generate a mapping population. In its simplest form, this requires the pre-characterization of donor material in order to identify contrasting parental lines and then generation time to develop RILs from biparental or back-cross populations. This approach can complement GWAS (Genome wide association study), allowing independent validation of the effect of an identified marker [11, 12]. Once the markers are identified in a diversity panel or a RIL population those need to be validated for further use. Please refer to Chap. 28 to learn about the validation and MAS (Marker-assisted selection) approaches [13].

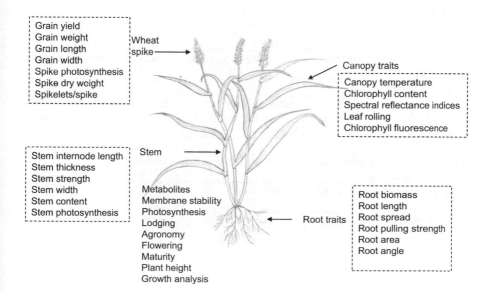

Fig. 25.3 A list of general traits in wheat used for pre-breeding

25.4.4   Trait Value and Prioritization. Which Traits and Why?

Phenotypic and genetic screening using markers in germplasm panels under relevant environments result in the identification of lines with high value traits/alleles, and can suggest trait combinations to boost yield and climate resilience. This step also identifies the heritability estimates of the trait, the genetic complexity of the trait, and if it is a simple or complex trait (Table 25.1). The traits fall within a continuum of simple to complex in nature, reflecting gene action from major effect to minor effect and therefore high heritability to low heritability. Even though heritability is assumed to be a genetic background and environment dependent parameter, some of the traits generally have high heritability, compared to others which are determined by multiple genes with minor effects. Large G × E, also makes it important to match traits with environments for breeding. Hence there is 'no one trait fits all environment' rule in pre-breeding. The main criteria for use of a trait in crossing and selection is its association with yield or other key performance trait -like biomass, kernel size or root capacity- and how easy it is to screen during generation advancement directly or using proxies like SRIs.

Table 25.1 Traits can be grouped based on number of genes involved, trait heritability, and the selection methods-phenotypic and genetic- that can be used for pre-breeding

Trait grouping	
Simple traits	**Complex traits**
Major genes	Multiple genes with small effects
Traits with high heritability	Traits with relatively low heritability
Categorical traits	Quantitative traits
Easily measurable	Time consuming
Low G × E	High G × E
Eg. Flowering time, plant height, grain weight	Eg. Photosynthesis, grain yield, radiation use efficiency, water use efficiency
High heritability	**Low heritability**
Plant height, flowering time, maturity; grain size and color, spike size	Grain yield, most physiological traits, metabolites, spectral reflectance indices that are influenced by environmental fluxes
Methods used	
Marker assisted selections/marker assisted backcrossing	Trait-based selections, genomic selection
Fine mapping and cloning of genes possible	Cloning of genes not possible
Single plants can be measured	Need to measure multiple plants in the field

25.4.4.1 Trait Integration

Incorporation of yield boosting, yield protecting or nutritional/quality trait(s) into elite backgrounds is the key goal of pre-breeding and to deliver proof of concepts of their value in appropriate target environments through trialing. Here breeding methods differ for simple vs. complex traits and if there is availability of molecular markers.

25.4.4.2 Pre-breeding for Simple Traits

Trait- and marker-based incorporation of simple traits is possible if based on the availability of robust linked markers. Simple traits or major gene-based traits are relatively easier to incorporate since their selection in subsequent generations can be done phenotypically or through marker assisted selection approaches. For example, reduced plant height, controlled by the famous green revolution alleles *Rht-B1b* and *Rht-D1b* is relatively easy to select visually. In addition, the identification of molecular markers with technologies such as *Kompetitive allele specific PCR* (KASP) and gene cloning is possible with relatively fewer generations and less time compared to complex traits. Another example of a trait that is easy to screen and incorporate is developmental traits. Although multiple genes related to flowering time have been identified in wheat, the different genes determining spring to winter growth habit remains a major screen in every pre-breeding activity. Other examples include genes for traits such as vernalization (*Vrn*), photoperiod *(Ppd)*, plant height *(Rht)*, *earliness per se* (*Eps*), thousand grain weight *(TaGW2)* and rust genes for leaf rust (*Lr*), stripe rust (*Sr*), and yellow rust (*Yr*) (For details on rusts refer to Chap. 8). Marker-assisted selection and marker-assisted backcrossing are normally used in selection for simple traits.

25.4.4.3 Pre-breeding for Complex Traits

(a) Phenomic approaches:

Strategic crossing, in which parents are selected to complement each other for 'source' and 'sink' related traits, is a successful pre-breeding strategy which has shown significant genetic gains in spring wheat for yield potential, heat and drought stressed target environments [14, 15]. In general, source refers to traits that are directly or indirectly associated with carbon assimilation (e.g. canopy architecture, radiation use efficiency, roots, above ground biomass, etc.). The sink is represented by grain number and potential size as well as the traits that enable yield formation such as spike architecture/fertility and traits showing negative trade-off with final spike dry weight such as specific internode growth [16]. Some traits and process may serve both source and sink roles such as spikes which also photosynthesize and sinks of labile carbohydrate -stored mainly in stems-, that are remobilized as sources of assimilate for grain filling especially under stress. The phenotype based approach to crossing occurs necessarily in the absence of sufficient genetic understanding of complex traits and how their alleles may interact. However, crossing among lines with complementary traits -backed by previous research- can stack the odds of accumulating favorable alleles in progeny selected for yield and complementary secondary traits.

In this scheme, progenies from F_2 to F_6 undergo a modified bulk method of selection (Fig. 25.4) employing selection for integrative traits like canopy temperature and NDVI (Normalized Difference Vegetation Index) on whole families for example, since the genetic value of individual plants cannot be measured accurately for complex traits. The resulting progenies represent well characterized, semi-elite lines that generally encompass alleles from diverse or exotic backgrounds such as synthetics, landraces and other genetic resources in a useful genetic background. These semi-elite materials can be used as parents by breeders aiming to achieve specific adaptation to their environments as mentioned in Chap. 3 [17, 18].

(b) Genomics based pre-breeding

The genetics of complex traits is not straight-forward to study through GWAS and QTL mapping. As trait complexity increases, the potential for $G \times E$ also increases. A typical QTL identified for a trait may be 15–20 cM in size, which may contain 1000s of genes, which need to be narrowed down through fine mapping for efficient use. If fine mapped, they still may not explain a high proportion of the phenotypic variance, so such QTLs and marker-trait associations (MTAs) need to be further refined and validated before applied in pre-breeding.

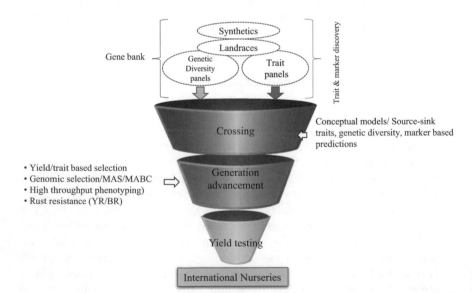

Fig. 25.4 Overview of the current IWYP and HeDWIC pre-breeding pipelines at CIMMYT

A more useful molecular breeding approach for complex traits like yield is genomic selection and prediction-based pre-breeding. Here genomic selection methods can be used for predicting parents, predicting the outcome from crosses, predicting the performance of progeny generations, and in selecting individual plants. However, these approaches are not routine in pre-breeding programs and are at the simulation or development stage. For example, rapid generation advance through speed-breeding combined with genomic selection may be advantageous for some traits [19].

The current scheme in CIMMYT uses four different approaches as shown in Fig. 25.5. One approach constitutes a fungicide (disease-free) pipeline where all generations are grown with fungicide to avoid the loss of high value alleles linked with rust susceptible backgrounds; this is most typical when neither parent is a modern, disease resistant line. Another stream is for simple traits, where MAS and MABC (Marker-assisted backcrossing) is used to incorporate genes/alleles. The third approach is based on speed-breeding where a rapid bulk-based approach is used to advance generations. In the fourth approach, lines in each generation are screened for rust (yellow rust and brown rust based on the shuttle breeding process of Dr. Normal Borlaug) to incorporate rust resistance into high value, semi-elite lines intended for breeders in countries where rust is an issue. The final products are distributed to public and private breeding programs globally through CIMMYT's International Wheat Improvement Network (IWIN) for yield testing. The better performing lines are used to cross, to reselect individual plants and incorporate locally important traits or disease resistance genes into their elite cultivars.

The collaborators in different countries share data back to IWYP and HeDWIC translational research and pre-breeding hubs, which is further used to select parents for breeding or pre-breeding or as semi-elite trait sources (http://orderseed.cimmyt.org/).

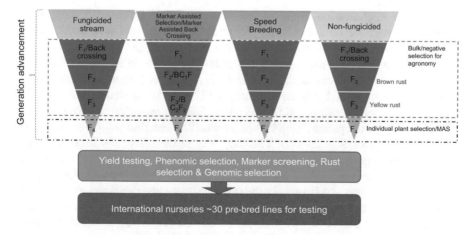

Fig. 25.5 Methods of pre-breeding practiced at CIMMYT

25.5   Proof of Concept -*in-silico* Approaches: Simulations

The cost and time required to run a pre-breeding program means that optimization and testing approaches may be best studied first by computer simulations and quantitative genetics theory. However, the process for major genes is well established. Backcrossing with selection on phenotype is effective if heterozygotes can be distinguished from the recurrent parent homozygous class. Selection on genome-wide markers can speed the process by reducing linkage drag and increasing the rate of recovery of the elite background: termed "background selection". For recessive major genes this is not possible without a slow process of progeny tested after each cross to the recurrent parent. Markers tagging the QTL are therefore required for "foreground selection". In the absence of a perfect marker for the trait, pairs of markers flanking the targeted QTL are desirable. Software to optimize the backcrossing the introgression of a major gene, including foreground and background selection has long been available (e.g. Popmin [20]) and more general-purpose software for genetic simulation (e.g. GeneDrop, AlphaSimR [21, 22]) can also be easily adapted to test alternative backcrossing strategies through gene-dropping approaches.

For quantitative traits, and if there are no tagged QTL to be introgressed, the situation is more complex and computer simulation is desirable to validate suitable strategies. If the desired phenotype or trait level is missing in the elite pool, then a cross between an elite and an exotic line can be followed by selection, either on phenotype, or through genomic selection, in the segregating generations. However, the trait to be introgressed will be in negative LD with the favorable traits already carried by the elite parent. Selection in an F_2 population or among F_2 derived lines may increase the frequency of favorable alleles carried by the resultant pre-breeding population, but the loss of the adapted background from the elite parent may be considerable. Equivalently, on making one or more backcrosses to the elite parent prior to selection, there is a strong risk that most favorable alleles in the exotic donor will be lost.

A further complication arises in instances where pre-breeding is intended to introduce novel variation for existing polygenic traits. In a cross, it is inevitable that most favorable alleles will be carried by the elite parent, but there may be novel variation in the exotic source, potentially at low frequency in an ancestral population but lost during domestication. In this case, there is a strong likelihood that favorable alleles carried by the exotic parent will be lost again, during selection. The chance of loss will be worse if selection takes place in generations derived from the backcross to the elite parent. This will occur whether selection is directly on phenotype or using genomic selection. The only way to unequivocally know that new variation has been introduced is to observe significant transgressive segregation over the elite parent and this is unlikely if the trait difference between the elite and exotic parents is large, as is usually the case. To overcome this "performance gap" other strategies have been tested in simulation. Gorjanc et al. [23] suggested establishing bespoke pre-breeding populations composed exclusively of exotic founders and then improving these rapidly through genomic selection, prior to introgression to elite lines through backcrossing with selection in the usual manner. Simulations showed this was more effective than the standard approach of direct backcrossing of elite to exotic. However, there is a chance that this approach does not capture novel variation but merely reintroduces alleles at a low frequency in the exotic pool which are already at a high frequency or fixed in the elite pool. It is unknown if this is a problem in practice. Simulating a process of crop domestication, breeding, then pre-breeding to recover lost variation could indicate its likelihood but has not been reported as far as we are aware. Such simulations should be possible with, for example AlphaSimR [22] which incorporates the coalescent based simulator into the package to generate founder haplotypes and already includes a models for wheat and maize domestication.

To circumvent the problem of the performance gap Yang et al. [24] tested the very simple approach of partitioning the genomic prediction equation into a component for markers for which the favorable allele was carried by the elite parent and one for which the favorable allele was carried by the exotic parent. Since most genomic prediction methods provide regression coefficients for every marker in the model, this amounts to partitioning regression coefficients into those with a positive sign and those with a negative sign, provided alleles carried by the two parents are coded consistently across markers (e.g. alleles from the elite parent are coded as 1 and those from the exotic as 0). In simulations they found, as expected, that genomic (or phenotypic) selection ignoring this partition would result in selection of a predominantly elite background and novel variation from the exotic would be lost. Partitioning the genomic prediction equation into two parts allowed a controlled approach to the introgression process, without excessive loss of novel exotic variation. They also tested this approach in barley and maize NAMs and found it effective.

Similar approaches have been developed by Allier [25] and tested in simulation: whereby the proportion of genome from the donor source is treated as a second trait. Simultaneous selection on two traits, the target trait for introgression and the proportion of donor genome, can then be used to ensure that the donor genome is not entirely lost, though there is no guarantee that the donor genome that is maintained in the selected lines is favorable. In practice, selection would be on an index of the two traits.

25.6 Pre-breeding Challenges

The primary challenges in practical pre-breeding are the identification of subsets of donor material which are likely to harbor novel and useful genetic variation for breeding and the scale of activities required to advance and assess material carrying diversity from pre-selected 'exotic' material. This creates complexity in delivering final products for uptake that meet core breeding objectives and can be smoothly integrated into established pipelines. Linkage drag is a major challenge when working with wild relatives and occurs between a high value allele of a primary trait

Fig. 25.6 Two main opportunities and challenges while synthetics are used for pre-breeding (**a**) new lines resistant to yellow rust and (**b**) necrosis of the new synthetics × elite crosses

associated with a high value allele of a secondary trait, for example a yield potential trait may increase lodging susceptibility. Yellow rust is one of the most devasting wheat diseases in the world and some pre-breeding wheat lines are susceptible to it. Cross incompatibility is another issue; when a new primary synthetic is crossed with an elite line it may not germinate or die after a few days (Fig. 25.6). Primary synthetic wheat especially performs differently, its spikelets are difficult to thresh and in some cases, shattering is an issue. Another issue associated with pre-breeding is the time required to develop the elite lines from the semi-elite material. It may require another full breeding cycle to come up with elite lines. The most critical step and challenge in pre-breeding is to know the genes identified in the genetic material are really novel and are not already present in the elite cultivars.

25.7 Technologies that Can Assist or Speed-Up Pre-breeding

Most of the new technologies mentioned in this volume (see Chaps. 27, 28, 29, 30, 31 and 32) will help to accelerate or increase precision of pre-breeding. Some are mentioned below.

(a) Trait screening methods

Some of the traits that are important to increase the yield potential of wheat are too complex to screen using normal visual selection [26]. This may need complicated equipment and a long processing time e.g. above ground biomass, root traits, harvest index. Development of genetic markers, prediction models, and genomic selection approaches can assist in pre-breeding of these traits. Please see Chap. 32 to learn about selection indices and their use in pre-breeding and breeding.

(b) Genetic markers

Genetic markers are highly useful for marker assisted selection and marker assisted back crossing for simple traits where the markers explain a large amount of variation. Some the traits that are routinely used and genes discovered are flowering time and plant height related genes. However genetic studies on RUE (radiation-use efficiency) and BM (biomass) also help to understand the genetic structure of the trait, its heritability estimate and complexity [16, 27, 28].

(c) Gene editing

This may work well for simple traits where the causative variants are known. It can be used to create new variation to test or to create a desired variant [29]. For complex traits this may still not work since edits to the causative genes may have only minor effects. Please see Chap. 29 for more details.

(d) Speed breeding

Rapid generation advancement can contribute to two different areas of pre-breeding (1) to develop RILs or BC populations to study the genetics of the traits (2) to advance the pre-breeding populations through bulking [30]. Genomic selection and prediction models together with rapid generation advance may be helpful once the training and testing populations are defined and well substituted when needed [31]. Refer to Chap. 30 for more details.

(e) Reference genomes

Reference genome helps to identify and cross check the novel alleles [32]. It also helps in comparison of different marker systems based on physical positions. In addition, it also assists with the prediction of candidate genes for further studies, cloning and studying haplotypes [33]. (See Chap. 28)

(f) Gene cloning

Even though gene cloning is not necessary for pre-breeding, having a cloned gene helps to fix them in elite cultivars and to identify the novel genes [34].

25.8 Linking Pre-breeding with Agronomy to Exploit G × M Synergies

Together with improved crop agronomy and management, pre-breeding has potential to deliver traits and understanding in exploiting opportunities in genotype × management interaction. Breeders carefully consider the target environment and farming system when selecting as adaptation and commercial success relies on varieties that perform reliably and at reduced cost to increase grower profitability. Among the most common considerations in modified management are changes in sowing date, reduced tillage including stubble retention, reduced herbicide-use through increased crop competitiveness, disease and insect resistance, and increased nutrient-use efficiency [27]. Opportunities exist in identifying traits that will support wider improvements in farm adaptation.

The gene pools typical of successful commercial breeding programs are fine-tuned (or 'co-adapted') for specific packages of alleles likely to deliver new varieties with improved performance across a wide range of disease, development, quality, and other key adaptation parameters. Key to the delivery of new traits/alleles is a greater understanding and access to a wider gene pool as described for pre-breeding in this chapter. For example, the green revolution has delivered improved grain yields through deployment of height-reducing *Rht-B1b* (syn. *Rht1*) and *Rht-D1b* (syn. *Rht2*) gibberellic acid (GA) insensitive dwarfing genes. However, while these genes reduce plant height, they also reduce seedling growth and particularly coleoptile length. Other GA-sensitive dwarfing genes have been identified that also reduce plant height to increase grain yield (Fig. 25.7). One dwarfing gene, *Rht18*, increases coleoptile length an average of 50% to increase field plant establishment 50 to 90% with deep sowing (Fig. 25.8). Genetic increases in coleoptile length will improve crop establishment with deep sowing to reach deep soil moisture, stubble retention and warmer soil temperatures.

Importantly, pre-breeding through improved physiological understanding of crop growth has permitted the identification and deployment of new dwarfing genes now being used in commercial breeding programs worldwide. Another example of physiological understanding is in the breeding of the polygenic early vigor trait important in drought tolerance and weed competitiveness.

Early vigor, defined as more rapid leaf area development following seedling emergence, is associated with wider leaves and greater biomass early in the season. As much as 60% of rainfall is evaporated from the soil representing a substantial loss in water needed for growth. Barley has greater early vigor to reduce soil evaporation loss and increase crop water-use efficiency. Barley is also more competitive with weeds owing to its shading of weeds early in the season. Wheat is very

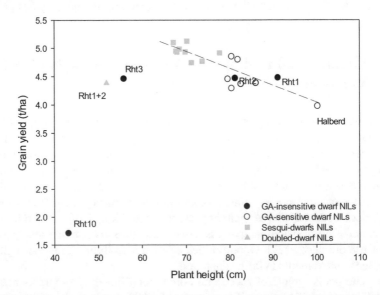

Fig. 25.7 Relationship of plant height and grain yield for gibberellic acid (GA) -insensitive and -sensitive single and doubled dwarfing gene near-isolines (NILs), and original tall parent Halberd at the Yanco Managed Environment Facility in 2018 (Line of best fit is Y = 7.061–0.031.X, $r^2 = 0.74$, $P < 0.01$). (Reprinted with permission from [35])

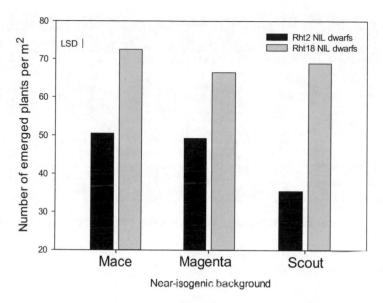

Fig. 25.8 Average numbers of emerged seedlings (per m²) for backcross three-derived *Rht2* and *Rht18* near-isogenic lines in Mace, Magenta and Scout genetic backgrounds when sown at 12 cm sowing depth at Merredin, Western Australia in 2018 [27]

conservative in its early growth, yet a large global screen of international wheats identified a set of genetically-unrelated landraces used in intermating in the development of a structured high vigor, recurrent selection population. Selection over six cycles produced progeny with 40–50% greater seedling leaf area than original parents and vigor equivalent to barley. Resulting high vigor progeny have been used as parents in the development of populations for selection of improved weed competitiveness, higher water-use efficiency and improved nutrient uptake [36].

25.9 Key Concepts

For a non-crop scientist, the distinction between pre-breeding and breeding may not be obvious, but while pre-breeding involves many of the same steps as breeding (in order to deliver adequate proof of concept) it focuses more on the identification of specific trait sources and achieving new trait combinations, as well as their selection where feasible in early progeny generations, to deliver well characterized germplasm for use as novel parents in breeding. The germplasm contains new sources of traits or alleles -and therefore increased genetic diversity- to underpin the requirement of future cultivars while broadening the wheat genepool.

Fig. 25.9 A general scheme of physiological pre-breeding pipeline. (Modified with permission from [5])

25.10   Conclusions

A newly emerging approach in pre-breeding is based on conceptual models for grain yield under yield potential (Chap. 21), heat stress (Chap. 22), and drought stress (Chap. 23). This approach divides grain yield into sub-component traits i.e., plant biomass (source) and harvest index (sink). These traits are further dissected into several sub-component traits [14]. This approach (Fig. 25.9) has gained more acceptance recently due to the challenge of identifying reliable markers for complex traits and with rapid advances in field based phenomics and genetic gains have been achieved using strategic crossing in this way.

In conclusion, pre-breeding is an essential activity in plant breeding to bring new traits and genetic diversity into elite germplasm. Many breeding programs perform this activity and a clear distinction between the breeding and pre-breeding may not exist. The key step in pre-breeding is to successfully incorporate novel genetic variation into the elite cultivar without linkage drag or disturbing the equilibrium of the genes in the elite genotype. Pre-breeding, especially if it involves discovery and translational research and possibly wide crossing with wild relatives, as well as strategic crossing and progeny selection, can be a long-term process but is necessary to exploit the full biological potential of crops.

References

1. Van Ginkel M, Ortiz R (2018) Cross the best with the best, and select the best: HELP in breeding selfing crops. Crop Sci 58:17–30. https://doi.org/10.2135/cropsci2017.05.0270
2. Sharma S, Upadhyaya HD, Varshney RK, Gowda CLL (2013) Pre-breeding for diversification of primary gene pool and genetic enhancement of grain legumes. Front Plant Sci 4. https://doi.org/10.3389/fpls.2013.00309

3. Sharma S (2017) Prebreeding using wild species for genetic enhancement of grain legumes at ICRISAT. Crop Sci 57:1132–1144. https://doi.org/10.2135/cropsci2017.01.0033
4. Moore G (2015) Strategic pre-breeding for wheat improvement. Nat Plants 1:15018. https://doi.org/10.1038/nplants.2015.18
5. Reynolds M, Langridge P (2016) Physiological breeding. Curr Opin Plant Biol 31:162–171. https://doi.org/10.1016/j.pbi.2016.04.005
6. Falconer DS, Mackay TFC (1961) Introduction to quantitative genetics
7. Reynolds M, Chapman S, Crespo-Herrera L, Molero G, Mondal S, Pequeno DNL, Pinto F, Pinera-Chavez FJ, Poland J, Rivera-Amado C, Saint-Pierre C, Sukumaran S (2020) Breeder friendly phenotyping. Plant Sci 295:110396. https://doi.org/10.1016/j.plantsci.2019.110396
8. Yu X, Li X, Guo T, Zhu C, Wu Y, Mitchell SE, Roozeboom KL, Wang D, Wang ML, Pederson GA, Tesso TT, Schnable PS, Bernardo R, Yu J (2016) Genomic prediction contributing to a promising global strategy to turbocharge gene banks. Nat Plants 2:16150. https://doi.org/10.1038/nplants.2016.150
9. Pask A, Pietragalla J, Mullan D (2012) Physiological breeding II: a field guide to wheat phenotyping. CIMMYT, Mexico
10. Sukumaran S, Yu J (2014) Association mapping of genetic resources: achievements and future perspectives. Genomics Plant Genet Resour:467–487. https://doi.org/10.1007/978-94-007-7575-6
11. Zhu C, Gore M, Buckler ES, Yu J (2008) Status and prospects of association mapping in plants. Plant Genome J 1:5
12. Tibbs Cortes L, Zhang Z, Yu J (2021) Status and prospects of genome-wide association studies in plants. Plant Genome:1–17. https://doi.org/10.1002/tpg2.20077
13. Dreisigacker S, Sukumaran S, Guzmán C, He X, Lan C, Bonnett D, Crossa J (2016) Molecular marker-based selection tools in spring bread wheat improvement: CIMMYT experience and prospects. In: Molecular breeding for sustainable crop improvement. Springer, Cham, pp 421–474
14. Reynolds MP, Pask AJD, Hoppitt WJE, Sonder K, Sukumaran S, Molero G, Saint PC, Payne T, Singh RP, Braun HJ, Gonzalez FG, Terrile II, Barma NCD, Hakim A, He Z, Fan Z, Novoselovic D, Maghraby M, Gad KIM, Galal EHG, Hagras A, Mohamed MM, Morad AFA, Kumar U, Singh GP, Naik R, Kalappanavar IK, Biradar S, Sai Prasad SV, Chatrath R, Sharma I, Panchabhai K, Sohu VS, Mavi GS, Mishra VK, Balasubramaniam A, Jalal-Kamali MR, Khodarahmi M, Dastfal M, Tabib-Ghaffari SM, Jafarby J, Nikzad AR, Moghaddam HA, Ghojogh H, Mehraban A, Solís-Moya E, Camacho-Casas MA, Figueroa-López P, Ireta-Moreno J, Alvarado-Padilla JI, Borbón-Gracia A, Torres A, Quiche YN, Upadhyay SR, Pandey D, Imtiaz M, Rehman MU, Hussain M, Hussain M, Ud-Din R, Qamar M, Kundi M, Mujahid MY, Ahmad G, Khan AJ, Sial MA, Mustatea P, von Well E, Ncala M, de Groot S, Hussein AHA, Tahir ISA, Idris AAM, Elamein HMM, Manes Y, Joshi AK (2017) Strategic crossing of biomass and harvest index—source and sink—achieves genetic gains in wheat. Euphytica 213:23. https://doi.org/10.1007/s10681-017-2040-z
15. Reynolds M, Manes Y, Izanloo A, Langridge P (2009) Phenotyping approaches for physiological breeding and gene discovery in wheat. Ann Appl Biol 155:309–320. https://doi.org/10.1111/j.1744-7348.2009.00351.x
16. Rivera-Amado C, Trujillo-Negrellos E, Molero G, Reynolds MP, Sylvester-Bradley R, Foulkes MJ (2019) Optimizing dry-matter partitioning for increased spike growth, grain number and harvest index in spring wheat. Field Crop Res 240:154–167. https://doi.org/10.1016/j.fcr.2019.04.016
17. Paper C, Co GS (2015) CIMMYT' s wheat breeding mega- environments
18. Braun H-J, Rajaram S, Ginkel M (1996) CIMMYT's approach to breeding for wide adaptation. Euphytica 92:175–183. https://doi.org/10.1007/BF00022843
19. Voss-Fels K, Herzog E, Dreisigacker S, Sukumaran S, Watson A, Frisch M, Hayes B, Hickey LT (2019) "SpeedGS" to accelerate genetic gain in spring wheat. 303–327. https://doi.org/10.1016/B978-0-08-102163-7.00014-4

20. Decoux G, Hospital F (2002) Popmin: A Program for the Numerical Optimization of Population Sizes in Marker-Assisted Backcross Programs. J Hered 93(5):383–384. https://doi.org/10.1093/jhered/93.5.383

21. Ladejobi O, Elderfield J, Gardner KA, Gaynor RC, Hickey J, Hibberd JM, Mackay IJ, Bentley AR (2016) Maximizing the potential of multi-parental crop populations. Appl Transl Genomics 11:9–17. https://doi.org/10.1016/j.atg.2016.10.002

22. Gaynor RC, Gorjanc G, Hickey JM (2021) AlphaSimR: an R package for breeding program simulations. G3 Genes|Genomes|Genetics 11. https://doi.org/10.1093/g3journal/jkaa017

23. Gorjanc G, Jenko J, Hearne SJ, Hickey JM (2016) Initiating maize pre-breeding programs using genomic selection to harness polygenic variation from landrace populations. BMC Genomics 17:30. https://doi.org/10.1186/s12864-015-2345-z

24. Yang CJ, Sharma R, Gorjanc G, Hearne S, Powell W, Mackay I (2019) Origin specific genomic selection: a simple process to optimize the favourable contribution of parents to progeny. bioRxiv 10:2445–2455. https://doi.org/10.1101/2019.12.13.875690

25. Allier A, Moreau L, Charcosset A, Teyssèdre S, Lehermeier C (2019) Usefulness criterion and post-selection parental contributions in multi-parental crosses: application to polygenic trait introgression. G3 Genes|Genomes|Genetics 9:1469–1479. https://doi.org/10.1534/g3.119.400129

26. Reynolds M, Van Ginkel M, Ribaut JM (2000) Avenues for genetic modification of radiation use efficiency in wheat. J Exp Bot:447–458. https://doi.org/10.1093/jexbot/51.suppl_1.447

27. Molero G, Joynson R, Pinera-Chavez FJ, Gardiner LL, Rivera-Amado C, Hall A, Reynolds MP (2018) Elucidating the genetic basis of biomass accumulation and radiation use efficiency in spring wheat and its role in yield potential. Plant Biotechnol J 52:1–13. https://doi.org/10.1111/pbi.13052

28. Rivera-Amado C, Molero G, Trujillo-Negrellos E, Reynolds M, Foulkes J (2020) Estimating organ contribution to grain filling and potential for source upregulation in wheat cultivars with a contrasting source-sink balance. Agronomy 10:1–21. https://doi.org/10.3390/agronomy10101527

29. Miao J, Guo D, Zhang J, Huang Q, Qin G, Zhang X, Wan J, Gu H, Qu LJ (2013) Targeted mutagenesis in rice using CRISPR-Cas system. Cell Res 23:1233–1236. https://doi.org/10.1038/cr.2013.123

30. Watson A, Ghosh S, Williams MJ, Cuddy WS, Simmonds J, Rey MD, Asyraf Md Hatta M, Hinchliffe A, Steed A, Reynolds D, Adamski NM, Breakspear A, Korolev A, Rayner T, Dixon LE, Riaz A, Martin W, Ryan M, Edwards D, Batley J, Raman H, Carter J, Rogers C, Domoney C, Moore G, Harwood W, Nicholson P, Dieters MJ, Delacy IH, Zhou J, Uauy C, Boden SA, Park RF, BBH W, Hickey LT (2018) Speed breeding is a powerful tool to accelerate crop research and breeding. Nat Plants 4:23–29. https://doi.org/10.1038/s41477-017-0083-8

31. Crossa J, Pérez-Rodríguez P, Cuevas J, Montesinos-López O, Jarquín D, de los Campos G, Burgueño J, Camacho-González JM, Pérez-Elizalde S, Beyene Y, Dreisigacker S, Singh R, Zhang X, Gowda M, Roorkiwal M, Rutkoski J, Varshney RK (2017) Genomic selection in plant breeding: methods, models, and perspectives. Trends Plant Sci xx:1–15. https://doi.org/10.1016/j.tplants.2017.08.011

32. Walkowiak S, Gao L, Monat C, Haberer G, Kassa MT, Brinton J, Ramirez-Gonzalez RH, Kolodziej MC, Delorean E, Thambugala D, Klymiuk V, Byrns B, Gundlach H, Bandi V, Siri JN, Nilsen K, Aquino C, Himmelbach A, Copetti D, Ban T, Venturini L, Bevan M, Clavijo B, Koo D-H, Ens J, Wiebe K, N'Diaye A, Fritz AK, Gutwin C, Fiebig A, Fosker C, Fu BX, Accinelli GG, Gardner KA, Fradgley N, Gutierrez-Gonzalez J, Halstead-Nussloch G, Hatakeyama M, Koh CS, Deek J, Costamagna AC, Fobert P, Heavens D, Kanamori H, Kawaura K, Kobayashi F, Krasileva K, Kuo T, McKenzie N, Murata K, Nabeka Y, Paape T, Padmarasu S, Percival-Alwyn L, Kagale S, Scholz U, Sese J, Juliana P, Singh R, Shimizu-Inatsugi R, Swarbreck D, Cockram J, Budak H, Tameshige T, Tanaka T, Tsuji H, Wright J, Wu J, Steuernagel B, Small I, Cloutier S, Keeble-Gagnère G, Muehlbauer G, Tibbets J, Nasuda S, Melonek J, Hucl PJ, Sharpe AG, Clark M, Legg E, Bharti A, Langridge P, Hall A, Uauy

C, Mascher M, Krattinger SG, Handa H, Shimizu KK, Distelfeld A, Chalmers K, Keller B, Mayer KFX, Poland J, Stein N, McCartney CA, Spannagl M, Wicker T, Pozniak CJ (2020) Multiple wheat genomes reveal global variation in modern breeding. Nature 2020:1–7. https://doi.org/10.1038/s41586-020-2961-x

33. Brinton J, Ramirez-Gonzalez RH, Simmonds J, Wingen L, Orford S, Griffiths S, Haberer G, Spannagl M, Walkowiak S, Pozniak C, Uauy C (2020) A haplotype-led approach to increase the precision of wheat breeding. Commun Biol 3:1–11. https://doi.org/10.1038/s42003-020-01413-2

34. Steuernagel B, Periyannan SK, Hernández-Pinzón I, Witek K, Rouse MN, Yu G, Hatta A, Ayliffe M, Bariana H, Jones JDG, Lagudah ES, Wulff BBH (2016) Rapid cloning of disease-resistance genes in plants using mutagenesis and sequence capture. Nat Biotechnol 34:652–655. https://doi.org/10.1038/nbt.3543

35. Rebetzke G, Ingvordsen C, Bovill W, Trethowan R, Fletcher A (2019) Breeding evolution for conservation agriculture. In: Pratley J, Kirkegaard J (eds) Australian agriculture in 2020: from conservation to automation. Agronomy Australia and Charles Sturt University, pp 273–287

36. Zhang L, Condon AG, Richards RA, Rebetzke GJ (2015) Recurrent selection for wider seedling leaves increases early leaf area development in wheat (Triticum aestivum L.). J Exp Bot 66:1215–1226. https://doi.org/10.1093/jxb/eru468

Chapter 26
Translational Research Networks

Matthew P. Reynolds, Hans-Joachim Braun, Richard B. Flavell,
J. Jefferson Gwyn, Peter Langridge, Jeffrey L. Rosichan, Mark C. Sawkins,
and Stephen H. Visscher

Abstract Without higher yielding and more climate resilient crop varieties, better agronomy and sustainable inputs, the world is on a course for catastrophes in food and nutritional security with all the associated social and political implications. Achieving food and nutritional security is one of the most important Grand Challenges of this century. These circumstances demand new systems for improving wheat to sustain current needs and future demands. This chapter presents some of the networks that have been developed over the years to help address these challenges. Networks help to: identify the most urgent problems based on consensus; identify and bridge knowledge silos; increase research efficacy and efficiency by studying state of the art germplasm and sharing common research environments/platforms so multiple strands of research can be cross-referenced; and creating communities of practice where the *modus operandi* becomes cooperation towards common goals rather than competition. Networks can also provide identity and visibility to research programs and their stakeholders, thereby lending credibility, increasing investment opportunities and accelerating outputs and dissemination of valuable new technologies.

M. P. Reynolds (✉) · H.-J. Braun
International Maize and Wheat Improvement Centre (CIMMYT), Texcoco, Mexico
e-mail: m.reynolds@cgiar.org

R. B. Flavell · J. J. Gwyn · M. C. Sawkins
International Wheat Yield Partnership (IWYP), Texas A&M AgriLife Research,
College Station, TX, USA
e-mail: iwypprogdirector@iwyp.org; iwypprogmanager@iwyp.org

P. Langridge
University of Adelaide, Adelaide, SA, Australia
e-mail: peter.langridge@adelaide.edu.au

J. L. Rosichan
Crops of the Future Collaborative, Foundation for Food & Agriculture Research,
Washington, DC, USA
e-mail: JRosichan@foundationfar.org

S. H. Visscher
Global Institute for Food Security, Saskatoon, SK, Canada
e-mail: steve.visscher@gifs.ca

© The Author(s) 2022
M. P. Reynolds, H.-J. Braun (eds.), *Wheat Improvement*,
https://doi.org/10.1007/978-3-030-90673-3_26

Keywords Research bottlenecks · Pre-breeding · Traits · Proofs of concept · Physiology

26.1 Learning Objectives

- Understanding the added value associated with translational and collaborative research networks.

26.2 The Research Continuum from Pure Science to Application

No one laboratory or organization can realistically encompass the full continuum of science from discovery research to delivery and adequate testing of new crop cultivars. There are many reasons for this including historical precedents of research organizations, the means by which science funding is allocated and the different specializations, research facilities and even locations required to achieve specific classes of research outputs. Nonetheless, for crop improvement to be dynamic enough to ultimately have impact in farm fields and meet societal demands, there needs to be a flow of knowledge from academia to applied crop science and breeding. Each area has its own specialization and demands, so rather than seeking conformity, linking them as a stepwise pipeline is a more likely approach to achieve synergy.

Plant scientists in academia are funded to work at the frontiers of understanding of genetics, physiology, cell biology etc., disciplines which even among themselves may not necessarily be interconnected or built upon. The use of model species and controlled environments -from petri dishes to growth rooms- maximize control and repeatability of treatments, as well as throughput since the cutting-edge of science is a highly expensive space. This approach furthers the understanding of specific processes but by definition is considered reductionist, since the different directions at the frontiers of science are not necessarily contiguous. Furthermore, much effort is invested in developing tools to further the research scope, a recent example is the use of tomography to study root growth and architecture, which while clearly of great potential in crop improvement is not tailored for application *per se*.

Crop scientists are typically trained in the academic approach and seek to apply discovery research in a real world context by applying treatments to understand specific growth and adaptive processes. If that understanding is intended to be used for genetic improvement, it can still take some time to move this from academic to applied research. For example, a textbook may explain what makes a cactus more stress tolerant than cabbage but to understand how two wheat genotypes, individuals of the same species, differ in their adaptive capacity is likely more challenging, requiring research approaches that may be quite different compared to academic precedents [1, 2]. A crop scientists may conduct experiments under more realistic growing environments, preferably under field conditions, posing additional challenges in obtaining controlled and accurate data (Table 26.1). Nonetheless, such approaches more likely lead to genetic improvement, being representative of growing conditions [3, 4]. However, the step from crop science to breeding is also significant.

Demand-driven breeding must establish priority traits that are better defined by high-throughput and application of well-established methods, than science *per se*. Examples are maintenance breeding to assure a crop does not become susceptible to new strains of diseases and pests [5] (see Chaps. 8 and 9) the need for diverse

Table 26.1 Main differences between field crop growing environments and controlled growth facilities

LIGHT	Light quality, intensity, and diurnal pattern are typically different in growth facilities, even in greenhouses where artificial light supplements may be employed for a variety of reasons, such as during dark winter months at high latitudes
AIR TEMPERATURE	Greenhouses are usually warmer than outside, notwithstanding use of costly cooling systems, while many growth-room facilities experience more abrupt changes in temperature than those experienced in the field.
SOIL TEMPERATURE	The impact of soil temperature is almost completely overlooked in growth facilities, where pots typically experience temperatures that are warmer, more uniform down the soil profile and less buffered to ambient air temperature than of field soil profiles
WATER & HUMIDITY	Both irrigation and relative humidity can mimic field conditions, though are costly/labor-intensive to control
SOIL	Soil from target environments can be used in pots, however, it is much harder to simulate the natural variation in bulk density, aeration and most importantly depth of field soil profiles
FERTILITY	Fertilizer is probably the easiest factor to control, notwithstanding the impact of differences in soil factors, including soil volume and temperature that may impact uptake by roots.
BIOTIC FACTORS	One of the advantages of the controlled environment is the relative ease with which pests and diseases can be identified and controlled compared to the field, though strict hygiene is necessary in the former to avoid infestation.
WIND & CO2	Wind patterns are not typically controlled in growth facilities; this has implications for boundary layers that affect transpiration and gas exchange (which in turn affect plant temperature), as well as local depletion of CO_2; wind can also modify plant mechanical strength
SCALE & COST	The biggest advantage of using the field as a laboratory is that in most situations field costs per unit area are much lower than in growth facilities, affecting experimental design and scale.
ROOT VOLUME	To maximize number of test pots and minimize costs of growth facilities, plants are typically grown in small pots. Resulting data show little correlation with field data, since roots can't develop normally; for example to depths where subsoil water may be present.

end-use quality requirements [6] (see Chap. 11), and increments in yield and yield stability to maintain a competitive edge in industry, and to ensure food security [7] (see Chap. 7). While many of these priority traits may be appropriate in 'upstream' research (see Part III of this book), a breeding program typically does not have the required resources, facilities or expertise to investigate novel, potentially better approaches, and at the same time develop improved new competitive varieties. To ensure food security and competitivity, a successful breeding program must put most of its resources into 'production line' efforts rather than research *per se*.

In a changing environment, consumer demand and the economic landscape require breeding methods to be continually fine-tuned and the occasional and necessary step-change in how science is applied in cultivar improvement. Hence a well-defined pathway involving networks of experts is needed to translate basic science to crop science, then to breeding and finally to farm level productivity increases.

26.3 Identifying and Prioritizing Opportunities that Represent Current Bottlenecks to Crop Improvement

Whether the threat to achieving adequate productivity is biotic or abiotic in nature, the principles of identifying and prioritizing opportunities for genetic improvement are similar. The literature is obviously a good place to begin, starting with wheat but also considering breakthroughs that may have been achieved in other crop species. It is more likely that a successful approach in another cereal or monocot species would be translatable to wheat [8], at least in the short term, than from a totally unrelated species. Nonetheless, many funding agencies encourage 'blue sky' or high risk-high return research, in which case the scope may be expanded to model species. While such research has pushed back the frontiers of understanding, there are few examples of translational research to crops [9, 10]. The problem should also be tackled from the bottom up. Experienced breeders can provide insights into what needs 'fixing'. The example often cited was the need for lodging resistance in wheat that sparked the Green Revolution. Another example was emphasis placed by breeders on retention of chlorophyll during grain-filling that arguably led to a body of research on the stay-green trait [11] and ways to measure it at high throughput [12]. Somewhere in between, crop physiologists, working with genetic resource experts, identified sources of a shorter, more upright leaf type that was introgressed from *T. timopheevii* (Zhuk.) Zhuk. in the 1970s. This trait is expressed in many modern wheat cultivars [13], improving light penetration into the canopy and inspiring further research for improving radiation use efficiency (RUE) via improved canopy architecture and photosynthesis [14, 15].

Ironically, some of the most important bottlenecks to improve productivity may be underrepresented in the literature and even in people's models of wheat 'ideotypes', for various reasons [16]. Such bottlenecks can become apparent in discussions among colleagues who share common goals. The practice of some funding agencies in issuing competitive calls for so-called disruptive research is a way to identify new opportunities in this space.

26.4 Establishing Collaborative Networks to Complement Skill Sets and Research Infrastructure

In the early 1950s, USDA initiated the first international wheat rust testing network followed by globally-coordinated research into a number of staple crops -including wheat, rice and maize- starting with the Green Revolution in the mid-1960s. This led to several international crop improvement networks linking national programs to the creation of CGIAR centers and beyond; one of the most impactful has been the International Wheat Improvement Network (IWIN). The scope and function of IWIN and other complementary networks are described below.

26.4.1 The International Wheat Improvement Network

The IWIN tests new bread and durum wheat and triticale lines at hundreds of sites in over 90 countries (Box 26.1). Breeding for traits of strategic importance, such as diseases that threaten entire regions or may cause pandemics, is conducted at strategic research hubs in order to develop and distribute approximately 1000 high yielding, disease-resistant lines targeted to major agro-ecologies each year, made freely available on request [17].

> **Box 26.1: Global Trialing Sites of the International Wheat Improvement Network**
> IWIN embraces a global collaboration of wheat scientists testing approximately 1000 new high yielding, stress adapted, disease resistant wheat lines each year as approximately 1800 sets of nurseries at around 250 locations annually, resulting in massive phenotypic data sets [18, 19].
>   To date, IWIN has collected over 10 million raw phenotypic data points and delivered germplasm that is estimated to be worth several billion dollars in extra productivity to more than 100 million farmers in less developed countries, annually [20] and by raising yields has saved more than 20 M ha of land from being brought under cultivation [21].

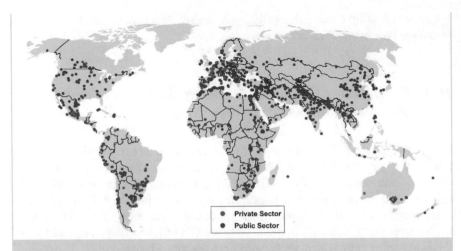

Public and private breeding programs that have received germplasm under the International Wheat Improvement Network (Figure drawn by Kai Sonder, CIMMYT).

National programs use the lines as new breeding material for new sources of traits i.e. parental lines for crossing; as candidates for release of new varieties; and for research. Data on new lines is shared within the network. Economic analysis of IWIN-related cultivars suggests that they are grown on over 50% of spring wheat area in less developed countries (Fig. 26.1), generating additional value (attributable to IWIN research) of US $2–$3 billion annually, spread among resource-poor farmers and consumers. The cost-benefit ratio of investment is estimated at ~100:1 [22].

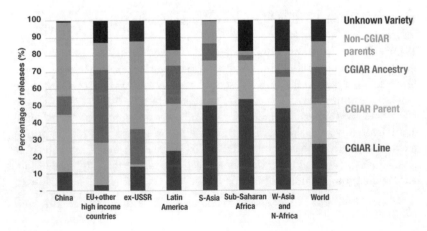

Fig. 26.1 Spring bread wheat released by region/origin through IWIN, 1994–2014. (Reprinted with permission from [22])

This does not even factor in the added value of avoiding rust and other disease epidemics by incorporating genetic disease resistance [5]. Further, IWIN curates a database containing millions of phenotypic and genotypic data points that have value for data mining and modelling (e.g. see Sect. 26.4.2).

26.4.2   The Heat and Drought Wheat Improvement Consortium (HeDWIC)

HeDWIC was formally established in 2020 to complement the IWIN by connecting translational research on climate resilience to mainstream wheat breeding through pre-breeding. HeDWIC's aims (https://hedwic.org/about/) are intended to add value to developing more climate resilient wheat varieties by:

- Facilitating global coordination of wheat research related to heat and drought stress with a special focus on countries in the Global South.
- Developing research and breeding technologies prioritized by stakeholders (researchers, breeders, farmers, seed companies, national programs, and funding organizations).
- Connecting geographically and agro-climatically diverse sites for rigorous testing of promising concepts.
- Curating data resources for use by the global wheat research community.
- Accelerating the deployment of new knowledge and strategies for developing more climate resilient wheat.
- Preparing a new generation of young scientists from climate-affected regions to tackle crop improvement challenges faced by their own countries.
- Building additional scientific capacity of wheat researchers in a coordinated fashion that enables a faster response to productivity threats associated with climate change.

Funding from the Foundation for Food and Agricultural Research (FFAR https://foundationfar.org/) is enabling HeDWIC to confront several research gaps (Fig. 26.2), in an effort led by CIMMYT in collaboration with many partners worldwide including IWIN and the International Wheat Yield Partnership (IWYP) (see Sect. 26.4.3).

HeDWIC inspired the Wheat Initiative (see Sect. 26.6) to establish the Alliance for Wheat Adaptation to Heat and Drought (AHEAD) program (*https://www.wheat-initiative.org/ahead*) which serves as an umbrella for HeDWIC and related projects, and brings into focus priorities for wheat improvement in the developed world, including partnerships between public and private sectors. The research goals of AHEAD and HeDWIC are broadly aligned and interactive, with the development of climate-resilient wheat as common goal.

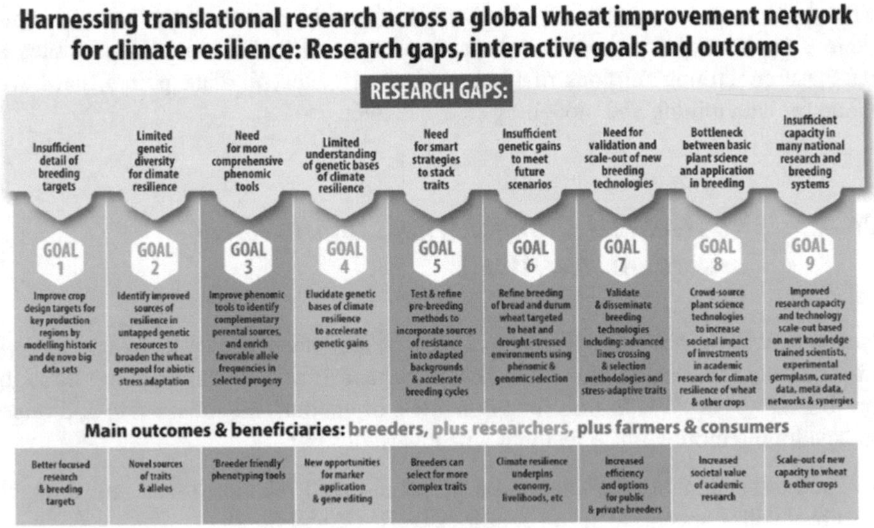

Harnessing translational research across a global wheat improvement network for climate resilience: Research gaps, interactive goals and outcomes

Fig. 26.2 Harnessing research across a global wheat improvement network for climate resilience: Research gaps, interactive goals and outcomes

26.4.3 The International Wheat Yield Partnership (IWYP)

The fact that much more food needs to be grown on essentially the same or less land amounts – to me correct is less of land in the coming decades is well established and accepted. This increase in productivity is compounded by changing diets, changing climates, and pests and diseases that will continue to undermine sustainable high crop production which puts more stress on food supplies and consumer prices. For these reasons, IWYP (https://iwyp.org/) was launched in late 2014 with the goal to increase the genetic yield potential of wheat (by 50% over 20 years was proposed). IWYP is a unique partnership of public and private institutions that deploy a highly efficient model for funding international research and coordinating and integrating the research into a holistic science and development program. IWYP complements the IWIN by linking research on yield potential to wheat breeding through translational research and pre-breeding.

IWYP was launched in late 2014 to increase the genetic yield potential of wheat closer to its biological limits (by 50% over 20 years was proposed) . IWYP is a unique partnership of public and private institutions that deploys a highly efficient model for funding international research, and coordinating and integrating the research into a holistic science and development program. IWYP complements the IWIN by linking research on yield potential to wheat breeding through translational research and pre-breeding.

IWYP operates as a not-for-profit voluntary collaborative partnership. The public sector funds and contributes high-quality research seeking breakthroughs to boost wheat yields around the world, and the private and public sectors across the world exploit the validated discoveries in their breeding pipelines, then test, scale and market better varieties for their respective markets. There is no significant duplication in public and private sectors because the environments and national markets for which their respective products are optimized are significantly different, and therefore all the locations where farmers grow wheat can benefit. IWYP exploits the best relevant science globally, is focused, operates with a sense of urgency, leverages outputs to generate added value and drives research outputs for delivery by both public and private wheat breeding programs worldwide, with the goal of generating significant yield improvements in farmer's fields. IWYP takes many steps to make certain its efforts are aligned with other current relevant research programs and initiatives worldwide (Fig. 26.3). All IWYP products are freely available.

Importantly, IWYP is product driven with focused scope and objectives. The basis of the IWYP strategy is:

- Deploy top quality scientific research from international teams with a united focus on potential yield boosting traits.
- Actively coordinate research projects around the world for greater efficiency.
- Achieve a succession of research breakthroughs in key traits.

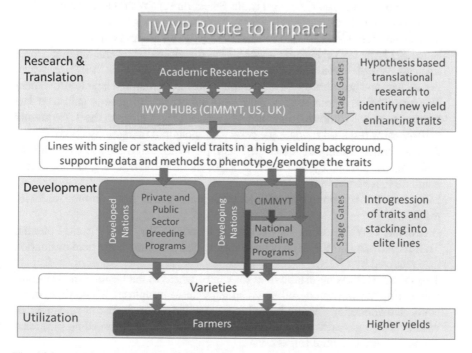

Fig. 26.3 IWYP deploys a model where a consortium of public funding organizations supports collaborative international research that feeds centralized development Hubs that deliver new traits and germplasm to breeding programs worldwide. These product pipelines further develop the IWYP innovations and deliver new higher yielding varieties to farmers worldwide

- Create added value by combining breakthroughs in elite germplasm.
- Utilize centralized downstream development platforms (Hubs) to deliver new higher yielding germplasm with novel traits in elite genetic backgrounds and push them toward deployment.
- Drive the improved germplasm into established breeding pipelines around the world, both public and private, which will deliver new higher yielding varieties to farmers in both the North and the South.

Details on IWYP research can be found in a series of monthly IWYP "Science Briefs" (https://iwyp.org/iwyp-science-briefs/) and are summarized in the IWYP Annual Reports (https://iwyp.org/annual-report/). Section 26.4.3.1 presents some key outputs of translational research.

26.4.3.1 Examples of Translational Research Outputs from Collaborative Platforms: The Case of IWYP

- Wheat lines developed in part by the IWYP Hub at CIMMYT through selection of IWYP target physiological traits have been released as varieties in Pakistan and Afghanistan.
- IWYP developed lines have shown higher yield than CIMMYT elite lines and local checks across multilocation trials. Selection for IWYP target traits such as biomass and radiation use efficiency, in combination with physiological and grain formation traits can lead to increased genetic yield potential.
- Since 2015, the IWYP Hub at CIMMYT has disseminated several hundred wheat lines as Wheat Yield Collaboration Yield Trial (WYCYT) "sets" to wheat researchers and breeders worldwide through the International Wheat Improvement Network (IWIN). From WYCYT data, the annual rate of genetic gain over last 5 years is ~1.3% (Sukumaran et al., Chap. 25). This is close to the 1.7% annual genetic gains required to meet the 50% yield potential increase by 2035.
- A better understanding of the contribution physiological traits such as biomass, radiation use efficiency and harvest index make to enhanced grain yield and combine these traits in new lines.
- Many sources for improvements to these traits come from unimproved/wild material.
- A dedicated IWYP testing network of 30 locations, the "IWYP Yield Potential Trait Experiment" (IYPTE) has been established to augment the field evaluation data received by IWIN.
- Two IWYP Hubs were established for winter wheat in the UK and the US complementing the work undertaken on spring wheat germplasm. These three interconnected validation and pre-breeding Hubs will develop the major categories of wheats grown globally, expanding IWYP's reach into more breeding programs and increasing potential impact.
- Early generation pre-breeding and experimental lines are made widely available.
- Information on genes/molecular genetic markers discovered by the IWYP Research Projects for source and sink traits is routinely collated and promoted for uptake by wheat breeders.

- Information on any novel phenotyping and genomics tools and protocols developed by IWYP are collated and promoted for use by third parties.
- Novel alleles for genes controlling grain size (width and length) and spike traits are routinely crossed into multiple wheat backgrounds, particularly wheat lines with high biomass, to develop germplasm with improved source sink balance and higher yield.
- Novel alleles have been identified for genes controlling wheat phenology along with the knowledge of which combinations of these alleles should be used to maximize yield and harvest index.
- Wheat parent lines transferred to public and private breeding programs with:

  - improved energy capture that leads to improved yield;
  - improved radiation use efficiency at the scale of the canopy;
  - high dry matter partitioning to the grain (increased harvest index) and lodging resistance;
  - rapid return to full photosynthetic efficiency following a short period of shading (sun-shade transition) ensuring optimal conversion of carbon dioxide to sugars;
  - chromosome segments introgressed from wild wheat relatives with increased photosynthetic efficiency relative to the wheat parents or variation in floral morphology;
  - favorable native alleles conferring enhanced shoot growth and biomass production backcrossed into multiple wheat lines.

- Identification of wheat landraces and other genetic resources with increased levels of photosynthetic efficiency compared to selected modern wheat varieties that serve as a resource for trait introgression in cultivated wheat.
- The identification of genes and molecular genetic markers that induce different wheat root phenotypes suitable for maximizing yield under different environments have been shared with public and private wheat breeding programs.

26.5 What It Takes to Establish and Fund an International Collaborative Platform; the Example of IWYP

26.5.1 Defining the Need

The need for a collaborative and coordinated international program or platform to address a specific global grand challenge requires a clear strategic purpose which sets out why such an approach is more likely to succeed than separate national programs. The key drivers for establishing IWYP were assuring food security for an increasing global population recognizing climate change impacts, a mismatch between supply and demand, risks of spikes in wheat prices and leveling off in the rate of yield growth. Forecasts indicated a substantial gap between projected demand and what wheat yield improvements could be achieved, at least in a business as usual scenario, i.e. continue with current incremental yield improvements. The overall analysis identified an urgent need to address this predicted shortfall forecast for the world's most widely grown crop. A strategy to deliver a step-change improvement in yield was therefore necessary.

26.5.2   Creating Awareness and Testing for Interest

The scale of the challenge demanded a collaborative approach that brought together the best researchers from around the world for both discovery and translational research. This could only succeed if stakeholders worked together as one team, sharing resources, results and implementing coordinated regional evaluations of new germplasm. An international conference convened by USAID at CIMMYT secured support for such an approach and program.

The next key step was to determine if support for funding could be secured in principle. Representatives from funding organizations from the UK, USA and Australia guided the development and brought in independent scientific input from world leaders in plant sciences who would not be directly involved with IWYP. A conference with key international development agencies and invited experts was convened and coordinated by the UK's BBSRC and hosted by the Government of Mexico. Here, the strategic need, proposed approach, key goals and an outline governance and review structure was presented. Support was quickly forthcoming from several funders, along with emphasis on the importance of involving the private sector in the partnership. This initial support proved to be a vital step to open up the detailed planning for the program and its scope.

26.5.3   Planning Governance and Operations

The next stage involved detailed planning and design of an effective governance structure, whilst addressing inevitable differences in national approaches and processes. To ensure acceptability, international best practice were adopted in areas such as independent international peer review and assessment criteria, program management, monitoring and evaluation of project milestones and key indicators of success. Regular meetings were agreed to assure discussions on new data and knowledge and to receiving advice and challenge from peers and partners.

26.5.4 Adding Value Through Program and Project Management

The major reason to have an international program is to generate more information, greater impact and added value beyond that originally envisaged. Project management needs to play a major role here by stimulating an open ambience of collaboration, belief in the common goals and the value of achieving more by additional collaborations. Such interactions can also lead to sharing of equipment, students, postdocs and skills, etc. This can be crucial where field work to assess outcomes is necessary but technical or staff support is not available in each institution. Project management should not only stimulate generation of added value but also monitor, track and communicate progress between the participating laboratories, Funders, and the scientific and other communities. This is important because individual scientists and Funders can then see the added value that comes with such partnerships. Management also needs to produce documents that increase transparency such as an Annual Report, a Strategic Plan and other papers that communicate the scientific novelty such as lists of publications, technologies and know-how. A high-level oversight Board of stakeholders is needed to assess progress, address problems, budget issues and for strategic planning.

26.5.5 Delivering Added Value

A worthwhile platform should generate added value and impact. IWYP therefore had to establish centralized "Hubs" to validate, develop, combine and test outputs from the discovery projects. This is important in agriculture because a discovery in a non-agricultural environment or a non-elite genotype may not be worthwhile taking further, and so learning this early is important to maximize efficiency. The principal IWYP Hub for spring wheat is currently based at CIMMYT, although other places are also involved in validation and testing. This Hub focuses on field-based testing of outputs, pre-breeding and testing of discoveries in elite genetic backgrounds. Two winter wheat IWYP Hubs in the US and the UK been established with additional financial inputs from private companies to stimulate uptake by breeding programs. It is important to recognize that the Hubs retain the responsibilities to bring added value to the discoveries made by other scientists and funding agencies.

Once an effective management system, oversight Board and the sharing, stimulation and oversight of science has been established this can be used for further exploitation and additional initiatives, both international and within a country, to generate added value.

26.6 Higher Level Networks

26.6.1 The Wheat Initiative's Expert Working Groups

The Wheat Initiative (WI) was established in 2011 following endorsement from the G20 agriculture ministers as part of a program to enhance global food security. The membership is made up of national research funding agencies, international research organizations and industry. The Wheat Initiative encourages and supports the development of a vibrant global wheat public-private research community sharing resources, capabilities, data and ideas to improve wheat productivity, quality and sustainable production around the world. The WI comprisess public and private researchers, educators and growers working on wheat to develop strong and dynamic national and transnational collaborative programs.

The current membership of the Wheat Initiative includes 14 countries, two CGIAR centers and six companies. A further five countries contribute as observers https://www.wheatinitiative.org/.

The Expert Working Groups (EWGs) are the scientific working force of the WI (Fig. 26.4). Currently there are ten scientific EWGs and one focusing on Funding. The EWGs bring together international experts in each field of expertise, to share ideas, knowledge, information, resources and data and identify international research priorities. There are presently 635 members from 47 countries in the EWGs from research organizations, universities, government and industry. A core task of

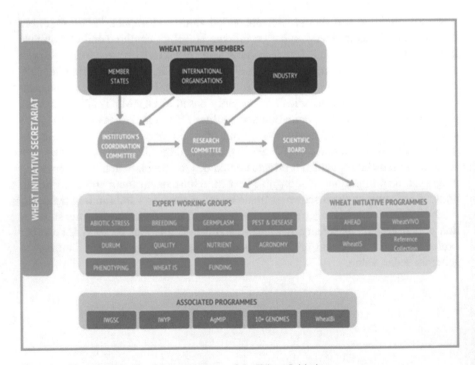

Fig. 26.4 The organization and management of the Wheat Initiative

the EWGs is to define global research priorities in their area of expertise and develop strategies to address these priorities. Each EWG also serves as a forum to bring researchers together to discuss major research advances and identify research gaps where increased investment or collaboration would be beneficial. A further role is to support and encourage the exchange of information, data, resources and explore opportunities to share capabilities.

The activities of the EWGs are supported by the secretariat and through a series of tools for information exchange. These include the Wheat Information System (WheatIS – http://wheatis.org), WheatVIVO (http://www.wheatvivo.org/), quarterly newsletters, weekly Media Briefs, and a biennial International Wheat Congress.

26.6.2   Multi-crop Networks

Traditionally, most private sector investment in agriculture has focused on a few, select large acre row crops, and high value vegetable crops, leaving many globally important food crops under-resourced. Given the urgent need to feed more people, there is increasing emphasis to produce nutritious, affordable food on thriving farms through efficient crops that increase yields with fewer inputs. Achievement of this ambitious goal requires an increase in both public and private investment to increase crop diversity and on farm profitability. Crop diversity creates greater economic security for farmers, offers environmental benefits and can increase food security. Farmers growing a range of crops may be able to sell to multiple markets and supply chains. Additionally, some crops can improve soil, filter water and reduce climate emissions.

A major hurdle toward meeting these needs is the significant decrease in public funding for agricultural research in the last decade (Fig. 26.5). In the Global South the problem has been seen for several decades. In the meantime, private sector investment has steadily increased. Much of the increase in private sector investment is driven by the acceleration of technology development and implementation of that technology into those major cash crops grown in the developed economies. The funding imbalance has left many, traditionally public funded crops under-resourced and technology poor. This gap leads to greater inequity for many important food security crops to meet the growing global demand for food, particularly in the face of climate change.

In this context we need to have a look at the situation in the poorest countries. While private sector Ag R&D investments in middle income countries have significantly increased, and investments of top investors are shifting to the private sector, the situation for the poorest countries is very different. No changes were observed since 1980, when for every dollar of AgR&D spent in high-income countries, just 3.5 cents was spent in the low-income countries. Thirty years later, the gap has widened. In 1980, high income countries spent on a per capita basis 13.25 $ vs 1.73$ in the poorest countries (7.7 fold difference) while in 2011, high income countries spent 17.73$ per capita compared with 1.51$ in the poorest country (a 11.7 fold

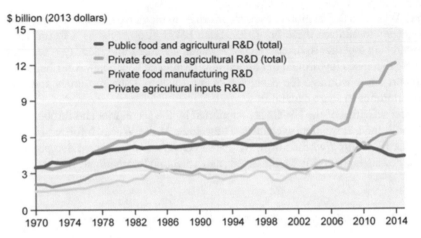

Fig. 26.5 A comparison of public and private crop research funding over time, real (inflation-adjusted) dollars, 1970–2015. Note: Private agriculture research funding data are through 2014; public agricultural research funding is available through 2015. (Reprinted with permission from [23])

difference). This is dramatic, since in the poorest countries' population growth is highest, agriculture plays a key role for economic development, but investments in Ag R& D are among the lowest in the world [24]. However, stagnant Federal funding and increased private research investments has led to new funding models and opportunities. To leverage this opportunity, the United States Congress created the Foundation for Food & Agriculture Research (FFAR) in the 2014 Farm Bill. FFAR was asked to build public-private partnerships that pioneer the next frontiers of agriculture research. Through public-private partnerships, federal investments are doubled – and often garner more than a 1:1 match. Public-private partnerships enable stakeholders both inside and outside the food and agriculture industry to convene. The convening capabilities of FFAR and the depth of relationships with wide-ranging stakeholders create an atmosphere of collaboration that is unique within the agriculture research community. It is not every day that competitors join forces to address a common challenge, but FFAR's mission helps unusual partners work together for the common good.

To date, FFAR has brought together over 500 diverse stakeholders to form these unique partnerships that support innovative science addressing today's food and agriculture challenges. Research focuses on imminent challenges where science is either filling a knowledge gap or developing a solution previously deemed impossible. When private-sector partners participate in research with FFAR the results likely are closer to application. At the same time, research is made available to the public so breakthroughs can be implemented widely and swiftly.

The Crops of the Future Consortium (COTF) is one of FFARs earliest consortia. The private sector participants of COTF represent seed companies and technology providers. Partners work closely together to identify key research gaps of common interest to the industry, define pre-competitive space and collectively de-risk new areas of research. This approach allows competitors to jointly fund the research, use the results in their internal R&D to develop products, and then compete in the market with those products. Key to the success is understanding and navigating how to deal with IP issues. It is critical to discuss data sharing and IP up front and get buy in from prospective private sector funders to make this model work. Consortia funding aligns well with non-exclusive access to technology. Further, using and leveraging public dollars requires that there be public benefit from the research. In COTF, mechanism to do this benefit the scientific community, the companies and the end users of the research.

For example, COTF also is participating in funding a large project with CIMMYT and the Bill and Melinda Gates Foundation focused on accelerating genetic gains in corn and wheat. This research is looking to shorten the breeding cycle and introduction of new varieties from what currently takes 8–10 years. This research also provides a path for translation to other crops once validated in corn and wheat.

26.7 Delivering Proofs of Concepts for Research Ideas Through Translational Research and Pre-breeding

Demographic and environmental factors stress the urgency to boost yield potential and climate resilience – yield stability. Ideas are suggested by academia, many stemming from studies with model species in controlled environments but satisfactory proofs of concept in a breeding context are prerequisite. The translational step is essential to ensure results will hold up under realistic field conditions [3]. Thus the last stage of translational research must ultimately show proofs of concept in the field, across an appropriate range of target environments, and using relevant, up to date germplasm whose genetic backgrounds encompass the collateral traits needed to make a new cultivar marketable [25] (see Fig. 25.9). In this way, translational research provides the link between more upstream research and crop breeding-through networks like IWYP, HeDWIC and IWIN- adding value to both.

26.8 Networking to Train the Next Generation of Crop Scientists

Networks of the type described here provide ample opportunity for capacity building, whether as part of a graduate degree or other opportunities for young scientists and technicians to learn about different methods and approaches in a new context. For example, CIMMYT's research platform in the Sonora Desert, jointly sponsored by the Mexican Government, IWYP and now FFAR, has helped train 12 PhDs over the last 10 years. The HeDWIC project formally initiated a Doctoral Training Program in 2020 which is already supporting 3 young scientists, to conduct novel research into: root imaging and growth analysis under heat and drought; identifying high throughput proxies for 'minimum data set' traits used in crop simulation modelling; and remote sensing to identify pigments associated with photoprotection at breeding scale, mentored by experts at Nottingham, Purdue and Hohenheim, respectively. The IWYP graduate program broke new ground in the areas of photosynthesis, partitioning and lodging research using realistic field conditions, e.g. [26–31]. In each case graduate committees comprised expertise from very different research fields, whose expertise and experience were complementary to producing results that not only demonstrated new science but also technologies ready for application in wheat improvement.

26.9 Key Concepts

Translational research capitalizes on prior large investments in upstream research; collaborative networks widen access to expertise, environments and infrastructure.

26.10 Conclusions

Co-authors of this chapter considered it important to emphasize that a continuum in breeding, from basic to applied research is vital since few scientists occupy the applied research space where validations of novel technologies and proofs of concept for crop improvement hypotheses are rigorously tested in a breeder-friendly context [32]. There is no scientific reason why these areas are neglected; perhaps partly because of the effort involved, funding constraints and perhaps due to silos that form for a variety of reasons. However, networking among scientists across the spectrum of research from pure to applied is an effective way to fill this space. Furthermore, the synergy that is created adds robustness to scientific conclusions while translational research and pre-breeding add societal value to investments made in science. Networks allow results from upstream plant science to have application in downstream problem-solving research. Given the increased demand from a growing population, the fact that new temperature records are being set annually and that water resources and soil fertility are on the decline in many parts of the world, science needs to become more efficient, and networking is a proven method for boosting modern plant breeding.

References

1. Molero G, Joynson R, Pinera-Chavez FJ, Gardiner LL, Rivera-Amado C, Hall A, Reynolds MP (2018) Elucidating the genetic basis of biomass accumulation and radiation use efficiency in spring wheat and its role in yield potential. Plant Biotechnol J 52:1–13. https://doi.org/10.1111/pbi.13052

2. Wilson PB, Rebetzke GJ, Condon AG (2015) Pyramiding greater early vigour and integrated transpiration efficiency in bread wheat; trade-offs and benefits. Field Crop Res 183:102–110. https://doi.org/10.1016/j.fcr.2015.07.002
3. Poorter H, Fiorani F, Pieruschka R, Wojciechowski T, Van Der Putten WH, Kleyer M, Schurr U, Postma J (2016) Pampered inside, pestered outside? Differences and similarities between plants growing in controlled conditions and in the field. New Phytol 212:838–855. https://doi.org/10.1111/nph.14243
4. Füllner K, Temperton VM, Rascher U, Jahnke S, Rist R, Schurr U, Kuhn AJ (2012) Vertical gradient in soil temperature stimulates development and increases biomass accumulation in barley. Plant Cell Environ 35:884–892. https://doi.org/10.1111/j.1365-3040.2011.02460.x
5. Singh RP, Hodson DP, Huerta-Espino J, Jin Y, Bhavani S, Njau P, Herrera-Foessel S, Singh PK, Singh S, Govindan V (2011) The emergence of Ug99 races of the stem rust fungus is a threat to world wheat production. Annu Rev Phytopathol 49:465–481. https://doi.org/10.1146/annurev-phyto-072910-095423
6. Guzmán C, Autrique E, Mondal S, Huerta-Espino J, Singh RP, Vargas M, Crossa J, Amaya A, Peña RJ (2017) Genetic improvement of grain quality traits for CIMMYT semi-dwarf spring bread wheat varieties developed during 1965–2015: 50 years of breeding. Field Crop Res 210:192–196. https://doi.org/10.1016/j.fcr.2017.06.002
7. Crespo-Herrera LA, Crossa J, Huerta-Espino J, Autrique E, Mondal S, Velu G, Vargas M, Braun HJ, Singh RP (2017) Genetic yield gains in CIMMYT'S international elite spring wheat yield trials by modeling the genotype × environment interaction. Crop Sci 57:789–801. https://doi.org/10.2135/cropsci2016.06.0553
8. Valluru R, Reynolds MP, Lafarge T (2015) Food security through translational biology between wheat and rice. Food Energy Secur 4:203–218. https://doi.org/10.1002/fes3.71
9. Harjes C, Rocherford T, Bai L, Brutnell T, Bermudez-Kandiannis C, Sowinski S, Stapleton A, Vallabhaneni R, Williams M, Wurtzel E, Yan J, Buckler E (2008) Natural genetic variation in lycopene epsilon cyclase tapped for maize biofortification. Science 319:330–333. https://doi.org/10.1126/science.1150255
10. Hu H, Scheben A, Edwards D (2018) Advances in integrating genomics and bioinformatics in the plant breeding pipeline. Agriculture 8:1–18. https://doi.org/10.3390/agriculture8060075
11. Tian F, Gong J, Zhang J, Zhang M, Wang G, Li A, Wang W (2013) Enhanced stability of thylakoid membrane proteins and antioxidant competence contribute to drought stress resistance in the tasg1 wheat stay-green mutant. J Exp Bot 64:1509–1520. https://doi.org/10.1093/jxb/ert004
12. Lopes MS, Reynolds MP (2012) Stay-green in spring wheat can be determined by spectral reflectance measurements (normalized difference vegetation index) independently from phenology. J Exp Bot 63:3789–3798. https://doi.org/10.1093/jxb/ers071
13. Richards RA, Cavanagh CR, Riffkin P (2019) Selection for erect canopy architecture can increase yield and biomass of spring wheat. Field Crop Res 244:107649. https://doi.org/10.1016/j.fcr.2019.107649
14. Long SP, Marshall-Colon A, Zhu XG (2015) Meeting the global food demand of the future by engineering crop photosynthesis and yield potential. Cell 161:56–66. https://doi.org/10.1016/j.cell.2015.03.019
15. Murchie EH, Townsend A, Reynolds M (2018) Crop radiation capture and use efficiency. In: Meyers RA (ed) Encyclopedia of sustainability science and technology. Springer, New York, pp 73–106
16. Reynolds M, Atkin OK, Bennett M, Cooper M, Dodd IC, Foulkes MJ, Frohberg C, Hammer G, Henderson IR, Huang B, Korzun V, McCouch SR, Messina CD, Pogson BJ, Slafer GA, Taylor NL, Wittich PE (2021) Addressing research bottlenecks to crop productivity. Trends Plant Sci 26:607–630. https://doi.org/10.1016/j.tplants.2021.03.011
17. Reynolds MP, Braun HJ, Cavalieri AJ, Chapotin S, Davies WJ, Ellul P, Feuillet C, Govaerts B, Kropff MJ, Lucas H, Nelson J, Powell W, Quilligan E, Rosegrant MW, Singh RP, Sonder

K, Tang H, Visscher S, Wang R (2017) Improving global integration of crop research. Science 357:359–360. https://doi.org/10.1126/science.aam8559

18. Braun HJ, Atlin G, Payne T (2010) Multi-location testing as a tool to identify plant response to global climate change. In: Reynolds M (ed) Climate change and crop production (CABI climate change series). CABI Publishing, Wallingford, pp 115–138

19. Gourdji SM, Mathews K, Reynolds M, Crossa J, Lobell DB (2012) An assessment of wheat breeding gains in hot environments. AGU Fall Meet Abstr 1:1075

20. Pingali PL (2012) Green revolution: impacts, limits, and the path ahead. Proc Natl Acad Sci 109:12302–12308. https://doi.org/10.1073/pnas.0912953109

21. Stevenson JR, Villoria N, Byerlee D, Kelley T, Maredia M (2013) Green revolution research saved an estimated 18 to 27 million hectares from being brought into agricultural production. Proc Natl Acad Sci 110:8363–8368. https://doi.org/10.1073/pnas.1208065110

22. Lantican MA, Braun H-J, Payne TS, Singh RP, Sonder K, Michael B, van Ginkel M, Erenstein O (2016) Impacts of international wheat improvement research, pp 1994–2014

23. Economic Research Service (2019) Agricultural research funding in the public and private sectors. In: U.S. Department Agric. https://www.ers.usda.gov/data-products/agricultural-research-funding-in-the-public-and-private-sectors/

24. Pardey PG, Chan-Kang C, Dehmer SP, Beddow JM (2016) Agricultural R&D is on the move. Nature 537:301–303. https://doi.org/10.1038/537301a

25. Reynolds MP, Pask AJD, Hoppitt WJE, Sonder K, Sukumaran S, Molero G, Saint PC, Payne T, Singh RP, Braun HJ, Gonzalez FG, Terrile II, Barma NCD, Hakim A, He Z, Fan Z, Novoselovic D, Maghraby M, Gad KIM, Galal EHG, Hagras A, Mohamed MM, Morad AFA, Kumar U, Singh GP, Naik R, Kalappanavar IK, Biradar S, Sai Prasad SV, Chatrath R, Sharma I, Panchabhai K, Sohu VS, Mavi GS, Mishra VK, Balasubramaniam A, Jalal-Kamali MR, Khodarahmi M, Dastfal M, Tabib-Ghaffari SM, Jafarby J, Nikzad AR, Moghaddam HA, Ghojogh H, Mehraban A, Solís-Moya E, Camacho-Casas MA, Figueroa-López P, Ireta-Moreno J, Alvarado-Padilla JI, Borbón-Gracia A, Torres A, Quiche YN, Upadhyay SR, Pandey D, Imtiaz M, Rehman MU, Hussain M, Hussain M, Ud-Din R, Qamar M, Kundi M, Mujahid MY, Ahmad G, Khan AJ, Sial MA, Mustatea P, von Well E, Ncala M, de Groot S, Hussein AHA, Tahir ISA, Idris AAM, Elamein HMM, Manes Y, Joshi AK (2017) Strategic crossing of biomass and harvest index—source and sink—achieves genetic gains in wheat. Euphytica 213:23. https://doi.org/10.1007/s10681-017-2040-z

26. González-Navarro OOE, Griffiths S, Molero G, Reynolds MP, Slafer GA (2015) Dynamics of floret development determining differences in spike fertility in an elite population of wheat. Field Crop Res 172:21–31. https://doi.org/10.1016/j.fcr.2014.12.001

27. Quintero A, Molero G, Reynolds MP, Calderini DF (2018) Trade-off between grain weight and grain number in wheat depends on GxE interaction: a case study of an elite CIMMYT panel (CIMCOG). Eur J Agron 92:17–29. https://doi.org/10.1016/j.eja.2017.09.007

28. Silva-Perez V, Molero G, Serbin SP, Condon AG, Reynolds MP, Furbank RT, Evans JR (2018) Hyperspectral reflectance as a tool to measure biochemical and physiological traits in wheat. J Exp Bot 69:483–496. https://doi.org/10.1093/jxb/erx421

29. Piñera-Chavez FJ, Berry PM, Foulkes MJ, Molero G, Reynolds MP (2016) Avoiding lodging in irrigated spring wheat. II. Genetic variation of stem and root structural properties. Field Crop Res 196:64–74. https://doi.org/10.1016/j.fcr.2016.06.007

30. Sierra-Gonzalez A, Molero G, Rivera-Amado C, Babar MA, Reynolds MP, Foulkes MJ (2021) Exploring genetic diversity for grain partitioning traits to enhance yield in a high biomass spring wheat panel. Field Crop Res 260:107979. https://doi.org/10.1016/j.fcr.2020.107979

31. Rivera-Amado C, Trujillo-Negrellos E, Molero G, Reynolds MP, Sylvester-Bradley R, Foulkes MJ (2019) Optimizing dry-matter partitioning for increased spike growth, grain number and harvest index in spring wheat. Field Crop Res 240:154–167. https://doi.org/10.1016/j.fcr.2019.04.016

32. Reynolds M, Chapman S, Crespo-Herrera L, Molero G, Mondal S, Pequeno DNLDNL, Pinto F, Pinera-Chavez FJ, Poland J, Rivera-Amado C, Saint Pierre C, Sukumaran S, Pinera-Chavez FJ, Poland J, Rivera-Amado C, Saint Pierre C, Sukumaran S (2020) Breeder friendly phenotyping. Plant Sci 295:110396. https://doi.org/10.1016/j.plantsci.2019.110396

Part IV
Rapidly Evolving Technologies & Likely Potential

Chapter 27
High Throughput Field Phenotyping

Jose Luis Araus, Maria Luisa Buchaillot, and Shawn C. Kefauver

Abstract The chapter aims to provide guidance on how phenotyping may contribute to the genetic advance of wheat in terms of yield potential and resilience to adverse conditions. Emphasis will be given to field high throughput phenotyping, including affordable solutions, together with the need for environmental and spatial characterization. Different remote sensing techniques and platforms are presented, while concerning lab techniques only a well proven trait, such as carbon isotope composition, is included. Finally, data integration and its implementation in practice is discussed. In that sense and considering the physiological determinants of wheat yield that are amenable for indirect selection, we highlight stomatal conduc tance and stay green as key observations. This choice of traits and phenotyping techniques is based on results from a large set of retrospective and other physiological studies that have proven the value of these traits together with the highlighted phenotypical approaches.

Keywords Genetic advance · High throughput · Modelling · Field phenotyping platforms · Remote sensing

27.1 Learning Objectives

- Understanding how phenotyping may contribute to wheat genetic advance and potential techniques to apply.

J. L. Araus (✉) · M. L. Buchaillot · S. C. Kefauver
Integrative Crop Ecophysiology Group, Faculty of Biology, University of Barcelona, Barcelona, Spain

AGROTECNIO (Center for Research in Agrotechnology), Lleida, Spain
e-mail: jaraus@ub.edu; sckefauver@ub.edu

495

27.2   Introduction

Phenotyping is nowadays considered a major bottleneck limiting the breeding efforts [1]. In fact, high throughput precision phenotyping is becoming more accepted as viable way to capitalize on recent developments in crop genomics (see Chaps. 28 and 29) and prediction models (see Chap. 31). However, for many breeders, the adoption of new phenotyping traits and methodologies only makes sense if they provide added value relative to current phenotyping practices. In that sense, a basic concern for many breeders is still the controlled nature of many of the phenotyping platforms developed in recent years and the perception that most of these platforms are unable to fully replicate environmental variables influencing complex traits at the scale of climate variability nor handle the elevated numbers of phenotypes required by breeding programs [2]. This does not exclude for example the interest of indoors (i.e., fully controlled) platforms for specific studies or traits to be evaluated, or even the need to developing special outdoor (i.e., near field) but still controlled facilities. This is the case of phenotyping arrangements aimed to evaluate resilience to particular stressors (e.g., diseases, pests, waterlogging…) or the performance of hidden plant parts (i.e., roots) or non-laminar photosynthetic organs (e.g., ears, culms). While this chapter will focus on the general aspects concerning wheat phenotyping, specific information about special setups is very abundant.

Phenotyping of simple traits (e.g., plant height) can be achieved even by untrained personnel within a manageable time frame. However, manual phenotyping of complex traits, which is often the case when focusing on drought or heat tolerance, requires experienced professionals and is time intensive. Another important point to consider is that phenotyping of the large genotype sets is generally only feasible if conducted by several persons. Moreover, in case phenotyping is conducted visually this results in an inflation of measuring error, which might be further increased by fatigue setting, and is prone to subjective appreciation of each person. A recent paper [3] has defined the *high throughput phenotyping* as "relatively new for most breeders and requiring significantly greater investment with technical hurdles for implementation and a steeper learning curve than the minimum data set," where visual assessments are often the preferred choice.

In what follows, this chapter will address crop phenotyping within the context of its implementation under real growing (i.e., field) conditions. Literature and examples included will refer as much as possible to wheat or other small grain cereals under field conditions. In that sense we will introduce the term high throughput field phenotyping (HTFP).

The aim of an efficient phenotyping method is to enhance genetic gain (Fig. 27.1), which is defined as the amount of increase in performance achieved per unit time through artificial selection (see Chap. 7), usually referred to the increase after one generation (or cycle) has passed. Continuing on, the potential contribution of phenotyping to wheat breeding is placed in context by taking the genetic-advance determinants as a framework of reference. Alternative ways to dissect the role of phenotypic on genetic gain have been assessed elsewhere [4].

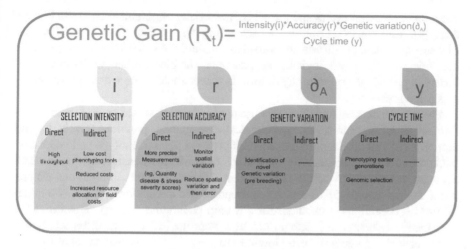

Fig. 27.1 Direct and indirect ways how high throughput precision field phenotyping may contribute to genetic gain in wheat

Accelerating genetic gain can be achieved by increasing selection intensity, accuracy and genetic variation, and/or reducing cycle time (see also Chap. 30). Phenotyping contributes both directly and indirectly to these variables [5]. Direct effects include *increasing selection intensity* by the development and deployment of more high throughput phenotyping techniques, evaluating larger populations eventually across different environments, which is actually the main purpose of this chapter, *improved selection accuracy*, which involves the repeatability and precision of the phenotyping techniques deployed, and *identifying new genetic variability for the targeted traits*, which, provided that it exists [6], may be secured through preselection, using very high throughput affordable approaches, even if they are not as accurate [4].

Indirect positive effects are diverse but also relevant. Low-cost phenotyping protocols allow breeders to increase selection intensity and identify new genetic variability, for example through the evaluation of larger populations. Phenotyping means more than just selecting the right traits and choosing the appropriate tools for evaluation, together with efficient data management. It also requires appropriate trial management and spatial variability handling [1, 5]. Improved trial management and field variation control will increase the selection accuracy -of phenotyping and thus the heritability of the trait being selected (see Chaps. 5, 6, 7 and 12). Therefore, selection accuracy is also improved through the deploying of phenotyping techniques to account for the growing conditions where plants are phenotyped (spatial variability in environmental factors, which also may involve the use of phenotyping techniques). While phenotyping does not directly contribute towards *decrease cycle time*, it is likely to play a more important role indirectly. For example, targeted HTFP will permit the reliable phenotyping of greater numbers of genetic resources derived from breeding lines by using smaller plot sizes and assessments obtained at earlier stages of population development. This allows breeders to reduce the duration of breeding cycles and the loss of potentially important alleles with linkage drag [4], therefore contributing to increasing the genetic gain (see Chap. 7). Moreover, while most efforts are considered toward direct selection for yield, indirect selection for physiological, morphological or biochemical yield-component traits can provide the opportunity to introduce new alleles from which genetic progress can be made [4].

Phenotypic expression is the response of genotypes to varied environmental conditions (GxE) or even to the agronomical management practices (GxExM), and therefore the full disentangling of the link between plant phenotype and its genetic background cannot be achieved (see Chap. 15) without considering the full and accurate quantification of the environmental and agronomical conditions experienced during growth [5]. Therefore, appropriate documentation of the environmental growth conditions is essential for any crop phenomics strategy. This implies a systematic collection and integration of meteorological data at different spatio-temporal scales, frequently using low-cost sensors [7]. Finally, new avenues for data management and exploitation are required in order to optimally capitalize on recent improvements in data capture and computation capacity.

Summarizing, the objective of practical phenotyping innovation is the implementation of high throughput precision phenotyping under real (i.e., field in most cases) conditions and preferably at an affordable cost. On the other hand, proper HTFP requires some basic uniform characteristics such as similar phenology of the whole set of varieties selected as well as the identification of the right growth stage (or stages) when phenotyping has to be conducted. In other words, a phenotypic trait may have a positive, negative or no relationship with grain yield or another target parameter depending on the growth stage at measurement. Such differential performance of a phenotypic trait may depend on different factors such as the phenological stage when it is measured or the growing conditions. The phenotypical performance may even be biased if the targeted germplasm is too diverse in terms of phenology (e.g., heading, anthesis or maturity dates). Therefore, in addition to choosing the optimal phenotypic traits, the time at which they are assessed, while avoiding too wide of a genotypic range in phenology, is also crucial. This applies for remote sensing traits such as vegetation indices as well as for lab traits such as the carbon isotope composition [6, 8].

27.3   Platforms: From Ground to the Sky

The drawbacks of most time-consuming phenotyping methods in terms of throughput and standardization can be overcome using image-based data collection. Remote sensing technologies, with the respective controllers and data loggers that complement the imaging systems, are usually assembled into what are termed as phenotyping platforms [5, 7, 9, 10]. The use of these platforms allows for a more efficient and accurate phenotyping with stable error across all genotypes, whether as single plants or in micro-plots. However, currently many of these platforms are costly and/or not applicable on a wide scale. Therefore, there is a strongly expressed need by the crop breeding community to develop both state-of-the-art and cost-effective, easy to use, and nonstationary HTFP platforms. These platforms may also represent tailored solutions to specific cases or a feasible formula on how to apply standard phenotyping tools in breeding programs with limited resources [11].

The concept of the phenotypic platform is wide and embraces a varied range of options in terms of placement: ground, aerial or even eventually (in the coming years) at the space level (Fig. 27.2; [10]). Within the category of ground phenotyping, platforms have quickly diversified, and the range of options is very wide: from a simple hand-held sensor, including for example monopods and tripods carrying any sensors from a simple yet effective RGB color camera, to complex unmanned ground vehicles of diverse nature, which are generically termed as "phenomobiles," and include tractor-mounted sensors, other tailored solutions (e.g., carts, buggies) or mobile cranes. Within the ground category one may also include highly complex stationary facilities. Cable-based robotics systems are becoming also an alternative for outdoor (i.e. field) phenotyping, which allow imaging platforms to move about a defined area [12]. Within the hand-held category of platforms, smartphones are becoming an alternative giving they may carry out different imagers (e.g. RGB and thermal), data management activities and geo-referencing functions [5, 7, 9, 10].

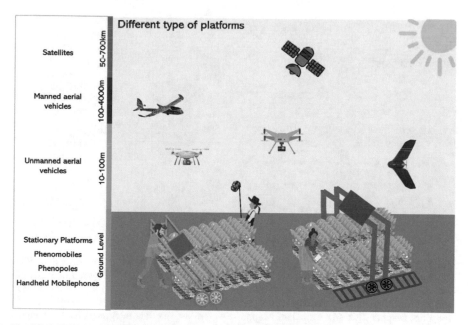

Fig. 27.2 Different Categories of Ground and Aerial Phenotyping Platforms. Ground level: these include from Handheld sensors (in this case just a person holding a mobile), to Phenopoles, Phenomobiles, Stationary Platforms. From 10 to 100 m: Unmanned Aerial Vehicles, as drones of different sizes and more or less compactness, fixed-wind drone. From 100 to 4000 m Manned Aerial Vehicles as airplanes or helicopters. In the near future different categories of satellites (Nanosatellite, Microsatellite and Satellites) from 50 to 700 km

Aerial platforms of different nature are being widely used, particularly more and more involving unmanned aerial vehicles (UAV), popularly known as drones. Proximal and remote sensing sensors are now able to be mounted on low flying multirotor UAVs, with image acquisition capabilities at spatial scales in centimeters, relevant to crop breeding [13]. The use of drones has popularized in the recent years [10] and even book manuals (even if mostly focused on crop management) have been produced. The remote sensing tools most frequently deployed in phenotyping platforms are RGB cameras, alongside multispectral and thermal sensors or imagers [14]. The increasing availability of compact drones which don't need to be assembled, bring the sensors embedded, and are affordable, reliable and easy to control, is popularizing more and more this option (Fig. 27.3). Nevertheless, other unmanned options offer appealing alternatives with contrasting capabilities, particularly fixed wing UAVs, where for example, the crop area to monitor area larger than a few hectares, or in related precision agriculture activities [15]. Other alternatives such as manned aircrafts are less used by crop breeders given the cost of this alternative, while the use of satellites on phenotyping are not yet a reality in practical terms due to the lack of free sub-meter resolution data, but they will surely be of increasing interest in the near future as these technologies advance [5, 10].

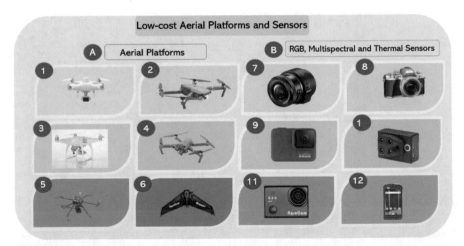

Fig. 27.3 Example of different types of affordable aerial platforms and sensors (less than 5000 USD). (**A**) Aerial Platforms: (*1*) Phantom 4 Multispectral (https://www.dji.com/es/p4-multispectral); (*2*) Mavic 2 Pro (https://www.dji.com/es/mavic-2); (*3*) & (*4*) are for a company named Sentera which add a multispectral camera to Phantom and Mavic (https://sentera.com); (*5*) AgroCam Mapper QC; (*6*) AgroCam Mapper FW the last one has integrated NDVI cameras (https://www.agrocam.eu/uav-system). (**B**) Affordable sensors that can be used on a phenopole or on a drone: (*7*) Sony Qx1 RGB (https://www.sony.es); (*8*) Olympus OM-D E-M10 MKII RGB (https://www.olympus.es); (*9*) GoPro, an RGB camera that can be modified to calculate the NDVI; (*10*) Parrot Sequoia multispectral Camera (https://www.parrot.com); (*11*) AgroCam NDVI camera, (https://www.agrocam.eu); (*12*) Smartphone CatS60 with RGB and Thermal camera (https://www.catphones.com)

Here we outline standards for the deploying simple stationary, cable-based robotics and ultimately UAVs as progressively more mobile and high throughput phenotyping platforms for the transport of the various proximal and remote sensing sensors/imagers. The primary selection criterion concerning the equipment to carry out the HTFP platforms concerns the choice of the most adequate sensors for the estimation of the specific biophysical traits of interest at an appropriate technology readiness level. Many of the simultaneous major technological advancements in HTFP platforms come from the impressive miniaturization of imaging and measurement technologies and on-board processing capacities, but perhaps even more importantly massive leaps and bounds in communications, compact and lightweight batteries, inertial sensors, electronic compasses, data storage and intelligent automated control algorithms. These have come together to enable the development of more compact and light weight scientific imaging sensors and at the same time improved indoor robotics systems and UAVs with increasing autonomy, carrying capacity, stability, and security. The result is that scientific quality of remote sensing platforms and sensors that only 5 years ago were nearly exclusively limited to very expensive indoor installations and manned airborne platforms (only able to provide ground spatial resolution a.k.a. pixel sizes on the order of 5–20 m and not amenable to phenotyping) are now available in more cost-effective unmanned systems. In fact, UAVs are nowadays the most popular mobile platform for phenotyping purposes [16].

27.4   Phenotyping Is More than Just Monitoring Techniques

Crop phenotyping is about collecting useful and meaningful data for integration into crop breeding programs. As such, a complete HTFP platform research protocol should include considerations for every part of the full process in order to ensure that no bottlenecks impede the throughput of the phenotyping activities. This includes but is not limited to (1) the equipment (sensors, platforms and software); (2) use operation (e.g. pilot permits and training, flight plans, and image acquisition in case of UAV); (3) proper storing and managing of experimental datasets for long term use; (4) image processing (pre-processing, calibration, mosaicking); (5) data generation (extraction from processed image to plot level data); (6) data analysis (index calculations, stats scripts) and database structure (storage, linkages, inventory indexing, ontologies, etc.); (7) specific case studies of bottlenecks to throughput, training requirements, costs, and optimization for specific crops (scalable/transferable traits) [7]. All the major components from sensors to platforms to software for each key processing step are intricately intertwined and need to be considered together such that pre-integrated systems or close attention to integration details will improve both data quality and data throughput. Some examples have been provided for the deployment of RGB images.

27.5   Data Integration: From Ideotype to Modelling and More

Connecting genomic and phenomic datasets remains challenging. Is in this context where plant phenotyping is creating new needs for data standardization, analysis and storage [17]. In addition, the value of phenotypic data is moving from empirical/descriptive context, where ideotype, understood as the fixed combination of traits that confers advantage to a given wheat genotype was the target, to the use of phenotypic data in a more mechanistic way, through simulation models aiming to predict genotype performance (see Chaps. 31 and 32). In that sense integration of phenotypic data into simulation models to predict trait value is of increasing importance [18]. Besides that, large amounts of phenotypic data are used in a statistically oriented manner, for marker-assisted and even more for genomic selection (see Chaps. 28 and 29). The future of crop breeding lies in the standardization of data collection across phenotyping platforms.

On the other hand, the development of specific software tools that meet the needs of the crop phenotyping community in terms of remote sensing data processing, extraction and analysis have been identified as potentially the greatest bottleneck for generating high quality phenotypic data [19]. This includes for example the development of intuitive, easy-to-use semiautomatic programs for microplot extraction encompassing also appropriate flight planning to capture images with sufficient quality, which implies relevant concepts such as view, sharpness and exposure calculations, in addition to consider ground control points (GCPs), viewing geometry and way-point flights [20]. These new software tools will need be integrated to include not only the assessment of crop growth performance (including for example crop establishment, stay green) and grain yield, but also the detection and quantification of phenological stages (heading or maturity times and even anthesis), agronomical yield components (ear density), total biomass, or identifying specific pests and diseases and further quantifying its impact.

Overall, there is a great need for new analytical approaches that can integrate multiple types of data or provide proper experimental design in observational contexts. This need will only grow with the development of imaging, sequencing, and sensing technologies. A recent push in this direction has been an emphasis on machine learning and artificial intelligence in phenotyping [21]. Concerning trait measurements, implementing machine learning methods on UAV data enhances the capability of data processing and prediction in various applications [16], such as wheat ear counting [22]. High spatial resolution UAV-based remote sensing imagery with a resolution between 0 and 10 cm is the most frequently employed data source amongst those utilized for machine learning approaches [16]. Classification and regression are two main prediction problems that are commonly used in UAV-based applications. Taking RGB images as a proximal remote sensing approach may increase the resolution of images and therefore the usefulness of these images when analyzed with machine learning methods. Thus, for example, using an RGB camera placed on a pole at 1.2 m from the ground provided a ground spatial resolution better than 0.2 mm, able to assess the thickness of the residual stems standing straight after the cutting by the combine machine during harvest. In that case, a faster Regional Convolutional Neural Network (Faster-RCNN) deep-learning model was first trained to identify the stems cross section [23]. Machine learning algorithms can be implemented using either open source or commercial software. Open source coding environments such as Python and R are freely available and may be redistributed and modified.

27.6   Affordable Phenotyping Approaches

Many of the desired phenotypic traits can be acquired using cost-effective and readily available RGB cameras, which are characterized as very high spatial resolution imaging sensors, with quality color calibration and PAR spectral coverage (Fig. 27.3). These are extensively addressed elsewhere (e.g. [7]). In short, several RGB vegetation indexes use the spectral concept for the estimation of biomass and canopy chlorophyll, while others are based on alternate color space transforms such as Hue Saturation Intensity (HSI), CIE-LAB and CIE-LUV [24]. Practical solutions exist for the calculation of these RGB vegetation indexes using free, open-source software. Thus, for example, our team at the University of Barcelona has developed open-source software tools for analyzing high resolution RGB digital images, with special consideration to cost-effectiveness, technology availability and computing capacity using digital cameras or smartphones for data acquisition. Besides the formulation of vegetation indices amenable to monitor crop growth, stay green, or quantify the impact of a given pest or disease which affect the green biomass, examples exist on the use of RGB images to specific purposes such as for example assessing ear density [22]. Recently methods have been proposed to phenotype early development of wheat, specifically to assess the rate of plant emergence, the number of tillers, and the beginning of stem elongation using drone-based RGB imagery. Moreover, the characteristics of the digital RGB images, together with the support of machine learning approaches, make feasible the automatic identification of plant deficiencies and biotic stressed based in the shape and pattern of leaf symptoms such as chlorosis, necrosis spots etc.

Besides the RGB sensors, in the last years a wide range of affordable multispectral imagers, and even thermal imagers, and dual multispectral/RGB or thermal/RGM imagers are available, making HTFP more feasible to, for example, small seed companies and national agricultural research organizations.

The main traits that can be measured in the field using affordable HTP-approaches is included, with the sensors/indices, as well as a qualitative assessment of their precision, in Table 27.1.

Table 27.1 Example indexes per trait and sensor type with a qualitative assessment of precision

Trait	Examples of Spectral Indexes	Sensors			Qualitative assessment of their precision	Reference
		Multispectral/ Hyperspectral	RGB	Thermal		
Growth Early Vigor Stay Green Quantification pest/disease Green Biomass Senescence	Normalized Difference Vegetation Index (NDVI)				Some indexes formulated using RGB and multispectral images may become saturated at medium to high levels of biomass, which implies a loss in accuracy. Saturating canopies are common between end of tillering to grain filling. Captured with a field sensor (points) or with cameras (images) at different heights such us in UAVs, phenopoles, etc. (Fig 27.2). The platform and the sensor determine the spectral resolution. Biomass estimation using thermal sensors is only related to fraction of vegetation cover, so accuracy is low.	[24, 25]
	Optimized Soil-adjusted Vegetation index (OSARI)					
	Greener Green Area (GGA)					
	a*					
	u*v*A					
	Crop Senescence Index (CSI)					
	Normalized Green-Red Difference Index (NGRDI)					
	Thermal bands					
Nitrogen content Leaf Pigments: Carotenoids, anthocyanins	Chlorophyll Content Index (CCI)				Some indices, like multispectral and RGB, estimate the different pigment content. Could be taking with a sensor (only points) or with cameras (images) at different height such us in drones, phenopoles (Fig 27.2). The platform and the cost of the sensor will determinate the spectral resolution.	[24, 25]
	Transformed Chlorophyll Absorption Ratio Index (TCARI) or (TCARI/OSAVI)					
	Anthocyanin Reflectance Index (ARI2)					
	Carotenoid Reflectance Index 2 (CRI2)					
	Triangular Green Index (TGI)					
Phenology (e.g heading, anthesis, maturity times)	Specific algorithms are required for specific stages such as anthesis. For another phenological stages, such as heading or maturity, even changes in the vegetation indices presented above, or an increase in canopy temperature may suffice				Depending of the phenological stage to assess resolution is key and therefore imager and distance of acquisition must be considered. This is the case for example of anthesis time which is determined based in appearance of extruded anthers. In such a case resolution in the range of mm are needed	[24]
Detection pest/disease, Agronomical yield components, Number of seedlings and spikes	Specific algorithms for detecting plant health symptoms, counting seedlings during crop emergence or the number of spikes during crop reproductive stage				Image resolution is key and varies depending on the imager used (RGG>multispectral>thermal) and the distance from where the image is acquired. For some traits, such as ear counting, only the ears above the canopy not overlapped by other tillers, will be accounted	[24]
Height/ Plant Architecture	RGB or multispectral 3D model				In the best scenarios, accuracy in the range of cm is achieved. It provides external canopy assessments, which means for additional assessment of crop yield or canopy architecture additional measurements using other approaches (or harvesting) should to be considered.	[26]
Light Use Efficiency	Photochemical Reflectance Index (PRI)				Powerful index at both, the single leaf level and for canopy level, even when its combination with gas exchange or chlorophyll fluorescence is recomended.	[27]
Crop water status, transpiration, Water content	Crop Water stress Index (CWSI)				Use NIR, SWIR or TIR bands to estimate canopy water content. Accuracy depends on the infrared water absorption bands selected.	[14, 25]
	Normalized Difference Water index (NDWI)					
	Thermal bands				Environmental factors such as wind, clouds or the presence of bare soil within the canopy may strongly affect thermal measurement accuracy. Moreover, using aerial platforms is recommended to avoid short-term time-related dynamics.	

27.7   Hyperspectral Imaging for Crop Phenotyping: Pros and Cons

Hyperspectral sensors and cameras are among the most promising for the phenotyping of advanced traits. The application of hyperspectral reflectance to proximal (i.e. ground level) plant phenotyping at high resolution range makes it possible to infer, in the case of wheat under field conditions, not only grain yield but for example the content of metabolites in leaves and ears [28], or photosynthetic capacities and quenching. Hyperspectral imaging techniques have been expanding considerably in recent years. The cost of current solutions is decreasing, but these high-end technologies are not yet available for moderate to low-cost outdoor or indoor applications. However new methodological developments, such as a single-pixel imaging setup [29], which do not require (as much of an investment) high computational capacity, may offer a more approachable alternative.

In spite of the recent availability of hyperspectral UAV sensors, both the sensor design and the resulting data result in several complications at the time of capture, pre-processing, calibration, and analysis [13]. Firstly, "hyper" literally means "too much" so hyperspectral sensors are and openly acknowledged as frequently capturing more data than is necessary for any specific given purpose. For that reason, they are and will continue to be considered as more exploratory and experimental rather than operational sensors. It is on the scientific community to take on the challenge of first acquiring what may be considered as excess data in order to later distil the "big data" down to the essential and prescribe the more specific and required measurements for any particular measurement goal, in this case the phenotyping of photosynthesis and biophysical traits relevant to the disentangling of genetic sequencing data and maximizing yield to feed the future [30].

Moreover, the use of hyperspectral images from moving platforms, such as those carried out from UAV, has additional challenges [13]. Unlike sensors that capture whole images in one instant like RGB (which captures three separate spectral regions in one image with its integrated Bayer filter) and the more common multispectral cameras (which capture each spectral region with a different sensor and are later corrected for parallax) most hyperspectral cameras are not "area array" type. Most hyperspectral imagers are of the "line scanning" type, which require a moving mirror and spectral prism to iteratively measure each wavelength over a single line of pixels as the UAV moves forward. This requires carefully programmed and timed internal sensor movements with the external robotics platform or UAV flights at specific forward movement velocities relative to the distance between the sensor and crop. The data is also thus more likely to be adversely affected by environmental conditions and gimbal instability. In turn, the carrying platform and hyperspectral camera system must be fully integrated as the inertial measurement unit (IMU) accelerometer (yaw, pitch and roll) and positioning, whether in local or GPS (geographical location) data in order to create a correct hyperspectral image. Ground topographical variability, if present, should also be optimally corrected for using a separately produced digital elevation model (DEM).

Still, despite these complications, adequately integrated hyperspectral sensors and platforms from stationary solutions to UAVs are available and may provide excellent data with in-depth knowledge and expertise in data interpretation and processing. More common UAV multispectral sensors are based precisely on the extensive data analysis from field spectroscopy and airborne hyperspectral imaging conducted by research laboratories over the past 40 years [31]. The best bands for measuring specific plant spectral properties that are associated with physiological traits of interest have been selected with regards to both their specific central wavelength and their bandwidth (range of wavelengths where radiation is measured) and designed accordingly. However, no full VNIR+SWIR hyperspectral sensors have been available for application as HTFPs with the specific purpose of crop breeding until very recently, due to the many technological barriers that impeded their deployment on UAVs, and as such the linkage between spectral wavelengths and breeding traits has not been completed.

27.8 Implementing Phenotyping in Practice

Some approaches for practical wheat phenotyping will be briefly presented taking grain yield as the breeding target (Fig. 27.4). A thorough set of examples of traits and conditions where phenotyping may be applied in practice at different levels (handy, high throughput, and precision phenotyping) may be accessed elsewhere [3].

Identifying the key traits for phenotyping may result in convergent approaches. On one hand, grain yield may be dissected into three main physiological components: the amount of resources (radiation, water, nitrogen…) captured by the crop, the efficient use of these resources and the dry matter partitioning (so called harvest index). The kind of resource considered depend in each case on what is the limiting factor (e.g. under drought conditions or low nitrogen fertility, water and nitrogen will be the relevant resources, respectively). On the other hand, retrospective studies, comparing cultivars developed through the last decades, also provide clues on the most successful, to date, physiological traits, involved in the genetic advance after Green Revolution. For this approach it is important to avoid confounding effects associated with the inclusion in the comparison, genotypes developed prior Green Revolution or even transitional ones. In that sense genetic advance in wheat for a wide range of environmental conditions has been associated with a higher stomatal conductance [32]. Remote sensing techniques such as infrared thermometry or thermography may be deployed as proxies for higher transpiration [1, 4, 5, 9]. An alternative is to use the stable carbon isotope, one of the few lab-phenotyping traits widely accepted. Usually a lower (i.e. more negative) carbon isotope composition ($\delta^{13}C$) or, alternatively, a higher carbon isotope discrimination ($\Delta^{13}C$), particularly when analyzed in mature kernels and confounding effects are avoided (such as differences in phenology [6]) is pursued, since it indicates a better water status and eventually more water captured by the crop, in spite the fact water use efficiency decreases. Another trait to consider is stay green, which may be relevant particularly under good agronomical conditions [33]. This trait may be assessed through

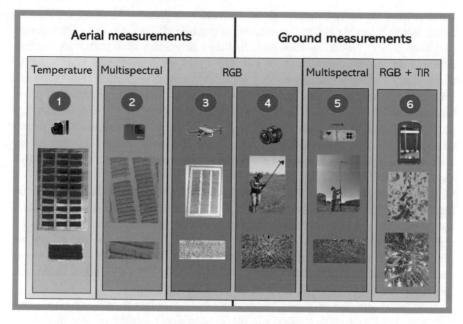

Fig. 27.4 Different examples from the University of Barcelona using different platforms and sensors; mostly of affordable nature: (*1*) Thermal Camera: FLIR Tau 640 with Thermal Capture; (*2*) Modified GoPro taking the NDVI, that was installed in a Mavic pro 2; (*3*) Mavic 2 Pro with an RGB camera; (*4*) Sony Qx1 used from ground to count spikes; (*5*) MultiSPEC 4C camera, which has 4 channels, from AIRINOV company, on a phenopole of 5 m; this multispectral camera is quite similar to Parrot Sequoia in capacities and cost; (*6*) Cat s60 mobile phone, which takes thermal and RGB pictures

multispectral of RGB-derived vegetation indices assessed during grain filling [24, 34]. The same category of indices may be used to assess early vigor and ground covering. The three categories of main remote sensing approaches (RGB, thermal or multispectral/hyperspectral) may be used to assess differences in phenology, particularly heading and anthesis nature.

Digital RGB imaging may allow to 3D surface reconstruction to provide estimations of plant height, and incidence of lodging, while image-pattern recognition may help to identify the presence of a pest or disease, which may be further quantified on their impact by RGB or multispectral vegetation indices [26].

Greater biomass is also considered as a key target trait for selection, particularly since harvest index is reaching theoretical maximum, while the increasing in biomass have been minor during the more the half century elapsed from the beginning of the Green Revolution to the present. HTFP, particularly when deployed from an aerial platform allows the assessment of biomass through different techniques in the full plot rather than in subsamples. Moreover, there is the capacity to undertake repetitive measurements which may improve the estimation [4]. A priory the most canonical way to assess biomass is using LiDAR (Light Detection and Ranging) mounted in an aerial platform or in a "phenomobile" [26]. However still today the most common way to assess green biomass is through vegetation indices, either multispectral or RGB-derived, given the common perception these approaches being more affordable and easier to use than the LiDAR [26]. However, an inherent limitation of the vegetation indices is that they saturate, which makes its use less effective during the central part of the crop cycle, even when still is of value to assess early stages of growth or stay green. Moreover, vegetation indices do not inform about canopy height. Nevertheless, a more accurate determination of green biomass than that associated to vegetation indices, together with plant height may be also achieved using RGB images; this time through three-dimensional reconstruction of the crop canopy. This evaluation may improve further if canopy height is combined with the number and thickness of the stems, evaluated through high-resolution RGB images [23].

Another potential target for current phenotyping, which has been traditionally neglected, is the photosynthetic contribution of the ear to grain filling. While a recent study has confirmed that genotypic variability exists for this trait and moreover showing the first examples of HTFP for this trait [35], the advent on remote sensing techniques based on the combination of RGB imaging for in situ organ detection, together with thermal and/or multispectral imaging may allow in the near future the evaluation of this trait from aerial platforms [36].

However, there still exist several areas not fully explored in terms of HTFP protocols, such as root phenotyping, just one among many hidden yet very important attributes to consider in new potential phenotyping target traits.

27.9 Key Concepts

HTFP may contribute to speed genetic advances in different ways. Nevertheless, phenotyping under controlled conditions may still have applicability in some cases. Usually, there is not a single technological solution, but rather different options in terms of throughput and even cost are available. In this sense, affordable phenotyping techniques, including various sensors and platforms, are more approachable than ever before. Remote sensing techniques are the most commonly used for phenotyping but other approaches, like the lab-based traits may be also useful. Eventually, hyperspectral techniques may even replace many lab-based approaches. Besides that, image processing and even more data analysis, including prediction models are the actual components of the phenotyping pipeline that will allow full exploitation of new technological developments, in terms of traits and platforms, for HTFP.

27.10 Conclusions

As a take-home message, phenotyping is evolving very fast in terms of throughput, the range of traits that can be assessed, and the adaptation of the costs of sensors and platforms to a growing market for these technologies. However, the computing and statistical components still remain as the most commonly perceived bottleneck that currently limits HTFP from reaching full operability. This includes a wide range of areas: from automation of data capturing and further data processing, to the use of the data produced to drive prediction models or even its integration and application in genomic selection.

In this sense remote sensing techniques will become more accessible to breeders if image analysis services were to become more widely available, affordable and automatized (i.e., customer friendly), providing curated phenotypic data in near real time. As examples, on board data pre-processing and 5G in-flight data transmission are two of the main paths forward for simplified processing and improved usability of remote sensing sensors. Both go hand-in-hand with improvements in sensor-platform integration, in which the sensor and platform have become more and more interconnected and thus are able to share GPS, altimetry, IMU, power sources and transmission capacities for improved efficiency and operability. Manual UAV flights and separate manual programming of sensor data capture are already in the past. In many modern commercial UAVs, smartphone connectivity already converts the UAV controller to an all-in-one command station for programming flight paths, viewing UAV and sensor details in-flight, and even limited data viewing and down-loading in real-time. Also, in smart sensors, such as the Tetracam MCAW system (https://www.tetracam.com/Products-Macaw.htm), images are calibrated to reflectance, corrected for parallax and combined into multiband TIFFs or even processed into programmable vegetation indexes in flight by the on-board micro-processor and fast solid state disk drives; these also include Wi-Fi to smartphone connectivity. Even though the current wireless connectivity of these can't keep pace with the onboard data capture and automated pre-processing, both of these, including even UAV hyperspectral data, should be both processable and transmissible in real-time with 5G Wi-Fi, enabling the automation of the rest of the pre-processing, from Structure-to-Motion orthomosaicking and on to micro-plot extraction (given the proper GIS metadata), either in PC or cloud-based services inter-connected to UAV functionality or specific sensors or as a third party solution, such as DroneMapper, Pix4Dcloud, AgisoftCloud, Micasense AtlasCloud, DroneDeploy, and many more (http://dronemapper.com, https://www.pix4d.com/product/pix4dcloud, https://cloud.agisoft.com, https://atlas.micasense.com, see also https://micasense.com/software-solutions). Given that there are already precision agriculture crop pest/disease UAVs that can detect specific pest or disease presence or absence and spray with onboard imaging and artificial intelligence decision support, the next step for plant phenotyping must be close behind.

On the other a routinely assessment under field conditions of particular traits, relevant for grain and fodder quality (e.g., contents of amino acids, micronutrients, provitamins), or for HTFP in frontier areas such as the breeding for higher and more efficient photosynthesis. This will be feasible through hyperspectral techniques, providing not only computing capabilities are optimized, but also cost of hyperspectral sensors and imagers decrease.

References

1. Araus JL, Cairns JE (2014) Field high-throughput phenotyping: the new crop breeding frontier. Trends Plant Sci 19:52–61
2. Pauli D, Chapman SC, Bart R, Topp CN, Lawrence-Dill CJ, Poland J, Gore MA (2016) The quest for understanding phenotypic variation via integrated approaches in the field environment. Plant Physiol 172:00592.2016. https://doi.org/10.1104/pp.16.00592
3. Reynolds M, Chapman S, Crespo-Herrera L, Molero G, Mondal S, Pequeno DNL, Pinto F, Pinera-Chavez FJ, Poland J, Rivera-Amado C, Saint Pierre C, Sukumaran S (2020) Breeder friendly phenotyping. Plant Sci 295:110396. https://doi.org/10.1016/j.plantsci.2019.110396
4. Rebetzke GJ, Jimenez-Berni J, Fischer RA, Deery DM, Smith DJ (2019) High-throughput phenotyping to enhance the use of crop genetic resources. Plant Sci 282:40–48. https://doi.org/10.1016/j.plantsci.2018.06.017
5. Araus JL, Kefauver SC, Zaman-Allah M, Olsen MS, Cairns JE (2018) Translating high-throughput phenotyping into genetic gain. Trends Plant Sci 23:451–466. https://doi.org/10.1016/j.tplants.2018.02.001
6. Araus JL, Slafer GA, Royo C, Serret MD (2008) Breeding for yield potential and stress adaptation in cereals. Crit Rev Plant Sci 27:377–412. https://doi.org/10.1080/07352680802467736
7. Araus JL, Kefauver SC (2018) Breeding to adapt agriculture to climate change: affordable phenotyping solutions (Review). Curr Opin Plant Biol 45:237–247. https://doi.org/10.1016/j.pbi.2018.05.003
8. Sanchez-Bragado R, Newcomb M, Chairi F, Condorelli GE, Ward R, White JW, Maccaferri M, Tuberosa R, Araus JL, Serret MD (2020) Carbon isotope composition and the NDVI as phenotyping approaches for drought adaptation in durum wheat: beyond trait selection. Agronomy 10:1679. https://doi.org/10.3390/agronomy10111679
9. Araus JL, Kefauver SC, Zaman-Allah M, Olsen MS, Cairns JE (2018) Phenotyping: new crop breeding frontier. In: Meyers R (ed) Encyclopedia of sustainability science and technology. Springer, New York
10. Jin X, Zarco-Tejada P, Schmidhalter U, Reynolds MP, Hawkesford MJ, Varshney RK, Yang T, Nie C, Li Z, Ming B, Xiao Y, Xie Y, Li S (2020) High-throughput estimation of crop traits: a review of ground and aerial phenotyping platforms. IEEE Geosci Remote Sens:1–33. https://doi.org/10.1109/MGRS.2020.2998816
11. Reynolds D, Baret F, Welcker C, Bostrom A, Ball J, Cellini F, Lorence A, Chawade A, Khafif M, Noshita K, Mueller-Linow M, Zhou J, Tardieu F (2019) What is cost-efficient phenotyping? Optimizing costs for different scenarios. Plant Sci 282:14–22. https://doi.org/10.1016/j.plantsci.2018.06.015
12. Andrade-Sanchez P, Gore MA, Heun JT, Thorp KR, Carmo-Silva AE, French AN, Salvucci ME, White JW (2014) Development and evaluation of a field-based high-throughput phenotyping platform. Funct Plant Biol 41. https://doi.org/10.1071/FP13126
13. Aasen H, Honkavaara E, Lucieer A, Zarco-Tejada PJ (2018) Quantitative remote sensing at ultra-high resolution with UAV spectroscopy: a review of sensor technology, measurement procedures, and data correction workflows. Remote Sens 10:1091. https://doi.org/10.3390/rs10071091
14. Gracia-Romero A, Kefauver SC, Fernandez-Gallego JA, Vergara-Díaz O, Nieto-Taladriz MT, Araus JL (2019) UAV and ground image-based phenotyping: a proof of concept with durum wheat. Remote Sens 11:1244. https://doi.org/10.3390/rs11101244
15. Maes WH, Steppe K (2019) Perspectives for remote sensing with unmanned aerial vehicles in precision agriculture. Trends Plant Sci 24:152–164. https://doi.org/10.1016/j.tplants.2018.11.007
16. Eskandari R, Mahdianpari M, Mohammadimanesh F, Salehi B, Brisco B, Homayouni S (2020) Meta-analysis of unmanned aerial vehicle (UAV) imagery for agro-environmental monitoring using machine learning and statistical models. Remote Sens 12:3511. https://doi.org/10.3390/rs12213511

17. Bolger M, Schwacke R, Gundlach H, Schmutzer T, Chen J, Arend D, Opperman M, Weise S, Lange M, Fiorani F, Spannagl M, Scholz U, Mayer K, Usadel B (2017) From plant genomes to phenotypes. J Biotechnol 261:46–52. https://doi.org/10.1016/j.jbiotec.2017.06.003

18. Brown TB, Cheng R, Siriault XRR, Rungrat T, Murray KD, Trtilek M, Furbank RT, Badger M, Pogson BJ, Borevitz JO (2014) TraitCapture: genomic and environment modelling of plant phenomic data. Curr Opin Plant Biol 18:73–79. https://doi.org/10.1016/j.pbi.2014.02.002

19. Furbank RT, Tester M (2011) Phenomics–technologies to relieve the phenotyping bottleneck. Trends Plant Sci 16:635–644. https://doi.org/10.1016/j.tplants.2011.09.005

20. Roth L, Hund A, Aasen H (2018) PhenoFly Planning Tool: flight planning for high-resolution optical remote sensing with unmanned aerial systems. Plant Methods 14:116. https://doi.org/10.1186/s13007-018-0376-6

21. Singh A, Ganapathysubramanian B, Singh AK, Sarkar S (2016) Machine learning for high-throughput stress phenotyping in plants. Trends Plant Sci 21:110–124. https://doi.org/10.1016/j.tplants.2015.10.015

22. Fernandez-Gallego JA, Lootens P, Borra-Serrano I, Derycke V, Haesaert G, Roldán-Ruiz I, Araus JL, Kefauver SC (2020) Automatic wheat ear counting using machine learning based on RGB UAV imagery. Plant J 103:1603–1613. https://doi.org/10.1111/tpj.14799

23. Jin X, Madec S, Dutartre D, de Solan B, Comar A, Baret F (2019) High-throughput measurements of stem characteristics to estimate ear density and above-ground biomass. Plant Phenomics 2019:1–10. https://doi.org/10.34133/2019/4820305

24. Fernandez-Gallego JA, Kefauver SC, Vatter T, Aparicio Gutierrez N, Nieto-Taladriz MT, Araus JL (2019) Low-cost assessment of grain yield in durum wheat using RGB images. Eur J Agron 105:146–156. https://doi.org/10.1016/j.eja.2019.02.007

25. Gracia-Romero A, Verdara-Diaz O, Thierfelder C, Cairns JE, Kefauver SC, Araus JL (2018) Phenotyping conservation agriculture management effects on ground and aerial remote sensing assessments of maize hybrids performance in Zimbabwe. Remote Sens 10:349. https://doi.org/10.3390/rs10020349

26. Madec S, Baret F, de Solan B, Thomas S, Dutartre D, Jezequel S, Hemmerlé M, Colombeau G, Comar A (2017) High-throughput phenotyping of plant height: comparing unmanned aerial vehicles and ground LiDAR estimates. Front Plant Sci 8:1–14. https://doi.org/10.3389/fpls.2017.02002

27. Garbulsky MF, Peñuelas J, Gamon J, Inoue Y, Filella I (2011) The photochemical reflectance index (PRI) and the remote sensing of leaf, canopy and ecosystem radiation use efficiencies. A review and meta-analysis. Remote Sens Environ 115:281–297. https://doi.org/10.1016/j.rse.2010.08.023

28. Vergara-Diaz O, Vatter T, Kefauver SC, Obata T, Fernie AR, Araus JL (2020) Assessing durum wheat ear and leaf metabolomes in the field through hyperspectral data. Plant J 102:615–630. https://doi.org/10.1111/tpj.14636

29. Ribes M, Russias G, Tregoat D, Fournier A (2020) Towards low-cost hyperspectral single-pixel imaging for plant phenotyping. Sensors 20:1132. https://doi.org/10.3390/s20041132

30. Verrelst J, Malenovský Z, Van der Tol C, Camps-Valls G, Gastellu-Etchegorry JP, Lewis P, North P, Moreno J (2019) Quantifying vegetation biophysical variables from imaging spectroscopy data: a review on retrieval methods. Surv Geophys 40:589–629. https://doi.org/10.1007/s10712-018-9478-y

31. Schaepman ME, Ustin SL, Plaza AJ, Painter TH, Verrelst J, Liang S (2009) Earth system science related imaging spectroscopy—an assessment. Remote Sens Environ 113:S123–S137. https://doi.org/10.1016/j.rse.2009.03.001

32. Roche D (2015) Stomatal conductance is essential for higher yield potential of C3 crops. Plant Sci 34:429–453. https://doi.org/10.1080/07352689.2015.1023677

33. Carmo-Silva E, Andralojc PJ, Scales JC, Driever SM, Mead A, Lawson T, Raines CA, Parry MAJ (2017) Phenotyping of field-grown wheat in the UK highlights contribution of light response of photosynthesis and flag leaf longevity to grain yield. J Exp Bot 68:3473–3486. https://doi.org/10.1093/jxb/erx169

34. Lopes MS, Reynolds MP (2012) Stay-green in spring wheat can be determined by spectral reflectance measurements (normalized difference vegetation index) independently from phenology. J Exp Bot 63:3789–3798. https://doi.org/10.1093/jxb/ers071
35. Molero G, Reynolds MP (2020) Spike photosynthesis measured at high throughput indicates genetic variation independent of flag leaf photosynthesis. Field Crop Res 255:107866. https://doi.org/10.1016/j.fcr.2020.107866
36. Sanchez-Bragado R, Vicente R, Molero G, Serret MD, Maydup ML, Araus JL (2020) New avenues for increasing yield and stability in C3 cereals: exploring ear photosynthesis. Curr Opin Plant Biol 56:223–234. https://doi.org/10.1016/j.pbi.2020.01.001

Chapter 28
Sequence-Based Marker Assisted Selection in Wheat

Marco Maccaferri, Martina Bruschi, and Roberto Tuberosa

Abstract Wheat improvement has traditionally been conducted by relying on artificial crossing of suitable parental lines followed by selection of the best genetic combinations. At the same time wheat genetic resources have been characterized and exploited with the aim of continuously improving target traits. Over this solid framework, innovations from emerging research disciplines have been progressively added over time: cytogenetics, quantitative genetics, chromosome engineering, mutagenesis, molecular biology and, most recently, comparative, structural, and functional genomics with all the related -omics platforms. Nowadays, the integration of these disciplines coupled with their spectacular technical advances made possible by the sequencing of the entire wheat genome, has ushered us in a new breeding paradigm on how to best leverage the functional variability of genetic stocks and germplasm collections. Molecular techniques first impacted wheat genetics and breeding in the 1980s with the development of restriction fragment length polymorphism (RFLP)-based approaches. Since then, steady progress in sequence-based, marker-assisted selection now allows for an unprecedently accurate 'breeding by design' of wheat, progressing further up to the pangenome-based level. This chapter provides an overview of the technologies of the 'circular genomics era' which allow breeders to better characterize and more effectively leverage the huge and largely untapped natural variability present in the Triticeae gene pool, particularly at the tetraploid level, and its closest diploid and polyploid ancestors and relatives.

Keywords Genetic diversity · Molecular marker · Mapping · QTLome · Cloning · Homoeologous loci

M. Maccaferri (✉) · M. Bruschi · R. Tuberosa
Alma Mater Studiorum–Università di Bologna, University of Bologna, Bologna, Italy
e-mail: marco.maccaferri@unibo.it; martina.bruschi3@unibo.it; roberto.tuberosa@unibo.it

© The Author(s) 2022
M. P. Reynolds, H.-J. Braun (eds.), *Wheat Improvement*,
https://doi.org/10.1007/978-3-030-90673-3_28

28.1   Learning Objectives

- Assessing the feasibility, benefits, and shortcomings of molecular techniques and especially marker-assisted selection in wheat breeding.

28.2   Introduction

Meeting the food demand of a population of 10 billion by 2050 will require a substantial increase in genetic gain presently achieved mostly by conventional breeding approaches (see Chap. 27). In wheat and other crops, gains from selection are tapering off, also in part due to climate change effects, and will not meet the estimated 70% increase in crop productivity required by 2050 to feed mankind (see Chap. 21). This worrisome trend can be mitigated through genomics-assisted breeding, particularly through marker-assisted selection (MAS) and genomic selection, two procedures increasingly adopted to accelerate gain from selection in breeding programs worldwide [1, 2].

The success of the Green Revolution that fueled the high selection gains in the 1960s–1980s was mainly due to the previous identification, followed by deployment, of the semi-dwarf *Rht* alleles in combination with photoperiod-insensitive *Ppd* alleles which allowed for the selection of short and early flowering cultivars able to escape heat and drought and take advantage of higher nitrogen fertilization regimes (see Chap. 10). These remarkable results highlight the key role played by the genotype x environment x management (GxExM) interaction.

Additional traits played an important role towards the release of novel cultivars provided with alleles able to mitigate the negative effects of biotic (e.g., rusts, fusarium head blight, root rot, septoria tritici blotch, etc.), and abiotic (e.g., drought, heat, nutrient deficiency and toxicity, etc.) stress on yield and its stability. In both cases, the identification of beneficial alleles at the loci (genes and mostly quantitative trait loci: QTLs) governing the resistance/tolerance to such factors and their selection through MAS are being increasingly adopted to accelerate the gain from selection (see Chaps. 5 and 6). The identification of QTLs with a major effect on the target traits has been more frequently reported for biotic stress (see Chap. 19; see also Fig. 28.5), though some notable examples have been reported for abiotic stress [3, 4], particularly when targeting morpho-physiological traits (e.g., early vigor, root system architecture, staygreen, isotope discrimination, etc.) with predictive value as proxies for biomass production, water-use efficiency, yield components, yield and its stability [5].

28.3   Genetic Resources, Mapping Approaches and Database

The key role played by germplasm collections for both gene discovery and pre-breeding purposes has been highlighted in hexaploid wheat with the Watkins collection [6] and in tetraploid wheat with the Global Durum wheat Panel (GDP; [7]) and the Tetraploid wheat Global Collection (TGC; [8]) (Fig. 28.1).

Linkage mapping and association mapping also known as genome wide association study (GWAS) in wheat have been conducted using various molecular marker sets and platforms [9]. Therefore, cross-referencing loci and QTL mapping results across experiments and genetic materials is cumbersome but otherwise essential for increasing the accuracy of mapping, as well as for mapping the allele/haplotype distribution in germplasm collections and breeding pools across the QTLome [10]. A valuable approach to prioritize the QTLs to focus on with MAS and eventually attempting their cloning is provided by meta-analyses compiling and comparing the results of multiple QTL studies, hence providing a more accurate mapping of QTLs and their overall value across environments [11, 12].

The wheat community shares the knowledge related to the various molecular marker sets used during the past 40 years, mainly through dedicated publications and the GrainGenes database (https://wheat.pw.usda.gov/GG3/). Widely used, common and high-quality molecular marker sets were first adopted for RFLPs and then for SSR markers. The genome density of SSR markers allowed for cross-referencing

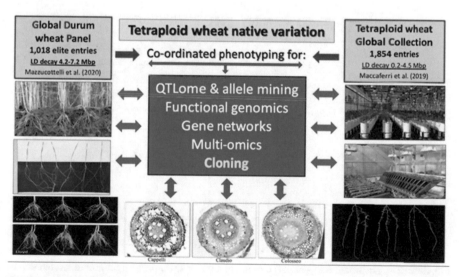

Fig. 28.1 The Global Durum wheat Panel (GDP; [7]) and the Tetraploid wheat Global Collection (TGC; [8]) are instrumental to mine the vast biodiversity present in the A and B tetraploid wheat genomes. The higher genetic variability coupled with lower linkage disequilibrium (LD) decay of the TGC indicates its suitability for QTL discovery and cloning while the GDP is more suitable for breeding purposes

across diverse linkage maps and highly polymorphic reference maps. Most important were the ITMI mapping population, a highly polymorphic map obtained from the cross of the bread wheat cv. Opata with a highly diverse wheat Synthetic line obtained from a cross between durum wheat and *Aegilops tauschii* and the Courtot x Chinese Spring intervarietal molecular marker linkage map. Subsequently, thanks to dedicated software, consensus maps providing higher genetic resolution and denser markers were assembled in both durum wheat [13] and bread wheat [14].

28.4   Dissecting the Wheat QTLome

The prevailing assumption has been that the variation in quantitative traits observed among wheat accessions is caused by the effects of multiple QTLs – mostly due to natural dominant mutations like insertion or deletion of bases (INDELs) in the regulatory gene regions – and the environment that inevitably limits our capacity for identifying QTLs, particularly under conditions of low heritability frequently present under abiotic stress ([5]; Chap. 13). Additionally, the wheat genome is huge and highly repetitive [8, 15], thus posing further difficulties in managing map-based cloning procedures that are implemented for the most interesting QTLs, clearly a very limited number (Fig. 28.2).

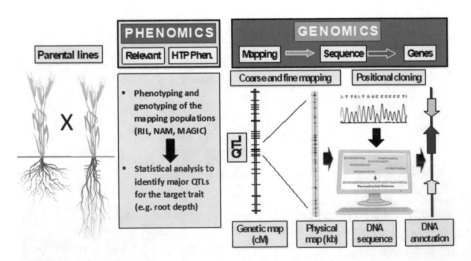

Fig. 28.2   The positional (map-based) cloning of a major QTL for a target trait (e.g., root depth) requires (1) the phenotyping and genotyping of an adequately large mapping population segregating for the trait, (2) the statistical analysis to map the QTLs and estimate their additive effect, (3) the fine mapping at high genetic resolution (possibly <0.1 cM) usually achieved with the phenotyping of a very large (from 1000 to 5000 F$_2$ plants depending on the heritability of the trait) population usually assembled from the cross of two near-isogenic lines contrasted for the QTL alleles. (Modified with permission from [16])

Enhancing genetic gain in wheat and other crops relies on the identification and, ideally, cloning of the loci governing the variability of the target traits followed by their selection via MAS and/or other genomic tools [2, 10].

More than three decades of dedicated experiments indicate that most QTL effects are small, as predicted by the so-called 'infinitesimal' model [17]. However, major QTLs (i.e., those accounting for >10% of the measured phenotypic variability) have also been reported and positionally cloned in wheat [18–20] which allows for designing the so-called 'perfect marker' for MAS (no recombination between the marker and the target locus) while advancing our understanding of the functional basis of variability of the target traits.

Once a QTL has been cloned via forward-genetics approaches, other reverse-genomics approaches (e.g., Targeting Induced Local Lesions in Genomes: TILLING, genetic engineering and gene editing, see Chap. 29) offer unprecedented opportunities to exploit native and/or artificially induced novel alleles. Considering the importance of quantitative traits for sustaining wheat performance under adverse conditions, increasing attention is being devoted to the mapping and cloning of major QTLs – hereafter defined 'QTLome' as a whole – which accounts for a sizeable portion of the variability targeted by breeders ([10]; Fig. 28.3).

Genomics-assisted wheat improvement is implemented in two complementary ways: (i) by targeting a limited number of well-characterized major QTLs via MAS (the tip of the iceberg in Fig. 28.3) and (ii) by leveraging the plethora of unknown

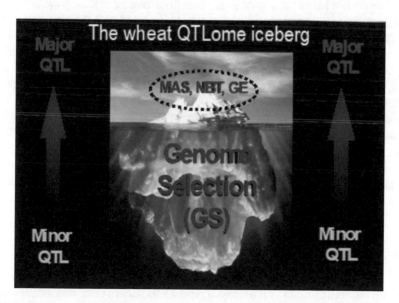

Fig. 28.3 The wheat QTLome represents the portion of QTLs with a sufficiently strong additive effect that makes their mapping and selection of the beneficial alleles via MAS possible. Only a minute fraction of these major QTLs can be cloned, hence allowing for the application of new breeding technologies (NBT; e.g., gene editing) and/or genetic engineering (GE). The vast majority of QTLs have additive effects too small to allow for their mapping. Their selection is possible through genome selection (GS)

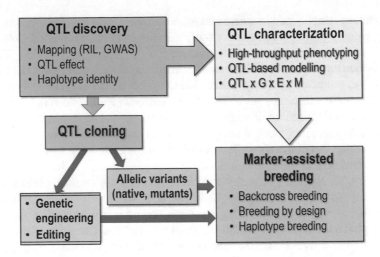

Fig. 28.4 How genomics-assisted breeding allows us to identify beneficial QTL alleles and deploys marker-assisted selection (MAS), genome editing, and/or genetic engineering (GE) to enhance the frequency of beneficial allelic variants in breeders' pools

QTLs with additive effects too small to be mapped (the submerged portion of the iceberg in Fig. 28.3) but otherwise indirectly selectable through genomic selection (GS; Chap. 30). The two approaches are complementary and their adoption – either as single approaches or in combination- should be based on a case-by-case evaluation, depending on the selection objectives, available genetic materials, and information on the genetic make-up and heritability of the target trait. The sequential or integrated adoption of both approaches (i.e., QTL-MAS followed by the application of GS models accounting for known genes/alleles as fixed effects) has been proven far more effective than GS alone [21].

Additionally, Fig. 28.4 indicates how, once a QTL has been cloned, the sequence information of the causative sequence (coding or non-coding) allows for the design of 'perfect' markers and the identification of rare native haplotypes present in the collection. Alternatively, the sequence of the QTL can be used to create novel alleles through gene editing (Chap. 29) and/or through genetic engineering, thereby enriching the MAS pipeline with novel alleles.

28.5   Selecting Traits and Loci for the MAS Pipeline

Choosing the traits suitable for the MAS pipeline requires a clear understating of the priorities and limiting factors of the breeding project based on (i) the prevailing environmental and phytosanitary conditions in the target environment and (ii) the concurrent effects on other traits (e.g., quality) of the targeted alleles/haplotypes *per se* caused by metabolic pleiotropy and/or caused by loci tightly linked to the allele/s targeted by MAS, the so-called 'linkage drag'.

An aspect of paramount importance for loci/QTL/allele proper exploitation in applied breeding programs is the thorough evaluation of the QTLxGxExM interaction that underpins the QTL effects [3]. This issue is often inadequately addressed, because an appropriate experimental design of field trials to achieve such a goal can be too expensive. Equally important when evaluating QTL effects is the concept of 'envirotyping' as a third 'typing' technology, complementing genotyping and phenotyping (Chap. 3). Envirotyping is a fundamental prerequisite to crop modeling and phenotype prediction through its functional components [22]. In this respect, modeling yield in wheat is particularly challenging due to its broad distribution across the globe and the contrasting environmental conditions under which wheat is grown (Chap. 31).

28.5.1   Loci for Phenology

The *Rht* and *Ppd* loci that fueled the Green Revolution are obvious "low-hanging fruit" for the application of MAS since heading date and height are primary determinants optimizing yield while ensuring its stability across environments. Data on the haplotype profiles at the *Rht* and *Ppd* loci are increasingly available for the founders and other modern genotypes that most frequently are used as parents to create novel segregating populations. Among the 46 currently known *Rht* genes and alleles (https://shigen.nig.ac.jp/wheat/komugi/), *RhtB1b* and *Rht-D1b* confer insensitivity to GA_3 and are the first two loci identified and used in the Green Revolution. However, taller and faster-growing wheat cultivars can be higher yielding than dwarf or semi-dwarf wheat genotypes under early and severe drought conditions, a finding likely related to effects of *Rht* alleles on coleoptile length and seedling/tiller vigor at early growth stages but also on root traits (e.g., root mass and depth) as shown by Beyer et al. [23]. The shorter wheat varieties are considered as better adapted to well-watered and nutrient-rich conditions rather than conditions of low soil moisture, a notable example of GxExM interaction, indicating how breeders can leverage MAS for *Rht* alleles to optimize yield and yield stability based upon the environmental conditions. As an example of wide differentiation of allelic distribution driven by adaptation and yield potential, we can consider the case of worldwide *RhtB1* allelic distribution in durum wheat. Most of the modern, highly productive durum varieties grown in the fertile and temperate areas under fall sowing and overwinter tillering are homozygous for the semi-dwarf *RhtB1b* allele while, on the contrary, this allele is rarely found in modern varieties bred for the Northern American prairies including North Dakota, Montana and Canada where extensive agriculture and short growing cycle are dominating.

Based on the environmental conditions (e.g., photoperiod, precipitation, temperature, etc.) of the target environment, breeding programs have been optimized for the alleles present at these loci in the parental lines and pre-breeding materials (Chaps. 3 and 25). Developmental regulatory networks include response to vernalization (*VRN loci;* Chap. 3) and response to day-length conditions, including *PHOTOPERIOD1*, *PHYB* or *PHYC*, *CO1*, and *CO2* as well as response to vernalization and freezing tolerance, including *CBF* and *COLD REGULATED* (*COR*) genes.

A key player for the fine tuning of flowering time in both durum and bread wheat is the *Vrn-1* locus that regulates the switch from vegetative to the reproductive mode based upon the duration of the exposure to a critical threshold of number of days with temperatures between −2 and 15 °C. The regulation of flowering time in response to environmental temperature and day-length conditions is further fine-tuned by partially redundant networks, including a vernalization responsive network with four *VRN* loci: *VRN1, VRN2, VRN3=FLOWERING LOCUS T (FT)*, and *VRN4*, all amenable to MAS.

A similar situation has been reported for the *Ppd1* locus, also present with three homeologs (*Ppd-A1, Ppd-B1,* and *Ppd-D1*) of different strength and with different alleles, including various *Ppd*-insensitive dominant mutations in the gene promoter regions at all the three homeologs that were rapidly selected by breeders due to their positive effects in temperate environments. Additionally, copy number variation is another major cause of natural allelic variation in *VRN* and *PPD* genes. The *VRN* and *PPD* allelic combinations consciously or unconsciously selected by breeders at the three *VRN* genes, *Ppd1*, and at their homeologs, respectively, have been surveyed in both tetraploid and hexaploid wheat [24].

28.5.2   Loci for the Root System Architecture (RSA)

Notwithstanding the well-demonstrated importance of the *Rht* loci, increasing attention is being devoted to the loci that control RSA, particularly root mass and root depth, both of which have been shown to play a pivotal role in capturing soil moisture and nutrients [25]. Selection and breeding for RSA traits have been documented to be effective under conditions where plants complete their cycle based on stored soil water, a condition where deeper roots allow the plants to access deeper soil layers where more residual moisture is available as compared to upper soil layers. A marker-assisted approach targeting plants enriched in alleles conferring deeper roots would expedite the release of drought-tolerant cultivars under such conditions when residual moisture is more likely available at depth around anthesis and grain-filling when surface layers become dry [26].

28.5.3   Loci for Disease Resistance

Nowadays, MAS for resistance to fungal diseases, mainly rusts, fusarium head blight and root rot, septoria tritici blotch, and powdery mildew accounts for the vast majority of the MAS activities, particularly marker-assisted backcrossing, routinely carried out in wheat breeding programs worldwide.

The release of new cultivars during the Green Revolution largely relied on three-way (top crosses) and less from simple crosses. At CIMMYT, the three-way crosses (top crosses) approach was mostly effective in introducing and immediately recombine new innovative and beneficial alleles at multiple loci for plant architecture (*Rht*), phenology (*Ppd*), and rust (*Lr*, *Yr*, and *Sr*) resistance. Importantly, this approach resembled the three-way cross already adopted by the early Italian wheat geneticist and breeder Nazareno Strampelli to develop a first series of innovative wheat varieties in the 1920s that in Italy supported the 'Battle for Grain' launched in 1925 and eventually allowed the country to become self-sufficient in wheat production. The many Strampelli's innovative varieties selected from the cross 'Rieti/Wilhelmina//Akakomugi' carried out in 1913, later spread worldwide, particularly in South America and China [27].

Increasing attention and effort are devoted to the identification of markers associated to loci for resistance to viruses (e.g., *SBCMV*) and/or insects (e.g., Hessian fly) whose diffusion and damaging effects are being increased by global warming. An example is provided by the search of markers linked to the loci for resistance to soil-borne cereal mosaic virus (SBCMV) which has been shown to reduce yield by 40–50% in susceptible commercial winter wheat cultivars in UK up to 70% in durum wheat in Italy [28].

28.6 Molecular Marker Technologies for MAS

A summary of marker technologies and their pros and cons is reported in Table 28.1. The 'first generation markers' developed at the onset of MAS in the late 1980s was based on RFLP, a very expensive and time-consuming technology. The advent of the PCR technique ushered in a number of much cheaper and faster 'second generation markers' such as random amplified polymorphic DNA (RAPD), and derived markers such as sequence characterized amplified regions (SCAR). Previous studies conducted to dissect the QTLome of soil-borne cereal mosaic virus (SBCMV) resistance in durum wheat were based on SSR and Diversity Arrays Technology (DArT) markers [28]. However, these marker classes present a series of constraints: low throughput (SSR markers), genome density insufficient for fine mapping (SSR and DArT markers [29]) and limited informativeness (DArT markers in their original version). In the past decade, efficient use of SNPs has become possible thanks to the development of arrays like the Illumina 90K [30]. Based on the wide use of the Illumina 90K wheat array worldwide, Maccaferri et al. [31] developed a consensus map for tetraploid wheat harboring 30,144 markers in which the high density of gene-derived SNPs provides useful anchor points for positional cloning. The abudance of SNPs in the wheat genome, together with the possibility of coupling them with high-throughput genotyping technologies, like KASP (Kompetitive Allele Specific PCR; Chap. 18) makes them suitable for fine mapping which requires the sampling of thousands of plants.

Table 28.1 Time-course progress in molecular marker technologies with their pros and cons

Marker type	Platform/Technology	Time interval	Time/assay	Marker numerosity	Pros and cons/Multiplexability
Hybridization-based markers					
Restriction fragment length polymorphism (RFLP)	Enzymatic restriction + Probe hybridization + Electrophoresis + Southern Blot + radioactive/chemioluminescent detection	'70–'90	Several days	Hundreds	Very laborious and expensive. Mainly codominant. Multiplex (yes, 2–3 assay)
Polymerase chain reaction (PCR)-based markers					
Simple PCR from expressed sequence tags (EST)	PCR + horizontal electrophoresis	'80-now	One day	Hundreds, or custom design	One day. Allows to detect insertion/deletions (INDEL). Multiplex (no). Codominant
Cleaved amplified polymorphisms (CAPS)	PCR + Enzymatic restriction + electrophoresis	'80-now	One day	Hundreds, or custom design	One day. Allows to detect single nucleotide substitutions. Multiplex (no) Codominant
Allele specific oligonucleotide PCR (ASO)	PCR + electrophoresis	'80-now	One day	Hundreds, or custom design	One day. Allows to detect substitutions or INDEL. Multiplex (no). Codominant
Random Amplified Polymorphic DNA (RAPD)	PCR + electrophoresis	'80–'90	One day	Hundreds	One day. Multiplex (yes). Robustness drawback. Dominant
Amplified Fragment Polymorphisms (AFLP)	PCR + vertical electrophoresis (polyacrylamide gel fragment analysis)	'80–'90	One day	Hundreds	One day. Multiplex (yes). Robust. Dominant
Simple Sequence Repeat (SSR)	PCR + vertical electrophoresis	'80-now	One day	Thousands	One day. Multiplex (yes, 3–6 assay). Robust. Codominant
Real-Time hydrolysis probes (TaqMan®)	SNP Fluorescent detection	'90-now	One day	Thousands. Custom design	One day. Multiplex (no). Robust. Codominant. Custom design

Marker type	Platform/Technology	Time interval	Time/ assay	Marker numerosity	Pros and cons/Multiplexability
Kompetitive Allele Specific PCR (KASP®)	SNP Fluorescent detection	2000-now	One day	Thousands. Custom design	One day. Multiplex (no). Robust. Codominant. Custom design
Single Nucleotide Polymorphism arrays (SNP arrays)	DNA microarrays. Illumina and Affimetrix technology	2000-now	Short turn around time	Highly flexible based on discovery panels	One week. High Multiplex of 10K-50K-90K-280K-660K-800K. Robust. Codominant. Suitable for whole genome analysis (GWAS, GS)
Targeted resequencing	Illumina Next Generation massive sequencing	2010	Short turn around time	Highly flexible based on discovery panels	One week. Multiplex of up to 1-10K Robust. Codominant. Suitable for whole genome analysis

The adoption of the SNP array technology and genotyping-by-sequencing (GBS) allowed for an unprecedented level of marker density and mapping quality [32]. A few arrays were quickly adopted by the wheat community, like the Illumina iSelect wheat 9K and 90K arrays [30] and the Affymetrix 35K array [33]. This allowed for the accumulation of mapping data sufficient to generate a newer, highly dense generation of reference and consensus maps. These maps reached a density of 1–10 marker/cM across the entire genome. Among those maps, the SNP-based durum consensus map assembled by Maccaferri et al. [31], joined all previous SSR- and SNP-based mapping information from tetraploid wheat. Reference consensus maps were quickly and widely adopted for (i) projecting QTL mapping results and QTL confidence intervals from multiple experiments into reference consensus maps/ assembled genomes and (ii) providing a framework for assisting the wheat genome sequence assembly procedures/pipelines.

28.7 Reference Genome Assembly

Gold-standard wheat genome assemblies have been obtained for the hexaploid wheat Chinese Spring [15] and the tetraploid wheat Svevo [8]. Second-generation, highly accurate, platinum-standard genome assemblies are being developed based on the integration of Optical Mapping (Bionano) and third generation long-read sequencing technology (PacBio), as recently shown with the release of the hexaploid wheat pangenome based on 10 high-quality genome assemblies from highly diverse and widely used cultivars worldwide (http://www.10wheatgenomes.com). The release of these highly contiguous wheat genomes allows to accurately project most of the molecular marker sets irrespectively of the marker technology adopted (DArT and SSR markers, SNP array, GBS, etc.) and represent the best reference for investigating the wheat QTLome [8].

28.8 Handling Sequence Data for Developing KASP Markers

For over a decade, SSR markers have provided a highly accurate and sufficiently dense marker framework that allowed for the development of many MAS protocols [34]. The drawback of SSR genotyping is that it required high-resolution polyacrylamide gel electrophoresis, no longer required with SNP technology where alternative alleles are discriminated by fluorescence. However, the entire molecular marker detection technology had to be revisited to adapt to the requirement of the SNP substitution detection, that does not involve differences in molecular weight between the alternative alleles. Discriminating the alternative SNP alleles requires 'allele-specific' recognition assays, with discrimination based on in-plate direct fluorescence reading, usually detected on real-time PCR (also known as quantitative PCR, qPCR) machines or plate fluorescence readers, which bypasses electrophoresis and

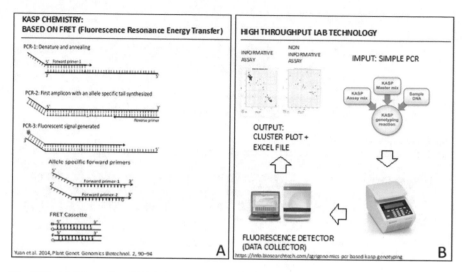

Fig. 28.5 (a) Schematic of KASP PCR (reprinted with permission from [36]). In evidence, the two allele-specific primer and the FRET cassette containing HEX and FAM fluorochromes. 1. The allele-specific primer anneals to the complementary sample DNA. 2. The first amplicon with allele-specific tail is synthesized. 3. The subsequent PCR cycles synthesize complements of the allele-specific tail sequence enabling the FRET cassette to bind the DNA and to emit allele-specific fluorescence based on the sample genotype formula. (b) Workflow of KASP genotyping technique. 1. Reagents required for KASP PCR. 2. Thermal cycler used to perform the reaction. 3. Detection of fluorescence during multiple amplifications performed in a Real-Time PCR instrument 4. Software output. See also https://info.biosearchtech.com/agrigenomics-pcr-based-kasp-genotyping

allows for the automation and high-throughput robotization of the assays. Together with the already well-established TaqMan technology, the KASP assay technology was progressively adopted due to the optimal combination of accuracy, easy implementation, and cost-effectiveness. Both technologies can accurately genotype SNPs based on either allele-specific probes (Taqman) or primers (KASP) (reviewed in [35]). Figures 28.5, 28.6, and 28.7 explain the main technical steps to design suitable KASP primers and implement the assays.

While in diploids the development of KASP assay is straightforward, this task poses several problems in tetra- and hexaploid wheat due to the high rate of similarity among the homeolog genome sequences adjacent to the varietal SNP. This entails a 'dilution effect' of the fluorescent signals that makes it progressively more difficult to accurately discriminate the target allelic variants that are genome-specific. Therefore, for allopolyploids, KASP primer design requires due attention to Mendelize the assay, i.e., making the assay as much genome-specific as possible, with primers being both allele- and genome-specific.

This requires multiple alignments of the two or three reference genomes in the SNP region in order to identify the position and sequence of both the varietal SNP (target- and genome-specific) and of the neighbor homeolog SNPs/INDELs that locally differentiate the genomes. Subsequently, an accurate design of the

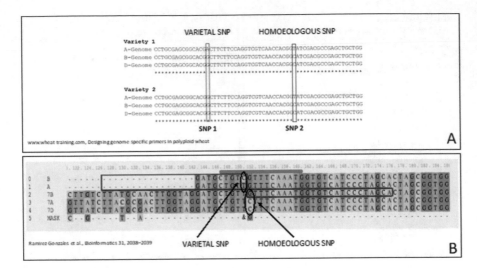

Fig. 28.6 (a) Example of hexaploid wheat sequence containing varietal (SNP 1) and homoeologous (SNP 2) SNPs from www.wheat-training.com. Varietal SNPs are polymorphisms between varieties while homoeologous SNPs are polymorphisms between genomes of a polyploid individual and typically non-polymorphic, though heterozygous, among varieties. A reliable genotype call can be obtained only by ensuring a sufficient NGS Illumina read depth on the polymorphic region (e.g., >8 reads). (b) Example of alignment performed by PolyMarker, a primer design pipeline for polyploids. KASP allele-specific primers are designed based on the varietal SNP, while the common primer is based on the homoeologous SNP and gives genome specificity to the KASP assay. (Modified with permission from [37])

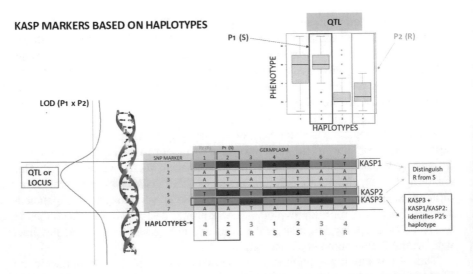

Fig. 28.7 Haplotype-based development of KASP assays for a disease resistance QTL. Haplotypes of resistant/susceptible parental lines can be used to develop diagnostic KASP assays that are predictive of multiallelic haplotypes (four haplotypes are represented). *P1* parental line 1, *P2* parental line 2, *R* resistant, *S* susceptible

allele-specific primers and of the genome-specific common primers is implemented with the support of dedicated primer-design software [38].

Additionally, the use of the reference genomes is relevant also to check genome-wide for off-target priming sites to prevent designing potentially non-specific primers on SNPs at loci other than those being targeted.

28.9 Examples of MAS

In wheat, protocols for tagging beneficial alleles suitable to MAS have been published in dedicated journals and made available to public and private research institutions and breeders worldwide since the late 1980s. Apart from specific literature searches using scientific publication browsers, effort has been made to provide access to this vast albeit fragmented knowledge. In particular, websites and databases specifically cataloguing MAS results are available at Graingenes (https://wheat.pw.usda.gov/GG3/), Komugi (https://shigen.nig.ac.jp/wheat/komugi/) as well as the Catalogue of Gene Symbols for Wheat. Even more focused websites are MAS-WHEAT (https://maswheat.ucdavis.edu/), the T3 Triticeae Toolbox website (https://wheat.triticeaetoolbox.org/), and CIMMYT publications (Laboratory Protocols, CIMMYT Applied Molecular Genetics Laboratory). In particular, MAS-Wheat provides a concise and informative report for each locus of breeding interest, assembled by the original study authors and including a short locus and allele description, molecular marker protocols, primer sequences, and expected amplification/hybridization allelic results. To date, 65 protocols are stored in MAS-wheat and more are expected. Additionally, databases are being developed to store, classify and manage the QTL results that are continuously published, either in the form of meta-QTL studies for several traits or more recently, as QTL databases: see T3 and WheatQTLdb [24].

Once the target locus/QTL has been identified, either through linkage or association mapping, geneticists and breeders develop one and preferably multiple user-friendly molecular marker assays useful for tracing the beneficial alleles through MAS. Due to the inherent difficulty in understanding the nature/localization of the causative gene and causative polymorphism (i.e., quantitative trait nucleotide, QTN), most molecular assays for MAS have been developed from the same original markers (SNP and/or INDEL) used in the mapping study, provided they are linked (<5 cM) or preferably tightly linked (<1 cM) to the locus/QTL peak.

These newly or re-designed single marker assays are immediately available for the MAS of plants with the desired allele/s. However, there are cases where these single assays are not acceptable for their weak diagnostic power and excess of false positives. This discrepancy is proportional to their distance from the target locus, since assayed markers still recombine with the causative gene, and to their capacity to discriminate the functional haplotypes at the causal loci. To limit the impact of recombination, it is always advisable to rely on at least a couple of markers flanking the target locus/QTL peak. In this case, the frequency of false positives can be

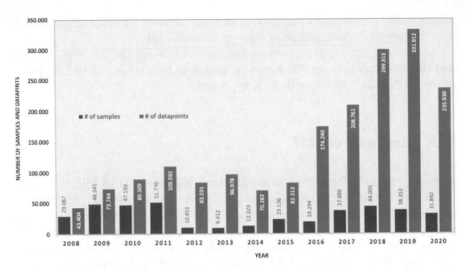

Fig. 28.8 Number of DNA samples and molecular marker assays used for MAS by CIMMYT's Global Wheat Program from 2008 to 2020. (Courtesy of the CIMMYT's Global Wheat Program)

predicted based upon the product of the distances of the two markers flanking the locus being selected. As an example, the rate of false positives of a MAS relying on two markers flanking the target locus at 2.2 and 1.6 cM will be slightly lower (due to negative crossover interference) than 2.2% × 1.6% = 0.35%.

A more subtle albeit critical aspect is that the natural variation found at causal genes occurs at multi-allelic haplotypes in a gene. Multiple mutations with diverse phenotypic effects occur at different times in the promoter, exons, and introns of causal genes and are typically organized in haplotypes and haplogroups. Notably, single bi-allelic markers do not have enough discriminant power to trace the haplotypes of interest and a single SNP usually pre-dates or post-dates the haplotypes of interest. Therefore, precise MAS applications require the use of haplotypes comprising multiple SNPs in the target regions rather than single SNPs. In durum wheat, haplotype discrimination was adopted for the MAS of *Lr14a* based on SSR markers [29] and it is now increasingly adopted thanks to the rapid accumulation of genotypic data.

At CIMMYT, MAS was introduced around 2006 to select parental lines with the beneficial alleles at key loci for phenology, mainly *Vrn*, *Rht*, *Ppd* for phenology, resistance to rust (*Lr34*) and fusarium head blight (*Fhb*), and quality (*Glu-1*). The program quickly scaled up to segregating materials once more user-friendly markers became available and were adopted for multiplexing multiple traits. Figure 28.8 clearly indicates this trend after 2012 when each DNA sample was probed, on average, for up to 7 loci.

Table 28.2 presents a synopsis of the main loci targeted to develop MAS protocols which are being implemented in pre-breeding and/or breeding programs in wheat. The details of the references reporting the loci targeted by MAS are reported in Gupta et al. [1], Kumar et al. [39], and King et al. (Chap. 18).

Table 28.2 Targets of marker-assisted selection in wheat for loci/QTL present in the primary gene pool, native alleles, and number of loci with markers and protocols available in the public domain. Loci reported in **bold** have been directly identified through positional cloning

Traits and loci amenable to marker-assisted selection	Bread wheat	Durum wheat	Loci
Plant architecture			
Reduced height (*Rht*)	***	**	***RhtB1, Rht12, Rht18***
Phenology, days to flowering			
Vernalization (*VRN*)	***	***	***VRN1, VRN2, VRN3, VRN4***
Sensitivity to photoperiod (*Ppd*)	***	***	***Ppd1***
Earliness *per se (Eps)*	***		***ELF3***
Resistance to foliar pathogens			
Biotrophs			
Yellow rust – *Yr*	***	***	80 > loci named; 17 reported on MAS-WHEAT: ***Yr5a and b***; ***Yr7; Yr10; Yr18***; ***YR36 = WHEAT KINASE START1 (WKS1)***; ***Yr46; Wtk1(Yr15, YrG303,*** and ***YrH52)***; ***YrAS2388***
Black or Stem rust –*Sr*	***	***	60 > loci named; 28 reported on MAS-WHEAT: ***Sr2; Sr13; Sr21; Sr22; Sr26; Sr33; Sr35; Sr45; Sr46; Sr50; Sr60/WTK2; Sr61; SrTA1662***
Brown rust –*Lr*	***	***	80 > loci named; 16 reported on MAS-WHEAT: ***Lr1; Lr10; Lr13; Lr14; Lr21; Lr22a; Lr34; Lr46; Lr67***
Powdery mildew – *Pm*	***	***	100 > loci named; KASP available for several loci in literature. ***Pm1a; Pm2; Pm3; Pm4; Pm5e***
Emi-biotrophs			
Septoria tritici blotch (*Stb*)	***	***	20 > loci named; ***Stb6-Stb16***
Parastagonospora nodorum (*Snn* and *Tsn-Tox*)	***	**	Nine host gene–necrotrophic effector interaction identified
P. tritici-repentis and *Bipolaris sorokiniana*	***	*	***Tsn1*** to ***Tsn5; ToxA*** to ***ToxD***
			Tsn1/SnToxA; Snn1/SnTox1;
			Snn3-D1/SnTox3; Snn5/SnTox5
Fusarium head blight (FHB)	***	***	***Fhb1*** to ***Fhb7. Fhb1, Fhb7. TaUGT6.*** Several QTLs reported including ***Qfhs.ifa-5A***
Fusarium root rot (*FRR*)	**	**	***FRR*** QTLs
Viruses			
Soil Borne Viruses	**	***	***Sbm1, Sbm2.*** QTLs for wheat yellow mosaic
Aphids-transmitted	***	**	QTLs for barley yellow dwarf virus
Pests			
Hessian Fly (*HF*)	***	***	35 HF resistance genes identified. *H* genes
Orange Wheat Blossom Midge	***	*	***Sm1*** QTLs
Wheat Stem Sawfly	***	***	***SSt1*** (*Solid stem locus*)

(continued)

Table 28.2 (continued)

Traits and loci amenable to marker-assisted selection	Bread wheat	Durum wheat	Loci
Grain yield components			
Spike development/shape	*	*	*AP2L2* and *Q. microRNA172, FRIZZY PANICLE (WFZP)*
Spike fertility (spikelet number per spike, fertile florets per spikelet)	**	*	Several QTLs. *WAPO-A1. GNI1, GNI2*
Grain size and weight	***	**	Grain Weight (*GW1* to *GW8*), Grain width, Grain Length, and Grain Size genes. *TaGW2-A1. TaGS3.* Several additional QTLs identified
Carbon metabolism	***	**	Trehalose 6-phosphate synthase and sucrose synthase. ADP glucose pyrophosphorylase (AGPase)
Abiotic stress tolerance			
Nitrogen use (NU) and nitrogen-use efficiency (NUE)	***	**	*GS. Fd-GOGAT.* QTLs
Resilience to low temperatures/frost	*	*	*DREB/CBF* factors. *COR. VRN* QTLs.
Resilience to water and heat stress	**	*	*DREB/CBF* factors. QTLs. Heat shock proteins
Root system traits	**	**	*TaVSR-B1. QTLs* Root Growth Angle (*RGA*), root biomass and root elongation, root/shoot ratio. QTLs
Resistance to herbicides (metribuzin and imazamox)	***	*	*AhasL-D1* and *AhasL-B1* (resistance to imidazolinones). QTLs
Resistance to boron	***	*	*Bo1*
Aluminium tolerance	***	*	*TaALMT1*
Salinity stress tolerance	***	**	*Kna1 = TaHKT1;5-D, Nax2 = TmHKT1;5-A*
Technological and nutritional quality			
Grain Storage proteins. Quality and quantity. Gluten strength.	***	***	*Glu-1, Bx7, Gpc-B1 (=TaNAC).* QTLs for GlutoPeak parameters and for gluten strength
Starch quality and quantity, amylose (*resistant starch*) content	**	***	*SBEII, GBSS (Wx1)*
Carotenoids and luteins synthesis and degradation (*grain yellow pigment content*)	*	***	*Psy1, Zds1, CYP, LCYE, LCYB, Lpx1-3, Ppo1-2.* QTLs and causative genes in the terpenoid pathway
Grain texture (*Hardness*)	***	*	*PINA, PINB*
Pre-Harvest Sprouting (*PHS*)	***	*	*Several **PHS** loci* and *QTLs tagged by KASP* *TaPHS1, TaMFT, Myb10-D; TaABI4*

(continued)

Table 28.2 (continued)

Traits and loci amenable to marker-assisted selection	Bread wheat	Durum wheat	Loci
Cadmium accumulation (reduced cadmium concentration)	*	***	*Cdu1*
Antinutritional factors	**	**	α-Amylase/Trypsin inhibitors

***wide interest and high relevance worldwide
**high interest at the local level, possible increase of relevance in the future
*low interest

The more tightly associated the markers are to the causative gene and the more markers that are being developed at the target locus, the more the combined SNP assays (haplotypes or 'haplo-markers') are diagnostic of the functional alleles at the causative gene and can be considered as 'diagnostic' or 'predictive' of the favorable alleles and phenotypes in various genetic backgrounds/crosses. These markers can be considered highly reliable and as such, widely recommended and used. Haplotype-based breeding is thus one of the most advanced areas for MAS [40].

Once the causative gene is cloned either by positional cloning or by functional analysis of candidate genes in the locus region and the causative polymorphisms are identified, it is possible to design the so-called 'perfect markers', i.e., one or more molecular assays diagnostic for the causative alleles and phenotypes, and coinciding with the functional haplotypes at the causative gene and as such not subjected to recombination.

Identifying the causal gene underlying a locus/QTL is a long-time, resource-demanding procedure, albeit highly rewarding in the case of loci of paramount breeding importance. In the past decade, due to the huge and complex wheat genome this goal has only been reached for few genes (*Bo1, GPC, Lr1, Lr10, Ppd, Q, Tsn1, VRN,* and *Yr36*). Importantly, the international efforts aiming at developing the genomic resources in wheat have shown an impressive acceleration in the last 5 years [8]. This recently led to the isolation of the causative genes for several loci in a few years, with *Fhb1* being one of the most relevant ones, followed by *Cdu1, Fhb7, MIWE18, Lr14, Pm4, SSt1,* and *Yr15* as well as several *Lr, Sr,* and *Yr* genes. First, the isolation of the causative genes at several loci of breeding interest allowed to develop so-called 'perfect molecular markers' designed rightly on the nucleotide polymorphisms causative of the phenotype, and therefore highly diagnostics and not subjected to recombination. Secondly isolation of causative genes allows us to better appreciate the range and the complexity of the mutations causing the functional native allelic diversity. A notable example is the *Fhb1* locus, an example of complex locus including natural variation at a causative gene for which the wheat reference genome Chinese Spring was uninformative [41]. Additionally, it has been shown that presence/absence variants (PAV) and copy number variation (CNV) as frequent causal polymorphism of native variation at *Bo1* and *Sst1*.

Importantly, the allopolyploid nature of wheat entails the presence of two and three copies of the same gene (called homeologous copies) in tetraploid and

hexaploid wheat, respectively. Gene functionality is usually retained, although differentiation is common in terms of genes silencing, sub-functionalization, and
even neo-functionalization [42], hence introducing a wider variation in effects and
a dosage effects not observed in diploids. QTLs can be found at the AA, BB, and
DD homeologs, as in the case of *VRN1* and *Ppd1*. Another side effect of allopoly-
ploidy is that most of the natural variation with an appreciable phenotypic effect is
mostly caused by dominant mutations, particularly in the regulative gene regions
(promoter region, first intron) while recessive mutation effects are frequently hidden
by the presence of at least one functional gene copy.

28.10   MAS for Transferring Beneficial Haplotypes
from Wheat Wild Relatives

Historically, wheat breeding has leveraged the wide diversity present in wheat wild
relatives (Chap. 18). The Triticeae tribe is huge, with many diverse species well
adapted to a wide range of environments, each showing specific peculiarities.
Targets for chromatin transfer from wheat wild relatives are (i) resistance to several
diseases, mainly rusts, powdery mildew, and fusarium head blight, (ii), grain quality, (iii) male sterility, (iv) resilience to abiotic stress, and (v) perenniality.

   Both close and distant (alien) relatives have been largely used across decades
(Chap. 18). Among the close relatives, *T. urartu, T. dicoccoides, T. monococcum,
Aegilops speltoides,* and *Ae. tauschii,* and among the distant relatives, *Ae. geniculata Ae. longissima, Ae. ventricosa, Haynaldia villosa, Secale cereale, Thinopyrum
elongatum, Th. intermedium,* and *Th. ponticum,* were more frequently used.

   The effective transfer and recombination of alien chromatin from distant wild
relatives heavily relied on chromosome engineering techniques, most exclusively
with the use of mutations at wheat *Ph* (*Pairing homoeologous*) genes, mainly *Ph1*.
Chromosome engineering programs and main results are reviewed in King et al.
(Chap. 18). While chromosome engineering holds great promises for transferring
traits absent in cultivated wheat and potentially of major breeding impact on a major
drawback is the linkage drag caused by the alien chromatin segments, often inducing negative features such as sterility, reduced seed germination, segregation distor-
tion, anomalies of plant growth habit, etc., which often reduce grain size and other
yield components.

   The linkage drag effects are proportional to the segment size of the transferred
chromatin. The transfer of alien chromatin in wheat through chromosome engineering generally involves first the transfer of single wide segments from the donor
species, mainly through translocation. Additional local recombinations are induced
to reduce the size of the alien chromatin around the target locus. This can be considered as a pre-breeding activity where a crucial role is played by the use of fluorescence *in situ* hybridization (FISH) PCR-based molecular markers functional in both

Triticum and the alien species and well distributed in the target region. The development of the high-density SNP array technology and subsequently, the KASP technology and the accumulation of massive genome sequence data allowed to specifically design probe sets for targeting and tracking introgressions from several wild relatives (Chap. 18).

28.11 Next Generation Sequencing (NGS) Technologies to Enhance MAS Effectiveness

The advancement in molecular technologies continuously affects how MAS can be implemented more cost-effectively. KASP panels covering the majority of the assays to target and trace beneficial alleles at the most relevant loci have been specifically designed for a first genome-wide characterization of the breeders' ready germplasm. At a higher throughput level, targeted Illumina-based re-sequencing of polymorphisms used for standard single-assay marker development has been proposed to streamline and increase the throughput of screening germplasm for the presence of beneficial alleles. The developed techniques are either targeted amplicon sequencing or direct multiplexed SNP interrogation, already offered by several private providers, combined with sample barcoding for efficient exploitation of the NGS sequencing capacity.

28.12 Integration Between MAS and Genomic Selection in Breeding Programs

While the concepts of MAS and Genomic Selection (GS) appear rather independent because they tap into two distinct portions of the wheat genome (see Fig. 28.3), the most successful genomic-assisted breeding programs combine both approaches in a synergic integration. Therefore, the role of MAS in pre-breeding is and will remain unique to rapidly introgress in breeding-compatible genetic stocks the new sources of variation made available through research and pre-breeding activities.

Once the novel beneficial alleles are introgressed and fixed in elite populations, this germplasm is ideal to implement genomic selection (GS) to efficiently tap into the plethora of minor QTLs. Due to its high efficiency, a well-managed GS program leads to selection and increase of the beneficial alleles more rapidly than conventional breeding programs. Hence, the importance of continuously refueling the program with novel beneficial allelic variants to be progressively cumulated into breeders' germplasm under selection (see Fig. 28.4).

28.13 Key Concepts

Future genetic gain in wheat will rely on a more effective application of marker-assisted selection and genomics approaches leveraging the fast-increasing capacity to sequence the entire wheat genome which will eventually provide a glimpse on the structural complexity of the wheat pangenome.

The effectiveness and success of genomics-assisted wheat breeding will depend on the following factors/issues:

- Availability of well-characterized germplasm collections capturing the biodiversity present worldwide in both tetraploid and hexploid wheat and closely related species.
- Capacity to accurately phenotype, preferably in high-throughput fashion, large populations under controlled and field conditions.
- Apply (i) linkage mapping based on sufficiently large segregating populations and (ii) genome-wide association (GWA) mapping based on germplasm collections with low linkage disequilibrium (LD) decay to accurately dissect the QTLome and fine-map major QTLs of key proxy traits for yield and yield stability.
- Based on the above, implement MAS with the closest markers flanking the target locus.
- Deploy forward- and reverse-genetics approaches to clone the functional sequences governing the target trait. Cloning allows for the design of 'perfect' markers ideal for an error-free MAS.
- The availability of high-quality genome assemblies greatly facilitates the identification of candidate genes and the design of high-throughput and precise KASP markers diagnostic for inter-varietal and homelog-SNPs.
- The two domestication bottlenecks undergone by the A and B wheat genomes make tetraploid and durum wheat germplasm resources a particularly suitable biodiversity source to identify novel, underesploited beneficial alleles.
- Overall, the development of organized, informative, and user-friendly dedicated genomic databases is relevant for all the above-mentioned activities. The number and variety of discovered, marker-tagged, and cloned loci are already huge and the available scientific information is fragmented and not filtered by quality parameters. Databasing and database-interconnection are crucial aspects to be addressed.

28.14 Conclusions

Marker-assisted selection (MAS) started in parallel with the earliest achievements in genetic mapping and isolation of the most relevant loci for wheat biology, genetics, and improvement. Today, wheat breeding benefits from a full range of techniques and genomic resources, including the recently completed wheat pangenome, for developing sequence-based molecular assays to enable high-throughput MAS. Importantly, the number of cloned wheat loci/QTLs, novel MAS protocols and genetic stocks developed in the last 5 years has grown steadily. As main achievements, the release of the reference gold standard wheat genome sequences paved the way to streamline genetic studies and MAS applications. Nowadays, the integrated and combined use of gene/QTL discovery, MAS in pre-breeding and breeding programs, together with genomic selection and gene editing are key for more effectively leveraging and bridging of biodiversity of the tetraploid with hexaploid A and B genomes while contributing to advance our knowledge in and understanding of wheat functional genomics.

References

1. Gupta PK, Balyan HS, Sharma S, Kumar R (2020) Genetics of yield, abiotic stress tolerance and biofortification in wheat (*Triticum aestivum* L.). Theor Appl Genet 133:1569–1602. https://doi.org/10.1007/s00122-020-03583-3
2. Langridge P, Reynolds M (2021) Breeding for drought and heat tolerance in wheat. Theor Appl Genet 134:1753–1769. https://doi.org/10.1007/s00122-021-03795-1
3. Collins NC, Tardieu F, Tuberosa R (2008) Quantitative trait loci and crop performance under abiotic stress: where do we stand? Plant Physiol 147:469–486. https://doi.org/10.1104/pp.108.118117
4. Gabay G, Zhang J, Burguener GF, Howell T, Wang H, Fahima T, Lukaszewski A, Moriconi JI, Santa Maria GE, Dubcovsky J (2021) Structural rearrangements in wheat (1BS)–rye (1RS) recombinant chromosomes affect gene dosage and root length. Plant Genome 14:e20079. https://doi.org/10.1002/tpg2.20079
5. Tuberosa R (2012) Phenotyping for drought tolerance of crops in the genomics era. Front Physiol 3:1–26. https://doi.org/10.3389/fphys.2012.00347
6. Winfield MO, Allen AM, Burridge AJ, Barker GLA, Benbow HR, Wilkinson PA, Coghill J, Waterfall C, Davassi A, Scopes G, Pirani A, Webster T, Brew F, Bloor C, King J, West C, Griffiths S, King I, Bentley AR, Edwards KJ (2016) High-density SNP genotyping array for hexaploid wheat and its secondary and tertiary gene pool. Plant Biotechnol J 14:1195–1206. https://doi.org/10.1111/pbi.12485
7. Mazzucotelli E, Sciara G, Mastrangelo AM, Desiderio F, Xu SS, Faris J, Hayden MJ, Tricker PJ, Ozkan H, Echenique V, Steffenson BJ, Knox R, Niane AA, Udupa SM, Longin FCH, Marone D, Petruzzino G, Corneti S, Ormanbekova D, Pozniak C, Roncallo PF, Mather D, Able JA, Amri A, Braun H, Ammar K, Baum M, Cattivelli L, Maccaferri M, Tuberosa R, Bassi FM (2020) The global durum wheat panel (GDP): an international platform to identify and exchange beneficial alleles. Front Plant Sci 11:2036. https://doi.org/10.3389/fpls.2020.569905
8. Maccaferri M, Harris NS, Twardziok SO, Pasam RK, Gundlach H, Spannagl M, Ormanbekova D, Lux T, Prade VM, Milner S, Himmelbach A, Mascher M, Bagnaresi P, Faccioli P, Cozzi P, Lauria M, Lazzari B, Stella A, Manconi A, Gnocchi M, Moscatelli M, Avni R, Deek J, Biyiklioglu S, Frascaroli E, Corneti S, Salvi S, Sonnante G, Desiderio F, Marè C, Crosatti C, Mica E, Özkan H, Kilian B, De Vita P, Marone D, Joukhadar R, Mazzucotelli E, Nigro D, Gadaleta A, Chao S, Faris JD, Melo A, Pumphrey M, Pecchioni N, Milanesi L, Wiebe K, Ens J, MacLachlan RP, Clarke JM, Sharpe AG, Koh CS, Liang K, Taylor GJ, Knox R, Budak H, Mastrangelo AM, Xu SS, Stein N, Hale J, Distelfeld D, Hayden MJ, Tuberosa R, Walkowiak S, Mayer KFX, Ceriotti A, Pozniak CJ, Cattivelli L (2019) Durum wheat genome highlights past domestication signatures and future improvement targets. Nat Genet 51:885–895. https://doi.org/10.1038/s41588-019-0381-3

9. Adhikari S, Saha S, Biswas A, Rana TS, Bandyopadhyay TK, Ghosh P (2017) Application of molecular markers in plant genome analysis: a review. Nucleus 60:283–297. https://doi.org/10.1007/s13237-017-0214-7

10. Salvi S, Tuberosa R (2015) The crop QTLome comes of age. Curr Opin Biotechnol 32:179–185. https://doi.org/10.1016/j.copbio.2015.01.001

11. Venske E, dos Santos RS, Farias D d R, Rother V, da Maia LC, Pegoraro C, de Oliveira A (2019) Meta-analysis of the QTLome of Fusarium head blight resistance in bread wheat: refining the current puzzle. Front Plant Sci 10:727. https://doi.org/10.3389/fpls.2019.00727

12. Shariatipour N, Heidari B, Richards CM (2021) Meta-analysis of QTLome for grain zinc and iron contents in wheat (*Triticum aestivum* L.). Euphytica 217:86. https://doi.org/10.1007/s10681-021-02818-8

13. Maccaferri M, Canè MA, Sanguineti MC, Salvi S, Colalongo MC, Massi A, Clarke F, Knox R, Pozniak CJ, Clarke JM, Fahima T, Dubcovsky J, Xu S, Ammar K, Karsai I, Vida G, Tuberosa R (2014) A consensus framework map of durum wheat (*Triticum durum* Desf.) suitable for linkage disequilibrium analysis and genome-wide association mapping. BMC Genomics 15:873. https://doi.org/10.1186/1471-2164-15-873

14. Quraishi UM, Pont C, Ain Q, Flores R, Burlot L, Alaux M, Quesneville H, Salse J (2017) Combined genomic and genetic data integration of major agronomical traits in bread wheat (*Triticum aestivum* L.). Front Plant Sci 8:1843. https://doi.org/10.3389/fpls.2017.01843

15. The International Wheat Genome Sequencing Consortium, Appels R, Eversole K, Stein N, Feuillet C, Keller B, Rogers J, Pozniak CJ, Choulet F, Distelfeld A, Poland J, Ronen G, Sharpe AG, Barad O, Baruch K, Keeble-Gagnère G, Mascher M, Ben-Zvi G, Josselin A-A, Himmelbach A, Balfourier F, Gutierrez-Gonzalez J, Hayden M, Koh C, Muehlbauer G, Pasam RK, Paux E, Rigault P, Tibbits J, Tiwari V, Spannagl M, Lang D, Gundlach H, Haberer G, Mayer KFX, Ormanbekova D, Prade V, Šimková H, Wicker T, Swarbreck D, Rimbert H, Felder M, Guilhot N, Kaithakottil G, Keilwagen J, Leroy P, Lux T, Twardziok S, Venturini L, Juhász A, Abrouk M, Fischer I, Uauy C, Borrill P, Ramirez-Gonzalez RH, Arnaud D, Chalabi S, Chalhoub B, Cory A, Datla R, Davey MW, Jacobs J, Robinson SJ, Steuernagel B, van Ex F, Wulff BBH, Benhamed M, Bendahmane A, Concia L, Latrasse D, Bartoš J, Bellec A, Berges H, Doležel J, Frenkel Z, Gill B, Korol A, Letellier T, Olsen O-A, Singh K, Valárik M, van der Vossen E, Vautrin S, Weining S, Fahima T, Glikson V, Raats D, Číhalíková J, Toegelová H, Vrána J, Sourdille P, Darrier B, Barabaschi D, Cattivelli L, Hernandez P, Galvez S, Budak H, Jones JDG, Witek K, Yu G, Small I, Melonek J, Zhou R, Belova T, Kanyuka K, King R, Nilsen K, Walkowiak S, Cuthbert R, Knox R, Wiebe K, Xiang D, Rohde A, Golds T, Čížková J, Akpinar BA, Biyiklioglu S, Gao L, N'Daiye A, Kubaláková M, Šafář J, Alfama F, Adam-Blondon A-F, Flores R, Guerche C, Loaec M, Quesneville H, Condie J, Ens J, Maclachlan R, Tan Y, Alberti A, Aury J-M, Barbe V, Couloux A, Cruaud C, Labadie K, Mangenot S, Wincker P, Kaur G, Luo M, Sehgal S, Chhuneja P, Gupta OP, Jindal S, Kaur P, Malik P, Sharma P, Yadav B, Singh NK, Khurana JP, Chaudhary C, Khurana P, Kumar V, Mahato A, Mathur S, Sevanthi A, Sharma N, Tomar RS, Holušová K, Plíhal O, Clark MD, Heavens D, Kettleborough G, Wright J, Balcárková B, Hu Y, Salina E, Ravin N, Skryabin K, Beletsky A, Kadnikov V, Mardanov A, Nesterov M, Rakitin A, Sergeeva E, Handa H, Kanamori H, Katagiri S, Kobayashi F, Nasuda S, Tanaka T, Wu J, Cattonaro F, Jiumeng M, Kugler K, Pfeifer M, Sandve S, Xun X, Zhan B, Batley J, Bayer PE, Edwards D, Hayashi S, Tulpová Z, Visendi P, Cui L, Du X, Feng K, Nie X, Tong W, Wang L (2018) Shifting the limits in wheat research and breeding using a fully annotated reference genome. Science 361:eaar7191. https://doi.org/10.1126/science.aar7191

16. Varshney RK, Tuberosa R (2007) Genomics-assisted crop improvement. Springer

17. Visscher PM, Goddard ME (2019) From R.A. Fisher's 1918 paper to GWAS a century later. Genetics 211:1125–1130. https://doi.org/10.1534/genetics.118.301594

18. Uauy C, Distelfeld A, Fahima T, Blechl A, Dubcovsky J (2006) A NAC gene regulating senescence improves grain protein, zinc, and iron content in wheat. Science 314:1298–1301. https://doi.org/10.1126/science.1133649

19. Pallotta M, Schnurbusch T, Hayes J, Hay A, Baumann U, Paull J, Langridge P, Sutton T (2014) Molecular basis of adaptation to high soil boron in wheat landraces and elite cultivars. Nature 514:88–91. https://doi.org/10.1038/nature13538

20. Rawat N, Pumphrey MO, Liu S, Zhang X, Tiwari VK, Ando K, Trick HN, Bockus WW, Akhunov E, Anderson JA, Gill BS (2016) Wheat Fhb1 encodes a chimeric lectin with agglutinin domains and a pore-forming toxin-like domain conferring resistance to Fusarium head blight. Nat Genet 48:1576–1580. https://doi.org/10.1038/ng.3706

21. Sallam AH, Conley E, Prakapenka D, Da Y, Anderson JA (2020) Improving prediction accuracy using multi-allelic haplotype prediction and training population optimization in wheat. G3 Genes|Genomes|Genetics 10:2265–2273. https://doi.org/10.1534/g3.120.401165

22. Rasheed A, Xia X (2019) From markers to genome-based breeding in wheat. Theor Appl Genet 132:767–784. https://doi.org/10.1007/s00122-019-03286-4

23. Beyer S, Daba S, Tyagi P, Bockelman H, Brown-Guedira G, Mohammadi M, IWGSC (2019) Loci and candidate genes controlling root traits in wheat seedlings—a wheat root GWAS. Funct Integr Genomics 19:91–107. https://doi.org/10.1007/s10142-018-0630-z

24. Singh K, Batra R, Sharma S, Saripalli G, Gautam T, Singh R, Pal S, Malik P, Kumar M, Jan I, Singh S, Kumar D, Pundir S, Chaturvedi D, Verma A, Rani A, Kumar A, Sharma H, Chaudhary J, Kumar K, Kumar S, Singh VK, Singh VP, Kumar S, Kumar R, Gaurav SS, Sharma S, Sharma PK, Balyan HS, Gupta PK (2021) WheatQTLdb: a QTL database for wheat. Mol Gen Genomics. https://doi.org/10.1007/s00438-021-01796-9

25. Tuberosa R, Frascaroli E, Maccaferri M, Salvi S (2021) Understanding and exploiting the genetics of plant root traits. In: Gregory PJ (ed) Understanding and improving crop root function. Burleigh Dodds Science Publishing, Cambridge

26. Ober ES, Alahmad S, Cockram J, Forestan C, Hickey LT, Kant J, Maccaferri M, Marr E, Milner M, Pinto F, Rambla C, Reynolds M, Salvi S, Sciara G, Snowdon RJ, Thomelin P, Tuberosa R, Uauy C, Voss-Fels KP, Wallington E, Watt M (2021) Wheat root systems as a breeding target for climate resilience. Theor Appl Genet 134:1645–1662. https://doi.org/10.1007/s00122-021-03819-w

27. Hao C, Jiao C, Hou J, Li T, Liu H, Wang Y, Zheng J, Liu H, Bi Z, Xu F, Zhao J, Ma L, Wang Y, Majeed U, Liu X, Appels R, Maccaferri M, Tuberosa R, Lu H, Zhang X (2020) Resequencing of 145 landmark cultivars reveals asymmetric sub-genome selection and strong founder genotype effects on wheat breeding in China. Mol Plant 13:1733–1751. https://doi.org/10.1016/j.molp.2020.09.001

28. Maccaferri M, Ratti C, Rubies-Autonell C, Vallega V, Demontis A, Stefanelli S, Tuberosa R, Sanguineti MC (2011) Resistance to soil-borne cereal mosaic virus in durum wheat is controlled by a major QTL on chromosome arm 2BS and minor loci. Theor Appl Genet 123:527–544. https://doi.org/10.1007/s00122-011-1605-9

29. Terracciano I, Maccaferri M, Bassi F, Mantovani P, Sanguineti MC, Salvi S, Šimková H, Doležel J, Massi A, Ammar K, Kolmer J, Tuberosa R (2013) Development of COS-SNP and HRM markers for high-throughput and reliable haplotype-based detection of Lr14a in durum wheat (Triticum durum Desf.). Theor Appl Genet 126:1077–1101. https://doi.org/10.1007/s00122-012-2038-9

30. Wang S, Wong D, Forrest K, Allen A, Chao S, Huang BE, Maccaferri M, Salvi S, Milner SG, Cattivelli L, Mastrangelo AM, Whan A, Stephen S, Barker G, Wieseke R, Plieske J, Consortium IWGS, Lillemo M, Mather D, Appels R, Dolferus R, Brown-Guedira G, Korol A, Akhunova AR, Feuillet C, Salse J, Morgante M, Pozniak C, Luo M-C, Dvorak J, Morell M, Dubcovsky J, Ganal M, Tuberosa R, Lawley C, Mikoulitch I, Cavanagh C, Edwards KJ, Hayden M, Akhunov E (2014) Characterization of polyploid wheat genomic diversity using a high-density 90 000 single nucleotide polymorphism array. Plant Biotechnol J 12:787–796. https://doi.org/10.1111/pbi.12183

31. Maccaferri M, Ricci A, Salvi S, Milner SG, Noli E, Martelli PL, Casadio R, Akhunov E, Scalabrin S, Vendramin V, Ammar K, Blanco A, Desiderio F, Distelfeld A, Dubcovsky J, Fahima T, Faris J, Korol A, Massi A, Mastrangelo AM, Morgante M, Pozniak C, N'Diaye A, Xu S, Tuberosa R (2015) A high-density, SNP-based consensus map of tetraploid wheat as a bridge to integrate durum and bread wheat genomics and breeding. Plant Biotechnol J 13:648–663. https://doi.org/10.1111/pbi.12288

32. van Poecke RMP, Maccaferri M, Tang J, Truong HT, Janssen A, van Orsouw NJ, Salvi S, Sanguineti MC, Tuberosa R, van der Vossen EAG (2013) Sequence-based SNP genotyping in durum wheat. Plant Biotechnol J 11:809–817. https://doi.org/10.1111/pbi.12072
33. Sun C, Dong Z, Zhao L, Ren Y, Zhang N, Chen F (2020) The Wheat 660K SNP array demonstrates great potential for marker-assisted selection in polyploid wheat. Plant Biotechnol J 18:1354–1360. https://doi.org/10.1111/pbi.13361
34. Gupta PK, Langridge P, Mir RR (2010) Marker-assisted wheat breeding: present status and future possibilities. Mol Breed 26:145–161. https://doi.org/10.1007/s11032-009-9359-7
35. Ayalew H, Tsang PW, Chu C, Wang J, Liu S, Chen C, Ma X-F (2019) Comparison of TaqMan, KASP and rhAmp SNP genotyping platforms in hexaploid wheat. PLoS One 14:1–9. https://doi.org/10.1371/journal.pone.0217222
36. Yuan J, Wen Z, Gu C, Wang D (2014) Introduction of high throughput and cost effective SNP genotyping platforms in soybean. Plant Genet Genomics Biotechnol 2:90–94
37. Ramirez-Gonzalez RH, Uauy C, Caccamo M (2015) PolyMarker: a fast polyploid primer design pipeline. Bioinformatics 31:2038–2039. https://doi.org/10.1093/bioinformatics/btv069
38. Makhoul M, Rambla C, Voss-Fels KP, Hickey LT, Snowdon RJ, Obermeier C (2020) Overcoming polyploidy pitfalls: a user guide for effective SNP conversion into KASP markers in wheat. Theor Appl Genet 133:2413–2430. https://doi.org/10.1007/s00122-020-03608-x
39. Kumar A, Saripalli G, Jan I, Kumar K, Sharma PK, Balyan HS, Gupta PK (2020) Meta-QTL analysis and identification of candidate genes for drought tolerance in bread wheat (*Triticum aestivum* L.). Physiol Mol Biol Plants 26:1713–1725. https://doi.org/10.1007/s12298-020-00847-6
40. Thudi M, Palakurthi R, Schnable JC, Chitikineni A, Dreisigacker S, Mace E, Srivastava RK, Satyavathi CT, Odeny D, Tiwari VK, Lam H-M, Bin HY, Singh VK, Li G, Xu Y, Chen X, Kaila S, Nguyen H, Sivasankar S, Jackson SA, Close TJ, Shubo W, Varshney RK (2021) Genomic resources in plant breeding for sustainable agriculture. J Plant Physiol 257:153351. https://doi.org/10.1016/j.jplph.2020.153351
41. Ma Z, Xie Q, Li G, Jia H, Zhou J, Kong Z, Li N, Yuan Y (2020) Germplasms, genetics and genomics for better control of disastrous wheat Fusarium head blight. Theor Appl Genet 133:1541–1568. https://doi.org/10.1007/s00122-019-03525-8
42. Adamski NM, Borrill P, Brinton J, Harrington SA, Marchal C, Bentley AR, Bovill WD, Cattivelli L, Cockram J, Contreras-Moreira B, Ford B, Ghosh S, Harwood W, Hassani-Pak K, Hayta S, Hickey LT, Kanyuka K, King J, Maccaferri M, Naamati G, Pozniak CJ, Ramirez-Gonzalez RH, Sansaloni C, Trevaskis B, Wingen LU, Wulff BBH, Uauy C (2020) A roadmap for gene functional characterisation in crops with large genomes: lessons from polyploid wheat. elife 9:e55646. https://doi.org/10.7554/eLife.55646

Chapter 29
Application of CRISPR-Cas-Based Genome Editing for Precision Breeding in Wheat

Wei Wang and Eduard Akhunov

Abstract Wheat improvement relies on genetic diversity associated with variation in target traits. While traditionally the main sources of novel genetic diversity for breeding are wheat varieties or various wild relatives of wheat, advances in gene mapping and genome editing technologies provide an opportunity for engineering new variants of genes that could have beneficial effect on agronomic traits. Here, we provide the overview of the genome editing technologies and their application to creating targeted variation in genes that could enhance wheat productivity. We discuss the potential utility of the genome editing technologies and CRISPR-Cas-induced variation incorporated into the pre-breeding pipelines for wheat improvement.

Keywords Genome editing · CRISPR-Cas-based technologies · Precision breeding · Wheat improvement

29.1 Learning Objectives

- Understanding the basics of CRISPR-Cas-based technology and learning how to apply it in wheat improvement.

29.2 Introduction to the Development of Genome Editing (GE) Technologies

Compared to the conventional random mutagenesis, GE tools provide opportunity to modify specific genomic regions of interest. The GE tools share two major common features: (1) a programable DNA binding domain, and (2) a DNA nuclease

W. Wang · E. Akhunov (✉)
Kansas State University, Manhattan, KS, USA
e-mail: wwang0604@ksu.edu; eakhunov@ksu.edu

© The Author(s) 2022
M. P. Reynolds, H.-J. Braun (eds.), *Wheat Improvement*,
https://doi.org/10.1007/978-3-030-90673-3_29

capable of introducing double or single strand breaks (DSB or SSBs) into the targeted DNA sites. The introduced DNA breaks could then be repaired either through (1) error-prone non-homologous end joining (NHEJ) process resulting in insertions, deletions, or single base substitutions, or (2) homology-directed repair (HDR) process, or (3) be utilized as sites for sequence replacement using prime editing (PE) technology. The HDR could be used for precise sequence insertion or replacement through recombination with the exogenously supplied "donor DNA".

Before the development of CRISPR-based GE technologies, targeted GE was performed using Zinc Finger Nuclease (ZFN) and Transcription Activator-Like Effector Nuclease (TALEN). ZFN, the first programable GE system, was based on a Zinc Finger Protein (ZFP) fused with the nuclease domain of restriction endonuclease FokI (Fig. 29.1a). ZFP comprised 3–6 tandemly repeated DNA binding units. Each unit contains 30 amino acids and recognizes a 3 base pair (bp) DNA sequence. The DNA targeting specificity of ZFNs is defined by the composition of DNA binding units. ZFNs are designed to work in pairs because the cleavage domain of FokI must dimerize to introduce DSBs (Fig. 29.1a). In the TALEN-based GE system, the DNA binding domain is composed of tandemly repeated DNA recognition units with additional N- and C- terminal domains derived from TALEs (Fig. 29.1b). Each unit contains 33–35 highly conserved amino acids except for the variable residues 12 and 13, which define the single nucleotide binding specificity. Both ZFN and TALEN systems have been broadly used for GE in multiple organisms. With the development of a more effective CRISPR-Cas-based technology, their application in research and biotechnology declined.

The CRISPR-Cas system is composed of a CRISPR-associated protein (Cas) and a mature transcript originating from a Clustered Regularly Interspaced Short Palindromic Repeats (CRISPR) locus. In 2012, the groups led by two biochemists, Jennifer A. Doudna from USA and Emmanuelle Charpentier from France, demonstrated the first application of the CRISPR-Cas9 programmable endonuclease for in vitro DNA editing [1]. The group led by a Lithuanian biochemist, Virginijus Siksnys, also independently achieved the CRISPR-Cas9 mediated in vitro DNA cleavage [2]. It took only a few years for the newly developed CRISPR-Cas-based system [3, 4] to become a major GE tool for studying eukaryotic genomes. Because

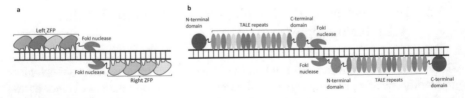

Fig. 29.1 The schematic illustration of ZFN and TALEN GE systems. (**a**) ZFN is composed of tandemly repeated DNA binding units fused to a nuclease domain of restriction endonuclease FokI. Each DNA binding units recognize 3 bp of DNA. Left and right ZFN work in pairs to make DSBs. (**b**) TALEN is composed of TALE repeats, the N- and C-terminal domains, and the fused with FokI nuclease. Each unit of TALE repeats recognize 1 bp of DNA. TALENs work in pairs to introduce DSBs

of the revolutionary changes brought into the basic genetic and genomic studies by the CRISPR-Cas-based technologies and new possibilities provided by this technology for curing diseases and improving crops, Jennifer A. Doudna and Emmanuelle Charpentier were awarded the 2020 Nobel Prize in Chemistry.

Since the first discovery of the CRISPR loci in 1987, it took scientists more than 20 years to understand that the CRISPR-Cas is a part of the bacteria/archaea immune system, which evolved to recognize and cleave invading DNA/RNA molecules [5]. The CRISPR-Cas complex is formed by Cas proteins and CRISPR RNAs (crRNAs) spliced from the RNA transcripts of the CRISPR loci. The CRISPR-Cas systems could be categorized into class 1 and class 2, which could adopt multiple and single Cas proteins as nucleases for target cleavage, respectively [6]. Class 2 is further divided into type II, V and VI. The two mostly widely used CRISRP-Cas-based genome editors are CRISPR-Cas9 and CRISPR-Cas12a, which belong to class 2 type II and V, respectively. The ease of target design, high GE efficiency and ability to simultaneously target multiple genomic regions made the CRISPR-Cas technology more popular than ZFN and TALEN. During the last decade, the power of CRISPR-Cas-based system has been harnessed to better understand function of genes underlying variation in major agronomic traits and to develop novel strategies for crop improvement.

29.3 CRISPR-Cas-Based GE Toolbox

29.3.1 CRISPR-Cas Variants and Their Basic Applications

The CRISPR-Cas9 from *Streptococcus pyogenes* (CRISPR-SpCas9) is one of the most commonly used CRISPR-Cas systems for GE. To induce targeted DSB, the SpCas9 needs a single-guide RNA (sgRNA) containing 20 nucleotide spacer complementary to the targeted DNA sequence, which is followed by the 3′-end NGG protospacer-adjacent motif (PAM) (Fig. 29.2a). The synthetic sgRNAs for CRISPR-Cas9 system are engineered based on the crRNAs and trans-activating crRNA (tracrRNA) [1]. In bacteria/archaea, the directed repeat sequences of the immature crRNA array form complexes with the tracrRNAs, which are processed into crRNA-tracrRNA duplexes by Cas9 and RNase III [5]. As part of the CRISPR-Cas9 complex, the sgRNA guides is responsible for guiding Cas9 to the target site. The two endonuclease domains of SpCas9, HNH and RuvC, will cleave the paired and non-paired DNA strands, respectively (Fig. 29.2a), and predominantly result in blunt end DSBs located 3 bp from the NGG PAM.

By introducing either aspartate-to-alanine substitution (D10A) in the RuvC domain or Histidine-to-alanine substitution (H840A) in the HNH domain, SpCas9 could be converted to the DNA nickase (nCas9). The inactive form of Cas9, also called dead Cas9 or dCas9, could be created by introducing both D10A and H840A substitutions simultaneously. The nCas9 and dCas9 variants are currently used in a

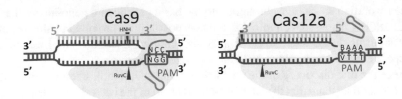

Fig. 29.2 The schematic illustration of CRISPR-Cas9 and CIRSPR-Cas12a systems. (**a**) The CRISPR-Cas9 nuclease cuts the double stranded DNA 3 bp upstream of the NGG protospacer adjacent motif (PAM) and forms blunt end DSB. The DNA cuts catalyzed by the two nuclease domains of Cas9 are shown by the purple arrows. The NGG PAM is shown with the blue rectangle. Amino acid changing mutations in Cas9 (D10A in RucV domain and H840A in HNH domain) result in variants that are either capable of making cuts on only one DNA strand (nCas9) or incapable of cutting DNA (dCas9). (**b**) The CRISPR-Cas12a has one RuvC nuclease domain and one novel nuclease domain, which are shown as purple arrows. CRISPR-Cas12a has different PAM sequence (TTTV [V = A, C, G] or TTV) and makes 4–5 nucleotide long staggered double stranded DNA breaks at the distal end of the protospacer

number of GE applications. The range of editable genomic targets was further expanded by creating the variants of Cas9 recognizing different PAMs. For example, the engineered Cas9NG [7], which predominantly recognizes 'NG' PAM, was shown to be effective for GE in wheat [8]. Additional information about the variants of Cas9 editors could be found in review by Alzalone et al. [9].

CRISPR-Cas12a, also known as Cpf1, is another broadly used CRIRSP-Cas system that has several features distinguishing it from CRISPR-Cas9 [10] (Fig. 29.2b). CRISPR-Cas12a (1) recognizes 5'-end T-rich PAMs (TTTV (V = A, C, G) for LbCas12a and AsCas12a, and TTV for FnCas12a); (2) generates staggered end DSBs with 4–5-nucleotide 5'-overhangs; (3) induces mutations on the distal end of protospacer, thereby preserving PAM and enabling multiple rounds of GE; (4) processes transcripts with tandem repeats into mature crRNAs and induces genome edits independent of tracrRNA, which simplifies the design of multiplex GE experiments. Similar to Cas9, a series of Cas12a variants recognizing various PAMs have been engineered [9]. Both the wild-type and engineered Cas12a have been shown to be effective in wheat [8].

The repair of DSBs created by Cas nucleases through the NHEJ process introduces stochastic mutations, mostly short DNA sequence insertions/deletions or base pair substitutions. This approach is commonly used to disrupt functional elements in genome, such as gene coding sequences or regulatory elements. In the earlier GE applications, the precise modifications in genome, including specific single-base mutations, gene replacements, targeted deletions or insertions could be achieved through HDR of DSBs by co-delivering a CRISPR-Cas reagent along with a donor DNA template. The donor DNA includes a DNA sequence with desired mutations flanked from both sides by sequences similar to the sequences around the CRISPR-Cas target site. This structure of a donor template promotes HDR and allows for replacing the original sequence with DNA carrying desired changes. Compared to NHEJ-mediated GE, the efficiency of HDR-based GE remains relatively low because it relies on DNA replication activity, which is initiated only at the S- and G2-phases of the cell cycle [3] One exception is precise DNA deletion, which could be achieved by the CRISPR-Cas targeting of a pair of sites flanking the region of interest [3].

29.3.2   Base-Editors and Prime-Editors

The low efficiency and precision of HDR-mediated DSB repair in the CRISPR-Cas applications aimed at sequence replacement necessitated the development of alternative approaches. A series of nonconventional GE systems, which do not rely on DSBs or donor DNA templates, greatly expanded the range of possible genome modifications [11–13].

In base editors BE3 (cytosine base editor or CBE) and ABE7.10 (adenine base editor or ABE), single-stranded DNA deaminases are fused to the N-terminal domain of nCas9 (Fig. 29.3a) [12, 13]. During base editing, after the hybridization of sgRNA with its cognate target, nucleotides in the editing window on the PAM-containing DNA strand (non-targeted) are exposed to deaminase.

In BE3, the editing window of cytidine deaminase (APOBEC1 from rat) spans nucleotides 4–8 (Fig. 29.3a), in which CBE converts cytosine (C) to uracil (U). The U will be recognized as thymine (T) by DNA polymerase, and adenosine (A) will be added to the nCas9-nicked DNA strand when it is repaired using the modified strand as a template. Finally, during DNA replication U will be replaced with T, resulting in transition mutations C → T and G → A on the non-targeted and targeted strands, respectively (Fig. 29.3a). The U introduced by CBE could be rapidly removed by uracil DNA glycosylase and reduce the efficiency of editing. To address this issue, an uracil glycosylase inhibitor (UGI) is fused to the C-terminal of nCas9 (D10A) in CBE to increase the half-life of U. In ABE7.10, the editing window of the heterodimer including both the wild and engineered TadA deoxyadenosine deaminase spans nucleotides 4–7 (Fig. 29.3b). The engineered TadA in ABE7.10 converts A to inosine (I), which will be read as G by DNA polymerases. Similar to CBE, after two rounds of DNA repair, the transition mutations A → G and T → C will be formed on the non-targeted and targeted strands, respectively (Fig. 29.3b).

Base editing is restricted to the four types of transition mutations (C → T, G → A, A → G, and T → C), shows high off-target (non-specific) GE and induces mutations within the protospacer sequence at positions different from the targeted base (bystander mutations). These limitations were resolved in the recently developed prime editor (PE) that was demonstrated to be more versatile and precise, though its editing efficiency was lower than that of the BEs [9, 11]. The most commonly used PE2 system is based on the combination of (1) an engineered reverse transcriptase (RT) from Moloney Murine Leukemia Virus fused to the N-terminal domain of nCas9, and (2) prime editing gRNA (pegRNA), which includes both a primer binding site (PBS) and a template for RT carrying the desired genome edits (Fig. 29.3c). During prime editing, nCas9 cuts the unpaired (non-targeted) DNA strand 3 bases

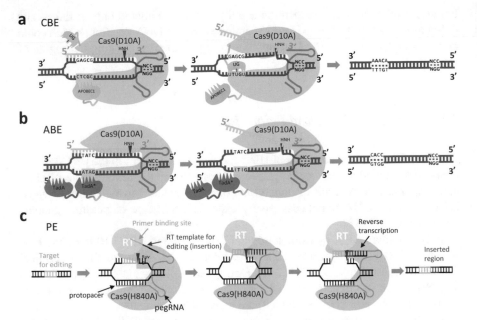

Fig. 29.3 The mechanisms of Base Editing and Prime Editing. The (**a**) Cytosine Base Editor (CBE) BE3 and (**b**) Adenine Base Editor (ABE) ABE7.10 are based on nickase Cas9(D10A) fused with deaminases. (**c**) Prime Editor (PE). This figure shows how PE inserts sequence (shown in red) downstream of the targeted locus (shown in blue). PE is a fusion of nCas9 with Reverse Transcriptase (RT). pegRNA contains targeting spacer sequence, primer binding sequence and template used by RT to introduce desired changes into a target site

upstream of PAM. The 3'-end of cut DNA strand hybridizes with PBS and directs RT using the template in the pegRNA. This reverse transcription step copies the desired sequence from the pegRNA to target DNA. The prime editor could induce all possible types of DNA point mutations and deletions and insertions as long as 10 nucleotides, and as far as 30 nucleotides from the cut site. Once issues with reduced GE efficiency are resolved, these features will make PE a powerful tool for basic and applied studies in agricultural crops. Recently, PE was successfully used to edit the rice and wheat genomes [14].

29.3.3   Gene Suppressors and Activators, Epigenomic Modifiers, and Others

The level of gene expression is defined by regulatory elements, transcription factors (TFs), and the chromatin accessibility of genomic regions with regulatory function. Since dCas9 could be easily programmed to target any region of genome using sgRNAs, it was adopted to deliver transcription factors or proteins involved in chromatin remodeling, histone modification or epigenetic reprogramming to promoters or other regulatory regions of genome. The catalytically inactive dCas9 protein was used to engineer TFs capable of activating or suppressing the expression of any gene, or to develop a broad array of gene regulation and epigenome modification tools by fusing with various proteins domains [15]. For example, the dCas9 fused with protein domains SRDX and VP64 was successfully used for respectively suppressing and activating gene expression in plants. Thus, dCas9 provides a unique opportunity to reversibly modulate gene expression and investigate its role in the regulation of biological pathways underlying phenotypic diversity.

29.4   Recent Application of GE for Improving Major Agronomic Traits and Breeding Technologies

Since its invention, CRISPR-Cas-based GE technologies have been widely applied to modify agronomically important traits in crops. The GE has been applied to improve crop productivity, nutritional quality, storage life, abiotic and biotic stress resistance. The improved understanding of the genetic basis of trait variation in many crops combined with CRISRP-Cas GE enabled novel breeding strategies. These strategies include inducing targeted genetic variation in genes controlling agronomic traits, de-novo domestication of novel crops, development of herbicide resistant crop varieties and male sterile lines, manipulation of hybrid incompatibility, hybrid vigor fixation, and development of haploid induction (HI) lines and haploid induction editing technology (HI-Edit) [16].

29.5   Genome Editing in Wheat

Compared to other main crops (e.g. rice or maize), the adoption of CRISRP-Cas system for wheat improvement lags behind. Among factors that contributed to this trend are (1) the low efficiency of wheat transformation, (2) until recently, the lack of a high-quality reference genome, and (3) the complexity of allopolyploid wheat genetics that complicates comparative genomic analyses and require additional efforts for inferring the biological role of duplicated genes. The recent advances in wheat transformation methods [17] and release of the high-quality reference genomes of wheat [18, 19] hold great promise to broaden application of CRISRP-Cas technologies in wheat genetics and breeding. This section introduces the progress in CRISRP-Cas-based GE in wheat, and describe procedures utilized for modifying the wheat genome, which include CRISPR-Cas system optimization, target gene selection, GE strategy selection, gene target design and validation, genetic transformation, CRISPR-Cas mutant screening, phenotypic validation of

CRISPR-Cas-based gene edits, and introgression of beneficial CRISPR-Cas-induced alleles into adapted wheat cultivars.

29.5.1 Optimization of the CRISRP-Cas System for Wheat Genome Editing

Since most of the CRISRP-Cas tools have originally been developed for human or mammalian cells, it was necessary to optimize these tools for crop genome editing. The optimization of CRISPR-Cas systems includes the selection of optimal codons for the effective Cas gene translation and promoters for the effective expression of Cas-encoding genes and sgRNAs. In the reported studies, both rice and maize codon-optimized Cas9 [20, 21] as well as the wheat codon-optimized Cas9 [22, 23] were successfully used for editing the wheat genome. The maize ubiquitin promoter is one of the most commonly used RNA polymerase II promoters to express the Cas genes in transgenic plants. The RNA polymerase III promoters, U3 or U6, are usually employed to drive the sgRNA expression. Recently, the ubiquitin promoter from switchgrass was shown could be used to express guide RNAs and support effective GE mediated by CRISPR-Cas12a [8].

The optimization of CRISPR-Cas systems for the multiplexed gene editing (multiplex GE or MGE) is also critical for the crop GE applications. In the MGE constructs, the expression of individual sgRNAs could be driven from independent promoters. Alternatively, multiple sgRNA units could be placed under the control of a single promoter and expressed as a precursor RNA molecule. The processing of this RNA into functional sgRNAs could be supported by the (1) self-cleaving ribozymes, (2) glycine tRNA, or (3) Csy4 recognition sites that separate the sgRNA units from each other [24].

29.5.2 Selection of Target Genes for CRISPR-Cas-Based GE

The application of GE for precision crop breeding depends on how well we understand the role of different genes and pathways in controlling phenotypic traits. Over the last decade, the development of novel sequence-based genotyping approaches and of new genetic and genomic resources including multiple annotated reference genomes, re-sequenced diversity panels, TILLING populations, gene expression atlases, and genetic mapping populations enabled quick validation of candidate genes underlying variation in important agronomic traits in wheat. These advances combined with the power of GE technologies opens unique opportunities to start improving wheat by introducing the new CRISPR-Cas-induced alleles of genes into the breeding process.

Using forward genetics approaches, the candidate gene(s) underlying quantitative trait loci (QTL) could be found by map-based cloning or genome-wide association mapping (GWAM). The function of a candidate gene could be validated by the CRISRP-Cas-induced knock-out. One such example is the CRISPR-Cas9-induced mutagenesis of the TaHRC gene, which was found to be a negative regulator of resistance to Fusarium head blight [25]. Thus, the CRISRP-Cas system could potentially be utilized for the large-scale functional screening of hundreds of genes identified by GWAM, comparative genomics or other genomic approaches.

The selection of candidate genes for GE could be guided by the bioinformatically or experimentally inferred gene interaction networks. For example, the experimental screening of protein-DNA interactions involving the promoter of the TaGlu-1Dy gene, which encodes high molecular weight glutenin contributing to the dough quality, helped to identify the TaNAC019 candidate gene [26]. The CRISPR-Cas9-induced knock-out mutants of this gene showed that TaNAC019 positively regulates the amount of seed storage protein and grain starch content by modulating the expression of grain quality genes.

Plant species showed parallel variation of their morphological and physiological traits, and the molecular mechanisms underlying these traits appear to be broadly conserved. The release of wheat genome references [18, 19] enabled identification of candidates genes based on the comparative analyses (e.g. using Ensembl Plants) involving closely and distantly related crop genomes. By integrating comparative genomics approach with the CRISPR-Cas-based GE, genes controlling grain number [27], grain size and weight [20, 23, 28], powdery mildew resistance [21], and herbicide resistance [29] have been functionally characterized. These findings indicate that comparative genomics is a powerful tool for the extrapolation of gene mapping information across related crop species.

29.5.3   Selection of GE Strategies

The choice of GE strategy and Cas Editors is primarily defined by the biological nature of alleles having positive effects on agronomic traits. The ease of the GE experiment design, and the genetic architecture and heritability of a trait are also taken into account. The loss of function mutations could be readily induced using the targeted random mutagenesis via the NHEJ process or point mutagenesis using BEs or PE. The DNA sequence insertions, deletions or replacements could be achieved by HDR-mediated GE or PE. However, among these strategies only random and point mutagenesis have successfully been used to create wheat lines with modified traits.

When a gene negatively regulates an agronomic trait, the most efficient GE strategy is the random mutagenesis of coding regions. For example, the loss-of-function mutants of the TaGW2 gene created by the CRISPR-Cas9-induced frame shift mutation in the coding region increased the grain size and weight (Fig. 29.4a) [20]. A point mutagenesis strategy was successfully used to develop herbicide-tolerant

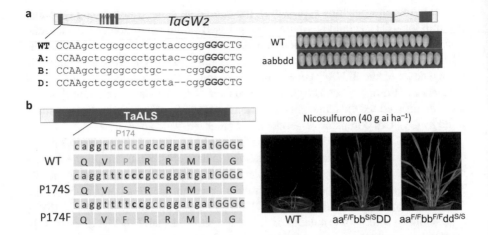

Fig. 29.4 Selection of GE strategies and target design. (**a**) An example of a gene negatively regulating agronomic trait. CRISPR-Cas9-induced frame shift mutations in the TaGW2 gene homoeologs increased the grain size and weight. (**b**) An example of CBE application to wheat GE. CBE was used to introduce point mutations in the TaALS gene and create herbicide resistant wheat lines [29]. The GE window of CBE and the targeted amino acid residue are shown in green, the CRISPR-Cas-induced mutations are shown in red

wheat [29]. Several plant herbicides target acetolactate synthase (ALS), which is essential for branched-chain amino acids synthesis and plant growth. A nonsense mutation at residue P174 of ALS was known to cause herbicide resistances in multiple plant species. In this case, the CBE-based GE strategy was adopted and successfully used to create herbicide resistant wheat with various ALS alleles (Fig. 29.4b).

29.5.4   GE Target Selection and Plasmid Construction

The selection of the CRISPR-Cas targets is influenced by the availability of PAMs and the base composition of the targeted regions. A number of web-based CRISPR-Cas target design tools facilitate this task. When target selection is not restricted to a small region, the overlap between the top-ranked targets chosen by multiple tools is recommended to ensure effective GE. However, if target selection is restricted to a small region, a target could be selected manually. In addition, target selection in polyploid wheat should consider the presence of multiple target copies in distinct genomes. To modify all homoeologous copies of genes in the wheat genome, a target site should be located in the conserved region shared by all gene copies. To specifically modify only one homoeologous copy of a gene, a target should include nucleotide sites unique to that copy of a gene. The proximity of a genome-specific mutation to PAM correlates well with the specificity of GE. At least one unique

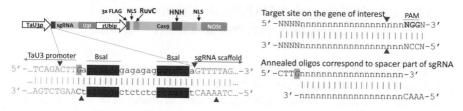

Fig. 29.5 Illustration of the CRISPR-Cas9 construct used for the wheat GE. The expression of sgRNA is driven by the U3 promoter from wheat (TaU3p) and terminated at the U3 terminator (U3t). Cas9 expressed from the maize ubiquitin promoter (zUbip) is terminated at the NOS terminator. A 3 x Flag tag, two nuclear location signals (NLSs), and the RuvC and HNH nuclease domains are marked. The sgRNA transcription start site is highlighted in green. The BsaI cut sites are used for inserting the spacer portion of sgRNA created by annealing the synthesized oligonucleotides. The Cas9 cut site on the target are shown with red triangles. The GE target site and the sequences of oligonucleotides complementary to the target site are shown in lowercase letter *n*

mutation less than 10 bp away from the PAM would be needed to make the CRISPR-Cas target specific.

Commonly, the CRISPR-Cas constructs have the sgRNA promoter and scaffold connected by two back-to-back Type IIS restriction enzyme cut sites, which could create 4 nucleotide 5′ overhangs at the ends of the promoter and the sgRNA scaffold (Fig. 29.5). The spacers matching the target site are synthesized as two reverse complimentary oligonucleotides that after annealing generate 5′ overhangs matching the sticky ends of the restriction enzyme cut sites. The annealed oligos are then inserted into the construct via Golden Gate reaction [22]. For MGE, a tandem array of multiple sgRNA separated by tRNA, ribozyme or Cys4 recognition site sequences could be synthesized and inserted into the CRISPR-Cas construct.

29.5.5 Validation of the Selected GE Targets

Though web-based tools substantially improved selection of optimal targets for GE, the efficiency of GE could still be influenced by factors whose effects are difficult to account for (e.g. high target duplication rate, chromosome state or epigenetic modifications at the targeted regions). As a result, the GE efficiency of selected targets could vary substantially, as was shown for the coding regions in the wheat genome (0.08%–8.33%) [23]. The transient expression of the CRISPR-Cas constructs in the wheat protoplasts was successfully used to experimentally assess the ability of sgRNAs to modify the selected targets. The GE efficiency could be estimated by restriction enzyme digestion (RED), if GE leads to mutation at the enzyme recognition site. Otherwise, the GE events could be validated by amplifying targeted regions followed by the Sanger sequencing of amplicons or by highly multiplexed next-generation sequencing (NGS) of pooled barcoded PCR products [23].

29.5.6   The Delivery of CRISPR-Cas Reagents
and Regeneration of Genome-Edited Wheat Lines

The success of GE heavily relies on the efficiency of wheat transformation, which previously was performed using the biolistic transformation and limited only to certain wheat cultivars. However, the recently reported methods of wheat regeneration and Agrobacterium-mediated transformation [17, 30] broaden the range of wheat varieties amenable to transformation. The usage of transgenic plants for inducing CRISPR-Cas-mediated mutations remains one of the major concerns for general public. To eliminate the time-consuming backcrossing steps aimed at removing the transgenic constructs, biolistic transformations using a CRISPR-Cas-based RNA [28] or ribonucleoprotein complex [31] have been developed for wheat. However, these methods remain labor-intensive because of the lack of selectable markers at the tissue culture and plant regeneration steps.

Haploid induction editing technology (HI-Edit) uses pollen from a haploid inducer line (e.g. maize) expressing the CRISPR-Cas reagents to fertilize a recipient line. The genome of the recipient line is edited before the elimination of the haploid inducer genome [32] and the haploid embryo with the edited copy of a gene is then used to produce double haploid (DH) plants. This approach was applied to create the edited DH lines of wheat using the pollen of maize line carrying active CRISPR-Cas9 [32]. HI-Edit overcomes not only the drawbacks of time-consuming crossing in wheat breeding aimed at generating inbred lines and reducing linkage drag around beneficial alleles, but also the issues related to the genotype-specific regeneration efficiency in wheat transformation. However, the necessity to generate transgenic maize, the low editing efficiency, the need for DH production step and freedom-to-operate related to this technology will likely affect the adoption of HI-Edit in wheat breeding.

29.5.7   Screening Plants Carrying GE Events

The same methods used for mutation screening in the protoplasts, RED, Sanger sequencing and NGS, could be utilized to detect the GE events in plants. Among these methods, however, only NGS provides a throughput necessary for large-scale mutation screening, and also is capable of generating deep coverage sequence data for the targeted regions that would allow for assessing the types and allelic dosage of new GE mutations. This information is important for detecting the GE plants with homozygous and heterozygous mutations, or plants that show the evidence of mosaic mutations in somatic tissues, which are usually not heritable [33].

In some cases, when it is required to edit a genomic target showing the low GE efficiency or to obtain plants with multiple targets modified, it was found useful to screen the GE events across several generations of plants derived from the original transgenic event. This approach provides an alternative to regenerating a large number of plants from independent transgenic events, and relies on the CRISPR-Cas9 transgenerational activity to produce mutations in the next generation [23, 33]. In addition, this transgenerational activity allows for inducing the CRISPR-Cas9 mutations on the homologous chromosomes of other cultivars crossed with the transgenic lines expressing the CRISPR-Cas constructs.

29.5.8 Phenotypic Evaluation of GE Wheat Lines

The phenotypic effects of GE depend on the genetic architecture of traits. In allo-polyploid wheat, the effects of GE are strongly influenced by polyploidy and were shown to be dosage-depend, with the strongest effects observed in lines carrying mutations in all copies of homoeologous genes [20]. The expansion of gene families in the wheat genome [18] represents another challenge for GE. The sub-functionalization or functional redundancy among duplicated genes would compli-cate estimating the phenotypic effects of GE, unless many gene family members are edited. The consequences of GE in genes selected based on the comparative genom-ics depend on the evolutionary conservation of gene function. For example, the GE of the wheat-rice orthologs TaGW7 and OsGW7 showed some differences in the phenotypic effects, although their main functions appear to be conserved [23].

One should be also cautious about the selection of a wild-type control for com-paring with the GE lines. The regeneration of transgenic plants could induce epi-genetic modifications in the genome, which by themselves could influence phenotypes. To reduce the effect of genetic background on phenotype, the best option is to use controls homozygous for the wild-type alleles, which are selected from a progeny of heterozygous mutants or cross between the wild-type genotype and the GE wheat line [20, 23].

29.5.9 Prospects of CRISRP-Cas Application in Wheat Improvement

Though the number of traits modified by GE in wheat is still limited and mostly based on the loss-of-function mutations in genes having a negative effect on agro-nomic traits, the CRISPR-Cas-based technologies has a great potential to accelerate wheat improvement. Combined with advances in wheat genome sequencing and the development of rich genetic resources for trait mapping, GE will accelerate identi-fication and functional analyses of genes and pathways controlling major agronomic traits. The range of phenotypic variation for these traits could potentially be further expanded by applying GE to engineer the novel allelic variants of genes with modi-fied regulatory regions or coding sequences. The development of herbicide resis-tant, haploid induction and male sterile lines using the GE is clear demonstration of how GE could be used to improve wheat breeding and production processes. The improvement of wheat transformation technologies holds promise to broaden the range of genotypes amenable to regeneration and in near future will likely allow for GE in a broad range of germplasm including wild relatives. The advent of HI-Edit technology is another step towards eliminating the dependence of GE from the genotype-specific variation in regeneration efficiency among different wheat cultivars. These advances will enable CRISPR-mediated induction of beneficial allelic diversity at multiple genes in diverse wheat cultivars, and identify the optimal combinations of CRISPR-induced and natural alleles with positive effects on agronomic traits by performing selection in the wheat breeding programs. In the future, the ability to transform wild relatives of wheat combined with the improved understanding of genes involved in domestication process will open possibilities for developing novel crops by the GE of domestication genes.

29.6   Key Concepts

The ability to engineer positive changes in agronomic traits using CRISPR-Cas genome editing technologies relies on decades of research performed to identify causal genes controlling phenotypic variation in agronomic crops. The CRISPR-Cas system uses guide RNAs designed to target specific regions of the genome to introduce precise modifications into coding or regulatory sequences of these causal genes. The genome editing projects start with the identification of these genes and choosing a CRISPR-Cas system and a genome editing strategy most suitable for performing desired modifications in the targeted genomic regions. The bioinformatically designed guide RNAs are experimentally validated and used to build plasmid constructs for genome editing. There are a number of methods exist for delivering the genome editing reagents into wheat plants including biolistic and Agrobacterium-mediated transformation, transformation with ribonucleoprotein complexes, and HI-Edit. The genome edited plants are screened for mutations in the targeted regions and phenotypically evaluated to assess the effect of newly engineered alleles on the traits of interest. These newly created variants of genes broaden the scope of genetic diversity available for selection in breeding programs and hold great potential to accelerate the improvement of agronomic traits.

29.7   Conclusions

CRISPR-Cas system is a powerful technology that could take full advantage of new genomic and genetic resources developed for wheat and related crops. Integration of this GE tool into the modern breeding practice may help to speed up the rate of genetic gain by accelerating the identification of novel agronomic genes, broadening genetic diversity of the identified genes, and reducing time required for transferring beneficial alleles into the adapted germplasm and evaluating their effects in the relevant environments. These advances will enable redesigning biological pathways underlying major agronomic traits in wheat by introducing the engineered genes into the breeding populations and selecting optimal allelic combinations to maximize target trait expression. The CRISPR-Cas-based tools will play important role in addressing many future wheat breeding challenges.

References

1. Jinek M, Chylinski K, Fonfara I, Hauer M, Doudna JA, Charpentier E (2012) A programmable dual-RNA-guided DNA endonuclease in adaptive bacterial immunity. Science 337:816–821. https://doi.org/10.1126/science.1225829
2. Gasiunas G, Barrangou R, Horvath P, Siksnys V (2012) Cas9-crRNA ribonucleoprotein complex mediates specific DNA cleavage for adaptive immunity in bacteria. Proc Natl Acad Sci 109:E2579–E2586. https://doi.org/10.1073/pnas.1208507109
3. Cong L, Ran FA, Cox D, Lin S, Barretto R, Habib N, Hsu PD, Wu X, Jiang W, Marraffini LA, Zhang F (2013) Multiplex genome engineering using CRISPR/Cas systems. Science 339:819–823. https://doi.org/10.1126/science.1231143
4. Mali P, Yang L, Esvelt KM, Aach J, Guell M, DiCarlo JE, Norville JE, Church GM (2013) RNA-guided human genome engineering via Cas9. Science 339:823–826. https://doi.org/10.1126/science.1232033
5. Wiedenheft B, Sternberg SH, Doudna JA (2012) RNA-guided genetic silencing systems in bacteria and archaea. Nature 482:331–338. https://doi.org/10.1038/nature10886
6. Makarova KS, Wolf YI, Iranzo J, Shmakov SA, Alkhnbashi OS, Brouns SJJ, Charpentier E, Cheng D, Haft DH, Horvath P, Moineau S, Mojica FJM, Scott D, Shah SA, Siksnys V, Terns MP, Venclovas Č, White MF, Yakunin AF, Yan W, Zhang F, Garrett RA, Backofen R, van der Oost J, Barrangou R, Koonin EV (2020) Evolutionary classification of CRISPR–Cas systems: a burst of class 2 and derived variants. Nat Rev Microbiol 18:67–83. https://doi.org/10.1038/s41579-019-0299-x
7. Nishimasu H, Shi X, Ishiguro S, Gao L, Hirano S, Okazaki S, Noda T, Abudayyeh OO, Gootenberg JS, Mori H, Oura S, Holmes B, Tanaka M, Seki M, Hirano H, Aburatani H, Ishitani R, Ikawa M, Yachie N, Zhang F, Nureki O (2018) Engineered CRISPR-Cas9 nuclease with expanded targeting space. Science 361:1259–1262. https://doi.org/10.1126/science.aas9129
8. Wang W, Tian B, Pan Q, Chen Y, He F, Bai G, Akhunova A, Trick HN, Akhunov E (2020) Expanding the range of editable targets in the wheat genome using the variants of the Cas12a and Cas9 nucleases. bioRxiv. https://doi.org/10.1101/2020.12.09.418624
9. Anzalone AV, Koblan LW, Liu DR (2020) Genome editing with CRISPR–Cas nucleases, base editors, transposases and prime editors. Nat Biotechnol 38:824–844. https://doi.org/10.1038/s41587-020-0561-9
10. Zetsche B, Gootenberg JS, Abudayyeh OO, Slaymaker IM, Makarova KS, Essletzbichler P, Volz SE, Joung J, van der Oost J, Regev A, Koonin EV, Zhang F (2015) Cpf1 is a single RNA-guided endonuclease of a class 2 CRISPR-Cas system. Cell 163:759–771. https://doi.org/10.1016/j.cell.2015.09.038
11. Anzalone AV, Randolph PB, Davis JR, Sousa AA, Koblan LW, Levy JM, Chen PJ, Wilson C, Newby GA, Raguram A, Liu DR (2019) Search-and-replace genome editing without double-strand breaks or donor DNA. Nature 576:149–157. https://doi.org/10.1038/s41586-019-1711-4
12. Gaudelli NM, Komor AC, Rees HA, Packer MS, Badran AH, Bryson DI, Liu DR (2017) Programmable base editing of A•T to G•C in genomic DNA without DNA cleavage. Nature 551:464–471. https://doi.org/10.1038/nature24644

13. Komor AC, Kim YB, Packer MS, Zuris JA, Liu DR (2016) Programmable editing of a target base in genomic DNA without double-stranded DNA cleavage. Nature 533:420–424. https://doi.org/10.1038/nature17946

14. Lin Q, Zong Y, Xue C, Wang S, Jin S, Zhu Z, Wang Y, Anzalone AV, Raguram A, Doman JL, Liu DR, Gao C (2020) Prime genome editing in rice and wheat. Nat Biotechnol 38:582–585. https://doi.org/10.1038/s41587-020-0455-x

15. Nakamura M, Gao Y, Dominguez AA, Qi LS (2021) CRISPR technologies for precise epigenome editing. Nat Cell Biol 23:11–22. https://doi.org/10.1038/s41556-020-00620-7

16. Gao C (2021) Genome engineering for crop improvement and future agriculture. Cell 184:1621–1635. https://doi.org/10.1016/j.cell.2021.01.005

17. Hayta S, Smedley MA, Demir SU, Blundell R, Hinchliffe A, Atkinson N, Harwood WA (2019) An efficient and reproducible Agrobacterium-mediated transformation method for hexaploid wheat (Triticum aestivum L.). Plant Methods 15(121) https://doi.org/10.1186/s13007-019-0503-z

18. Appels R, Eversole K, Stein N, Feuillet C, Keller B, Rogers J, Pozniak CJ, Choulet F, Distelfeld A, Poland J, Ronen G, Sharpe AG, Barad O, Baruch K, Keeble-Gagnère G, Mascher M, Ben-Zvi G, Josselin A-A, Himmelbach A, Balfourier F, Gutierrez-Gonzalez J, Hayden M, Koh C, Muehlbauer G, Pasam RK, Paux E, Rigault P, Tibbits J, Tiwari V, Spannagl M, Lang D, Gundlach H, Haberer G, Mayer KFX, Ormanbekova D, Prade V, Šimková H, Wicker T, Swarbreck D, Rimbert H, Felder M, Guilhot N, Kaithakottil G, Keilwagen J, Leroy P, Lux T, Twardziok S, Venturini L, Juhász A, Abrouk M, Fischer I, Uauy C, Borrill P, Ramirez-Gonzalez RH, Arnaud D, Chalabi S, Chalhoub B, Cory A, Datla R, Davey MW, Jacobs J, Robinson SJ, Steuernagel B, van Ex F, Wulff BBH, Benhamed M, Bendahmane A, Concia L, Latrasse D, Bartoš J, Bellec A, Berges H, Doležel J, Frenkel Z, Gill B, Korol A, Letellier T, Olsen O-A, Singh K, Valárik M, van der Vossen E, Vautrin S, Weining S, Fahima T, Glikson V, Raats D, Číhalíková J, Toegelová H, Vrána J, Sourdille P, Darrier B, Barabaschi D, Cattivelli L, Hernandez P, Galvez S, Budak H, JDG J, Witek K, Yu G, Small I, Melonek J, Zhou R, Belova T, Kanyuka K, King R, Nilsen K, Walkowiak S, Cuthbert R, Knox R, Wiebe K, Xiang D, Rohde A, Golds T, Čížková J, Akpinar BA, Biyiklioglu S, Gao L, N'Daiye A, Kubaláková M, Šafář J, Alfama F, Adam-Blondon A-F, Flores R, Guerche C, Loaec M, Quesneville H, Condie J, Ens J, Maclachlan R, Tan Y, Alberti A, Aury J-M, Barbe V, Couloux A, Cruaud C, Labadie K, Mangenot S, Wincker P, Kaur G, Luo M, Sehgal S, Chhuneja P, Gupta OP, Jindal S, Kaur P, Malik P, Sharma P, Yadav B, Singh NK, Khurana JP, Chaudhary C, Khurana P, Kumar V, Mahato A, Mathur S, Sevanthi A, Sharma N, Tomar RS, Holušová K, Plíhal O, Clark MD, Heavens D, Kettleborough G, Wright J, Balcárková B, Hu Y, Salina E, Ravin N, Skryabin K, Beletsky A, Kadnikov V, Mardanov A, Nesterov M, Rakitin A, Sergeeva E, Handa H, Kanamori H, Katagiri S, Kobayashi F, Nasuda S, Tanaka T, Wu J, Cattonaro F, Jiumeng M, Kugler K, Pfeifer M, Sandve S, Xun X, Zhan B, Batley J, Bayer PE, Edwards D, Hayashi S, Tulpová Z, Visendi P, Cui L, Du X, Feng K, Nie X, Tong W, Wang L (2018) Shifting the limits in wheat research and breeding using a fully annotated reference genome. Science 361. https://doi.org/10.1126/science.aar7191

19. Walkowiak S, Gao L, Monat C, Haberer G, Kassa MT, Brinton J, Ramirez-Gonzalez RH, Kolodziej MC, Delorean E, Thambugala D, Klymiuk V, Byrns B, Gundlach H, Bandi V, Siri JN, Nilsen K, Aquino C, Himmelbach A, Copetti D, Ban T, Venturini L, Bevan M, Clavijo B, Koo D-H, Ens J, Wiebe K, N'Diaye A, Fritz AK, Gutwin C, Fiebig A, Fosker C, Fu BX, Accinelli GG, Gardner KA, Fradgley N, Gutierrez-Gonzalez J, Halstead-Nussloch G, Hatakeyama M, Koh CS, Deek J, Costamagna AC, Fobert P, Heavens D, Kanamori H, Kawaura K, Kobayashi F, Krasileva K, Kuo T, McKenzie N, Murata K, Nabeka Y, Paape T, Padmarasu S, Percival-Alwyn L, Kagale S, Scholz U, Sese J, Juliana P, Singh R, Shimizu-Inatsugi R, Swarbreck D, Cockram J, Budak H, Tameshige T, Tanaka T, Tsuji H, Wright J, Wu J, Steuernagel B, Small I, Cloutier S, Keeble-Gagnère G, Muehlbauer G, Tibbets J, Nasuda S, Melonek J, Hucl PJ, Sharpe AG, Clark M, Legg E, Bharti A, Langridge P, Hall A, Uauy C, Mascher M, Krattinger SG, Handa H, Shimizu KK, Distelfeld A, Chalmers K, Keller B, Mayer KFX, Poland J, Stein

N, McCartney CA, Spannagl M, Wicker T, Pozniak CJ (2020) Multiple wheat genomes reveal global variation in modern breeding. Nature 2020:1–7

20. Wang W, Simmonds J, Pan Q, Davidson D, He F, Battal A, Akhunova A, Trick HN, Uauy C, Akhunov E (2018) Gene editing and mutagenesis reveal inter-cultivar differences and additivity in the contribution of TaGW2 homoeologues to grain size and weight in wheat. Theor Appl Genet 131:2463–2475. https://doi.org/10.1007/s00122-018-3166-7

21. Wang Y, Cheng Z, Shan W, Zhang Y, Liu J, Gao C, Qiu J (2014) Simultaneous editing of three homoeoalleles in hexaploid bread wheat confers heritable resistance to powdery mildew. Nat Biotechnol 32:947–951. https://doi.org/10.1038/nbt.2969

22. Wang W, Akhunova A, Chao S, Akhunov E (2016) Optimizing multiplex CRISPR/Cas9-based genome editing for wheat. bioRxiv. https://doi.org/10.1101/051342

23. Wang W, Pan Q, Tian B, He F, Chen Y, Bai G, Akhunova A, Trick HN, Akhunov E (2019) Gene editing of the wheat homologs of TONNEAU1-recruiting motif encoding gene affects grain shape and weight in wheat. Plant J 100:251–264. https://doi.org/10.1111/tpj.14440

24. Čermák T, Curtin SJ, Gil-Humanes J, Čegan R, Kono TJY, Konečná E, Belanto JJ, Starker CG, Mathre JW, Greenstein RL, Voytas DF (2017) A multipurpose toolkit to enable advanced genome engineering in plants. Plant Cell 29:1196–1217. https://doi.org/10.1105/tpc.16.00922

25. Su Z, Bernardo A, Tian B, Chen H, Wang S, Ma H, Cai S, Liu D, Zhang D, Li T, Trick H, St. Amand P, Yu J, Zhang Z, Bai G (2019) A deletion mutation in TaHRC confers Fhb1 resistance to Fusarium head blight in wheat. Nat Genet 51:1099–1105. https://doi.org/10.1038/s41588-019-0425-8

26. Gao Y, An K, Guo W, Chen Y, Zhang R, Zhang X, Chang S, Rossi V, Jin F, Cao X, Xin M, Peng H, Hu Z, Guo W, Du J, Ni Z, Sun Q, Yao Y (2021) The endosperm-specific transcription factor TaNAC019 regulates glutenin and starch accumulation and its elite allele improves wheat grain quality. Plant Cell. https://doi.org/10.1093/plcell/koaa040

27. Zhang Z, Hua L, Gupta A, Tricoli D, Edwards KJ, Yang B, Li W (2019) Development of an Agrobacterium-delivered CRISPR/Cas9 system for wheat genome editing. Plant Biotechnol J 17:1623–1635. https://doi.org/10.1111/pbi.13088

28. Zhang Y, Liang Z, Zong Y, Wang Y, Liu J, Chen K, Qiu J-L, Gao C (2016) Efficient and transgene-free genome editing in wheat through transient expression of CRISPR/Cas9 DNA or RNA. Nat Commun 7:12617. https://doi.org/10.1038/ncomms12617

29. Zhang R, Liu J, Chai Z, Chen S, Bai Y, Zong Y, Chen K, Li J, Jiang L, Gao C (2019) Generation of herbicide tolerance traits and a new selectable marker in wheat using base editing. Nat Plants 5:480–485. https://doi.org/10.1038/s41477-019-0405-0

30. Debernardi JM, Tricoli DM, Ercoli MF, Hayta S, Ronald P, Palatnik JF, Dubcovsky J (2020) A GRF–GIF chimeric protein improves the regeneration efficiency of transgenic plants. Nat Biotechnol 38:1274–1279. https://doi.org/10.1038/s41587-020-0703-0

31. Liang Z, Chen K, Li T, Zhang Y, Wang Y, Zhao Q, Liu J, Zhang H, Liu C, Ran Y, Gao C (2017) Efficient DNA-free genome editing of bread wheat using CRISPR/Cas9 ribonucleoprotein complexes. Nat Commun 8:14261. https://doi.org/10.1038/ncomms14261

32. Kelliher T, Starr D, Su X, Tang G, Chen Z, Carter J, Wittich PE, Dong S, Green J, Burch E, McCuiston J, Gu W, Sun Y, Strebe T, Roberts J, Bate NJ, Que Q (2019) One-step genome editing of elite crop germplasm during haploid induction. Nat Biotechnol 37:287–292. https://doi.org/10.1038/s41587-019-0038-x

33. Wang W, Pan Q, He F, Akhunova A, Chao S, Trick H, Akhunov E (2018) Transgenerational CRISPR-Cas9 activity facilitates multiplex gene editing in allopolyploid wheat. CRISPR J 1:65–74. https://doi.org/10.1089/crispr.2017.0010

Chapter 30
Accelerating Breeding Cycles

Samir Alahmad, Charlotte Rambla, Kai P. Voss-Fels, and Lee T. Hickey

Abstract The rate of genetic gain in wheat improvement programs must improve to meet the challenge of feeding a growing population. Future wheat varieties will need to produce record high yields to feed an anticipated 25% more inhabitants on this planet by 2050. The current rate of genetic gain is slow and cropping systems are facing unprecedented fluctuations in production. This instability stems from major changes in climate and evolving pests and diseases. Rapid genetic improvement is essential to optimise crop performance under such harsh conditions. Accelerating breeding cycles shows promise for increasing the rate of genetic gain over time. This can be achieved by concurrent integration of cutting-edge technologies into breeding programs, such as speed breeding (SB), doubled haploid (DH) technology, high-throughput phenotyping platforms and genomic selection (GS). These technologies empower wheat breeders to keep the pace with increasing food demand by developing more productive and robust varieties sooner. In this chapter, strategies for shortening the wheat breeding cycle are discussed, along with the opportunity to integrate technologies to further accelerate the rate of genetic gain in wheat breeding programs.

Keywords Speed breeding · Doubled haploid · Shuttle breeding · Genomic selection · Genetic gain · Breeding technologies

S. Alahmad · C. Rambla · K. P. Voss-Fels · L. T. Hickey (✉)
The University of Queensland, Queensland Alliance for Agriculture and Food Innovation, Brisbane, Australia
e-mail: s.alahmad@uq.edu.au; c.rambla@uq.edu.au; k.vossfels@uq.edu.au; l.hickey@uq.edu.au

557

30.1 Learning Objectives

- Realise the challenges associated with increasing the rate of genetic gain in wheat.
- Explore technologies that can reduce the length of a breeding cycle.
- Understand the challenges associated with adapting new technologies to breeding programs.
- Explore opportunities for integrating technologies to improve breeding efficiency.

30.2 Introduction

Wheat breeding programs are designed to prioritise the release of cultivars better adapted to drought and heat, and resistant or tolerant to pests and diseases, with the ultimate goal of increasing yield potential [1]. The estimated rate of genetic gain in wheat is 0.9% per year and an overall agronomic and genetic yield improvement of over 2% per year is needed to meet the increased demand. A typical wheat breeding cycle can take over 12 years for crossing, inbreeding, testing, and selection. Broadly, a breeding cycle involves three distinctive phases, starting and finishing with parental lines used for crossing. These phases include: (i) The crossing and inbreeding phase, which requires the progenies to go through six generations of self-pollination to reach homozygosity and minimise segregation of the traits of interest; (ii) The testing phase, which involves screening for biotic and abiotic traits followed by multi-environment trial evaluation. Lines that are stable across a wide range of environments and carry desirable traits can be recycled and used as parents for making new crosses; (iii) Bulking up seed of the most successful line and releasing a new cultivar available to growers. Due to the lengthy process of cultivar development, breeding programs must adopt new technologies to reduce the time required for completing breeding cycles. Therefore, efforts must focus on speeding up the rate of genetic gain by optimising the components that fit in the breeder's equation. This includes increasing the intensity and precision of selecting individuals within a population that has a high level of useful genetic variation for traits of interest, while also accelerating breeding cycles by reducing the time required to achieve a single cycle [2].

Modern wheat breeding programs have adopted several technologies and breeding strategies that are instrumental in increasing the rate of genetic gain. For example, high-throughput phenotyping platforms can be used to screen large numbers of selected candidates for target traits more precisely, which leads to enhanced selection accuracy and selection intensity [3]. The evolution of next-generation sequencing platforms has resulted in cost-effective genotyping services that have increased the efficiency and accuracy of selection in breeding programs [4]. GS in combination with high-throughput phenotyping platforms shows promise for enhancing predictive breeding approaches, particularly for complex traits such as grain quality and yield [5]. Furthermore, adopting GS in wheat breeding programs can lead to an increase in selection accuracy and intensity, while reducing the breeding cycle concurrently [6]. Shuttle breeding, and most recently 'speed breeding' (SB), has transformed breeding programs by shortening the crossing and inbreeding phase [7, 8]. The faster generation turnover enables evaluation of selection candidates sooner, ultimately reducing the length of the breeding cycle (Fig. 30.1). Likewise, DH technology enables the development of homozygous lines in only two plant generations

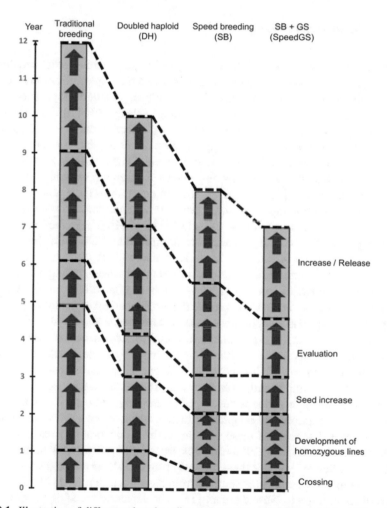

Fig. 30.1 Illustration of different wheat breeding technologies (DH, SB, and SpeedGS) and their impact on the length of the breeding cycle. Combining speed breeding and genomic selection (i.e. SpeedGS) further reduces the length of the breeding cycle by reducing the need for extensive phenotyping. Black arrows indicate a single plant generation. Green indicates steps performed under conventional growing conditions, whereas pink indicates steps performed under speed breeding conditions. (Modified with permission from Ref. [8])

and has dramatically reduced the length of breeding cycles, particularly for winter wheat programs [9].

This chapter provides an overview of the technologies and strategies available to wheat breeders for reducing the length of the breeding cycle. Furthermore, the opportunity to combine technologies to accelerate genetic gain is discussed.

30.3   Strategies to Shorten Breeding Cycles in Wheat

30.3.1   Shuttle Breeding

Shuttle breeding is an off-season field-testing technique whereby genetic material is grown in contrasting environments to turn over two plant generations per year. By implementing this simple, yet effective technique, breeders have successfully reduced the time required to complete a breeding cycle by 50% [10]. In this method, segregating populations are subject to screening, selection and simultaneous generation advancement [11]. The strategy was first developed by Norman Borlaug at the International Maize and Wheat Improvement Centre (CIMMYT) Mexico in 1946. The goal was to speed up breeding cycles and develop varieties faster for the Mexican wheat farmers [12]. To achieve the two generations each year, the material is grown at two contrasting locations in terms of precipitation, altitude and latitude (Fig. 30.2). Today, the technique is still adopted by CIMMYT. In the winter season (November to April), breeding populations are grown in the Sonora Desert (Yaqui Valley, North-Western Mexico) at 39 m.a.s.l. under short days, and selection is applied for yield, agronomic type, leaf and stem rust disease, grain quality and photoperiod insensitivity. In summer (May – October), populations are grown at the Toluca station at 2649 m.a.s.l.), to ensure the crop experiences cooler temperatures during grain filling, and selection is applied for resistance to yellow rust (*Puccinia striiformis* f. sp. *tritici*) and speckled leaf blotch (*Septoria tritici*) disease [7]. In addition to the key advantage of shortening the breeding cycle, the strategy enables selection of breeding materials exposed to different soil types, temperatures, disease pressure and most importantly, photoperiod. The semi-dwarf, rust-resistant and photoperiod insensitive varieties developed by Borlaug and his colleagues resulted in widespread adoption and adaptation to the wheat mega environments of the world. The global success of these varieties is the foundation of what is known as the Green Revolution of the 1960s and 70s, transitioning CIMMYT to internationalisation [12]. Several winter wheat breeding programs have adopted shuttle breeding, for example, the material is shuttled from breeding programs in North America and Northern Europe to New Zealand [13]. The Japanese breeding program initiated shuttle breeding in the 1970s, taking advantage of the wide variation in latitude [14]. Shortly after the emergence of Ug99, a modified shuttle breeding program was implemented by CIMMYT and partners to incorporate screening of the highly virulent race of stem rust in Noro, Kenya [15]. A similar application of shuttle breeding

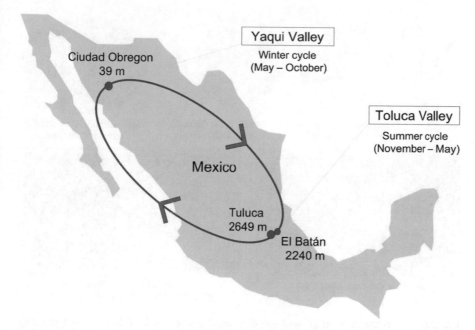

Fig. 30.2 Overview of the shuttle breeding strategy developed by Dr. Norman Borlaug at CIMMYT, Mexico. The arrows indicate the shuttling of genetic material between contrasting North-West and South-East Mexican environments over the winter and summer seasons

that incorporates selection for biotic and abiotic stresses has been implemented at the International Centre for Agricultural Research in the Dry Areas (ICARDA). For instance, winter × spring crosses are generated at Terbol Station, Lebanon (34° N; 36° E, 900 m.a.s.l.) during winter and summer and the segregating F_2 material is shuttled to Sids station (29° N; 31° E, 32.2 m.a.s.l.) in Egypt during winter for early yield potential trials, the F_3 material is then shuttled to Kulumsa (08° N; 39° E, 2220 m.a.s.l.) in Ethiopia during summer for rust screening, and finally, the F_4 segregating populations are shuttled to Merchouch station (33.6° N; 6.7° W, 430 m.a.s.l) in Morocco during the winter season and screened for insect tolerance and disease resistance [11]. This process can generate broadly adapted inbred lines enriched with target traits prior to yield testing, thus assisting the development of robust cultivars for farmers.

30.3.2   Doubled Haploid Technology

A doubled haploid genotype is created when the chromosomes in haploid cells (n) are doubled. DH wheat was first developed by culturing immature haploid pollen and generating diploid homozygous plants using Colchicine (Fig. 30.3). However, this technique is limited due to several factors related to growth conditions of the

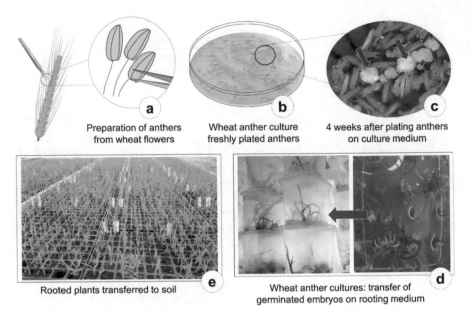

Fig. 30.3 Overview of the steps involved in generating wheat doubled haploid lines using the anther culture method, including (**a**) harvesting anthers from the selected wheat plants; (**b**) culturing anthers; (**c**) callus initiation; (**d**) transferring embryos to rooting medium; and (**e**) transplanting rooted plants into the soil

donor plants, their genetic stability and anther-culture ability, resulting in a low success rate to regenerate healthy and green DH plants [9]. This has led to the search for alternative and high-throughput routes for developing DH plants more suited to applications into breeding programs. Chromosome elimination of wide crosses, such as the cross between wheat and maize (*Zea mays L.*), was first trialled by Zenkteler and Nitzsche [16]. Following pollination, maize chromosomes are eliminated during cell division and the haploid wheat is induced and the chromosomes are doubled using Colchicine. This technique requires less effort and tripled the success rate, providing a DH production system that is more suitable for breeding purposes. The success of this technique is underpinned by high efficiency and low genotype specificity in comparison to the anther-culture technique. The technology of DH has been integrated into many global largely winter wheat breeding programs which required the upscaling of DH production and the development of a large number of individuals necessary to increase the rate of genetic gain. The advantage of this technology is that the breeding cycle is significantly shortened due to the efficiency in developing inbred lines. The DH homozygous lines could be available for evaluation and breeding within two generations in comparison to six generations or more when using traditional self-pollination methods in the field [11] (Fig. 30.1). Despite the breakthrough in accelerating breeding cycles by rapidly developing homozygous lines, DH technology has drawbacks when implemented in breeding programs. For example, it does not allow evaluation and selection for important traits during the early segregating generations of a classical inbred line development approach. The technique requires specialised labs, is quite labour intensive and requires a high level of experience. Furthermore, there is often significant variability in successfully producing DH lines, due to the genotype dependency. However, the main limitation for DH technology to become a mainstream tool in wheat breeding programs is cost. For example, if a large-scale wheat breeding program were to generate 40,000 DH lines, at a very conservative rate of US $15 per line, it would cost US $600,000. Notably, the cost of DH production for wheat can be even higher, up to US $50 per line (see Chap. 5). Despite this, DH technology is common in breeding programs in Europe and the UK, due to the strong vernalisation requirement of winter wheat that extends generation turnover time. On the other hand, for spring wheat DH technology has been largely adopted in research and pre-breeding programs for QTL mapping studies.

30.3.3   Speed Breeding

SB involves growing plant populations under controlled environmental conditions that are conducive to early flowering and generation advance. The concept was inspired by research funded by the National Aeronautics and Space Administration (NASA) at Utah State University in the 1980s, which explored the possibility of growing a fast crop of wheat on space stations [8]. This research resulted in the development of 'USU-Apogee' wheat which can flower within 25 days after sowing when grown under 23°C and continuous light [17]. In 2003, researchers at the University of Queensland coined the name SB, and the technique was first applied to wheat breeding for fast-tracking the introgression of yellow spot resistance. With advances in science and technology over the last few decades, techniques for SB crops have evolved [8]. In particular, advances in light-emitting diodes (LED) technology have led to the widespread availability of affordable grow-lights that provide healthy and efficient plant growth in controlled environment facilities. SB regimes that use LED lighting systems have been developed for a range of long-day or short-day crops [18]. Furthermore, LED lights with 'tunable' wavelengths are now available, opening the door for optimising wavelengths to further manipulate plant growth, such as plant height and flowering [19]. This technology has reduced energy consumption and costs, but also facilitated the delivery of optimal light quality in fully controlled SB facilities, resulting in improved plant growth, increased quality and enhanced seed production. SB facilities can take many shapes and forms, including growth cabinets for small-scale research, and modified glasshouses or warehouses with multi-tiered growing spaces for large-scale programs. In glasshouses fitted with supplemental lighting systems, sensor technologies coupled with automated systems can adjust the light intensity depending on the cloud cover, sky luminance and radiation. This provides an integrated light control system to promote flowering, maximise growth rate, and importantly, minimise costs associated with generation turnover.

Speed breeding enables wheat researchers and breeders to grow up to 4–6 generations per year instead of 1–3 generations in the field or under regular glasshouse conditions. Protocols have been made available [20], including a step-by-step guide for establishing large-scale facilities [18]. Exposing plants to extended photoperiods (22 h light, 2 h day) and controlled temperatures (22/17 °C, day/night) respectively, promotes early flowering and seed formation. A common feature in many rapid generation advance methods is embryo rescue, which is a laborious process. However, it is possible to avoid the need for embryo rescue using a pretreatment to break seed dormancy (e.g. 4 °C for 3 days) and achieve high germination rates [18]. For example, to short-cut generation time, spikes can be harvested green (just 2 weeks post flowering) and placed into an air-forced dehydrator at 35 °C for 3 days to artificially mature and dry the seed prior to re-sowing. The premature harvest technique is very effective when applied on a small scale, making it suitable for research and pre-breeding activities. However, for large-scale wheat breeding programs, harvest at maturity is preferred because it avoids the need for multiple harvests for populations that are typically segregating for flowering time, and as such minimises labour (Fig. 30.4). To hasten maturity, water supply can be reduced after flowering to enable harvest approximately 4 weeks later [18].

In addition to accelerating breeding cycles, SB can be used to fast-track research and pre-breeding outcomes. The tool is particularly useful to accelerate the development of populations suitable for trait dissection and mapping QTL for important traits. For example, nested-association mapping populations (NAM) suitable for dissecting drought adaptive traits were rapidly generated using SB at The University of Queensland, requiring only 18 months from crossing to development of

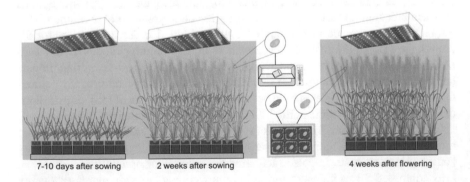

7-10 days after sowing   2 weeks after sowing   4 weeks after flowering

Fig. 30.4 Illustration of a wheat breeding population growing at high-density under speed breeding conditions. Seven to ten days after sowing, wheat seedlings reach 2–3 leaf growth stage. Time to flowering typically ranges between 4 and 6 weeks for most spring wheat genotypes. To complete the plant generation within eight weeks, wheat heads can be harvested prematurely two weeks after flowering and dried down in an air-forced dehydrator at 35 °C for three days. The slightly shrivelled seeds germinate well provided a cold treatment is applied, and can be used to fast-track generation turnover for pre-breeding and research applications. Alternatively, for breeding applications that involve larger and more diverse populations, after flowering water supply can be reduced to accelerate maturity and enable harvest of mature/well-filled seed four weeks later

F4-derived lines [21]. To support trait screening and selection for pre-breeding applications, a number of protocols have been developed for disease resistance traits, such as wheat rusts [22], yellow spot [23] and crown rot [24]. These techniques incorporate rapid generation advance and enable screening of large segregating populations all year round. Selected plants can be advanced to develop inbred lines or backcrossed to elite parents for trait introgression. When selection is applied in early generations, the resulting inbred lines are enriched with desirable allelic combinations, which enables more targeted and efficient field testing. To support breeding, SB can reduce the number of years required for trait dissection and introgression of new traits into elite genetic backgrounds.

The key to integrating SB into a large-scale wheat breeding program is establishing cost-effective facilities and streamlining operations. With this in mind, SB facilities have been established by private wheat breeding companies, as well as public breeding and research centres, such as CIMMYT and ICARDA. Reducing the cost per plant is important, which ultimately reduces the cost per line development through the SB facility. Ghosh et al. [18] provide detailed protocols and advice for scaling-up wheat SB protocols, including growing plants at high density (e.g. 1000 plants/m^2). Such techniques can enable wheat breeding programs to generate large populations in a cost-effective manner. Other important considerations include designing low-cost and energy efficient infrastructure, adopting automation where possible (e.g. automated irrigation systems), and streamlining operations.

30.3.4 Genomic Selection

The implementation of GS is outlined in Chap. 5. This method was initially developed and applied to accelerate genetic gain in animal breeding. The method is based on statistical models that can predict the genetic merit of individuals based on high-resolution genome profiles, before they are tested in the field. The genetic merit of a genotype, mostly referred to as the genomic estimated breeding value (GEBV), is calculated by simultaneously estimating genome-wide DNA marker effects and then summing them for each individual. Individuals with the highest GEBV can then be selected either for extensive field evaluation or to be used as parents in the next breeding cycle [25]. The method requires a training population where the individuals within the population are phenotyped for the traits of interest and genotyped using high-density DNA markers. By combining the genetic makeup of individuals in the training population and their phenotypes, a statistical model that can estimate the association between markers and quantitative traits is initiated. It is then used to predict GEBV of genotypes in the 'selection candidates' that have only been genotyped using the same genome-wide DNA markers [26].

Several growing seasons are usually required for testing breeding material under different environments in conventional breeding methods, with a risk of inaccurate selection due to genotype-by-environment interaction [27]. One strategy by which GS models can help overcome this issue is to incorporate proxy traits that explain an additional amount of the phenotypic variation observed for the target trait. Incorporating such traits in GS models has been shown to improve GEBV prediction [5]. This enables a more accurate selection of superior genotypes with the desired combination of alleles without the need for phenotyping. Besides prioritising genotypes for further field testing based on their GEBV, a recurrent GS method could be applied during early generations of the line development phase. For example, recurrent GS could be used at F_2 or even F_1 generation to select candidates for intercrossing. This would enable more rapid enrichment of populations with desired alleles (e.g. [26, 28]). Thereafter, further shortcutting the testing phase and accelerating the breeding cycle is possible, unlike in conventional breeding methods. The shift from phenotypic selection to GS has resulted in saving resources and time. For example, GEBVs can be obtained in 2 years when GS is applied, instead of 5 years in conventional breeding [4]. In addition to shortening the breeding cycle and improving selection accuracy, GS has increased the rate of genetic gain in breeding programs by increasing selection intensity. This is possible when the number of tested lines is boosted due to reduction in replication [8]. This approach has been implemented in wheat breeding programs around the globe [27]. However, application of GS in plant breeding programs has experienced a significant shift from single-trait/single-site prediction models to multi-trait/multi-site models which incorporate genotype-by-environment effects. The shift to more complex models has enabled more accurate predictions of phenotypes which have been sparsely collected in multiple environments [29]. GS is usually applied to quantitative traits governed by a large number of genes, such as drought adaptive traits [5], grain quality [30] and grain yield [31]. Furthermore, GS has been applied to proxy traits such as flowering time, which has high heritability, is easy to measure, and has a significant impact on yield [32]. GS has been shown to be powerful for improving resistance to fusarium head blight and stem rust in wheat using large historical datasets [33]. In the early stages of GS application, research largely focused on improving selection accuracy based on optimised prediction models by accounting for variation caused by genotype-by-environment interaction in the models [34]. More recently, high-throughput phenotyping platforms that generate very large datasets, such as unmanned aerial vehicle (UAV) platforms, have been used for improving GS models with enhanced accuracies [35]. GS has shifted from being the focus of research to being widely adopted into wheat breeding programs in both private and public sectors. This is due to several factors including advances in molecular biotechnology, a dramatic decrease in genotyping cost and the development of bioinformatic tools that can efficiently calculate GEBVs for very large datasets [4].

30.4 Integrating Breeding Technologies

A radical change and redesign of breeding programs incorporating advanced technologies holds the key to improving yield potential for our future wheat varieties. Breeding tools such as, DH, SB and GS have great potential for accelerating breeding cycles, however, their implementation on an industrial scale in mainstream wheat breeding programs is yet to see the light of day. As detailed in Sects. 30.2 and 30.3, these breeding tools can play an instrumental role in accelerating breeding cycles. Implementation of these technologies in early stages of breeding cycles enables rapid production of homozygous lines enriched with desired allele combination for field testing.

The process of DH line development usually requires two years and an additional year of self-pollination for increasing sufficient seed necessary for field trials. The rapid development of DH lines could be possible when SB is integrated at different stages of DII production systems. For example, parental lines used for haploid induction could be grown under SB. Furthermore, SB technology could be used during the self-pollination stage following chromosome doubling and the additional generation for seed bulking. The integration of SB into the DH system facilitates shortcutting time to further accelerate breeding cycles.

The adoption of GS enables the identification of superior individuals based on their GEBV, either for advanced yield testing or as parents used for crossing in the next breeding cycle [8]. The integration of GS in breeding programs has thus resulted in the emergence of highly productive wheat cultivars in a shorter time. Notably, combining these technologies targeting different stages of the breeding cycle may have an additional impact on the rate of genetic gain. For example, integrating high throughput phenotyping with GS methods has been shown to improve efficiency and outcomes for both methods [5]. GS is also employed for detecting and stacking the best haplotypes using parents with optimal genetic variation used for DH line development [36]. Implementation of SB and GS has demonstrated significant increases in the rate of genetic gain when applied separately in the breeding programs. However, combining these two technologies may result in a larger effect on shortening time required for completing a breeding cycle (Fig. 30.1). Despite the potential advantages of combining these technologies, it is yet to be employed in existing breeding programs.

To explore the potential, simulation studies have been performed. The study by Voss-Fels et al. [37] compared the rate of genetic gain for four breeding strategies, including traditional phenotypic selection, GS, the combination of SB and GS (SpeedGS), and SpeedGS with introduced diversity. Overall, genetic merit for grain yield was used to determine the amount of gain that could be achieved by implementing the different breeding tools in isolation or in combination. The results concluded that significant gains were possible using all strategies, however, breeding schemes that implemented SpeedGS displayed a 34% increase in the rate of genetic gain per unit of time when compared to conventional breeding [37]. In these simulations GS was only used to identify improved lines in the breeding cycle for advanced field testing, however, the use of GS in a rapid recurrent selection framework has been shown to hold huge potential for substantially increasing genetic gain [19, 28]. Despite these promising results from combining SB and GS, adoption of these tools by breeding programs needs further investigation. Further empirical and simulation studies are needed to determine a suitable pathway in which these technologies

could be applied in the most efficient and cost-effective way to maximise investment return.

30.5 Key Concepts

This chapter has outlined (1) key strategies and advanced technologies that can
reduce the length of a breeding cycle, (2) opportunities to integrate these technologies to further accelerate genetic gain. Widespread adoption of these enabling tools
in public and private wheat breeding programs would enhance breeding efficiency
and support global food security.

30.6 Conclusions

Crop production must increase by 50% by 2050 to meet the future demand for food.
Adoption of precision farming systems that use cutting-edge technology to optimise
management practices, such as fertiliser, pest and disease management will assist in
meeting this goal. Most importantly, plant breeding aimed at improving the genetics
of crop varieties with high performance in the face of abiotic and biotic challenges
will help secure increments in crop production at a global scale. However, the development of resilient crop varieties requires a huge investment and is notoriously
slow. In order to speed up the process, redesign and transformation of crop breeding
programs is required. Fortunately, plant breeders now have at their disposal technologies that could become a game-changer and transform traditional plant breeding to being significantly more cost-effective and efficient in releasing high yielding
and stable varieties. Employing SB, DH, high-throughput phenotyping platforms
and GS in breeding programs has advantages, but could be further exploited by
integrating these tools at different stages of research and breeding programs; from
trait discovery to rapid introgression and population improvement, field testing and
release of new cultivars. Fusing these technologies into our existing breeding programs will play a key role in rapidly improving future wheat crops.

Acknowledgments The authors give thanks to the Grain Research and Development Corporation
of Australia for a Postdoctoral Research Fellowship (9177334) to SA, and a PhD scholarship
(9176855) to CR. The authors would like to acknowledge Jon Falk from SU Biotec for providing
images used to create Fig 30.3. and BioRender (BioRender.com), which was used to create
Fig. 30.4.

References

1. Reynolds MP, Hays D, Chapman S (2010) Breeding for adaptation to heat and drought stress. In: Climate change and crop production. CABI Publishing, Wallingford, pp 71–91
2. Bentley A, Mackay I (2017) Advances in wheat breeding techniques. In: Langridge P (ed) Achieving sustainable cultivation of wheat: breeding, quality traits, pests and diseases. Burleigh Dodds Science Publishing, Cambridge, pp 53–76
3. Araus JL, Kefauver SC, Zaman-Allah M, Olsen MS, Cairns JE (2018) Translating high-throughput phenotyping into genetic gain. Trends Plant Sci 23:451–466. https://doi.org/10.1016/j.tplants.2018.02.001
4. Bassi FM, Bentley AR, Charmet G, Ortiz R, Crossa J (2016) Breeding schemes for the implementation of genomic selection in wheat (Triticum spp.). Plant Sci 242:23–36. https://doi.org/10.1016/j.plantsci.2015.08.021
5. Rutkoski J, Poland J, Mondal S, Autrique E, Pérez LG, Crossa J, Reynolds M, Singh R (2016) Canopy temperature and vegetation Indices from high-throughput phenotyping improve accuracy of pedigree and genomic selection for grain yield in wheat. G3 6:2799–2808. https://doi.org/10.1534/g3.116.032888
6. Edwards SM, Buntjer JB, Jackson R, Bentley AR, Lage J, Byrne E, Burt C, Jack P, Berry S, Flatman E, Poupard B, Smith S, Hayes C, Gaynor RC, Gorjanc G, Howell P, Ober E, Mackay IJ, Hickey JM (2019) The effects of training population design on genomic prediction accuracy in wheat. Theor Appl Genet 132:1943–1952. https://doi.org/10.1007/s00122-019-03327-y
7. Ortiz R, Trethowan R, Ferrara GO, Iwanaga M, Dodds JH, Crouch JH, Crossa J, Braun H-J (2007) High yield potential, shuttle breeding, genetic diversity, and a new international wheat improvement strategy. Euphytica 157:365–384. https://doi.org/10.1007/s10681-007-9375-9
8. Hickey LT, Hafeez AN, Robinson H, Jackson SA, Leal-Bertioli SCM, Tester M, Gao C, Godwin ID, Hayes BJ, Wulff BBH (2019) Breeding crops to feed 10 billion. Nat Biotechnol 37:744–754. https://doi.org/10.1038/s41587-019-0152-9
9. Srivastava P, Bains NS (2018) Accelerated wheat breeding: doubled haploids and rapid generation advance. In: Gosal SS, Wani SH (eds) Biotechnologies of crop improvement, volume 1: cellular approaches. Springer, Cham, pp 437–461
10. Borlaug NE (2007) Sixty-two years of fighting hunger: personal recollections. Euphytica 157:287–297. https://doi.org/10.1007/s10681-007-9480-9
11. Tadesse W, Sanchez-Garcia M, Assefa SG, Amri A, Bishaw Z, Ogbonnaya FC, Baum M (2019) Genetic gains in wheat breeding and its role in feeding the world. Crop Breed Genet Genomics 1:e190005. https://doi.org/10.20900/cbgg20190005
12. Rajaram S (1995) Wheat breeding at CIMMYT: commemorating 50 years of research in Mexico for global wheat improvement. In: Wheat breeding at CIMMYT. CIMMYT, Ciudad Obregón, p 162
13. Ikeguchi S (2002) Present state of study for spring wheat breeding in canada and USA. Hokunou
14. Nonaka S (1979) Present state and problem of wheat and barley breeding with rapid generation advance. Recent Adv Breed 20:102–107
15. Singh RP, Herrera-Foessel S, Huerta-Espino J, Singh S, Bhavani S, Lan C, Basnet BR (2014) Progress towards genetics and breeding for minor genes based resistance to Ug99 and other rusts in CIMMYT high-yielding spring wheat. J Integr Agric 13:255–261. https://doi.org/10.1016/S2095-3119(13)60649-8
16. Zenkteler M, Nitzsche W (1984) Wide hybridization experiments in cereals. Theor Appl Genet 68:311–315. https://doi.org/10.1007/BF00267883
17. Bugbee B, Koerner G (1997) Yield comparisons and unique characteristics of the dwarf wheat cultivar 'USU-Apogee'. Adv Sp Res 20:1891–1894. https://doi.org/10.1016/S0273-1177(97)00856-9
18. Ghosh S, Watson A, Gonzalez-Navarro OE, Ramirez-Gonzalez RH, Yanes L, Mendoza-Suárez M, Simmonds J, Wells R, Rayner T, Green P, Hafeez A, Hayta S, Melton RE, Steed A, Sarkar A, Carter J, Perkins L, Lord J, Tester M, Osbourn A, Moscou MJ, Nicholson P, Harwood W,

Martin C, Domoney C, Uauy C, Hazard B, Wulff BBH, Hickey LT (2018) Speed breeding in growth chambers and glasshouses for crop breeding and model plant research. Nat Protoc 13:2944–2963. https://doi.org/10.1038/s41596-018-0072-z

19. Bhatta M, Sandro P, Smith MR, Delaney O, Voss-Fels KP, Gutierrez L, Hickey LT (2021) Need for speed: manipulating plant growth to accelerate breeding cycles. Curr Opin Plant Biol 60:101986. https://doi.org/10.1016/j.pbi.2020.101986

20. Watson A, Ghosh S, Williams MJ, Cuddy WS, Simmonds J, Rey MD, Asyraf Md Hatta M, Hinchliffe A, Steed A, Reynolds D, Adamski NM, Breakspear A, Korolev A, Rayner T, Dixon LE, Riaz A, Martin W, Ryan M, Edwards D, Batley J, Raman H, Carter J, Rogers C, Domoney C, Moore G, Harwood W, Nicholson P, Dieters MJ, Delacy IH, Zhou J, Uauy C, Boden SA, Park RF, Wulff BBH, Hickey LT (2018) Speed breeding is a powerful tool to accelerate crop research and breeding. Nat Plants 4:23–29. https://doi.org/10.1038/s41477-017-0083-8

21. Alahmad S, El Hassouni K, Bassi FM, Dinglasan E, Youssef C, Quarry G, Aksoy A, Mazzucotelli E, Juhász A, Able JA, Christopher J, Voss-Fels KP, Hickey LT (2019) A major root architecture QTL responding to water limitation in Durum wheat. Front Plant Sci 10:436. https://doi.org/10.3389/fpls.2019.00436

22. Riaz A, Hickey LT (2017) Rapid phenotyping adult plant resistance to stem rust in wheat grown under controlled conditions. In: Periyannan S (ed) Wheat rust diseases: methods and protocols. Springer, New York, pp 183–196

23. Dinglasan E, Godwin ID, Mortlock MY, Hickey LT (2016) Resistance to yellow spot in wheat grown under accelerated growth conditions. Euphytica 209:693–707. https://doi.org/10.1007/s10681-016-1660-z

24. Alahmad S, Dinglasan E, Leung KM, Riaz A, Derbal N, Voss-Fels KP, Able JA, Bassi FM, Christopher J, Hickey LT (2018) Speed breeding for multiple quantitative traits in durum wheat. Plant Methods 14:36. https://doi.org/10.1186/s13007-018-0302-y

25. Crossa J, Pérez-Rodríguez P, Cuevas J, Montesinos-López O, Jarquín D, de los Campos G, Burgueño J, Camacho-González JM, Pérez-Elizalde S, Beyene Y, Dreisigacker S, Singh R, Zhang X, Gowda M, Roorkiwal M, Rutkoski J, Varshney RK (2017) Genomic selection in plant breeding: methods, models, and perspectives. Trends Plant Sci xx:1–15. https://doi.org/10.1016/j.tplants.2017.08.011

26. Heffner EL, Sorrells ME, Jannink J-L (2009) Genomic selection for crop improvement. Crop Sci 49:1–12. https://doi.org/10.2135/cropsci2008.08.0512

27. Larkin DL, Lozada DN, Mason RE (2019) Genomic selection—considerations for successful implementation in wheat breeding programs. Agronomy 9. https://doi.org/10.3390/agronomy9090479

28. Gaynor RC, Gorjanc G, Bentley AR, Ober ES, Howell P, Jackson R, Mackay IJ, Hickey JM (2017) A two-part strategy for using genomic selection to develop inbred lines. Crop Sci 57:2372–2386. https://doi.org/10.2135/cropsci2016.09.0742

29. Ward BP, Brown-Guedira G, Tyagi P, Kolb FL, Van Sanford DA, Sneller CH, Griffey CA (2019) Multienvironment and multitrait genomic selection models in unbalanced early-generation wheat yield trials. Crop Sci 59:491–507. https://doi.org/10.2135/cropsci2018.03.0189

30. Hayes BJ, Panozzo J, Walker CK, Choy AL, Kant S, Wong D, Tibbits J, Daetwyler HD, Rochfort S, Hayden MJ, Spangenberg GC (2017) Accelerating wheat breeding for end-use quality with multi-trait genomic predictions incorporating near infrared and nuclear magnetic resonance-derived phenotypes. Theor Appl Genet 130:2505–2519. https://doi.org/10.1007/s00122-017-2972-7

31. Sun J, Poland JA, Mondal S, Crossa J, Juliana P, Singh RP, Rutkoski JE, Jannink J-L, Crespo-Herrera L, Velu G, Huerta-Espino J, Sorrells ME (2019) High-throughput phenotyping platforms enhance genomic selection for wheat grain yield across populations and cycles in early stage. Theor Appl Genet 132:1705–1720. https://doi.org/10.1007/s00122-019-03309-0

32. Thavamanikumar S, Dolferus R, Thumma BR (2015) Comparison of genomic selection models to predict flowering time and spike grain number in two hexaploid wheat doubled haploid populations. G3 5:1991–1998. https://doi.org/10.1534/g3.115.019745

33. Rutkoski J, Singh RP, Huerta-Espino J, Bhavani S, Poland J, Jannink JL, Sorrells ME (2015) Efficient use of historical data for genomic selection: a case study of stem rust resistance in wheat. Plant Genome 8. :plantgenome2014.09.0046. https://doi.org/10.3835/plantgenome2014.09.0046

34. Lopez-Cruz M, Crossa J, Bonnett D, Dreisigacker S, Poland J, Jannink J-L, Singh RP, Autrique E, de los Campos G (2015) Increased prediction accuracy in wheat breeding trials using a marker × environment interaction genomic selection model. G3 5:569–582. https://doi.org/10.1534/g3.114.016097

35. Lozada DN, Carter AH (2020) Genomic selection in winter wheat breeding using a recommender approach. Genes (Basel) 11:779. https://doi.org/10.3390/genes11070779

36. Daetwyler HD, Hayden MJ, Spangenberg GC, Hayes BJ (2015) Selection on optimal haploid value increases genetic gain and preserves more genetic diversity relative to genomic selection. Genetics 200:1341–1348. https://doi.org/10.1534/genetics.115.178038

37. Voss-Fels KP, Herzog E, Dreisigacker S, Sukumaran S, Watson A, Frisch M, Hayes B, Hickey LT (2019) "SpeedGS" to accelerate genetic gain in spring wheat. In: Miedaner T, Korzun V (eds) Applications of genetic and genomic research in cereals. Elsevier, Amsterdam, pp 303–327

Chapter 31
Improving Wheat Production and Breeding Strategies Using Crop Models

Jose Rafael Guarin and Senthold Asseng

Abstract Crop simulation models are robust tools that enable users to better understand crop growth and development in various agronomic systems for improved decision making regarding agricultural productivity, environmental sustainability, and breeding. Crop models can simulate many agronomic treatments across a wide range of spatial and temporal scales, allowing for improved agricultural management practices, climate change impact assessment, and development of breeding strategies. This chapter examines current applications of wheat crop models and explores the benefits from model improvement and future trends, such as integration of G × E × M and genotype-to-phenotype interactions into modeling processes, to improve wheat (*Triticum* spp.) production and adaptation strategies for agronomists, breeders, farmers, and policymakers.

Keywords Crop simulation model · Genetic improvement · Wheat yield · Adaptation strategies · Food security

31.1 Learning Objectives

- To showcase the importance, functionality, and advantages of utilizing wheat crop models for wheat production decision making and assistance in the development of breeding strategies.
- Outline future trends and areas of improvement needed in wheat crop models to mitigate future agricultural challenges.

J. R. Guarin (✉)
Department of Agricultural and Biological Engineering, University of Florida, Gainesville, FL, USA

Center for Climate Systems Research, Columbia University, New York, NY, USA
e-mail: j.guarin@columbia.edu

S. Asseng
Department of Life Science Engineering, Technical University of Munich, Freising, Germany
e-mail: senthold.asseng@tum.de

© The Author(s) 2022
M. P. Reynolds, H.-J. Braun (eds.), *Wheat Improvement*,
https://doi.org/10.1007/978-3-030-90673-3_31

31.2   Introduction

The global demand for food is continuously increasing due to the growing population and agricultural production strategies must continue to improve to ensure future global food security (see Chap. 4). Wheat (*Triticum* spp.) is the most important food crop in the world due to its high nutrition content contributing to approximately 20% of calories and protein in the human diet [1]. This chapter discusses potential agricultural strategies for improving wheat production by using crop simulation models as instruments for adaptation. Crop simulation models are computational tools used to determine crop growth and development using quantitative knowledge and input data from agronomic systems. Crop models produce dynamic simulations of crop phenology and growth based on the combined principles of crop physiology, soil science, and agrometeorology. Most crop models use daily maximum and minimum temperature, rainfall, solar radiation, atmospheric CO_2 concentration, soil attributes, cultivar characteristics, and crop management as inputs to simulate crop phenological development, biomass and yield accumulation, water use, and nutrient uptake.

Wheat crop modeling emerged in the 1980s based on the computational integration of mathematical relationships between environmental interactions and wheat growth developed in the 1960s and 1970s. In the 1990s, models began merging into crop modeling platforms, i.e., software that combines multiple models of various crops to facilitate evaluation and application for users [2]. Over the past decades, individual crop models and crop modeling platforms have received many technological and functional improvements which has promoted the development of new modeling approaches and new models. Currently, there are more than 30 wheat crop models used by modeling groups across the world [3]. These present-day crop models have progressed to become helpful tools in understanding how crops develop, grow, and yield in various agricultural scenarios which makes them advantageous for projecting how internal (e.g., genes, traits) and external (e.g., climate, crop management) factors impact crop production [4]. Thus, crop models are often integrated into new agricultural research activities, such as the improvement of breeding strategies. However, as knowledge and understanding of agricultural systems increases, model improvement is necessary to address the new challenges that arise when simulating complex agricultural systems interactions.

This chapter outlines the benefits and limitations of current wheat crop models for applications of wheat production regarding environmental sustainability, climate change, and breeding.

31.3 Assisting Breeding with Crop Modeling

Crop models can extrapolate data beyond field experimentation through simulation of different agronomic treatments with many variables over extended periods, allowing for agricultural systems studies across a wide range of spatial and temporal scales [5]. For example, wheat crop models can simulate the effects between climate and natural resources management (NRM) using various inputs (e.g., sowing dates, N fertilizer amounts, irrigation rates, etc.) to assist in the development of sustainable best management practices. This allows model users to maximize economic return at specified locations while managing spatial variability, environmental impact, and natural resources availability. Wheat crop models can also use seasonal forecasts and historical weather data to simulate a wide range of management practices for optimization of management strategies for predicted season types (e.g., increasing N fertilization in a predicted wet season) to improve efficiency and profitability of wheat production [5]. Additionally, wheat crop models are often paired with Global Climate Models (GCMs) to assess the impact of future climate change on wheat production under different Representative Concentration Pathway (RCP) scenarios of projected rainfall, temperatures (including extreme events), elevated atmospheric CO_2 concentrations, and increased tropospheric ozone (O_3) concentrations at local and global scales [6].

In addition to adaptive crop management, another strategy to increase global wheat production is through wheat breeding and genetic engineering of new stress resilient cultivars with improved resource use efficiencies [7, 8]. Breeding new wheat cultivars with desired traits such as heat, drought, or O_3 tolerance can mitigate the negative effects of climate change and improve overall agricultural productivity (see Chap. 7); however, producing a new wheat cultivar requires about 8–12 years to develop [9]. Wheat crop models can assist breeders to accelerate the development of improved cultivars by simulating many treatments of breeding programs in current and future environments with various crop management practices [10]. Figure 31.1 illustrates the interactions between breeding and crop modeling. Combining wheat breeding strategies with wheat models allows for the (1) characterization of wheat growing environments to better understand environmental variability, (2) assessment of physiological trait performance in targeted environments to focus on favorable traits, (3) evaluation of the potential effects of genetic controls on wheat yield in specified environments, (4) improved understanding of genotype × environment (G × E) interactions in statistical models, and (5) utilization of high-throughput phenotyping to identify traits of interest [4].

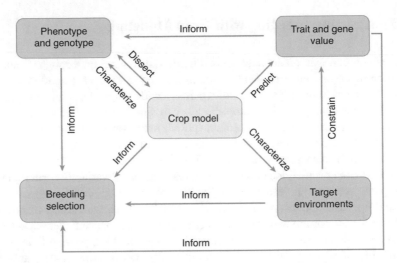

Fig. 31.1 Diagram describing the role of crop modeling in breeding. Crop models can be used to (1) dissect and characterize complex traits into simpler components traits, (2) simulate the impact of traits and genetic controls on crop growth and yield, (3) characterize environments of the target population of environments to identify the main environment types and their frequency, and (4) inform breeding via integrated analyses with breeding system models. (Reprinted with permission from Ref. [4])

31.3.1   Cultivars and Traits

Interactions between physiological traits within a wheat cultivar cause seasonal variations in wheat fecundity and nutrition. Wheat crop models can estimate the effect of single and combined agronomic traits on wheat growth and yield and can approximate the degree of additivity or reduction of the traits. Models can simulate multiple trait combinations of wheat grown in various *in silico* environments to observe if the traits will have positive or negative effects on crop growth, yield, or protein content [10]. The use of wheat models in breeding facilitates the testing of new wheat ideotypes, i.e., sets of wheat cultivar parameters that mimic the genotype for an "ideal wheat type." Wheat models can simulate potential ideotypes under various treatments so that they are optimized for specified locations and future adaptation needs, while using less time and resources than field experiments [11]. Wheat models have been used to simulate grain yields for possible wheat ideotypes under future climate scenarios and found combinations of specific traits that could lead to large yield increases in certain environments [12]. However, appropriate understanding and caution should be exercised when using wheat models to simulate ideotypes because (1) models cannot account for all the interactions among traits, (2) variations in the simulated traits may not represent existing genetic variability, and (3) simulated combinations of traits may not be physiologically or genetically possible in the field [4].

31.3.2   Simulating Genotype × Environment × Management (G × E × M) Interactions

For agricultural productivity to increase in a changing climate, the interactions between genotype, environment, and management (G × E × M) on crop development, growth, and yield must be considered. Wheat crop models can simulate wheat growth, development, and yield using G × E × M interactions to assess the benefits and risks of different adaptation strategies in different environmental scenarios (Fig. 31.2). Wheat models are ideal tools for determining the benefits of various genetic improvement strategies, but models are often limited by uncertainties related to processes and parameters when simulating genetic variations [11]. This is because modeling wheat phenology requires algorithms with cultivar-specific parameters, but parameter estimation for large numbers of genotypes can be time-consuming and costly [13], limiting the rate at which new genotypes are incorporated into models. Some studies have linked parameters with genetics, such as modifying a wheat model to incorporate gene effects into the estimation of wheat heading time [14]. This resulted in a gene-based model that showed new longer-season cultivars may

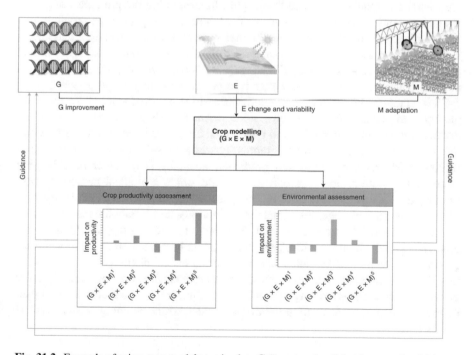

Fig. 31.2   Example of using crop models to simulate G (genotype) × E (environment) × M (management) interactions for climate change impact assessment. When optimizing existing and designing new agricultural adaptation strategies, lessons from both crop productivity and environmental sustainability assessments should be used as guidance. Brown lines and arrows indicate guidance from crop productivity assessment and green lines and arrows indicate guidance from environmental assessment. (Reprinted with permission from Ref. [8])

flower later when sown early in a season leading to potential yield increases. Recently, a multiscale (gene to globe) modeling framework has been developed to assess different adaptation strategies, including genetic improvement, under climate change at larger scales [8].

31.3.3   Integrating Genotype-to-Phenotype Interactions

A crop phenotype is the observable expression of the genes, i.e., specific traits/ characteristics such as crop structure, development, physiological properties, phenology, or behavior, resulting from the interaction of the crop genotype (genetic structure) and the environment (G × E interactions). Understanding wheat genotypes and phenotypes allows breeders to develop new and/or improved wheat cultivars, such as abiotic stress-tolerant cultivars, for improved agricultural productivity. Linking genotype-to-phenotype relationships in crop modeling is a novel area, but there is considerable potential to agricultural and breeding improvement strategies because wheat crop models can simulate multiple agronomic traits in various environmental and management conditions [14]. Incorporating the principles of genetics and genomics into wheat models enables genotype-to-phenotype prediction so agronomists and breeders can select the ideal cultivar for target locations and needs. The advantages to modeling with integrated physiological traits are that (1) model parameters may be more closely linked to Quantitative Trait Loci (QTLs) and genes, and (2) complex traits may be better represented as emergent properties arising from interactions between component traits and the environment [4]. To integrate genotype-to-phenotype interactions, some wheat models attempted a 'top-down' approach to simulate dynamics and phenotypic outcomes from genetic variation at the whole-crop scale with course granularity while considering phenotyping capabilities, measurement errors, and prediction accuracy [15]. Other models have attempted a 'bottom-up' approach to integrate single biological processes (e.g., photosynthesis) operating at different temporal and spatial scales by explicitly simulating gene network regulation, metabolic reactions, and metabolite transport to be scaled to the whole-crop level [16]. However, before scaling the 'bottom-up' approach to the whole-crop level, challenges must be addressed in the model integration of component modules and representation of phenotypic responses to G × E × M interactions. Studies have suggested multiscale modeling framework for combining these two modeling approaches to simulate the whole-crop level impact of trait manipulation at finer (gene-to-cell) scales while retaining model predictability [8].

31.3.4   Improvements for G × E × M and Genotype-to-Phenotype Interactions in Crop Models to Assist Agronomists and Breeders

The ability of wheat crop models to simulate G × E × M interactions and support genotype-to-phenotype predictions offers large potential to facilitate genetic improvement (e.g., dissection and understanding of complex traits) and the development of breeding and management adaptation strategies. However, challenges remain when simulating G × E × M and genotype-to-phenotype interactions in current wheat crop models because of (1) uncertainty in the representation of key physiological processes leading to accumulated uncertainty in simulated resource transport, growth, and yield, (2) differences between model parameters and underlying genetic interactions, (3) difficulty in quick and accurate phenotyping for model parameters, (4) limited availability of detailed quantitative data of the interaction between the genetic controls (genotype) and physiological processes in response to environmental and management changes, and (5) limited information on gene/QTL function, the genotypes characterized for specific gene/QTL, and their interactions [17, 18].

Current wheat crop models can be generalized into different levels to show similarities and differences in model processes and parameterization across the levels (Fig. 31.3) [17]. Most wheat models are considered 'level 3' cultivar models, where models describe cultivar differences of species using cultivar specific parameters. Models that link model parameters to gene effects and QTL for limited genotypes or traits are considered 'level 4' genotype models [14]. There have been several attempts to represent gene effects on phenological development as a 'level 5' model, e.g., 'top-down' or 'bottom-up' approaches or through capturing the interactive effects of specific genes on wheat vernalization [19]. Models that are considered 'level 4' or 'level 5' provide a better representation of genetic interaction with the environment and can assist in the development of breeding strategies. To increase the number of wheat models in 'level 4' and 'level 5', Wang et al. [18] outlined three stages of model improvement: (1) improving physiological understanding and modification of physiological algorithms to simulate subprocesses for trait prediction and evaluation, (2) linking model parameters to phenotypic responses of genetic variation using genomic data and identification of QTLs, and (3) modifying model structure to better represent physiological feedback and gene-level understanding. Figure 31.3 illustrates the incorporation of these areas of improvement at each model level. In order to incorporate these improvements, a new software design, improved crop model (that accurately represents crop responses to different environmental and management conditions), and statistical model (that links crop model parameters to genetic information) are recommended.

Improving genotype-to-phenotype prediction requires understanding and development of algorithms that represent the underlying genetic structure to generate a phenotype of a crop based on simulated dynamics. Recent development of new technologies for high-throughput phenotyping in both controlled environments and

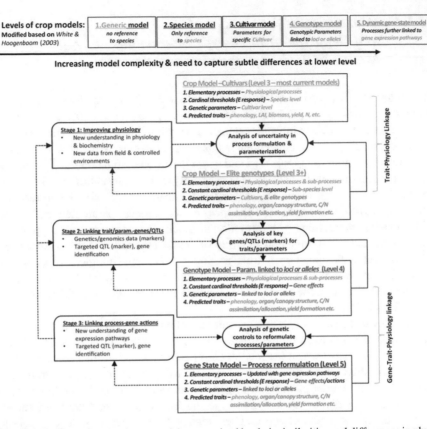

Fig. 31.3 Different levels of crop models categorized by their similarities and differences in physiological processes and parameters, with additional details on the three interactive stages to improve current process-based wheat crop models for simulation of gene effects and G × E × M interactions. Dotted lines with arrows indicate information feeding loops, and solid lines with arrows indicate model development loops. (Modified with permission from Ref. [18])

in the field (see Chap. 27) will open additional possibilities to improve genotype-to-phenotype prediction in models [20]. Additionally, breeding support systems, such as the Genomic Open-source Breeding informatics initiative (GOBii) (http://cbsug-obii05.biohpc.cornell.edu/wordpress/) or the Integrated Breeding Platform (IBP) (https://www.integratedbreeding.net/), are developing databases and software tools to maintain and organize large quantities of genomic data to facilitate efficient crop cultivar selection which will also benefit genotype-to-phenotype model development. However, the added rigor to improve genotype-to-phenotype predictions should balance simplicity and model complexity so that parameterization does not become problematic [15].

31.3.5 Identifying Target Regions for Breeding

Climate change is likely to shift the target population of environments (TPE), i.e., the areas and seasons in which cultivars produced by breeding programs will be grown (see Chap. 3), for wheat cultivars which could limit the genetic gain from breeding programs [4]. Integrated modeling approaches that characterize TPE for projected future climates would help breeders target genotypes, traits, and regions of interest for new cultivars. Wheat crop models paired with GCMs can simulate the G × E × M interactions of wheat cultivars for TPE under future climate scenarios. Additionally, wheat models can characterize TPE at large-scales and/or over long periods of time to estimate spatial and temporal variability. For example, studies in the North China Plain using wheat models found that the addition of new winter wheat cultivars could prolong the growing period which reduced the negative warming effects from climate change in this region [21]. Also, dynamic in-season modeling can target specific stress patterns of interest to improve relevant field phenotyping [22].

31.4 Limitations and Improvements in Crop Model Performance

The expanding ability of wheat crop models to assess agricultural management, environmental, climate change, and breeding adaptation strategies on various scales (e.g., points-regions-global) has advanced agricultural science and opened new areas of interdisciplinary research. This expansion of research has raised many novel questions and challenges, such as the issue of climate change and food security on a global scale. Currently, the main challenges facing wheat modeling are improving model development to (1) incorporate the phenotypic effects of genotypes on physiological processes, (2) enhance simulations of management consequences (e.g., leaching of nitrate and pesticides) and soil constraints (e.g., acidity, salinity, access water), (3) enhance simulations of physiological responses to compound climate factors (e.g., CO_2 × temperature interactions), (4) enhance simulations of the impacts from extreme climatic events (e.g., heat shocks, drought, elevated CO_2, frost), (5) incorporate biotic stresses such as weeds, pests, and diseases, and (6) incorporate grain quality aspects (e.g., grain protein content and composition) and nutrition (e.g., Zn, Fe). Additionally, incorporation of detailed physiological processes (e.g., respiration costs, role of hormones in signaling environmental factors, and partitioning of carbon, especially to the roots) may help to improve current and future model performance; however, inclusion of additional model parameters does not always improve model precision [23]. Therefore, these challenges require improvements in model processes and interactions determined through model testing and evaluation with comprehensive and detailed observations. Addressing these challenges will help wheat models to continue providing a key role in guiding future adaptation advancements of wheat crop systems.

31.4.1 $CO_2 \times$ Temperature Interactions

Additional understanding of the interactions between climate factors with other environmental factors, extreme events, and crop feedback (e.g., source-sink relationships) is needed to simulate wheat production. Before assessing the impacts of these combined interactions, the impact of climate variables on wheat yield should be separated and tested individually. It is well known that elevated atmospheric CO_2 concentrations stimulate crop growth through improved photosynthetic capacity and transpiration efficiency (TE), and that higher temperatures decrease wheat productivity (see Chaps. 10 and 22). Most wheat crop models can simulate the interactions between atmospheric CO_2 and temperature on wheat growth; however, multi-model grain yield simulations have been shown to diverge under higher temperatures [3] or elevated atmospheric CO_2 [24], which highlights the need for model improvement in response to both high temperatures and elevated atmospheric CO_2. A challenge for many wheat models is the lack of sensitivity to short-term stresses of one to two days related to extreme events that can affect yield-determining processes. Therefore, it is necessary to test and improve wheat model processes with data from field experimentation examining the impacts of climate factors at different developmental stages. In addition to simulating combined climate factors, it is important for wheat models to consider single or infrequent extreme events such as hailstorms, floods, or wind gusts as these can severely limit wheat yields. Climate change adds an additional modeling challenge because predicting the frequency of extreme events is difficult and extreme events are projected to become more variable [25]. Improving climate model projections of extreme events and the simulated physiological effects of individual and combined climatic factors on wheat growth within wheat models is necessary for developing future adaptation strategies.

31.4.2   Frost Stress

Severe yield losses in wheat production can be caused by extreme low temperatures or frost/chilling stress, an extended period of low temperatures (<2 °C), especially at the reproductive stage. Several wheat crop models include a "frost-kill" threshold, where simulated wheat crops start to fail when daily minimum temperatures fall below a critical low temperature threshold (e.g., <-10 °C) [2]. However, many models do not account for the effects of accumulated frost stress under low temperatures. It is necessary to incorporate these effects from frost stress into wheat models as future climate change can cause high variability in seasonal and daily temperatures adding potential risk to certain wheat producing areas [25]. Models that include frost stress functionality have been able to assess the potential impact from frost stress, such as quantifying the risk from frost in the major wheat growing areas of China [26].

31.4.3 O_3 Stress

O_3 is a ubiquitous secondary pollutant that can negatively impact wheat development and yield, especially since wheat is the most sensitive crop to O_3 stress [27]. Future global O_3 concentrations are projected to increase due to increased amounts of O_3 precursor emissions [25]. Recently, crop modeling studies have shown that the negative impact from future O_3 concentrations on wheat production can be comparable to, or larger than, the combined climate change impact from atmospheric CO_2, temperature, and rainfall depending on location [6]. However, many wheat models do not consider the effects from O_3 stress (or the combined interactions between O_3 and water and/or CO_2) on development and growth. These effects are often not included in models because of limited O_3 data availability. However, the new Tropospheric Ozone Assessment Report (TOAR) database contains the world's most extensive collection of global O_3 observations from 1970 to present and could alleviate O_3 data limitations [28]. Additionally, the Agricultural Modeling Intercomparison and Improvement Project (AgMIP) Ozone modeling community facilitates O_3 data collection and multi-model ensemble studies, which can help incorporate O_3 effects into models [27]. The inclusion of O_3 effects in wheat models will improve climate change impact assessment and the development of future adaptation strategies [6].

31.4.4 Weeds, Pests, and Diseases

Biotic factors such as weeds, pests, and diseases can severely limit wheat health and yield quality causing global losses in wheat production of approximately 28% [29]. A major challenge in global wheat production is limiting the negative effects and spread of weeds/pests/diseases such as Fusarium head blight (FHB, also known as scab). Wheat is highly susceptible to infection from FHB and other diseases, especially in warm and wet climates experiencing frequent precipitation, high humidity, or heavy dews (see Chaps. 8 and 9). Additionally, the increased weather variability caused by climate change creates another challenge when determining disease risk areas [30]. Combining wheat crop models with weed/pest/disease models provides a method to simulate the effects of weeds/pests/diseases on crops while accounting for future climatic changes. However, modeling the occurrence, movement, and dynamics of weeds/pests/diseases in relation to crops and their dynamic interactions is still a challenge which is why few crop models estimate the effects of weeds/pests/diseases.

Several studies have been conducted to link weed/pest/disease models to wheat crop models through modification of crop model processes and algorithms. Wheat crop models have been combined with weed competition models to evaluate model performance and estimate the effect of weeds in dryland and irrigated treatments in Australia [31]. Wheat models have been linked with disease models to estimate the seasonal consequences from FHB on wheat production under variable climatic conditions in southern Brazil [32]. In addition to FHB, framework for linking wheat crop models to other pest population models has been developed to estimate impacts from other major pests and diseases such as grain aphids (*Sitobion avenae*), eyespot, and rust (*Puccinia striiformis*) [33]. To improve simulated impacts caused by weeds/pests/diseases in crop models, several steps have been suggested: (1) improvement in weed/pest/disease data quality and availability for model inputs and evaluation, (2) improvement in integration of weed/pest/disease and crop physiological interactions, (3) development of standard criteria for weed/pest/disease model evaluation, and (4) development of a community for weed/pest/disease modelers for sharing of information and resources [30]. Although there are still many challenges in linking weed/pest/disease population models with wheat crop models, the ability of a wheat model to simulate G × E × M interactions while also accounting for weed/pest/disease risk provides an opportunistic goal to assist breeders in developing pest or disease resistant wheat cultivars for the future.

31.4.5   Grain Quality

Climate change will affect grain yield quality (e.g., grain protein concentration; see Chap. 11) which poses a major challenge for global food security [34]. Grain protein concentration, the ratio of grain protein amount to grain yield, is an important characteristic for evaluating the nutritional quality of wheat yield since wheat contributes about 20% of protein for global human consumption [1]. Grain protein concentrations depend on a combination of G × E factors, e.g., grain protein concentrations are negatively affected by elevated atmospheric CO_2 concentrations but increase under higher temperatures and drought stress due to lower starch accumulation. Wheat crop models could help determine the nutritional quality of grain yield under different G × E × M conditions. Some wheat models can simulate grain protein, but many models require a better understanding and/or integration of the physiology of yield quality components [10]. Recently, studies using wheat models have shown that climate change adaptation strategies that benefit grain yield may not be beneficial for grain quality depending on the environmental and input conditions [7].

31.5 Collaborative Global Crop Modeling Networks

Individual wheat crop models are powerful tools and provide a useful method to determine wheat growth in various environments and conditions as previously described. However, models are abstract representations of reality and many uncertainties and limitations exist within the processes and simulated interactions of each model. The use of multiple crop models in multi-model ensemble studies can improve overall accuracy and reduce uncertainty of simulated results [35]. In addition to improved accuracy and reduced uncertainty, another major benefit of using multi-model ensembles is that the comparison and evaluation of multiple models simulating the same scenario can highlight limitations or issues within individual crop models leading to model improvement. This improvement of individual models will then improve the overall accuracy of future multi-model ensemble studies.

Using multiple crop models in model intercomparison programmes for climate impact assessments is an auspicious method for projecting future crop productivity and for comparing results between modeling groups. The AgMIP (www.agmip.org) is a major international collaborative effort to combine interdisciplinary modeling communities with state-of-the-art information technology for the goal of significantly improving climate impact projections, crop models, and economic models for agricultural advancement and sustainability at local and global scales [36]. The AgMIP initiative established detailed protocols for simulating crop models, emissions scenarios, and GCMs on a global scale to facilitate the use of multi-model ensemble studies in climate change impact assessments. The multi-model impact assessment studies produced by global collaborative efforts, such as AgMIP, allow farmers, scientists, stakeholders, and policymakers to make improved decisions and adaptation strategies for future agricultural challenges.

31.6 Case Study – Using Crop Models to Determine the Effects of Genetic Adaptations

An increasing amount of agricultural studies are examining the impacts of climate change on grain yield, but few focus on the impacts on the nutritional content of the grain [34]. As mentioned earlier, grain protein concentration is an important characteristic for evaluating the nutritional quality of wheat yield (Sect. 31.3.5). The recent study by Asseng et al. [7] estimated the effects of climate change on wheat protein concentration for the main wheat producing areas across the world as part of the AgMIP. The study used a multi-model ensemble of 32 wheat crop models to simulate the combined effects of temperature, CO_2, water, and nitrogen (N) on wheat protein concentration while including a trait adaptation option of delayed anthesis with an increased grain filling rate. Sixty major wheat-growing locations were examined, thirty high-rainfall or irrigated locations (simulated with no N or water limitations) and thirty low-rainfall locations, to represent total global wheat production. The wheat models used projected climate data from 2040 to 2069 provided by 5 GCMs under RCP8.5. Model testing was done using outdoor chamber and Free-Air Carbon dioxide Enrichment (FACE) experiments to evaluate model performance of heat shocks, increased temperature, and elevated atmospheric CO_2 concentrations. The trait adaptation option of delayed anthesis with increased grain filling rate was determined from a wide range of observed field experiments at different locations across the world.

After confirming that the multi-model ensemble median produced acceptable results, the multi-model median impact of climate change on grain and protein yield at the sixty locations was determined with and without the genotypic adaptation option (Fig. 31.4). The study found that grain yields were improved in most locations with the trait combination of delayed anthesis and increased grain filling rate (Fig. 31.4b). However, the response of grain protein concentration was more variable and dependent upon the growing season and location. It was found that climate change and the combined trait adaptation could lead to an increase in grain protein concentration at low-rainfall locations, particularly where yield was projected to decline. After aggregating the sixty locations to the global scale, it was determined that the inclusion of the trait combination of delayed anthesis and increased grain filling rate could increase global yield 7% and protein yield 2% by 2050. However, this inclusion of the trait combination would decrease grain protein concentration by a relative change of −4%. This study shows the complex relationship of climate change and genetic adaptations on crop yield and illustrates the robust benefits of linking crop modeling and breeding disciplines for the development of adaptation strategies.

31.7 Key Concepts

- Utilization of wheat crop models can assist agronomists and breeders in the development of new wheat ideotypes and breeding strategies through dynamic simulations of many agronomic treatments across various spatial and temporal scales.
- Incorporation of G × E × M and genotype-phenotype interactions into wheat crop models is an emerging area of interest with high potential agricultural benefits.
- Wheat crop models are robust instruments for agricultural adaptation that have steadily improved over recent decades, but further improvement is needed to address future agricultural challenges.

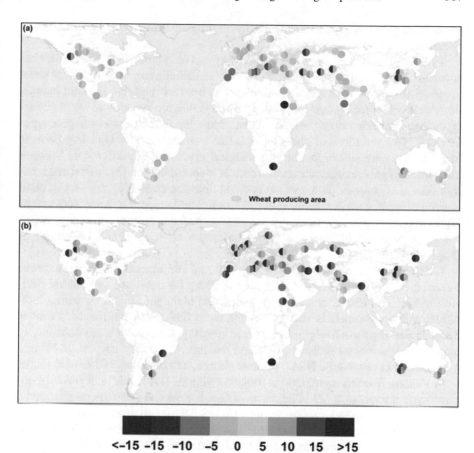

Fig. 31.4 Simulated multi-model ensemble projection under climate change of global wheat grain yield (left half) and protein yield (right half), (**a**) without genotypic adaptation and (**b**) with genotypic adaptation. Relative climate change impacts for 2036–2065 under RCP8.5 compared with the 1981–2010 baseline. Impacts were calculated using the medians across 32 models (or 18 for protein yield estimates) and five GCMs (circle color) and the average over 30 years of yields using region-specific soils, cultivars, and crop management. (Reprinted with permission from Ref. [7])

31.8   Conclusion

The challenge of supplying food to the expanding global population in an increasingly hostile climate with decreasing resources and cropping area will require advancement in many agricultural practices. Crop simulation models are powerful tools that produce dynamic simulations of crop growth and environmental interactions across various spatial and temporal scales using the principles of crop physiology, soil science, and agrometeorology. These simulations of agronomic treatments help farmers, scientists, and policymakers develop adaptation strategies to increase agricultural productivity and sustainability for projected climatic and environmental scenarios.

The emerging linkage between the breeding and crop modeling communities has accentuated the ability of crop models to assist in defining target ideotypes for novel breeding strategies. Wheat models simulate environment and management interactions well, and some models have incorporated genotypic variation of select phenotypes, e.g., flowering time [14]. However, simulating differences across genotypes and genotypic variations of phenotypes is still a challenge. This is mainly because of (1) uncertainty within model physiological processes, (2) differences between integration of genetic interactions and current model structure, (3) limited availability of detailed genomic data, and (4) cost and time constraints for large-scale, rapid phenotyping of complex traits (Sect. 31.3.4). The burgeoning improvements of high-throughput phenotyping allow researchers to focus on many new physiological traits, which will improve model development, testing, and evaluation of G × E × M interactions.

Collaborative global networks, like AgMIP, and multi-model ensembles improve overall model performance and accuracy allowing for improved agricultural decision making. Multi-model ensemble studies can highlight limitations within individual wheat crop models and addressing these limitations will further enhance decision making for wheat production and breeding. Crop models are successfully applied in agricultural research, decision making, and policy support, but future applications in agronomy, NRM, climate change, and breeding will require model improvement through improved (1) understanding of G × E × M and genotype-to-phenotype interactions, (2) integration of G × E × M and genotype-to-phenotype interactions into model processes, and (3) availability of high quality, detailed genomic, climatic, and management data sets. Tackling these challenges will improve crop model performance and assist farmers, agronomists, breeders, and policymakers to develop improved agricultural adaptation strategies.

References

1. Shiferaw B, Smale M, Braun HJ, Duveiller E, Reynolds M, Muricho G (2013) Crops that feed the world 10. Past successes and future challenges to the role played by wheat in global food security. Food Secur 5:291–317. https://doi.org/10.1007/s12571-013-0263-y
2. Jones JW, Hoogenboom G, Porter CH, Boote KJ, Batchelor WD, Hunt LA, Wilkens PW, Singh U, Gijsman AJ, Ritchie JT (2003) The DSSAT cropping system model. In: European Journal of Agronomy. Elsevier, Amsterdam, pp 235–265
3. Asseng S, Ewert F, Martre P, Rötter RP, Lobell DB, Cammarano D, Kimball BA, Ottman MJ, Wall GW, White JW, Reynolds MP, Alderman PD, Prasad PVV, Aggarwal PK, Anothai J, Basso B, Biernath C, Challinor AJ, De Sanctis G, Doltra J, Fereres E, Garcia-Vila M, Gayler S, Hoogenboom G, Hunt LA, Izaurralde RC, Jabloun M, Jones CD, Kersebaum KC, Koehler AK, Müller C, Naresh Kumar S, Nendel C, O'leary G, Olesen JE, Palosuo T, Priesack E, Eyshi Rezaei E, Ruane AC, Semenov MA, Shcherbak I, Stöckle C, Stratonovitch P, Streck T, Supit I, Tao F, Thorburn PJ, Waha K, Wang E, Wallach D, Wolf J, Zhao Z, Zhu Y (2015)

Rising temperatures reduce global wheat production. Nat Clim Chang 5:143–147. https://doi. org/10.1038/nclimate2470

4. Chenu K, Porter J, Martre P, Basso B, Chapman S, Ewert F, Bindi M, Asseng S (2017) Contribution of crop models to adaptation in Wheat. Trends Plant Sci 22:472–490. https://doi. org/10.1016/j.tplants.2017.02.003

5. Guarin JR, Asseng S (2017) Wheat crop modelling to improve yields. In: Langridge P (ed) Achieving sustainable cultivation of wheat volume 2: cultivation techniques, vol 2. Burleigh Dodds Science Publishing, Cambridge

6. Guarin JR, Emberson L, Simpson D, Hernandez-Ochoa IM, Rowland D, Asseng S (2019) Impacts of tropospheric ozone and climate change on Mexico wheat production. Clim Change 155:157–174. https://doi.org/10.1007/s10584-019-02451-4

7. Asseng S, Martre P, Maiorano A, Rotter R, O'Leary G, Fitzgerald G, Girousse C, Motzo R, Giunta F, Babar M, Reynolds M, Kheir A, Thorburn P, Waha K, Ruane A, Aggarwal P, Ahmed M, Balkovic J, Basso B, Biernath C, Bindi M, Cammarano D, Challinor A, De Sanctis G, Dumont B, Rezaei E, Fereres E, Ferrise R, Garcia-Vila M, Gayler S, Gao Y, Horan H, Hoogenboom G, Izaurralde R, Jabloun M, Jones C, Kassie B, Kersebaum K, Klein C, Koehler A, Liu B, Minoli S, San Martin M, Muller C, Kumar S, Nendel C, Olesen J, Palosuo T, Porter J, Priesack E, Ripoche D, Semenov M, Stockle C, Stratonovitch P, Streck T, Supit I, Tao F, Van der Velde M, Wallach D, Wang E, Webber H, Wolf J, Xiao L, Zhang Z, Zhao Z, Zhu Y, Ewert F (2019) Climate change impact and adaptation for wheat protein. Glob Chang Biol 25:155–173. https://doi.org/10.1111/gcb.14481

8. Peng B, Guan K, Tang J, Ainsworth E, Asseng S, Bernacchi C, Cooper M, Delucia E, Elliott J, Ewert F, Grant R, Gustafson D, Hammer G, Jin Z, Jones J, Kimm H, Lawrence D, Li Y, Lombardozzi D, Marshall-Colon A, Messina C, Ort D, Schnable J, Vallejos C, Wu A, Yin X, Zhou W (2020) Towards a multiscale crop modelling framework for climate change adaptation assessment. Nat Plants 6:338–348. https://doi.org/10.1038/s41477-020-0625-3

9. Fischer RA, Edmeades GO (2010) Breeding and cereal yield progress. Crop Sci 50;S85–S98. https://doi.org/10.2135/cropsci2009.10.0564

10. Martre P, He JQ, Le Gouis J, Semenov MA (2015) In silico system analysis of physiological traits determining grain yield and protein concentration for wheat as influenced by climate and crop management. J Exp Bot 66:3581–3598. https://doi.org/10.1093/jxb/erv049

11. Rotter RP, Tao F, Hohn JG, Palosuo T (2015) Use of crop simulation modelling to aid ideotype design of future cereal cultivars. J Exp Bot 66:3463–3476. https://doi.org/10.1093/jxb/erv098

12. Semenov MA, Stratonovitch P, Alghabari F, Gooding MJ (2014) Adapting wheat in Europe for climate change. J Cereal Sci 59:245–256. https://doi.org/10.1016/j.jcs.2014.01.006

13. Parent B, Tardieu F (2014) Can current crop models be used in the phenotyping era for predicting the genetic variability of yield of plants subjected to drought or high temperature? J Exp Bot 65:6179–6189. https://doi.org/10.1093/jxb/eru223

14. Zheng BY, Biddulph B, Li DR, Kuchel H, Chapman S (2013) Quantification of the effects of VRN1 and Ppd-D1 to predict spring wheat (Triticum aestivum) heading time across diverse environments. J Exp Bot 64:3747–3761. https://doi.org/10.1093/jxb/ert209

15. Hammer GL, van Oosterom E, McLean G, Chapman SC, Broad I, Harland P, Muchow RC (2010) Adapting APSIM to model the physiology and genetics of complex adaptive traits in field crops. J Exp Bot 61:2185–2202. https://doi.org/10.1093/jxb/erq095

16. Marshall-Colon A, Long SP, Allen DK, Allen G, Beard DA, Benes B, von Caemmerer S, Christensen AJ, Cox DJ, Hart JC, Hirst PM, Kannan K, Katz DS, Lynch JP, Millar AJ, Panneerselvam B, Price ND, Prusinkiewicz P, Raila D, Shekar RG, Shrivastava S, Shukla D, Srinivasan V, Stitt M, Turk MJ, Voit EO, Wang Y, Yin XY, Zhu XG (2017) Crops in silico: generating virtual crops using an integrative and multi-scale modeling platform. Front Plant Sci 8. https://doi.org/10.3389/fpls.2017.00786

17. White JW, Hoogenboom G (2003) Gene-based approaches to crop simulation: past experiences and future opportunitie. Agron J 95:52–64. https://doi.org/10.2134/agronj2003.0052

18. Wang EL, Brown HE, Rebetzke GJ, Zhao ZG, Zheng BY, Chapman SC (2019) Improving process-based crop models to better capture genotype × environment × management interactions. J Exp Bot 70:2389–2401. https://doi.org/10.1093/jxb/erz092
19. Brown HE, Jamieson PD, Brooking IR, Moot DJ, Huth NI (2013) Integration of molecular and physiological models to explain time of anthesis in wheat. Ann Bot 112:1683–1703. https://doi.org/10.1093/aob/mct224
20. Reynolds MP, Chapman S, Crespo-Herrera L, Molero G, Mondal S, Pequeno DNL, Pinto F, Pinera-Chavez FJ, Poland J, Rivera-Amado C, Saint Pierre C, Sukumaran S (2020) Breeder friendly phenotyping. Plant Sci 295:110396. https://doi.org/10.1016/j.plantsci.2019.110396
21. Liu YA, Wang EL, Yang XG, Wang J (2010) Contributions of climatic and crop varietal changes to crop production in the North China Plain, since 1980s. Glob Chang Biol 16:2287–2299. https://doi.org/10.1111/j.1365-2486.2009.02077.x
22. Rebetzke GJ, Chenu K, Biddulph B, Moeller C, Deery DM, Rattey AR, Bennett D, Barrett-Lennard EG, Mayer JE (2013) A multisite managed environment facility for targeted trait and germplasm phenotyping. Funct Plant Biol 40:1–13. https://doi.org/10.1071/fp12180
23. Challinor A, Martre P, Asseng S, Thornton P, Ewert F (2014) COMMENTARY: making the most of climate impacts ensembles. Nat Clim Chang 4:77–80. https://doi.org/10.1038/nclimate2117
24. Ahmed M, Stockle CO, Nelson R, Higgins S, Ahmad S, Raza MA (2019) Novel multimodel ensemble approach to evaluate the sole effect of elevated CO2 on winter wheat productivity. Sci Rep 9. https://doi.org/10.1038/s41598-019-44251-x
25. IPCC (2019) Climate Change and Land: an IPCC special report on climate change, desertification, land degradation, sustainable land management, food security, and greenhouse gas fluxes in terrestrial ecosystems
26. Xiao LJ, Liu LL, Asseng S, Xia YM, Tang L, Liu B, Cao WX, Zhu Y (2018) Estimating spring frost and its impact on yield across winter wheat in China. Agric For Meteorol 260:154–164. https://doi.org/10.1016/j.agrformet.2018.06.006
27. Emberson L, Pleijel H, Ainsworth E, van den Berg M, Ren W, Osborne S, Mills G, Pandey D, Dentener F, Buker P, Ewert F, Koeble R, Van Dingenen R (2018) Ozone effects on crops and consideration in crop models. Eur J Agron 100:19–34. https://doi.org/10.1016/j.eja.2018.06.002
28. Schultz MG, Schroder S, Lyapina O, Cooper OR, Galbally I, Petropavlovskikh I, von Schneidemesser E, Tanimoto H, Elshorbany Y, Naja M, Seguel RJ, Dauert U, Eckhardt P, Feigenspan S, Fiebig M, Hjellbrekke AG, Hong YD, Kjeld PC, Koide H, Lear G, Tarasick D, Ueno M, Wallasch M, Baumgardner D, Chuang MT, Gillett R, Lee M, Molloy S, Moolla R, Wang T, Sharps K, Adame JA, Ancellet G, Apadula F, Artaxo P, Barlasina ME, Bogucka M, Bonasoni P, Chang L, Colomb A, Cuevas-Agullo E, Cupeiro M, Degorska A, Ding AJ, FrHlich M, Frolova M, Gadhavi H, Gheusi F, Gilge S, Gonzalez MY, Gros V, Hamad SH, Helmig D, Henriques D, Hermansen O, Holla R, Hueber J, Im U, Jaffe DA, Komala N, Kubistin D, Lam KS, Laurila T, Lee H, Levy I, Mazzoleni C, Mazzoleni LR, McClure-Begley A, Mohamad M, Murovec M, Navarro-Comas M, Nicodim F, Parrish D, Read KA, Reid N, Nrl R, Saxena P, Schwab JJ, Scorgie Y, Senik I, Simmonds P, Sinha V, Skorokhod AI, Spain G, Spangl W, Spoor R, Springston SR, Steer K, Steinbacher M, Suharguniyawan E, Torre P, Trickl T, Lin WL, Weller R, Xu XB, Xue LK, Ma ZQ (2017) Tropospheric ozone assessment report: database and metrics data of global surface ozone observations. Elem Anthr 5:26. https://doi.org/10.1525/elementa.244
29. Oerke EC (2006) Crop losses to pests. J Agric Sci 144:31–43
30. Donatelli M, Magarey RD, Bregaglio S, Willocquet L, Whish JPM, Savary S (2017) Modelling the impacts of pests and diseases on agricultural systems. Agric Syst 155:213–224. https://doi.org/10.1016/j.agsy.2017.01.019
31. Deen W, Cousens R, Warringa J, Bastiaans L, Carberry P, Rebel K, Riha S, Murphy C, Benjamin LR, Cloughley C, Cussans J, Forcella F, Hunt T, Jamieson P, Lindquist J, Wang E

(2003) An evaluation of four crop: weed competition models using a common data set. Weed Res 43:116–129. https://doi.org/10.1046/j.1365-3180.2003.00323.x
32. Del Ponte EM, Jmc F, Pavan W, Baethgen WE (2009) A model-based assessment of the impacts of climate variability on Fusarium Head Blight seasonal risk in Southern Brazil. J Phytopathol 157:675–681. https://doi.org/10.1111/j.1439-0434.2009.01559.x
33. Whish JPM, Herrmann NI, White NA, Moore AD, Kriticos DJ (2015) Integrating pest population models with biophysical crop models to better represent the farming system. Environ Model Softw 72:52–64. https://doi.org/10.1016/j.envsoft.2014.10.010
34. Haddad L, Hawkes C, Webb P, Thomas S, Beddington J, Waage J, Flynn D (2016) A new global research agenda for food. Nature 540:30–32. https://doi.org/10.1038/540030a
35. Martre P, Wallach D, Asseng S, Ewert F, Jones JW, Rotter RP, Boote KJ, Ruane AC, Thorburn PJ, Cammarano D, Hatfield JL, Rosenzweig C, Aggarwal PK, Angulo C, Basso B, Bertuzzi P, Biernath C, Brisson N, Challinor AJ, Doltra J, Gayler S, Goldberg R, Grant RF, Heng L, Hooker J, Hunt LA, Ingwersen J, Izaurralde RC, Kersebaum KC, Mueller C, Kumar SN, Nendel C, O'Leary G, Olesen JE, Osborne TM, Palosuo T, Priesack E, Ripoche D, Semenov MA, Shcherbak I, Steduto P, Stoeckle CO, Stratonovitch P, Streck T, Supit I, Tao F, Travasso M, Waha K, White JW, Wolf J (2015) Multimodel ensembles of wheat growth: many models are better than one. Glob Chang Biol 21:911–925. https://doi.org/10.1111/gcb.12768
36. Rosenzweig C, Jones JW, Hatfield JL, Ruane AC, Boote KJ, Thorburn P, Antle JM, Nelson GC, Porter C, Janssen S, Asseng S, Basso B, Ewert F, Wallach D, Baigorria G, Winter JM (2013) The agricultural model intercomparison and improvement project (AgMIP): protocols and pilot studies. Agric For Meteorol 170:166–182. https://doi.org/10.1016/j.agrformet.2012.09.011

Chapter 32
Theory and Practice of Phenotypic and Genomic Selection Indices

José Crossa, J. Jesús Cerón-Rojas, Johannes W. R. Martini,
Giovanny Covarrubias-Pazaran, Gregorio Alvarado, Fernando H. Toledo,
and Velu Govindan

Abstract The plant net genetic merit is a linear combination of trait breeding values weighted by its respective economic weights whereas a linear selection index (LSI) is a linear combination of phenotypic or genomic estimated breeding values (GEBV) which is used to predict the net genetic merit of candidates for selection. Because economic values are difficult to assign, some authors developed economic weight-free LSI. The economic weights LSI are associated with linear regression theory, while the economic weight-free LSI is associated with canonical correlation theory. Both LSI can be unconstrained or constrained. Constrained LSI imposes restrictions on the expected genetic gain per trait to make some traits change their mean values based on a predetermined level, while the rest of the traits change their values without restriction. This work is geared towards plant breeders and researchers interested in LSI theory and practice in the context of wheat breeding. We provide the phenotypic and genomic unconstrained and constrained LSI, which together cover the theoretical and practical cornerstone of the single-stage LSI theory in plant breeding. Our main goal is to offer researchers a starting point for understanding the core tenets of LSI theory in plant selection.

Keywords Canonical correlation · Linear regression · Selection response

Gregorio Alvarado: To the memory of 'Goyito' our forever dear friend, and unique colleague.

J. Crossa
Colegio de Postgraduados (COLPOS), Montecillos, Mexico

International Maize and Wheat Improvement Center (CIMMYT), Texcoco, Mexico
e-mail: j.crossa@cgiar.org

J. J. Cerón-Rojas (✉) · J. W. R. Martini · G. Covarrubias-Pazaran · G. Alvarado (Deceased) ·
F. H. Toledo · V. Govindan
International Maize and Wheat Improvement Center (CIMMYT), Texcoco, Mexico
e-mail: g.covarrubias@cgiar.org; f.toledo@cgiar.org; velu@cgiar.org

593

32.1   Learning Objectives

- To understand the advantages of linear selection index (LSI) theory for making selection decisions.
- To understand and apply the unconstrained and constrained LSI in plant breeding.
- To understand how to estimate LSI parameters.

32.2   Introduction

The linear phenotypic selection index (LPSI) theory was first described in the plant breeding context [1] and later in the animal [2] breeding phenotypic selection context. When the phenotypic and genotypic covariance matrices of the traits are known, the LPSI is the best phenotype-based linear predictor of the individual net genetic merit. In LPSI theory, it is assumed that the genotypic values that define the net genetic merit are composed entirely of the additive effects of genes and that the LPSI and the net genetic merit have joint bivariate normal distribution [3]. The main objectives of using a selection index (LSI) are (i) to predict the unobservable net genetic merit values of the candidates for selection, (ii) to maximize the expected genetic gain per trait or multi-trait selection response, and (iii) to provide the breeder with an objective rule for evaluating and selecting for several traits simultaneously. The advantages of an LPSI are that it modifies the predefined economic weights according to the trait heritability, that it considers indirect selection effects resulting from the genetic correlation between traits, and that it is relatively easy to use. Its disadvantages are that it may be difficult to assign economic weights to some traits, and that it requires large amounts of information to reliably estimate the genetic covariance between traits. This may cause a large sampling error.

Because economic weights are difficult to assign to some traits, several modified indices, such as the base index, the modified base index, the non-weighted multiplicative index [4] and the eigen selection index method (ESIM) [5, 6], have been proposed. The main LSI theory was developed assuming that the economic weights are fixed and known, which is, for instance, not the case for the ESIM.

In the LPSI structure, each trait has an economic weight. This could also imply that, for each trait, a directional change is desired. This may not be suitable to achieve a breeding objective in which some traits should remain unchanged. Assuming that the breeder is interested in keeping a certain trait within a range, we could either combine the use of an LPSI with independent culling for the restricted traits, or we could incorporate the fact that not changing the trait is desired. This was the main idea of the restricted LPSI (RLPSI) [3] which solves the usual LPSI equations subject to the restriction that the covariance between the LPSI and some linear function of the genotypes involved equals zero, thus preventing selection on the index from causing any genetic change in the expected genetic advance of the restricted traits.

Later the RLPSI results were extended [7] to a selection index called constrained LPSI (CLPSI) that attempts to make some traits change their mean based on a predetermined level while the rest of them are unrestricted. The CLPSI equals the covariance between the LPSI and some linear functions of the genotypes to a constant or genetic gain predetermined by the breeder. Some authors [5] developed a constrained index ESIM (CESIM) that does not use economic weights. The CLPSI (CESIM) is the most general LPSI and includes the LPSI (ESIM) and the RLPSI as particular cases.

In a similar manner, in the marker-assisted selection (MAS) context, a linear marker selection index (LMSI) was proposed [8] that uses phenotypic and marker score values jointly to predict the net genetic merit. The LMSI combines information on markers linked to quantitative trait locus (QTLs) and the phenotypic values of the traits to predict the net genetic merit of the candidates for selection because it is not possible to identify all QTLs affecting the economically important traits. Several authors [9, 10] have criticized the LMSI approach because it makes inefficient use of the available data. In addition, because the LMSI is based on only a few large QTL effects, it violates the selection index assumptions of multivariate normality and small changes in allele frequencies. We shall not describe the LMSI. Readers interested in the LMSI can see [11] for details.

The linear genomic selection index (LGSI) and constrained linear genomic selection index (CLGSI) were developed in the genomic selection (GS) context in which animals and plants are selected based on the GEBV of the candidates for selection [12, 13]. In the LGSI context, all marker effects of the genotyped individuals in the training population are estimated using marker and phenotypic data. These estimated effects are used in subsequent selection cycles to obtain predictors (GEBVs) of the individual breeding values in the testing population for which there is only marker information about the candidates for selection.

It has been shown [9] that GS increased the accuracy of predicting the breeding values of the candidates for selection, and reduced the intervals between selection cycles and the costs of the breeding programs (See Chaps. 5, 6 and 30). Because GS decreases the generation interval, it leads to a much higher genetic gain per year. Some authors [14] indicated that GS could replace traditional progeny testing when maximizing the genetic gain per year, as long as the accuracy of GEBV is higher than or equal to 0.45.

The expected selection response of the net genetic merit and the expected genetic gain per trait are the main quantities to consider when comparing different LSI. These parameters give breeders an objective basis to compare different selection methods. We describe the practical applications of the phenotypic and genomic LSI using real wheat data. Readers unfamiliar with LSI theory should read the Appendix of this work first and then return to the manuscript. A complete exposition of LSI theory is in [15].

32.3 Definitions

Breeding Value the value of an individual measured by the mean phenotype of its progeny obtained by random mating with the population. It is also the sum of the average additive effects of the genes of the individual.

Economic Weight the increase in profit achieved by improving a particular trait by one unit.

Expected Genetic Gain Per Trait (Multi-trait Selection Response) a vector of expected genetic gains associated with the traits of the offspring of the selected parents.

GEBV the sum of additive whole genome allele effects of an individual. Allele effects are estimated by a regression of the phenotypic values on the whole genome DNA markers. It is used to predict breeding values of individuals in animal and plant breeding programs in the genomic selection context.

Genomic Selection the selection of parents based on the higher GEBV values or on a linear combination of them (e.g., LGSI or CLGSI).

Genotypic Value the average of the phenotypic values across a (large) population of environments.

Linear Selection Index (LSI) a linear combination of phenotypic and/or GEBV values, or marker scores. In addition, it can be unconstrained or constrained.

Net Genetic Merit a linear combination of breeding values of the individual traits of interest, each of them weighted by its respective economic value. It is also called the total economic value of one individual.

Phenotype Value the sum of genotypic (or breeding) value, environment value, and genotype-by-environment interaction.

Quantitative Traits plant and animal characteristics (or phenotypic expression) that exhibit continuous variability, which is the result of many gene effects interacting among themselves and with the environment.

Selection Response the expectation of the net genetic merit of the selected individuals when the mean of the original population is zero. It is also defined as the difference between the mean phenotypic values of the offspring of the selected parents and the mean of the entire parental generation before selection.

32.4   Key Points

- Selection indices are fundamental tools for modern plant breeding.
- The use of selection indices is a key to better estimate the net genetic merits of candidates for selection. Selection indices will ensure that wheat improvement research maximizes its impact.
- New breeding technologies like genomic assisted breeding and rapid cycle selection has to be combined with the use of selection indices to maximize response to selection.

32.5   Phenotypic and Genomic Selection Indices Theoretical Results

32.5.1   The Net Genetic Merit and the LPSI

The net genetic merit $(H = \mathbf{w}'\mathbf{g})$ is related to the vector of trait phenotypic $(\mathbf{y})$ values as

$$H = \mathbf{b}'\mathbf{y} + e = I + e, \tag{32.1}$$

where $\mathbf{g}' = [G_1 \ G_2 \ \dots \ G_t]$ and $\mathbf{y}' = [Y_1 \ Y_2 \ \dots \ Y_t]$ are vector $1 \times t$ (t=number of traits) of true unobservable breeding values and observable trait phenotypic values, respectively, $I = \mathbf{b}'\mathbf{y}$ is the LPSI, and $\mathbf{w}' = [w_1 \ w_2 \dots w_t]$ is the vector of economic weights. In Eq. 32.1, we assume that e has normal distribution with expectation $E(e) = 0$ and variance σ_e^2, and that I and e are independent; thus $\sigma_H^2 = \sigma_I^2 + \sigma_e^2$ is the variance of H, $\sigma_I^2 = \mathbf{b}'\mathbf{P}\mathbf{b}$ is the variance of I, $\mathbf{P}$ is the phenotypic covariance matrix, and $\sigma_e^2 = \sigma_H^2 - \sigma_I^2$ is the residual variance.

The LPSI $(I = \mathbf{b}'\mathbf{y})$ can be written as

$$I = \mathbf{w}'\mathbf{C}\mathbf{P}^{-1}\mathbf{y}, \tag{32.2}$$

where $\mathbf{b} = \mathbf{P}^{-1}\mathbf{C}\mathbf{w}$, $\mathbf{C}$ is the genotypic covariance matrix, $Cov(H, \mathbf{y}) = \mathbf{C}\mathbf{w}$ is the covariance among $H = \mathbf{w}'\mathbf{g}$ and $\mathbf{y}$, and $\mathbf{P}^{-1}$ is the inverse matrix of $\mathbf{P}$.

32.5.2   Economic Weights for LPSI

A method for assigning economic weights to the traits [1] is as follows. Suppose that in a wheat-selection program we are required to consider the vector $\mathbf{y}' = [Y_1 \ Y_2 \ \dots \ Y_t]$ of t traits. Let us evaluate each in terms of Y_1. Suppose that Y_1 denotes grain yield, Y_2 baking quality and Y_3 denotes resistance to flag smut.

Suppose that an advance of 10 in baking score (Y_2) is equal in value to an advance of 1 bushel per acre in yield (Y_1) and that a decrease of 20% infection (Y_3) is worth 1 bushel of yield (Y_1), and so on. Then, taking Y_1 as standard and units as indicated, $w_1 = 1.0$, $w_2 = 0.1$, $w_3 = -0.05$, etc., will be the economic values of each trait.

One additional method for assigning economic weights to the traits (which we have used in this work) is based on the expected genetic gain per trait (Appendix, Eq. 32.A5). Let us consider the real data HarvestPlus Association Mapping (HPAM) panel, which consists of 330 wheat lines from CIMMYT, and assume that the objective of the selection is to increase the mean value of Zn content in the grain (Zn), the Fe content in the grain (Fe), and grain yield (GY, t/h), while decreasing or maintaining the same plant height (PHT, cm). We found that the vector $\mathbf{w} = \begin{bmatrix} 0.1 & 0.5 & 2.8 & -0.6 \end{bmatrix}$ (see Sect. 32.10) is adequate for obtaining the expected genetic gain per trait described in the Results Section of this work. This method is by assay and error and requires the evaluation of Eq. 32.A5 until we obtain the desired results.

32.5.3   The Maximized Correlation and the Maximized LPSI Selection Response

The maximized correlation between H and I (ρ_{HI}) and the maximized LPSI selection response are

$$\rho_{HI} = \frac{\sqrt{\mathbf{b'Pb}}}{\sqrt{\mathbf{w'Cw}}},$$ (32.3)

$$R = k\sqrt{\mathbf{b'Pb}},$$ (32.4)

respectively, where $\mathbf{b} = \mathbf{P}^{-1}\mathbf{Gw}$ (Appendix, Eq. 32.A3). Equation 32.4 predicts the mean improvement in H due to indirect selection on $I = \mathbf{b'y}$. Here, k is the intensity of selection. The heritability of $I = \mathbf{b'y}$ is $h_I^2 = \dfrac{\mathbf{b'Cb}}{\mathbf{b'Pb}}$.

32.6   The Retrospective Index

This index is useful when, instead of the index values, the breeder observes only the vector of selection differentials ($\mathbf{s}$). In this case, the index that would give the same observed $\mathbf{s}$ is called the retrospective index and its vector of coefficients can be obtained as $\mathbf{b} = \mathbf{P}^{-1}\mathbf{s}$ [16].

32.7   Constrained LPSI (CLPSI)

The CLPSI vector of coefficients is

$$\beta = \mathbf{Kb}, \tag{32.5}$$

where $\mathbf{K} = [\mathbf{I}_t - \mathbf{Q}], \mathbf{Q} = \mathbf{P}^{-1}\mathbf{M}(\mathbf{M}'\mathbf{P}^{-1}\mathbf{M})^{-1}\mathbf{M}'$, $\mathbf{M}' = \mathbf{D}'\mathbf{U}'\mathbf{C}$, $\mathbf{I}_t$ is a $t{\times}t$ identity matrix and $\mathbf{b} = \mathbf{P}^{-1}\mathbf{Cw}$.

32.7.1   The Maximized CLPSI Selection Response and Expected Genetic Gain Per Trait

The maximized CLPSI selection response and expected genetic gain per trait are

$$R_C = k\sqrt{\beta'\mathbf{P}\beta}, \tag{32.6}$$

$$\mathbf{E}_C = k\frac{\mathbf{C}\beta}{\sqrt{\beta'\mathbf{P}\beta}}, \tag{32.7}$$

respectively, where k is the selection intensity.

32.8   The ESIM and CESIM Theory

32.8.1   The Maximized ESIM Selection Response and the Maximized ρ_{HI_1}

The maximized ESIM selection response (R_E) and the maximized correlation between $I_{E_1} = \beta'_{E_1}\mathbf{y}$ and $H_{E_1} = \mathbf{w}'_{E_1}\mathbf{g}$ (ρ_{HI_1}) are

$$R_E = k\sqrt{\beta'_{E_1}\mathbf{P}\beta_{E_1}}, \tag{32.8}$$

$$\rho_{HI_1} = \sqrt{\frac{\beta'_{E_1}\mathbf{C}\beta_{E_1}}{\beta'_{E_1}\mathbf{P}\beta_{E_1}}}, \tag{32.9}$$

respectively, where $\beta_{E_1} = \mathbf{Fb}_{E_1}$ is the first eigenvector of equation $(\mathbf{T}_2 - \rho^2_{HI_j}\mathbf{I})\beta_{E_j} = \mathbf{0}$ (Appendix, Eqs. 32.A7 and 32.A8). When $\mathbf{F}$ is not used, Eq. 32.8 is equal to $R_E = k\sqrt{\mathbf{b}'_{E_1}\mathbf{P}\mathbf{b}_{E_1}}$ (Appendix, Eq. 32.A7), whereas Eq. 32.9

is the square root of the first eigenvalue of Eq. 32.A7, *i.e.*, $\rho_{HI_1} = \sqrt{\rho^2_{HI_1}}$. The

heritability of $I_{E_1} = \beta'_{E_1} \mathbf{y}$ is $h^2_E = \dfrac{\beta'_{E_1} \mathbf{C} \beta_{E_1}}{\beta'_{E_1} \mathbf{P} \beta_{E_1}}$

32.8.2 The Maximized CESIM Selection Response and Expected Genetic Gain Per Trait

The maximized CESIM selection response (R_{CE}) and expected genetic gain per trait ($\mathbf{E}_{CE}$) are

$$R_{CE} = k\sqrt{\mathbf{b}'_{CE_1} \mathbf{P} \mathbf{b}_{CE_1}}, \qquad (32.10)$$

$$\mathbf{E}_{CE} = k\frac{\mathbf{C}\mathbf{b}_{CE_1}}{\sqrt{\mathbf{b}'_{CE_1} \mathbf{P} \mathbf{b}_{CE_1}}}, \qquad (32.11)$$

respectively, where all the terms were defined earlier.

32.9 The Unconstrained and Constrained Linear Genomic Selection Index Theory

The LGSI and the CLGSI are, respectively, an application of the LPSI and CLPSI to the genomic selection context. Thus, the LGSI and the CLGSI theoretical results are very similar to the LPSI and CLPSI theoretical results.

32.9.1 The Unconstrained Linear Genomic Selection Index (LGSI)

Let $\mathbf{z}' = \begin{bmatrix} GEBV_1 & GEBV_2 & \cdots & GEBV_t \end{bmatrix}$ be a vector of GEBVs for t traits. The individual LGSI is

$$I_G = w_1 GEBV_1 + w_2 GEBV_2 + \ldots + w_t GEBV_t = \mathbf{w}'\mathbf{z}, \qquad (32.12)$$

where $\mathbf{w}$ is the vector of economic weights for t traits.

32.9.2   The CLGSI Vector of Coefficients

The CLGSI vector of coefficients is

$$\beta_G = K_G w, \tag{32.13}$$

where w is the vector of economic weights, $K_G = [I_t - Q_G]$, $Q_G = UD(D'U\Gamma UD)^{-1}D'U\Gamma$, $\Gamma = Var(z)$ is the covariance matrix of GEBV, and I_t is an identity matrix of size $t \times t$, whereas D and U are the matrices described in Eq. 32.A6 (Appendix). When $d = 0$, $D = U$ and matrix K_G can be written as $K_G = [I_t - Q_G]$, where $Q_G = U(U'\Gamma U)^{-1}U'\Gamma$. In this case, the CLGSI is a null restricted LGSI. When $D = U$ and U' is a null matrix, $\beta_G = w$. Thus, the CLGSI includes the null restricted and the unrestricted LGSI as particular cases.

32.9.3   Maximized CLGSI Selection Response and Expected Genetic Gain Per Trait

The maximized CLGSI selection response and expected genetic gain per trait are

$$R_{CG} = k\sqrt{\beta_G'\Gamma\beta_G}, \tag{32.14}$$

$$E_{CG} = k\frac{\Gamma\beta_G}{\sqrt{\beta_G'\Gamma\beta_G}}, \tag{32.15}$$

respectively. The methods to estimate the index parameters are in [15].

32.9.4   The Genomic Estimated Breeding Values (GEBV)

To obtain the GEBV, we used a multi-trait genomic best linear unbiased predictor (GBLUP) described in [12, 13].

32.10   Real Wheat Data

We used the HarvestPlus Association Mapping (HPAM) panel, which consists of 330 wheat lines from CIMMYT and four traits: Zn content in the grain (Zn), Fe content in the grain (Fe), grain yield (GY, t/h), and plant height (PHT, cm). The objective of the selection was to increase the mean values of Zn, Fe, and GY while PHT decreased or stayed the same.

Using CLPSI, CESIM, and CLGSI, we constrained traits Zn, Fe and GY with the vector of constraints $\mathbf{d}' = \begin{bmatrix} 1.5 & 1.6 & 0.45 \end{bmatrix}$ and matrices $\mathbf{U}' = \begin{bmatrix} 1 & 0 & 0 & 0 \\ 0 & 1 & 0 & 0 \\ 0 & 0 & 1 & 0 \end{bmatrix}$ and

$\mathbf{D}' = \begin{bmatrix} 0.45 & 0 & -1.5 \\ 0 & 0.45 & -1.6 \end{bmatrix}$. Each element of vector $\mathbf{d}$ is the standard deviation of the

genotypic variance of Zn, Fe, and GY, respectively. The vector of economic weights for LPSI, CLPSI, LGSI, and CLGSI was $\mathbf{w} = \begin{bmatrix} 0.1 & 0.5 & 2.8 & -0.6 \end{bmatrix}$, whereas for

ESIM and CESIM, matrix $\mathbf{F}$ was $\mathbf{F} = \begin{bmatrix} -0.5 & 0 & 0 & 0 \\ 0 & 1.0 & 0 & 0 \\ 0 & 0 & 2.0 & 0 \\ 0 & 0 & 0 & -0.5 \end{bmatrix}$ and

$\mathbf{F} = \begin{bmatrix} 1.0 & 0 & 0 & 0 \\ 0 & -1.0 & 0 & 0 \\ 0 & 0 & 2.0 & 0 \\ 0 & 0 & 0 & -0.8 \end{bmatrix}$, respectively. The total proportion (p) retained was

6% (k=1.98) for the phenotypic indices and 12.45% (k=1.65) for the genomic indices. The estimated phenotypic ($\hat{\mathbf{P}}$) and genotypic ($\hat{\mathbf{C}}$) covariance matrices among the four traits were

$$\hat{\mathbf{P}} = \begin{bmatrix} 4.2 & 1.65 & -0.37 & 0.24 \\ 1.65 & 4.95 & 0.27 & 2.36 \\ -0.37 & 0.27 & 0.58 & 1.14 \\ 0.24 & 2.36 & 1.14 & 14.40 \end{bmatrix} \text{ and } \hat{\mathbf{C}} = \begin{bmatrix} 2.22 & 0.95 & -0.19 & 0.06 \\ 0.95 & 2.57 & 0.15 & 2.08 \\ -0.19 & 0.15 & 0.20 & 0.80 \\ 0.06 & 2.08 & 0.80 & 6.97 \end{bmatrix}$$

With the data described above, we obtained the estimated matrix $\mathbf{\Gamma}$ ($\hat{\mathbf{\Gamma}}$) for three cases denoted as G, G-COP and COP, where $\hat{\mathbf{\Gamma}} = \begin{bmatrix} 0.47 & 0.11 & -0.16 & -0.09 \\ 0.11 & 0.82 & 0.72 & 0.17 \\ -0.16 & 0.72 & 1.82 & 0.24 \\ -0.09 & 0.17 & 0.24 & 0.13 \end{bmatrix}$,

$\hat{\mathbf{\Gamma}} = \begin{bmatrix} 0.87 & 0.35 & -0.03 & -0.10 \\ 0.35 & 1.01 & 0.89 & 0.17 \\ -0.03 & 0.89 & 2.51 & 0.35 \\ -0.10 & 0.17 & 0.35 & 0.17 \end{bmatrix}$, and $\hat{\mathbf{\Gamma}} = \begin{bmatrix} 0.77 & 0.38 & 0.03 & -0.08 \\ 0.38 & 0.91 & 0.78 & 0.12 \\ 0.03 & 0.78 & 2.26 & 0.31 \\ -0.08 & 0.12 & 0.31 & 0.14 \end{bmatrix}$,

respectively.

32.11   Results

32.11.1   Phenotypic Results

Figure 32.1 presents the averages for four traits of the 20 selected individuals (with LPSI and ESIM) with a proportion of 6% ($k = 1.985$). In this case, those averages were very similar. We found similar results when we made selections using CLPSI and CESIM for this real dataset.

Table 32.1 presents the estimated LPSI, ESIM, CLPSI, and CESIM selection response, coefficient of correlation and heritability. The estimated ESIM and CESIM selection response, correlation and heritability were higher than the estimated LPSI and CLPSI selection response, correlation and heritability. Thus, ESIM and CESIM efficiency for predicting the net genetic merit was higher than LPSI and CLPSI efficiency.

Table 32.2 presents the estimated LPSI, ESIM, CLPSI, and CESIM expected genetic gain for four traits selected with a proportion of 6% ($k = 1.985$). The estimated CLPSI and CESIM expected genetic gains per trait were constrained by vector $\mathbf{d'} = \begin{bmatrix} 1.5 & 1.6 & 0.45 \end{bmatrix}$ values. Thus, the estimated expected genetic gains of traits Zn, Fe, and GY should be similar to the $\mathbf{d}$ values. The estimated CLPSI and CESIM expected genetic gain values were lower than the $\mathbf{d}$ values. This means that to reach $\mathbf{d}$ values, breeders will need to select once again using CLPSI and CESIM. However, the estimated CESIM expected genetic gain values were higher than the estimated CLPSI expected genetic gain values.

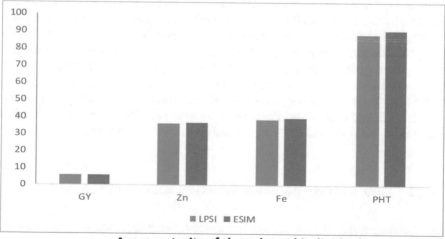

Average traits of the selected individuals

Fig. 32.1 Averages for four traits of 20 selected individuals with LPSI (linear phenotypic selection index) and ESIM (eigen selection index method)

Table 32.1 Estimated unconstrained and constrained linear phenotypic selection indices (LPSI and CLPSI, respectively) and eigen selection index methods (ESIM and CESIM, respectively) selection response, coefficient of correlation and heritability

	Estimated parameters		
Index	Selection response	Correlation	Heritability
LPSI	0.98	0.45	0.47
ESIM	3.02	0.72	0.52
CLPSI	0.93	0.43	0.20
CESIM	2.80	0.74	0.55

Table 32.2 Estimated unconstrained and constrained linear phenotypic selection indices (LPSI and CLPSI, respectively) and eigen selection index methods (ESIM and CESIM, respectively) expected genetic gain for four traits selected with a proportion of 6% ($k = 1.985$)

	Estimated expected genetic gain per trait			
Index	Zn	Fe	GY	PHT
LPSI	0.18	0.39	0.01	−0.41
ESIM	0.52	1.27	0.09	−0.01
CLPSI	0.09	0.10	0.03	−0.56
CESIM	1.24	0.66	0.09	−0.05

32.11.2 Genomic Selection Index Results

For datasets G, G-COP and COP, in Table 32.3 we present the estimated LGSI and CLGSI selection response and expected genetic gain for four traits with a selected proportion of 12.45% ($k = 1.65$). In this case, the estimated CLGSI expected genetic gains per trait were constrained by vector $\mathbf{d'} = \begin{bmatrix} 1.5 & 1.6 & 0.45 \end{bmatrix}$ values. Thus, the estimated expected genetic gain of traits Zn, Fe, and GY should be similar to the $\mathbf{d}$ values. The estimated CLGSI expected genetic gain values were lower than the $\mathbf{d}$ values. This means that to reach $\mathbf{d}$ values, breeders will need to select once again using CLGSI. Note, however, that the estimated LGSI and CLGSI selection response and expected genetic gain values were higher than the estimated LPSI and CLPSI expected genetic gain values. This means that for the predicted data, LGSI and CLGSI efficiency was higher than LPSI and CLPSI efficiency. In addition, the estimated LGSI and CLGSI selection response was not affected by the restriction imposed on the LGSI and CLGSI expected genetic gain, as we would expect.

32.12 How to Incorporate a Selection Index in Practice?

Incorporating a selection index requires a step-by-step approach to ensure its successful implementation. Most of the time, breeders use a customized procedure to select individuals based on independent culling that comprises multiple steps.

Table 32.3 Unconstrained and constrained estimated linear genomic selection indices (LGSI and CLGSI) selection response and expected genetic gain for four traits with a proportion of 12.45% ($k = 1.65$) for three datasets: G, G-COP and COP

Index	Response	Zn	Expected genetic gain per trait				Data
			Fe	GY	PHT		
LGSI	1.75	0.10	0.73	0.47	−0.11		G
LGSI	1.83	0.00	0.72	0.50	−0.12		G-COP
LGSI	1.63	0.03	0.61	0.44	−0.15		COP
CLGSI	1.75	0.42	0.44	0.12	−0.73		G
CLGSI	1.83	0.59	0.63	0.18	−0.70		G-COP
CLGSI	1.63	0.55	0.59	0.16	−0.66		COP

The *first step* consists of understanding the selection procedure executed by the program. The steps in the selection procedure can be mapped back to a set of reduction and selection steps applied to a selection unit (*i.e.*, lines, families, etc.); each step consists of trait conditions (value and directionality). Each selection step can consist of meeting more than one trait condition (Table 32.4).

The *second step* identifies which parts of the selection process can be replaced by an index. For example, by looking at Table 32.4, you can decide to pick a single step and replace independent culling or replace multiple steps with a single selection index. Here we will replace steps 2–11 with a selection index.

32.13   Retrospective Index

A *third step* consists of building the index. Indices that depend on economic weights are difficult to implement. Instead, a retrospective index is the best way to start implementing an index. For example, assume that the matrix of estimates for the traits indicated above is available together with an indicator column in which the material was selected by the breeder using the steps indicated in Table 32.4. The formula $\hat{\mathbf{b}} = \hat{\mathbf{P}}^{-1}\mathbf{s}$ is then used to infer the weights.

Suppose that $\hat{\mathbf{P}}$ and $\hat{\mathbf{P}}^{-1}$ are as follow:

$$\hat{\mathbf{P}} = \begin{bmatrix} 1 & -0.151 & -0.108 & -0.279 & 0.321 & -0.026 \\ -0.151 & 1 & -0.228 & -0.076 & 0.054 & -0.045 \\ -0.108 & -0.228 & 1 & 0.083 & -0.080 & -0.035 \\ -0.279 & -0.076 & 0.083 & 1 & -0.099 & -0.083 \\ 0.321 & 0.054 & -0.080 & -0.099 & 1 & 0.064 \\ -0.026 & -0.045 & -0.035 & -0.083 & 0.064 & 1 \end{bmatrix}$$

Table 32.4 Example of the type of independent culling selection steps carried out by a breeding program in a preliminary yield trial

step	selectionUnit	stepType	trait	value	directionality	trait	value	directionality	valuesUsedAs	useTraitCovariance	TotalSelected	TotalCumulative
1	Line	reduce	YLD_rel	0.4	>	ZNC_rel	0.3	>	culling	–	891	891
2	Line	select	Xa1	R	=	YLD_BV	6.6	>	culling	–	20	20
3	Line	select	Xa1	R	=	ZNC_BV	14	>	culling	–	20	30
4	Line	select	Xa2	R	=	YLD_BV	6	top	culling	–	8	34
5	Line	select	Xa2	R	=	ZNC_BV	6	top	culling	–	8	36
6	Line	select	Pi1	R	=	YLD_BV	6	top	culling	–	8	40
7	Line	select	Pi1	R	=	ZNC_BV	6	top	culling	–	8	45
8	Line	select	Pi2	R	=	YLD_BV	6	top	culling	–	6	50
9	Line	select	Pi2	R	=	ZNC_BV	6	top	culling	–	6	52
10	Line	select	YLD_BV	12	top				culling	–	6	60
11	Line	select	ZNC_BV	12	top				culling	–	6	62
12	Line	Reduce&select	PHT_BV	115	<	FLW_BV	100	<	culling	–	40	40
13	Family	reduce	ZNC_BV	3	random				culling	–	18	18

$$\hat{\mathbf{P}}^{-1} = \begin{bmatrix} 1.27 & 0.272 & 0.144 & 0.321 & -0.382 & 0.048 \\ 0.272 & 1.12 & 0.267 & 0.122 & -0.118 & 0.064 \\ 0.144 & 0.267 & 1.08 & -0.033 & 0.019 & 0.055 \\ 0.321 & 0.122 & -0.033 & 1.10 & 0.003 & -0.080 \\ -0.382 & -0.118 & 0.019 & 0.003 & 1.13 & -0.087 \\ 0.048 & 0.064 & 0.055 & -0.080 & -0.087 & 1.01 \end{bmatrix}.$$

Then, according to Table 32.5 values,

$$\hat{\mathbf{b}} = \hat{\mathbf{P}}^{-1}\mathbf{s} = \begin{bmatrix} 1.27 & 0.272 & 0.144 & 0.321 & -0.382 & 0.048 \\ 0.272 & 1.12 & 0.267 & 0.122 & -0.118 & 0.064 \\ 0.144 & 0.267 & 1.08 & -0.033 & 0.019 & 0.055 \\ 0.321 & 0.122 & -0.033 & 1.10 & 0.003 & -0.080 \\ -0.382 & -0.118 & 0.019 & 0.003 & 1.13 & 0.087 \\ 0.048 & 0.064 & 0.055 & -0.080 & -0.087 & 1.01 \end{bmatrix} \begin{bmatrix} 0.65 \\ 0.40 \\ -0.19 \\ -0.06 \\ 0.04 \\ -0.08 \end{bmatrix} = \begin{bmatrix} 0.873 \\ 0.561 \\ -0.004 \\ 0.208 \\ -0.251 \\ -0.032 \end{bmatrix}.$$

The obtained weights can be confusing if they have a different direction than the desired direction. For example, the 4th weight for Xa2 (Table 32.5) resistance is positive, when we would expect it to be negative. This is because covariances among traits are expected to account for taking the trait in the right direction despite the value of the weight.

To show that these weights are better than the current approach, we can calculate what would be the selected individuals and the selection differentials using the index and compare them to the selection differentials obtained with the current approach. As can be seen in Table 32.6, if these weights are considered the real weights (even economic weights), the index can select better individuals than the breeder's eyeball method. This example shows how the selection index theory can provide higher selection differentials than the breeder.

32.14   Discussion

32.14.1   The Unconstrained LSI Theory

The LSI theory includes, as particular cases, the unconstrained LPSI and LGSI, and any other unconstrained LSI associated with this theory that is based on the quantitative genetics and the multivariate normal distribution theory. The LSI theory is based on multivariate normal distribution theory because this distribution allows the

Table 32.5 Example of a trait matrix used for building a retrospective index. An indicator column is used to derive the selection differentials and phenotypic covariance matrix required for the calculation of the retrospective index

Select	Family	Line	YLD_BV	YLD_rel	FLW_BV	PHT_BV	ZNC_BV	ZNC_rel	Xa1	Xa2	Pi1	Pi2
	Cross1	Line1	7.54	0.38	80.28	114.71	11.76	0.29	S=2	R=1	S=2	S=2
	Cross1	Line2	7.50	0.46	81.53	110.57	12.76	0.27	S=2	R=1	R=1	S=2
YES	Cross1	Line3	7.48	0.50	81.64	107.79	13.04	0.38	S=2	R=1	R=1	S=2
	Cross1	Line4	7.43	0.46	80.00	113.60	12.03	0.37	S=2	R=1	S=2	S=2
	Cross2	Line5	7.42	0.43	81.40	107.74	12.83	0.37	S=2	R=1	R=1	S=2
	Cross2	Line6	7.39	0.43	85.41	113.91	12.29	0.37	S=2	R=1	R=1	S=2
	Cross2	Line7	7.39	0.51	82.10	108.09	12.97	0.45	S=2	R=1	R=1	S=2
	Cross2	Line8	7.39	0.49	82.19	107.09	13.11	0.43	S=2	R=1	R=1	S=2
YES	Cross2	Line9	7.38	0.42	79.61	109.82	11.96	0.35	S=2	S=2	R=1	S=2
	Cross3	Line10	7.37	0.48	81.72	106.58	13.41	0.36	S=2	R=1	R=1	S=2
	Cross4	Line11	7.37	0.18	80.71	100.42	11.81	0.13	S=2	S=2	R=1	S=2
	Cross4	Line12	7.36	0.43	81.96	107.32	12.91	0.37	S=2	R=1	R=1	S=2
	Cross4	Line13	7.36	0.55	82.26	110.00	12.98	0.48	#N/A	R=1	R=1	S=2
...	...	...	...	...	...	...	...	...	...	...	...	...
μ_{sel}			5.78				13.93		1.76	1.26	1.21	1.89
μ_{pop}			5.13				13.52		1.95	1.32	1.17	1.97
s			0.65				0.40		−0.19	−0.06	0.04	−0.08

Table 32.6 Selection differentials for six traits involved in selection steps 2–11 using two selection methods, the independent culling normally applied by breeders versus the selection index based on a retrospective analysis

Select	YLD_BV	ZNC_BV	Xa1	Xa2	Pi1	Pi2	Total gain
$S_{current}$	0.65	0.40	−0.19	−0.06	0.04	−0.08	0.776
S_{index}	0.44	2.20	−0.32	−0.06	−0.12	0.02	1.647

LSI to be completely described using only means, variances and covariances. When the phenotypic traits and GEBV values have multivariate normal distribution, linear combinations of phenotypic traits and GEBV are normal. Even if the phenotypic traits and GEBV values do not have multivariate normal distribution, this distribution serves as a useful approximation, especially in inferences involving sample mean vectors, which, by the central limit theorem, have multivariate normal distribution [17]. By this reasoning, a fundamental assumption in LSI theory is that the LSI and the net genetic merit have joint bivariate normal distribution. Under the latter assumption, the regression of the net genetic merit on any linear function of the phenotypic or GEBV values is linear [3].

The selection response and the expected genetic gain per trait were the main parameters of the LSI and the criteria to compare LSI efficiency and predict the net genetic merit of any linear index. These parameters give breeders a clearer base on which to objectively validate the effectiveness of the adopted selection method.

The LPSI was the first LSI used to predict the net genetic merit and has good statistical properties when the phenotypic and genotypic covariances matrices are known. The LGSI is the most recent LSI and has the advantage of reducing the intervals between selection cycles by more than two thirds.

32.14.2 The Constrained LSI

The constrained LPSI (CLPSI) and the constrained LGSI (CLGSI) impose constraints on the expected genetic gain per trait. These indices include the unconstrained indices as particular cases. There are two types of CLPSI and CLGSI: the null restricted index and the predetermined proportional gain index. The null restricted index allows imposing restrictions equal to zero on the expected genetic gain of some traits, while the expected genetic gain of other traits increases (or decreases) without imposing any restrictions. In a similar manner, the constrained index attempts to make some traits change their expected genetic gain values based on a predetermined level, while the rest of the traits remain without restrictions. The objective of both types of selection indices is to predict the net genetic merit and select parents for the next generation. The CLPSI and CLGSI are projections of the vector coefficients of the LPSI and LGSI, respectively, to a different space, and the constraining effects are observed on the CLPSI and CLGSI expected genetic gains per trait where each restricted trait has an expected genetic gain according to the constrained values imposed by the breeder.

32.14.3   Statistical Properties of the LSI

Both the unconstrained and constrained indices have the same statistical properties when the phenotypic and genotypic covariance matrices and the economic weights are known. For example, they have maximum correlation with the net genetic merit and the variance of the predicted error is minimal; however, when the phenotypic and genotypic covariance matrices and the economic weights are unknown, the statistical sampling properties of the indices described in this work are difficult to know. Assuming that the estimated LSI have normal distribution, some authors [18] found the statistical sampling properties of the LSI selection responses in the phenotypic and genomic selection context while others [15] reported the statistical sampling properties of ESIM and CESIM.

32.15   Key Concepts

- Using a selection index in plant breeding maximizes the expected genetic gain per trait or multi-trait selection response and provides an objective rule for evaluating and selecting for several traits simultaneously.
- The advantages of a selection index is that it considers indirect selection effects resulting from the genetic correlation between traits. Main disadvantages are that it may be difficult to assign economic weights to some traits. Several modified indices exists to overcome this problem.
- Recently genomic selection indices have been developed and used based on the genomic estimated breeding.

32.16   Conclusions

Our main goal was to offer researchers a starting point for understanding the core tenets of LSI theory in plant selection. We provided the unconstrained and constrained LSI theory associated with phenotypic and genomic selection. We validated the LSI phenotypic and genomic theoretical results in the wheat breeding context using a real wheat dataset with four traits.

Appendix

This Appendix is a brief review of the LSI theory. Readers interested in this theory should see [15], who describe the complete LSI theory.

Breeding and Trait Phenotypic Values

Let $\mathbf{g}' = [G_1 \quad G_2 \quad \ldots \quad G_t]$ be a vector $1 \times t$ (t= number of traits) of true unobservable breeding values associated with the observable vector of trait phenotypic values $\mathbf{y}' = [Y_1 \quad Y_2 \quad \ldots \quad Y_t]$, such that the jth (j= 1, 2, ..., t) individual trait phenotypic value for one environment is

$$Y_j = G_j + \varepsilon_j, \tag{32.A1}$$

where G_j is composed entirely of additive genetic effects and includes all types of gene and interaction values, whereas ε_j denotes the deviations of Y_j from the G_j values. In Eq. 32.A1, G_j and ε_j are independent unobservable random variables, have normal distribution with expectation $E(G_j) = 0$ and $E(\varepsilon_j) = 0$, and variance $\sigma_{G_j}^2$ and $\sigma_{\varepsilon_j}^2$, respectively. In addition, Y_j is an observable random variable, with normal distribution, expectation $E(Y_j) = 0$ and variance $\sigma_{G_j}^2 + \sigma_{\varepsilon_j}^2$. Finally, $cov(Y_j, G_j) = \sigma_{G_j}^2$ is the covariance between Y_j and G_j.

The Unconstrained Linear Phenotypic Selection Index (LPSI)

The random vectors $\mathbf{g}' = [G_1 \quad G_2 \quad \ldots \quad G_t]$ and $\mathbf{y}' = [Y_1 \quad Y_2 \quad \ldots \quad Y_t]$ (Equation 32.A1) have joint multivariate normal distribution with mean $\mu' = [\mathbf{0} \quad \mathbf{0}]$ and covariance matrix $Var\begin{pmatrix}\mathbf{y}\\\mathbf{g}\end{pmatrix} = \begin{bmatrix}\mathbf{P} & \mathbf{C}\\\mathbf{C} & \mathbf{C}\end{bmatrix}$, where $\mathbf{P}$ and $\mathbf{C}$ are $t \times t$ covariance matrices of trait phenotypic ($\mathbf{y}$) and breeding ($\mathbf{g}$) values, respectively. The joint distribution of the linear combination of $\mathbf{y}(I = \mathbf{b}'\mathbf{y}$, called LPSI) and $\mathbf{g}(H = \mathbf{w}'\mathbf{g}$, called "net genetic merit") values is bivariate normal distribution with mean $\mathbf{m}' = [m_H \quad m_I]$ and covariance matrix

$$Var\begin{bmatrix}H\\I\end{bmatrix} = \begin{bmatrix}\mathbf{w}'\mathbf{Cw} & \mathbf{w}'\mathbf{Cb}\\\mathbf{w}'\mathbf{Cb} & \mathbf{b}'\mathbf{Pb}\end{bmatrix} = \begin{bmatrix}\sigma_H^2 & \sigma_{HI}\\\sigma_{HI} & \sigma_I^2\end{bmatrix}, \tag{32.A2}$$

where $\mathbf{b}' = [b_1 \quad b_2 \quad \ldots \quad b_t]$ is an unknown vector of coefficients associated with $\mathbf{y}$, and $\mathbf{w}' = [w_1 \quad w_2 \ldots w_t]$ is a vector of known economic weight values associated with $\mathbf{g}$. In Eq. 32.A2, $\sigma_H^2 = \mathbf{w}'\mathbf{Cw}$ and $\sigma_I^2 = \mathbf{b}'\mathbf{Pb}$ are the variance of H and I, respectively, whereas $\sigma_{HI} = \mathbf{w}'\mathbf{Cb}$ is the covariance between H and I.

The Best Linear Predictor of the Mean Value of H

Suppose that $m_H = 0$ and $m_I = 0$; then the conditional expectation of H given $\mathbf{y}$ $(H/\mathbf{y})$ is

$$E\left(H \mid \mathbf{y}\right) = \mathbf{b}'\mathbf{y} = \mathbf{w}'\mathbf{CP}^{-1}\mathbf{y}, \qquad (32.A3)$$

where $\mathbf{b} = \mathbf{P}^{-1}\mathbf{Cw}$, $I = \mathbf{b}'\mathbf{y} = \mathbf{w}'\mathbf{CP}^{-1}\mathbf{y}$; $Cov(H, \mathbf{y}) = \mathbf{Cw}$ is the covariance among H and $\mathbf{y}$, and $\mathbf{P}^{-1}$ is the inverse matrix of $\mathbf{P}$. Eq. 32.A3 is the best linear predictor of the mean value of H.

The Selection Response

The selection response (R) is the expectation of H for a proportion p (Fig. 32.A1) of individuals selected and can be written as

$$R = k\sigma_H \rho_{HI}, \qquad (32.A4.1)$$

where k is the intensity of selection, σ_H is the standard deviation of H and ρ_{HI} is the correlation between H and I.

Equation 32.A4.1 is the same for all LSI; the only change is the type of information (phenotypic or genomic) and restrictions used when the index vector of coefficients is obtained to predict H and to maximize Eq. 32.A4.1, which is the main objective of any LSI. The genetic gain in Eq. 32.A4.1 will be larger as p becomes

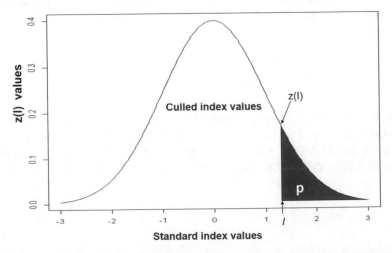

Fig. 32.A1 Relationship between the standard LSI (linear selection index) values (I), the proportion retained (p) and the density values [z(I)] of LSI

smaller—i.e., as the selection intensity becomes more intense (Fig. 32.A2). For example, in the LGSI context, the maximized selection response (Eq. 32.A4.1) can be written as

$$R_G = \frac{k}{L}\sqrt{\mathbf{w'\Gamma w}},\qquad(32.A4.2)$$

where k is the selection intensity, L denotes the interval between selection cycles, $\mathbf{\Gamma} = Var(\mathbf{z})$ is the covariance matrix of GEBV, and $\sigma_I = \sqrt{\mathbf{w'\Gamma w}}$ is the standard deviation of $I_G = \mathbf{w'z}$.

Constrained LPSI (CLPSI)

The main objective of the CLPSI is to maximize Eq. 32.A4 under some restrictions imposed on the expected genetic gain per trait (**E**), which can be written as

$$\mathbf{E} = k\frac{\mathbf{Cb}}{\sqrt{\mathbf{b'Pb}}}.\qquad(32.A5)$$

The type of restriction imposed on Equation 32.A5 can be a null restriction (RLPSI) or a predetermined constraint (CLPSI). Thus, let $\mathbf{d'} = \begin{bmatrix} d_1 & d_2 & \dots & d_r \end{bmatrix}$ be a vector of r constraints and assume that μ_q is the population mean of the qth trait $(q=1,2,\dots,r$, and r is the number of constraints) before selection. The CLPSI changes μ_q to $\mu_q + d_q$, where d_q is a predetermined change in μ_q imposed by the

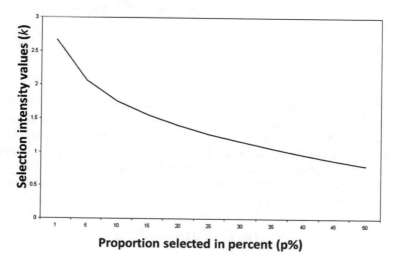

Fig. 32.A2 Values of the selection intensity (k) for different total proportion (p) values, in percentages

breeder. The restriction effects will be observed on the CLPSI expected genetic gains per trait (Equation 32.A5), where each restricted trait will have an expected genetic gain according to the **d** values imposed by the breeder.

The CLPSI Vector of Coefficients

Let $\mathbf{D}' = \begin{bmatrix} d_r & 0 & \dots & 0 & -d_1 \\ 0 & d_r & \dots & 0 & -d_2 \\ \vdots & \vdots & \ddots & \vdots & \vdots \\ 0 & 0 & \dots & d_r & -d_{r-1} \end{bmatrix}$ be a Mallard matrix [7] $(r-1) \times r$ of predeter-

mined proportional gains, where d_q ($q=1, 2\dots, r$) is the qth element of vector **d**, and let $\mathbf{U}'$ be a matrix of 1's and 0's, where 1 indicates that the traits are restricted and 0 that the traits are not restricted [3]. To obtain the CLPSI vector of coefficients, we minimized the mean squared difference between I and H, $E[(H-I)^2]$, with respect to **b** under the restriction $\mathbf{D}'\mathbf{U}'\mathbf{Cb} = \mathbf{0}$, where **C** is the covariance matrix of genotypic values.

The CLPSI vector of coefficients is

$$\beta = \mathbf{Kb}, \tag{32.A6}$$

where $\mathbf{K} = [\mathbf{I}_t - \mathbf{Q}], \mathbf{Q} = \mathbf{P}^{-1}\mathbf{M}(\mathbf{M}'\mathbf{P}^{-1}\mathbf{M})^{-1}\mathbf{M}', \mathbf{M}' = \mathbf{D}'\mathbf{U}'\mathbf{C}, \mathbf{I}_t$ is a $t \times t$ identity matrix and $\mathbf{b} = \mathbf{P}^{-1}\mathbf{Cw}$. When $\mathbf{d} = \mathbf{0}$, $\mathbf{D} = \mathbf{U}$, $\mathbf{Q} = \mathbf{P}^{-1}\mathbf{CU}(\mathbf{U}'\mathbf{CP}^{-1}\mathbf{CU})^{-1}\mathbf{U}'\mathbf{C}$, and CLPSI=RLPSI. When $\mathbf{D} = \mathbf{U}$ and $\mathbf{U}'$ is a null matrix, $\beta = \mathbf{b}$. Thus, the CLPSI is the most general index and includes the LPSI and the RLPSI as particular cases.

The Eigen Selection Index Method (ESIM)

The ESIM maximizes the correlation between $H = \mathbf{w}'\mathbf{g}$ and $I = \mathbf{b}'\mathbf{y}$, does not require a vector of economic weights **w**, and is associated with the canonical correlation theory [17]. In ESIM, I and H are *canonical variables*, whereas **b** and **w** are *canonical vectors*. The correlation between I and H (ρ_{HI}) is the *canonical correlation*. Thus, the measure of association between the jth linear combination of $\mathbf{y}(I_E = \mathbf{b}'_{E_j}\mathbf{y}$) and the jth linear combination of $\mathbf{g}(H_E = \mathbf{w}'_{E_j}\mathbf{g})$ is the jth canonical correlation (ρ_{HI_j}) value obtained from Eq. 32.A7.

$$\left(\mathbf{P}^{-1}\mathbf{C} - \rho^2_{HI_j}\mathbf{I}\right)\mathbf{b}_{Ej} = \mathbf{0}, \tag{32.A7}$$

where $\mathbf{b}_{E_j}$ is the jth *canonical vector* ($j=1,2,\ldots,t$) of matrix $\mathbf{P}^{-1}\mathbf{C}$, and $\mathbf{w}_{E_j} = \mathbf{C}^{-1}\mathbf{Pb}_{E_j}$. The first eigenvector ($\mathbf{b}_{E_1}$) of matrix $\mathbf{P}^{-1}\mathbf{C}$ is used in $I_E = \mathbf{b}'_{E_1}\mathbf{y}$ and in the maximized ESIM selection response.

Let $\mathbf{T}=\mathbf{P}^{-1}\mathbf{C}$; then Eq. 32.A7 can be written as $\mathbf{TIb}_{E_j} = \rho^2_{HI_j}\mathbf{Ib}_{E_j}$, where $\mathbf{I} = \mathbf{F}^{-1}\mathbf{F}$ is an identity matrix of size $t\times t$, and $\mathbf{F}=\mathrm{diag}\{f_1,f_2,\ldots,f_t\}$ is a diagonal matrix with values equal to any real number, except zero values. Thus, Eq. 32.A7 is equivalent to

$$\left(\mathbf{T}_2 - \rho^2_{HI_j}\mathbf{I}\right)\beta_{E_j} = \mathbf{0}, \qquad \text{(Eq. 32.A8)}$$

where $\mathbf{T}_2=\mathbf{FTF}^{-1}$ and $\beta_{E_j} = \mathbf{Fb}_{E_j}$; $\mathbf{T}$ and $\mathbf{T}_2=\mathbf{FTF}^{-1}$ are similar matrices and both have the same eigenvalues but different eigenvectors. Matrix $\mathbf{T}_2=\mathbf{FTF}^{-1}$ is called the *similarity transformation,* and matrix $\mathbf{F}$ is called the *transforming matrix* [19]. When the $\mathbf{F}$ values are only 1's, vector $\mathbf{b}_{E_j}$ is not affected; when the $\mathbf{F}$ values are only -1's, vector $\mathbf{b}_{E_j}$ will change its direction, and if the $\mathbf{F}$ values are different from 1 and -1, matrix $\mathbf{F}$ will change the proportional values of $\mathbf{b}_{E_j}$. In practice, $\mathbf{b}_{E_j}$ is first obtained from Eq. 32.A7 and then multiplied by matrix $\mathbf{F}$ to obtain β_{E_j}, that is, β_{E_j} is a linear transformation of $\mathbf{b}_{E_j}$. When vector β_{E_j} substitutes , the ESIM index should be written as $I_{E_1} = \beta'_{E_1}\mathbf{y}$.

The Constrained Eigen Selection Index Method (CESIM)

The CESIM is a constrained ESIM and its vector of coefficients is the first eigenvector of Eq. 32.A9

$$\left(\mathbf{KP}^{-1}\mathbf{C} - \rho^2_{HI}\mathbf{I}_t\right)\mathbf{b}_{CE_1} = \mathbf{0}, \qquad (32.A9)$$

where matrix $\mathbf{K}$ was described in Eq. 32.A6, and $\mathbf{b}_{CE_1}$ is the first eigenvector of matrix $\mathbf{KP}^{-1}\mathbf{C}$. When $\mathbf{D}' = \mathbf{U}'$, $\mathbf{b}_{CE_1} = \mathbf{b}_{R_1}$ (the vector of coefficients of RESIM), and when $\mathbf{U}'$ is a null matrix, $\mathbf{b}_{CE_1} = \mathbf{b}_{E_1}$.

References

1. Smith HF (1936) A discriminant function for plant selection. Ann Eugen 7:240–250. https://doi.org/10.1111/j.1469-1809.1936.tb02143.x
2. Hazel LN (1943) The genetic basis for constructing selection indexes. Genetics 28:476–490
3. Kempthorne O, Nordskog AW (1959) Restricted selection indices. Biometrics 15:10–19
4. Baker RJ (1986) Selection indices in plant breeding. CRC Press Inc., Boca Raton
5. Cerón-Rojas JJ, Sahagún-Castellanos J, Castillo-González F, Santacruz-Varela A, Crossa J (2008) A restricted selection index method based on eigen analysis. J Agric Biol Environ Stat 13:421–438. https://doi.org/10.1198/108571108X378911

6. Cerón-Rojas JJ, Crossa J, Toledo FH, Sahagún-Castellanos J (2016) A predetermined proportional gains eigen selection index method. Crop Sci 56:2436–2447. https://doi.org/10.2135/cropsci2015.11.0718
7. Mallard J (1972) The theory and computation of selection indices with constraints: a critical synthesis. Biometrics 28:713–735
8. Lande R, Thompson R (1990) Efficiency of marker-assisted selection in the improvement of quantitative traits. Genetics 124:743–756
9. Meuwissen TH, Hayes BJ, Goddard ME (2001) Prediction of total genetic value using genome-wide dense marker maps. Genetics 157:1819–1829
10. Dekkers JCM (2007) Prediction of response to marker-assisted and genomic selection using selection index theory. J Anim Breed Genet 124:331–341. https://doi.org/10.1111/j.1439-0388.2007.00701.x
11. Céron-Rojas JJ, Crossa J (2018) Linear marker and genome-wide selection indices. In: Linear selection indices in modern plant breeding. Springer, Cham, pp 71–98
12. Cerón-Rojas JJ, Crossa J (2019) Efficiency of a constrained linear genomic selection index to predict the net genetic merit in plants. G3 9:3981–3994. https://doi.org/10.1534/g3.119.400677
13. Cerón-Rojas JJ, Crossa J, Arief V, Basford K, Rutkoski J, Jarquín D, Alvarado G, Beyene Y, Semagn K, DeLacy I (2015) A genomic selection index applied to simulated and real data. G3 5:2155–2164. https://doi.org/10.1534/g3.115.019869
14. Börner V, Reinsch N (2012) Optimising multistage dairy cattle breeding schemes including genomic selection using decorrelated or optimum selection indices. Genet Sel Evol 44:11. https://doi.org/10.1186/1297-9686-44-1
15. Cerón-Rojas JJ, Crossa J (2018) Linear selection indices in modern plant breeding. Springer, Cham
16. Dickerson GE, Blunn CT, Chapman AB, Kottman RM, Krider JL, Warwick EJ, Whatley J, Baker ML, Lush JL, Winters LM (1954) Evaluation of selection in developing inbred lines of swine. Research Bulletin 551. University of Missouri, College of Agriculture, Agricultural Experiment Station
17. Rencher AC (2002) Methods of multivariate analysis, 2nd edn. Wiley-Interscience, New York
18. Cerón-Rojas JJ, Crossa J (2020) Expectation and variance of the estimator of the maximized selection response of linear selection indices with normal distribution. Theor Appl Genet. https://doi.org/10.1007/s00122-020-03629-6
19. Céron-Rojas JJ, Crossa J (2018) Linear phenotypic eigen selection index methods. In: Linear selection indices in modern plant breeding. Springer, Cham, pp 149–176

Correction to: Insect Resistance

Wuletaw Tadesse, Marion Harris, Leonardo A. Crespo-Herrera, Boyd A. Mori, Zakaria Kehel, and Mustapha El Bouhssini

Correction to:
Chapter 20 in: M. P. Reynolds, H.-J. Braun (eds.), *Wheat Improvement*, **https://doi.org/10.1007/978-3-030-90673-3_20**

The book was inadvertently published with an incorrect name information for one of the Chapter author as Body Mori, instead it should be "Boyd A. Mori" in the front matter and Chapter 20.

The updated original version of this chapter can be found at
https://doi.org/10.1007/978-3-030-90673-3_20

Correction to: Experimental Design for Plant Improvement

Ky L. Mathews and José Crossa

Correction to:
Chapter 13 in: M. P. Reynolds, H.-J. Braun (eds.), *Wheat Improvement*, **https://doi.org/10.1007/978-3-030-90673-3_13**

The book was inadvertently published with an inappropriate figure for Fig. 13.5 featured in Chapter 13. This has now been replaced with the appropriate figure as represented below.

The updated version of this chapter can be found at
https://doi.org/10.1007/978-3-030-90673-3_13

© The Author(s) 2023
M. P. Reynolds, H.-J. Braun (eds.), *Wheat Improvement*,
https://doi.org/10.1007/978-3-030-90673-3_34

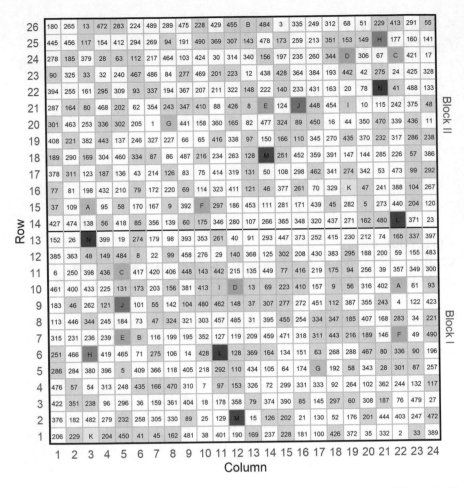

Fig. 13.5 Partially replicated design for $v = 504$ varieties in 624 plots, arranged in 26 rows by 24 columns. The bold horizontal line delineates the Blocks. Colors represent different check lines. The gray shaded plots are those allocated with 2 replicate varieties

Index

Printed in the United States
by Baker & Taylor Publisher Services